D1489281

Elementary Statistics
2nd Edition

Elementary Statistics
2nd Edition

Roger E. Kirk
Baylor University

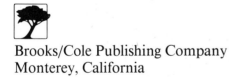

Brooks/Cole Publishing Company
Monterey, California

Brooks/Cole Publishing Company
A Division of Wadsworth, Inc.

Printed in the United States of America

10 9 8 7 6 5 4 3 2 1

Library of Congress Cataloging in Publication Data

Kirk, Roger E.
    Elementary statistics.

    Bibliography:
    Includes index.
    1. Statistics. I. Title
QA276.12.K58   1984   519.5   84-1871

ISBN 0-534-03213-3

Sponsoring Editor: C. Deborah Laughton
Production Service: Unicorn Production Services, Inc.
Manuscript Editor: Karin Hammann
Permissions Editor: Carline Haga
Interior Design: Jamie Sue Brooks
Cover Design: Charles Carter Design
Cover Illustration: © 1983 BLOCPIX Division, Watson-Manning, Inc.
Interior Illustration: Syntax International
Typesetting: Syntax International
Cover Printing: Lehigh Press Lithographers, Pennsauken, New Jersey
Printing and Binding: The Maple Press Company, York, Pennsylvania

Preface

Elementary Statistics was written for students in the behavioral sciences and education who are taking their first course in statistics. Its goals are threefold: to provide a sound introduction to descriptive and inferential statistics, to enable students to effectively use statistical procedures, and to develop an appreciation for the role of statistics in the scientific enterprise. In this age of sophisticated hand-held calculators and microcomputers, I believe that students are better served by a book that fosters an understanding of statistical logic than by one that presents statistics as a series of cookbook procedures to be mechanically applied. Accordingly, the emphasis throughout the book is on the purpose and rationale of the various statistical procedures.

In keeping with the intended audience, I have relied heavily on verbal rather than mathematical explanations. To be sure, the reader will encounter many formulas and a few proofs, but the level of mathematics used is quite elementary. A familiarity with the thinking patterns learned in arithmetic and high school algebra is sufficient for understanding the text. For those whose mathematical skills are rusty, Appendix A provides a review of elementary mathematics. The diagnostic mathematics skills test in Appendix A can be used for testing one's knowledge or identifying concepts that should be reviewed.

Today many students have access to statistical calculators that will compute complex statistics at the press of a key. And an increasing number are getting hands-on experience with microcomputers and statistical software packages. For those who want to learn about computers, I have included in Chapter 2 an optional introduction to computing hardware and software. Another optional feature is the inclusion of 13 annotated printouts from the major statistical packages and two microcomputer packages. A careful study of these printouts will go a long way toward developing skill in interpreting the output of computers.

A microcomputer package for Apple II's, DEC Rainbows, and IBM Personal Computers was written especially for use with this book. This package, which is available from the author, contains both statistical programs and a tutorial demonstration. The tutorial demonstration is particularly effective in illustrating the central limit theorem.

Other learning aids include a chapter outline at the beginning of each chapter, the use of boldface type to emphasize important definitions, an extensive glossary of statistical terms (Appendix B), chapter summaries, and 519 review exercises following chapter sections. These exercises (1) indicate which concepts and computational procedures are most important, (2) present interesting real-life examples, many based on recent journal articles, of the way statistics are used, and (3) provide practice in applying what has been learned. Answers are given in Appendix C for 210 of the exercises.

It is a pleasure to express my appreciation to Dennis Bonge, University of Arkansas, Robert Breckenridge, Northwestern State University of Louisiana, Richard Brown, California State University at Fresno, James Chumbley, University of Massachusetts, Stanley Cohen, West Virginia University, Steve Falkenberg, Eastern Kentucky University, Arthur Gutman, Florida Institute of Technology, Margaret Kerbaugh, North Texas State University, Michael Knight, Central State University, and Larry Ringer, Texas A & M University, for reading the manuscript and for their thoughtful comments. Also, it is a pleasure to express my appreciation to Jack Baty and to Sharon O'Connor, who contributed exercises, and to René Powers, who typed the manuscript. Brad Wilson deserves special mention for developing the microcomputer package for use with the book. C. Deborah Laughton of Brooks/Cole and Michael Michaud of Unicorn Production Services also deserve special recognition.

I am grateful to the Literary Executor of the late Sir Ronald A. Fisher, F.R.S., to Frank Yates, F.R.S., and to Longman Group Ltd., London, for permission to reprint Tables D.1, D.2, D.3, D.4, D.6, and D.8 from their book *Statistical Tables for Biological, Agricultural and Medical Research* (6th edition, 1974).

I am also grateful to E. S. Pearson and H. O. Hartley, editors of *Biometrika Tables for Statisticians,* Vol. 1, and to the *Biometrika* trustees for permission to reprint Table D.5; and to the editor of *Annals of Mathematical Statistics* for permission to reprint Table D.7.

Portions of this book were written while on sabbatical from Baylor University. I am grateful to the administration of Baylor University and in particular to Herbert H. Reynolds, John S. Belew, and William G. Toland for providing an environment that encourages creative and scholarly activity.

And finally I want to express my appreciation to my statistics classes for what I trust has been a mutually rewarding learning experience.

Roger E. Kirk

Contents

1

Introduction to Statistics

Section 1.1   Introduction

Some Misconceptions

It is widely believed that statistics can be used to prove anything—which implies, of course, that it can prove nothing. Furthermore, the word "statistics" conjures up visions of numbers piled upon numbers, uninterpretable charts, and computers cranking out gloomy predictions. To the ordinary person, besieged from all sides by advertising claims, statistics is hocus-pocus with numbers. It was Benjamin Disraeli who said "there are three kinds of lies—lies, damned lies, and statistics." In primitive cultures, exaggeration was common. One writer, with tongue in cheek, reasoned that, since primitive people did not have a science of statistics, they were forced to rely on exaggeration, which is a less effective form of deception. Another writer remarked that "if all the statisticians in the world were laid end to end— it would be a good thing." Regardless of its public image, statistics endures as a required course, and my students continue to refer to it, affectionately no doubt, as Sadistics 2402.

What Is Statistics?

In spite of frequent misuse, statistics can be an elegant and powerful tool for making decisions in the face of uncertainty. The word "statistics" comes from the Latin *status*, which is also the root for our modern term "state" or political unit. Statistics was a necessary tool of the state, since to levy a tax or to wage war a ruler had to know the number of subjects in the state and the amount of their wealth. Gradually the meaning of the term expanded to include any type of data. *Today the word **statistics** has four distinct meanings. Depending on the context, it can mean: (1) data; (2) functions of data, such as the mean and range; (3) techniques for the collection, analysis, and interpretation of data for subsequent decision making; and (4) the science of creating and applying such techniques.*

Why Study Statistics?

A knowledge of statistics yields more than the obvious benefits. For example, it generates new ways of thinking about questions and provides effective tools for answering them. It takes only a cursory examination of the professional literature in your own field to see the inroads made by statistical techniques and ways of thinking. Statistics is undoubtedly an indispensable tool for experimenters, but its usefulness is not limited to research. In many fields, it is virtually impossible to keep up with new developments without an understanding of elementary statistics. Also, statistics is an interesting subject—some people find it even fascinating.

In all likelihood you are reading this book because it was assigned in your required statistics course. You have been told that the study of statistics is necessary, and there is a strong implication that it will be good for you. At this point you may be skeptical. Just what can you expect to learn by studying statistics? A quick scanning of this book will give you an idea. You will acquire a new vocabulary, since in many ways learning statistics is like learning a foreign language, and you will learn to manipulate numbers according to symbolic instructions. But more important, you will learn when and how to apply statistics to research problems in the behavioral sciences and in education. Your study of statistics should enable you to read the literature in your professional field with greater understanding, and it will prepare you to learn more complex procedures in the design and analysis of experiments. You will probably become more critical of statistical presentations in your field and in the mass media. And, you should gain a greater appreciation of the probabilistic nature of scientific knowledge. Statistics involves a special way of thinking that can be used not only in research but also in daily life. I hope that you will add this mode of thinking to your conceptual ability.

Kinds of Statisticians

Users of statistics fall into four categories: (1) those who must be able to understand statistical presentations of findings in their fields; (2) those who select, apply, and interpret statistical procedures in their work; (3) professional (applied) statisticians; and (4) mathematical statisticians. This book is addressed to those in the first two categories, including psychologists, educators, sociologists, engineers, speech therapists, biologists, counselors, business people, physicians, medical researchers, political scientists, geologists, and city, state, and federal government officials, to mention only a few. In each case the person's primary interest is in his or her own field, be it sociology or city planning; he or she is interested in statistics because it is a useful tool for answering questions in that field. Such persons are both consumers and users of statistics. Their knowledge of statistics may range from meager to expert.

The professional statistician helps professionals in substantive areas to use statistics effectively.[1] He or she may work for industry or a government agency, engage in a private consulting practice, or teach in a university. Unlike individuals in the first two categories, a professional statistician usually has advanced degrees in mathematics and/or statistics.

The mathematical statistician is not interested in applied statistics but, instead, in pure (mathematical) statistics and probability theory. Most likely this

---

[1] A pamphlet, *Careers in Statistics*, which describes the work of statisticians, is available from the American Statistical Association, 806 15th St. N.W., Washington, D.C. 20005.

statistician teaches in a university and makes contributions to the theoretical foundations of statistics that may ultimately be used by those with applied interests.

Section 1.2   Studying Statistics

Read More Slowly

Statistics cannot be read like assignments in history, English, or political science. Statistics uses a specialized vocabulary that must be learned, and ideas and computational procedures are presented in a highly symbolic form. Consequently, a 30-page assignment may take three or four times as long to read as a comparable assignment in history. You will understand many sections of the book on a first reading; others will require two or more readings, lots of concentration, and perhaps some time between readings for the ideas to sink in.

Don't Worry If You Weren't an Ace in Math

If you're concerned about the level of mathematics required to understand statistics, stop worrying. Most statistical procedures in this book involve nothing more complicated than addition, subtraction, multiplication, and division. Although some use is made of high school algebra, the level is very elementary. The essential arithmetic and algebra are reviewed in Appendix A.

Resolve to Review Often

Frequent reviews of the material are a must, or it will slip away. The exercises at the end of each section should be a helpful review because they provide (1) feedback about what you know and what you don't, (2) an indication of which concepts and computational procedures are the most important, (3) numerous examples of how statistics is used, and (4) practice in applying what you are learning. Answers to the starred exercises are given in Appendix C. The chapter summaries should also be useful for reviewing, since they briefly present the major points and place the topics in perspective.

Master Foundation Concepts before Going on to New Material

In statistics, as in mathematics or a foreign language, the material presented first is the foundation for what follows. Each chapter should be mastered before going on to the next. Cramming is effective in some subjects, at least as far as tests are concerned. But in statistics it inevitably results in superficial understanding of basic concepts and subsequent learning problems. Periodic reviews require considerable discipline, but they pay off. Also, different teaching approaches are effective with different students. If you are having difficulty with a topic, ask your professor to recommend a book that gives an alternative explanation.

Strive for Understanding

This book contains hundreds of formulas. I have not memorized all of them and neither should you. Some, such as the one for the arithmetic mean, $\bar{X} = \sum X/n$, appear so often that you really can't help but learn them; the others aren't worth the effort. I decided a long time ago, when faced with my own inability to remember telephone numbers, addresses, and lock combinations, that books are better repositories than my head for such things. Instead of memorizing formulas, strive to understand concepts, and think about their applications. In what situations is a statistic useful? How is the statistic interpreted? What assumptions must be fulfilled to interpret the statistic? When you read about an experiment in your field, consider how you would have designed it and analyzed the data. And talk about your research ideas with your professor.

Section 1.3   Basic Concepts

Population and Sample Defined

Many statistical terms are a legacy from the time when statistics was concerned only with the condition of the state. "Population," for example, originally meant, and still means, the total number of inhabitants of the state. Its meaning in statistics is broader. *A **population** is the collection of all objects or observations having one or more specified characteristics.* The population is identified when we specify the common characteristics. All the people listed in a telephone directory constitute a population, as does the number of heads and tails obtained in tossing a coin for eternity. *A single object or observation is called an **element** of the population.* The population of telephone book listees contains a *finite* number of elements; the population resulting from tossing the coin contains an *infinite* number. A population is either concrete or conceptual. The population of telephone book listees is *concrete*—given sufficient time we could contact each person, since the number is finite and well defined. The population of heads and tails is *conceptual*—no matter how long we work at it, we can't record all the possible results of tossing a coin.

The elements of a population can be people, objects, events, or, as is usually the case, observations based on measurable characteristics of people, objects, or events. For example, a population could consist of all the students in a university, their cars, their pep rallies, or their IQ scores. In identifying a population for a particular experiment, an investigator considers the nature of the research problem as well as such practical matters as the availablilty of population elements.

*A **sample** is a proper subset of a population.* That is, a sample can contain a single element or all but one of the population elements. For practical reasons, such as limited resources and time, or because the population is infinite in size, most research is carried out with samples rather than with populations. It is assumed that the study of a sample will reveal something about the population. This leap of faith often appears to be justified, as when a laboratory technician analyzes a sample of a patient's blood or when an automobile manufacturer crash tests a sample of bumpers. Occasionally, however, samples lead us astray. Later we'll see how and why.

Descriptive and Inferential Statistics

It is convenient to divide statistical techniques into two categories—descriptive and inferential. ***Descriptive statistics** are tools for depicting or summarizing data so that they can be more readily comprehended.* When we say that a player's lifetime batting average is 0.420 or when we graph the gas consumption of American and foreign cars, we are using descriptive statistics. A computer printout listing the IQs of all college students in California would boggle our minds—not so, a statement that the mean IQ is 116. Large masses of data are difficult to comprehend. Descriptive statistics reduce data to some form, usually a number, that is easily comprehended. We will go into the details of descriptive statistics in the first half of this book.

We saw that it is usually impossible for experimenters to observe all the elements in a population. Instead they observe a sample of elements and generalize from the sample to all the elements—a process called *induction*. They are aided in this process by ***inferential statistics**, which are tools for inferring the properties of one or more populations from an inspection of samples drawn from them.* Inferential statistics were developed to improve decision making in cases where successive observations exhibit some degree of variation although they are taken under conditions that appear to be identical. The variation may be due to inherent variability in the phenomenon or in the subjects, errors of measurement, undetected changes in conditions, or a combination of these. In the behavioral sciences and education, the variability among subjects is the major stumbling block to induction. Suppose that a physiologist wants to know if a new drug will arrest the development of cancer in humans. It is impossible to administer the drug to the population of all cancer victims, but it is possible to administer it to a sample. The physiologist would probably attempt to control attitudinal and other extraneous factors by administering an inert druglike substance, a *placebo*, to half the sample and the new drug to the other half. Suppose that remission of cancer occurred in 100% of the sample receiving the new drug and in only 8% of those receiving the placebo. The difference, 100% versus 8%, between the drug and placebo samples is dramatic. The physiologist would probably conclude that if the drug had been administered to the population of all cancer victims, the remission rate would have been higher than if the population had received the placebo. But what if the remission rate were 12% for the new drug and 8% for the placebo? We know that chance factors can produce a difference between two samples even though the samples are taken from the same population and receive identical treatments. Is the difference, 12% versus 8%, greater than would be expected by chance? Stated another way, if the experiment were repeated many, many times, could the physiologist predict with confidence that the difference would consistently favor the sample receiving the drug? This is the kind of question that can be answered using inferential statistics. We will deal with procedures for answering such questions in the latter half of this book.

Random Sampling

Some samples provide a sound basis for drawing conclusions about populations, while others do not. The difference lies in the method by which the samples are selected. *The method of drawing samples from a population such that every possible sample of a particular size has an equal chance of being selected is called **random sampling**, and the resulting samples are **random samples**.* Sampling methods based on haphazard or purposeless choices, such as soliciting volunteers, using students enrolled in introductory psychology, or selecting every tenth person in an alphabetical listing of names, produce *nonrandom samples.* Such samples, unlike random samples, do not provide a sound basis for deducing the properties of populations. Hence, whenever sampling is mentioned in this book, it will refer to random sampling. A detailed discussion of random sampling must await the development of other basic concepts. At this point, we will simply illustrate several characteristics of random samples.

Consider a box containing 300 balls, each identified by a number stamped on its surface. Two hundred of them are red (*R*), and 100 are black (*B*). If you didn't know the ratio of red to black balls (2:1), you could estimate it from a random sample of balls. You close your eyes, shake the box, reach in, withdraw a ball, note its color and number, and replace it. You do this six times and obtain the following sample: R_{102}, R_{75}, B_{39}, R_{62}, B_{37}, R_{50}; the subscripts, 102, 75, and so on, denote the numbers stamped on the balls. From this sample you would infer that the box contains more red than black balls—in fact, twice as many red balls. Suppose you drew four more samples, each time replacing the balls drawn, and obtained R_{154}, B_{62}, R_{35}, R_{143}, R_4, R_{29} (sample 2); R_{104}, B_{41}, B_{21}, R_{50}, R_{192}, R_{67} (sample 3); B_{28}, B_{41}, R_{150}, B_{61}, R_{88}, R_{148} (sample 4); and R_{152}, R_{120}, B_{88}, R_{33}, R_{36}, B_5 (sample 5). The results of the five random samples are summarized in Table 1.3-1. This simple experiment illustrates several points about random samples. First, the elements obtained (and the ratio of *R* to *B* balls) differ from sample to sample. This is referred to as *sampling fluctuation* or *chance variability.* Second, the characteristics of a sample do not necessarily correspond to those in the population. It turns out, however, that the larger a random sample, the more likely it is to resemble closely the population. Hence, experimenters prefer to work with large samples if it is economically feasible. Although there is no guarantee that large random samples will resemble the population, in the long run they are more likely to do so than small ones.

Table 1.3-1. Outcomes of drawing five random samples

Color of balls	Sample				
	1	2	3	4	5
Number of red balls	4	5	4	3	4
Number of black balls	2	1	2	3	2
Ratio of red to black	2:1	5:1	2:1	1:1	2:1

Review Exercises for Section 1.3[2]

*1. Users of statistics fall into four categories.

   *a. List the categories.

   b. Considering your vocational goals, into which category do you fall? Why?

2. How does the original meaning of the term "population" differ from today's statistical definition?

3. (a) In your local newspaper, find two examples of populations and two examples of samples. (b) Classify the populations according to whether they are finite or infinite and whether they are concrete or conceptual.

*4. For each of the following statements, indicate (a) the population, (b) the element, and (c) the datum (observation) to be recorded.

   a. At least 50% of White female students in this university are ambivalent about having a career.

   b. Tequila Tech students are involved in more automobile accidents than other drivers in their age group.

   c. At least 80% of the homes in Chickasha, Oklahoma, have color televisions.

   d. Students at Ginebra University who hold outside jobs have higher grade point averages than those who don't hold outside jobs.

5. For each of the following statements, indicate (a) the population, (b) the element, and (c) the datum to be recorded.

   a. Thirty-six percent of 18-to-24-year-olds voted in the 1980 presidential election.

   b. In 1982, approximately 16% of all children under 18 were members of families whose incomes were below the poverty level.

   c. In 1982, approximately 42% of all prison inmates were 14 to 24 years old.

   d. In 1982, approximately 32% of all high school graduates 18 to 24 years old were enrolled in college.

   e. In 1982, four out of ten Americans were under 25 years old.

*6. What are the lower and upper limits on the size of a sample?

7. (a) Why is most research conducted on samples rather than populations? (b) How is sample size related to the resemblance between a random sample and the population?

8. Distinguish between descriptive statistics and inferential statistics.

*9. Indicate whether each of the following procedures would produce a random sample (R) or a nonrandom sample (NR) of students in an introductory sociology class.

   a. Write each student's name on a slip of paper, place the slips in a hat, shake the hat, and draw out 10 names.

   b. Place the blindfolded instructor in the middle of a circle made up of all the class members. Have the instructor point to 10 people around the circle. The student nearest the position pointed to becomes an element of the sample.

   c. For each student, flip a coin. If the coin lands heads, the student is in the sample.

   d. Line up the students from the tallest to the shortest. The third, fifth, seventh, . . . , twenty-first students become members of the sample.

10. Find two newspaper stories citing survey or poll results. Do the articles mention the method of sampling? Do they reveal the source of the financial support for the surveys? Why is this information—sampling method and source of financial support—important?

11. Terms to remember:

   a. Population

   b. Element

---

[2] Problems or portions thereof for which answers are given in Appendix C are denoted by *.

c. Sample
d. Random sample
e. Sampling fluctuation

Section 1.4   Describing Characteristics by Numerals

People, objects, and events have many distinguishable characteristics. Early in the design of an experiment, two key decisions must be made: What characteristics should be observed? And how should the characteristics be measured? The answer to the first question is determined by the experimenter's research problem and interests. Suppose an experimenter is interested in the IQs of male and female college students. College students differ in sex, age, IQ, major, hair color, family income, and so forth, but only two characteristics are of interest in this example—sex and IQ. These are the characteristics that are measured. The others are ignored. The second question, concerning how the characteristics should be measured, is outside the scope of this book. We are interested here in the more basic issues surrounding the assignment of numerals to characteristics. In the process of examining these issues, we will discuss variables and constants and see how mathematicians classify variables.

Variables and Constants

*A **variable** is a symbol that is used to stand for an unspecified element of a set.* The set of elements for which the variable stands is called the *range* of the variable, and each element of the range is called a *value*. When we replace a variable by one of the elements in its range, we say that the variable "takes" this value. For example, the variable of sex can take one of two values, male or female. *A **constant** is a symbol with a range that consists of a single element.* The ratio of the circumference of a circle to its diameter, denoted by π, is a constant because its range consists of the single value 3.1415926536. . . .

Perspectives on Numbers

We observed that selection of the characteristics to measure is relatively straightforward once the research problem has been identified. The second key decision, deciding how the characteristics should be measured, isn't as simple. For example, the variable of student IQ could be measured by (1) ranking or ordering IQ scores from highest to lowest and assigning each student the number of his or her rank; (2) assigning each student a label such as average, high average, or superior, based on IQ test performance; or (3) assigning each student her or his actual IQ score. Depending on the measuring scheme adopted, Jonathan Whiz would be designated 3, superior, or 132. The number 3 would represent his rank among students in the sample and 132 would be his IQ score. Sex can be classified by assigning a unique symbol such as M, 2, or ♂ to males and F, 1, or ♀ to females.

The assignment of numerals to characteristics of people, objects, or events and the accuracy of the representation are central concerns of researchers but not of mathematicians and mathematical statisticians. Only during its formative years was mathematics tied to the real world. Then it seemed perfectly natural to prove mathematical theorems by recourse to counting and measuring. But in recent times mathematics has shed its real-world ties—mathematicians are now free to manipulate symbols that are totally void of empirical meaning. They are interested in the formal properties of the systems they create; applications in the real world are left to other specialists. The theoretical work of mathematicians and mathematical statisticians laid the foundation for statistical tools, but the scientist who uses them must decide whether a particular tool is appropriate for his or her research applications and whether the numbers assigned to variables accurately represent the characteristics of interest. This division of interests between the developers and users of statistics has led to considerable confusion about the correct uses of statistical tools and also to two ways of thinking about numerals.

Classification of Variables in Mathematics

Mathematicians classify variables as qualitative or quantitative. *A **qualitative variable** is one whose range consists of attributes or nonquantitative characteristics of people, objects, or events, for example, sex (male, female), race (Caucasian, Negro, Oriental, other), and grade in a course (A, B, C, D, F).* The categories of a qualitative variable are nonoverlapping and exhaustive and may or may not suggest an order or rank. Grades in a course, A, B, C, D, F, clearly order academic achievement from highest to lowest, but no order is suggested by the categories for sex, race, religious preference, or blood type. Course grade is an example of an *ordered* qualitative variable; sex, race, religious preference, and blood type are examples of *unordered* qualitative variables.

A **quantitative variable** *is one whose range consists of a count or a numerical measurement of a characteristic.* Quantitative variables can be discrete or continuous. A variable is *discrete* if its range can assume only a finite number of values or an infinite number of values that is countable. (That is, the infinite number of values can be placed in a one-to-one correspondence with the counting or natural numbers.) Family size is an example of a variable with a finite range. It can assume values 1, 2, 3, 4, and so on, but not 200, 8000, or any noninteger value such as 0.5 and 4.3. The rational numbers—numbers that can be expressed as the ratio of two integers, for example, $\frac{2}{2}$, $-\frac{2}{3}$, $\frac{7}{4}$—illustrate countably infinite numbers. There is no largest number and no smallest number, and between, say, 1 and 2, an infinite number of rationals can be inserted, for example, $\frac{3}{2}$, $\frac{4}{3}$, $\frac{5}{4}$, ... Other examples of discrete quantitative variables are the number of parking tickets received, the number of trials required to learn a list of nonsense syllables, and one's score on a standardized achievement test. In each of these examples, the value assigned to the variable is obtained by counting, and the counting units—family members, parking tickets, learning trials, or achievement test items—are equivalent in arriving at the total count.

By contrast, a variable is *continuous* if its range is uncountably infinite. Such a range can be likened to points on a line that have no interruptions or intervening spaces between them. Examples of continuous variables are temperature in Bangor during January, length of fish caught off the Florida Keys, and speed of cars on the New Jersey Turnpike. Although a variable is continuous, our measurement of it is by necessity discrete because of limitations in the measuring instrument. For example, the thermometer is usually calibrated in $1°$ steps, the ruler in $\frac{1}{16}$ in., and the speedometer in 1 mph. Consequently, our measurement of continuous variables is always approximate. Discrete variables, on the other hand, can be measured exactly. A husband and wife with two children are a family of exactly four, but a temperature of $20°C$ refers to a range of temperatures—$20°C$ give or take $\frac{1}{2}°$.

The classification scheme for variables is summarized in Table 1.4-1. It is useful to mathematicians and statisticians because different mathematical tools and assumptions are required in derivations and proofs, depending on the classification of the variables. The classification scheme is a convenience; it was not devised to mirror characteristics in the real world. When we use statistical methods to answer real-world questions, we must remember that they were developed to analyze numbers as numbers. If the numbers analyzed bear no relation to the characteristics in which we are interested, the resulting answers will be meaningless.

Measuring Operations in the Behavioral Sciences and Education

Numbers are used for a variety of purposes, three of which are of particular interest to behavioral scientists and educators: (1) to serve as labels, (2) to indicate rank in a series, and (3) to represent quantity. For example, a football player is identified by the number 10 on his uniform, a team is ranked number two in the UPI poll, and the winning touchdown play covered 20 yd. Without thinking, we treat these numbers differently. It doesn't take a football fan to know that player 30 is not three times player 10 and that the number two team is not necessarily twice as good as the number four team, but that a 20-yd touchdown play did indeed move the ball twice as far down the field as a 10-yd play. We intuitively treat the numbers differently because they involve different levels of measurement. *Measurement is the process of assigning numerals to characteristics of people, objects, or events according to a set of rules.* We shall see that the rules used to

Table 1.4-1.   Mathematical classification of variables

Type of variable	Characteristics
Qualitative variable	Range consists of nonoverlapping and exhaustive categories that represent attributes or nonquantitative characteristics.
Unordered	Categories do not suggest an order or rank.
Ordered	Categories suggest an order or rank.
Quantitative variable	Range consists of a count or a numerical measurement of a characteristic.
Discrete	Range consists of only a finite number of values or an infinite number of values that is countable.
Continuous	Range consists of an uncountably infinite number of values.

assign the numerals determine the level of measurement. S. S. Stevens (1946), a behavioral scientist, identified four levels of measurement: nominal, ordinal, interval, and ratio.

Nominal Measurement

Nominal measurement *is the simplest of the four levels. It consists of assigning observations to mutually exclusive and exhaustive **equivalence classes** so that those in the same class are considered to be equivalent, whereas those in different classes are not equivalent. The classes are then denoted by unique labels.* The set of labels constitutes a *nominal scale*. Assignment of women to one equivalence class called "women" and men to the other called "men" is nominal measurement. Numbers may be used instead of words to identify the two classes, for example, 1 for women and 2 for men. Numbers used in this way are simply alternative names for the classes. We could just as well have assigned the numbers 9 and 6. A transformation of the scale in which 9 is substituted for 1 and 6 for 2 would be as useful for distinguishing between the classes as any other one-to-one transformation.[3] The numbers in a nominal scale could be added, subtracted, averaged, and so on, but the resulting numbers would tell us nothing about the equivalence classes represented by the numbers. For example, $1 + 2 = 3$ and $9 + 6 = 15$, but neither 3 nor 15 corresponds to any characteristic of men or women. This follows since we didn't utilize the properties of size and order of numbers when we assigned them to the classes. The only property of numbers that we utilized is that 1 is distinct from 2, 3, ... There are many examples of nominal scales in psychology and education, for example, Eysenck's four personality types (stable–extrovert, stable–introvert, unstable–extrovert, unstable–introvert), the primary taste qualities (sweet, sour, salty, bitter), and categories of psychoses (organic, functional). We can see a similarity between nominal measurement and one of the mathematician's types of variables. The nominal scale corresponds to the range of an unordered qualitative variable.

Ordinal Measurement

*The next higher level is **ordinal measurement.** It consists of assigning observations to equivalence classes that are ranked or ordered with respect to each other. The classes can be identified by numbers or any other ordered symbols that reflect the rank of the classes, such as letters.* Political candidates, for example, can be ranked with respect to voter appeal. If candidate Jane is judged to have the greatest appeal, followed by Keith, Lewis, and then Marvin, we could assign Jane the number 1; Keith, 2; Lewis, 3; and Marvin, 4. We have no reason to believe that Keith, ranked second, is half as appealing to voters as Jane or that the difference in appeal between Jane and Keith, 1 versus 2, is the same as the difference between Keith and Lewis, 2 versus 3. The numbers indicate rank order but not magnitude

---

[3] A one-to-one transformation associates with each element in one set, one and only one element in a second set, and vice versa. For example, if the two sets are $a_1, a_2, a_3, \ldots$, and $b_1, b_2, b_3, \ldots$, each a is paired with one and only one b, and each b is paired with one and only one a.

or difference in magnitude between classes. The numbers assigned to the equivalence classes can be subjected to any transformation that preserves the order-relationship of the numbers and classes.[4] For example, the set of ordered numbers 2, 16, 39, 40 would serve just as well to rank the candidates since only the order and not the distance between any two numbers is important. Alternatively, we could assign the letter *A* to Jane, *B* to Keith, *C* to Lewis, and *D* to Marvin.

Some characteristics, such as people's heights, can be measured in several ways, for example, ranking from tallest to shortest or recording actual feet and inches. The latter procedure assigns numbers that represent the magnitudes of the equivalence classes and therefore has several advantages over ordinal measurement, as we shall see later. For the moment we simply note that ordinal measurement is most often used when applying a more refined procedure is difficult or impossible.

Numerous examples of ordinal scales can be found in the behavioral sciences and education. Mentally subnormal children sometimes are classified as borderline, educable, trainable, or profoundly retarded. Other examples are rank in high school class and a supervisor's ranking of employees. Such ordinal scales correspond, in the language of the mathematician, to the range of an ordered qualitative variable.

Interval Measurement

Interval measurement is much more sophisticated than nominal and ordinal measurement. In ***interval measurement,*** *the numbers assigned to equivalence classes have the properties of distinctness and order, and in addition, equal differences between numbers reflect equal magnitude differences between the corresponding classes. The measurement procedure consists of defining a unit of measurement, such as a calendar year or 1°F, and determining how many units are required to represent the difference between equivalence classes.* In our measurement of calendar time, the same amount of time elapsed between 1970 and 1971 as between 1971 and 1972, and similarly the temperature difference between 70 and 75°F is the same as that between 80 and 85°F. A given numerical interval, say 1 yr or 5°F, represents the same difference in the characteristic measured, irrespective of the location of that interval along the measurement scale. In other words, numerically equal distances along the measurement continuum represent empirically equal differences among the corresponding equivalence classes—that is, the measured characteristic.

Since the units along interval scales are empirically equal, we meaningfully can perform most arithmetic operations on the numbers. For example, we can say that the difference between 80 and 60°F is twice as great as that between 60 and 50°F. That is, the ratio of intervals $(80 - 60°F)/(60 - 50°F) = 2$ has meaning with respect to temperature. However, not all operations are permissible, because the starting point or origin of an interval scale is always arbitrarily defined and does not necessarily correspond to an absence of the measured characteristic. In the case

---

[4] Such a transformation is called a *monotonic transformation.*

of the Fahrenheit scale, zero corresponds to the temperature produced by mixing equal quantities by weight of snow and salt. It does not indicate an absence of molecular action and hence an absence of heat. Therefore, although $80°F/40°F = 2$, we cannot say that $80°F$ is twice as hot as $40°F$. The ratio of temperatures is meaningless with respect to temperature. The situation is the same with calendar time, which is calculated from the birth of Christ, and altitude, which is calculated from sea level.

The numbers in an interval scale can be subjected to any positive linear transformation.[5] Degrees Fahrenheit, F, can be transformed into degrees Celsius, C, by means of the linear transformation

$$C = \frac{5}{9}(-32) + \frac{5}{9}F.$$

Although the variable represented by an interval scale may be continuous, our measurement of it is always discrete, since measuring instruments are calibrated in discrete steps. Thus, in practice an interval scale corresponds to the range of a discrete quantitative variable.

Ratio Measurement

Ratio scales constitute the highest level of measurement. *The numbers assigned to equivalence classes have the properties of distinctness, order, and equivalence of intervals, and, in addition, the origin of the scale represents the absence of the measured characteristic.* Thus, ratio scales have all the properties of interval scales plus an absolute zero. Most scales in the physical sciences are ratio scales—height in inches, weight in pounds, temperature on the Kelvin scale, and elapsed time such as age in years, years of experience, and reaction time in seconds.

Not only is the difference between 5 and 6 inches the same distance as that between 10 and 11 inches, but also 10 inches is twice as long as 5 inches. Ratio scales permit us to make meaningful statements about the ratio of two numbers, for example, 10 in./5 in. = 2; hence 10 inches is twice as long as 5 inches. The properties of a ratio scale permit us to perform all arithmetic operations on the numbers. However, the only transformation of a ratio scale that preserves these properties is multiplication by a positive constant. For example, we can transform inches into centimeters by multiplying inches by the constant 2.54: 10 in. = 25.4 cm and 5 in. = 12.7 cm. Ten inches is twice as long as 5 in. and, similarly, 25.4 cm is twice as long as 12.7 cm. As we move from the simplest measurement (nominal scales) to the most sophisticated (ratio scales), more and more constraints are placed on the transformations that can be meaningfully applied. This is because the numbers in higher-measurement scales contain more information that can be altered or destroyed. In practice, a ratio scale, like the interval scale, corresponds to the range of a discrete quantitative variable. The major characteristics of the four scales are summarized in Table 1.4-2.

---

[5] A positive linear transformation of a variable X consists of multiplying X by a positive constant b and adding a constant a to the product. The equation for a transformed value thus is $X' = a + bX$.

Table 1.4-2.   Overview of levels of measurement

Level of measurement	Characteristics
Nominal	Symbols serve as labels for mutually exclusive and exhaustive equivalence classes. The symbols have the property of distinctness. Appropriate transformation: any one-to-one substitution. Examples: sex, eye color, racial origin, personality types, and primary taste qualities.
Ordinal	Ordered symbols, usually numbers, indicate rank order of equivalence classes. The symbols have the properties of distinctness and order. Differences between ordered symbols provide no information about differences between equivalence classes. Appropriate transformation: monotonic. Examples: military rank, classification of mentally retarded children, rank in high school, and a supervisor's ranking of employees.
Interval[a]	Equal differences among numbers reflect equal magnitude differences among equivalence classes, but the origin or starting point of the scale is arbitrarily determined. Numbers have the properties of distinctness, order, and equivalence of intervals. Appropriate transformation: positive linear. Examples: Fahrenheit and Celsius temperature scales, calendar time, and altitude.
Ratio[a]	All the properties of interval scales apply, and, in addition, the origin of the scale reflects the absence of the measured characteristic. Appropriate transformation: multiplication by a positive constant. Examples: height, weight, Kelvin temperature scale, and measures of elapsed time.

[a] These two levels are sometimes referred to collectively as *metric measurement* or *numerical measurement*.

Implications of the Two Ways of Thinking about Numerals

Two ways of thinking about numerals have been described—one that reflects the concerns of mathematicians and statisticians and the other, the concerns of behavioral scientists and educators. Statistical methods were developed for analyzing numbers as numbers, whether or not they are true measures of some characteristic. If the assumptions associated with the statistical methods are fulfilled, they will produce answers that are formally correct as numbers. And this is true regardless of the degree of isomorphism or correspondence between the numbers and the characteristic they represent. The problem comes in translating statistical results into statements about the real world. If numbers representing a nominal scale are manipulated arithmetically, the result will be numbers that are numerically correct but uninterpretable. If nonsense is put into the equation, nonsense indeed will come out.

Experimenters concerned with problems of people and nature are very sensitive to the difficulties of interpreting numbers produced by statistical procedures—and rightfully so. Some authors have even gone so far as to prescribe the statistical procedures that can be used with each level of measurement.[6] Except in the physical sciences, few scales have equal intervals, so the number of statistical techniques on the approved list is relatively small. This position, it seems, fails to recognize that measurement of many variables in the behavioral sciences and in education lies somewhere between the ordinal and interval levels. The IQ scale is a good example. The 10-point difference between IQs of 100 and 110 is probably not the same as that between 130 and 140, but the scores contain some magnitude information, since most test administrators would agree that the intellectual difference between children with IQs of 100 and 110 is less than that between children with IQs of 110 and 130. Even though the measurement units along the scale are not equal, those familiar with IQ tests believe that the magnitude of the difference between scores does provide useful information about differences in intellectual aptitude. Should we avoid performing arithmetic operations on IQ scores because the measurement is below interval level? Or will mathematical manipulations yield interpretable answers and enable us to make better predictions concerning intelligence? Such questions have been heatedly debated.[7] We cannot look to mathematicians and statisticians for answers because these questions are outside their province. The answers must come from users of statistics who are acquainted with the problems of translating numerical answers into statements about the real world. An examination of the professional literature reveals that most experts in the behavioral sciences and education do apply arithmetic operations to numbers even though the measurement is somewhere between the ordinal and interval levels; further, they interpret the results as if differences between the numbers reflect something about the differences in the measured characteristics. Apparently, experts prefer to utilize whatever magnitude information the numbers contain even though differences among numbers only grossly approximate true magnitude differences. If an experimenter feels that any order-preserving transformation of a set of numbers adequately represents the equivalence classes, he or she should treat the scale as ordinal. If the experimenter's intuition is correct, the numbers contain no magnitude information, and they should not be treated as though they do. It is the experimenter, the person most familiar with the data, who must decide how much information the numbers contain.

If all our measurements were nominal or ratio, interpreting the outcomes of research would be fairly straightforward. Unfortunately, ratio measurement and even interval measurement are rare in the behavioral sciences and education. At best our measurement falls somewhere between the ordinal and interval levels. We can avoid some obvious and not-so-obvious interpretation errors by being sensitive to the degree of correspondence between a set of numbers and the charac-

---

[6] Examples can be found in Senders (1958), Siegel (1956), and Stevens (1946, 1951).

[7] The major issues in this debate have been presented by Anderson (1961), Boneau (1961), Gaito (1960), and Stevens (1968). These articles are reproduced in Kirk (1972a, chap. 2) along with suggestions for further reading on the issue. Also of interest are articles by Gardner (1975) and Wainer (1976).

teristic they represent. Suppose that on a standardized arithmetic-achievement test, Mortimer received a score of 0, Dude a score of 30, and Reginald a score of 60. Can we conclude that Mortimer knows nothing about arithmetic? Obviously not; a score of 0 means that he couldn't answer any questions on the test, but easier questions may exist that he could answer. Achievement tests, as well as many other tests, have arbitrary rather than absolute zero points, and therefore fall short of ratio measurement. It follows that although Reginald's score of 60 is twice as high as Dude's 30, Reginald's arithmetic achievement isn't necessarily twice Dude's.

The interpretation problem that results from lack of equal intervals is more subtle. Suppose we compare the effectiveness of two methods of teaching arithmetic. Students in a class using method A gained on the average 10 points; those in a class using method B, 7 points. The results seem straightforward, but suppose that originally the two groups were not equal in arithmetic achievement. Let the average score for group A be 40 and the score for group B be 10. Is it possible that a 7-point change from 10 to 17 represents more improvement in arithmetic achievement than a 10-point change from 40 to 50? Unless we know that, say, a 10-point change anywhere on the measurement scale represents the same empirical change, the interpretation of the experiment is equivocal. The greater the differences between the groups' initial performances, the more the problem is accentuated.

Because numbers don't always mean what they appear to mean, they should be carefully scrutinized. Consider a test that doesn't have enough ceiling to differentiate adequately among high-scoring subjects. Suppose that two individuals make the top score of 60. For one subject, this may represent maximum capability, but the other person may be capable of a much higher performance. The measuring instrument is simply incapable of showing it.

It is evident that the experimenter must be guided by two sets of rules. When the tools of statistics are used, the mathematician's and statistician's rules must be followed. When the numbers are interpreted as statements about the real world, measurement rules must be used as a guide. Of course, this calls for considerable informed judgment, since the application of measurement rules is the subject of unresolved controversy.

Review Exercises for Section 1.4

12. Mathematicians and behavioral scientists have somewhat different interests in numbers. Discuss these differences.

*13. Ignoring for the moment limitations of measuring instruments, classify measures of the following according to the mathematician's scheme (unordered qualitative, U; ordered qualitative, O; discrete quantitative, D; continuous quantitative, C).

  *a. Size of family
  *b. Race
  *c. Marital compatibility
  *d. Seeding of tennis players
   e. Employee production on an assembly line
   f. Creativity

        g. Political party affiliation
        h. Final standing of football teams in the Southwest Conference
        i. Weight loss after jogging for 3 miles
        j. Number of reported suicides in 1983
        k. Major in college
        l. Religious preference
       m. Grading scale in school (A, B, C, D, F)
        n. Amount of rainfall

14. Due to limitations of measuring instruments, measurement of some variables is of necessity approximate. Classify the variables in Exercise 13 according to whether our measurement is exact (E) or approximate (A).

*15. Reclassify the variables in Exercise 13 according to the mathematician's scheme, taking into account limitations in our ability to measure some of the variables.

16. (a) In what three ways do behavioral scientists use numbers in measurement? (b) Give two examples of each use.

*17. Classify the variables in Exercise 13 with respect to level of measurement, taking into account limitations in our ability to measure some of the variables.

18. For each level of measurement, list the properties that characterize the numbers assigned to the equivalence classes.

*19. For each level of measurement, indicate the appropriate transformation that can be performed on the numbers.

*20. What level of measurement is most often achieved (a) in the physical sciences and (b) in the behavioral sciences and education?

21. Who is in the best position to determine the degree of correspondence between a set of numbers and the corresponding equivalence classes and hence to determine the arithmetic operations that can meaningfully be applied?

22. What does a score of 0 on an aptitude test mean?

*23. Suppose that achievement test scores for a control group increased from 62 to 65 while those for the experimental group increased from 68 to 74. What must be true to conclude unequivocally that the experimental group improved twice as much as the control group?

24. Terms to remember:
        a. Variable
        b. Constant
        c. Qualitative variable, ordered and unordered
        d. Quantitative variable, discrete and continuous
        e. Measurement
        f. One-to-one transformation
        g. Monotonic transformation
        h. Positive linear transformation

Section 1.5   Historical Development of Statistics[8]

National Statistics

The science of statistics grew out of an attempt to solve practical problems associated with raising taxes, producing insurance tables, and determining the odds

---

    [8] The development of this section was strongly influenced by Dudycha and Dudycha (1972). Their article provides a more complete introduction to the history of statistics with special emphasis on the behavioral sciences.

in games of chance. Its subject matter was shaped by three lines of development—national statistics, probability theory, and experimental statistics. The oldest of these three is *national statistics*, which was enumerative and descriptive in character; it can be traced to the beginning of recorded history. David numbered his people, and the Egyptians and Romans kept detailed records of taxes and other state resources. Caesar Augustus simplified the enumerative process by ordering all citizens to report to the nearest statistician, better known as the tax collector. The descriptive use of statistics came of age in the work of an English army captain, John Graunt (1620–1674), who published in 1662 a small book of birth and death statistics for London from 1604 to 1661. Unlike earlier works, such as William the Conqueror's *Domesday Book*, which simply contained data compiled for purposes of taxation and military service, Graunt's book summarized and interpreted the data. His was the first work to shed light on the regularity of social phenomena. It marked the beginning of a theory of annuities and led to the founding of insurance societies.

Probability Theory

A second and independent line of development in statistics is *probability theory*. The earliest traces of probability, found in the Orient around 200 B.C., concerned whether an expected child would be male or female. However, the real impetus for development of probability came not from prospective parents but from gamblers who wanted to know the odds of winning at various games of chance. Leading mathematicians and scientists of the day—Pierre de Fermat (1601–1665), Blaise Pascal (1623–1662), Christianus Huygens (1629–1695), and James Bernoulli (1654–1705)—responded to the problem. Gradually they chiseled out the foundation of a theory of probability. A milestone in this development was the discovery of the normal curve of errors by Abraham de Moivre (1667–1754), a mathematics tutor who supplemented a meager income by calculating odds for gamblers at the coffeehouses he frequented. Apparently, de Moivre did not appreciate the significance of his discovery; it was published in 1733 only obscurely as a supplement written in Latin to a limited reprinting of a book he had published three years earlier. Therefore, it remained for others to demonstrate the pervasiveness of the *normal distribution*. For more than a century it was attributed to a later discoverer, Carl Friedrich Gauss (1777–1885), one of the greatest mathematicians of all time. It was also discovered independently by Pierre-Simon de Laplace (1749–1827), who forsook a cleric's robe for his lifework in celestial mechanics and probability. Both Laplace and Gauss used the normal distribution in investigating errors of observation in astronomy. Lambert Adolphe Jacques Quetelet (1796–1874), who is considered the father of social science, saw that the normal distribution and probability theory could be applied to all observational sciences—astronomy, anthropology, physics, the census, and the statistics of mental and moral traits. He used the normal curve, for example, to predict the number and type of crimes commited. His work integrated national statistics and probability theory and paved the way for the third line of development—experimental statistics.

Experimental Statistics

The emerging interest in the sciences in the early 1800s created a need for new statistical procedures and principles to guide the design of experiments. The result was *experimental statistics*. Its development was dominated by such intellectual giants as Sir Francis Galton (1822–1911). Terman, the developer of the Stanford–Binet intelligence test, estimated Galton's IQ at about 200. Galton, more than anyone before, used statistics in investigating problems of people and nature. His major statistical contributions were regression and correlation procedures (see Chapters 5 and 6), which he used to unravel mysteries of heredity. Karl Pearson (1857–1936) refined the mathematical theory of regression and made an astonishing number of other contributions to statistical theory and practice. Perhaps his greatest was the development in 1900 of the chi-square test for goodness of fit (see Chapter 17), which is used in testing the significance of differences between observed data and those expected on the basis of some hypothesis.

The modern era in experimental statistics was ushered in by William Sealey Gosset (1876–1937), who derived the *t* distribution (see Chapter 12) in 1908. Thus began the development of exact inductive procedures appropriate for both large and small samples. Heretofore experimenters had relied on large-sample statistical procedures. Gosset, who published under the pseudonym of *Student*, was a brewer for Messrs. Guinness. His discovery, like others in statistics, resulted from a practical need—in this case the need for inductive procedures appropriate for small samples. He was involved in brewing research, where variable materials and susceptibility to temperature changes precluded the use of large samples.

The modern era matured in the work of Sir Ronald A. Fisher (1890–1962), whose contributions to statistics are legion. He is best remembered for his derivation of the *F* distribution, contributions to the design and analysis of experiments, and heated exchanges about statistical theory with Jerzy Neyman (1894–1981) and Egon Pearson (1895–1981). Fisher's work was a unique blend of the rigor of the mathematician with a common-sense approach; the latter was undoubtedly due to his applied work in agriculture, biology, and genetics.

Neyman and Pearson carefully consolidated the work of Fisher and others while developing their own theory of statistical inference. The bulk of the statistical arsenal of today's researcher can be traced to Fisher, Neyman, and Pearson. But in response to changing research needs, there have been many new developments. The electronic computer has made possible the solution of problems that were heretofore intractable and has sparked new lines of inquiry. It seems unlikely, however, that a new era could be dominated to the extent that Fisher, Neyman, and Pearson dominated the one from 1920 to the present.

Review Exercises for Section 1.5

*25. *(a) What three lines of development shaped the subject matter of contemporary statistics? (b) Briefly summarize the major characteristics of each line.

26. For each of the following men, give at least one major contribution to statistics.
    a. Abraham de Moivre (1667–1754)
    b. Lambert Adolphe Jacques Quetelet (1796–1874)

   c. Francis Galton (1822–1911)
   d. Karl Pearson (1857–1936)
   e. William Sealey Gosset (1876–1937)
   f. Ronald A. Fisher (1890–1962)
   g. Jerzy Neyman (1894–1981)
   h. Egon Pearson (1895–1981)
*27. What distinguishes the modern era in experimental statistics from the previous period?

Section 1.6   Summary

This book is addressed to consumers and users of statistics and, more specifically, to students in the behavioral sciences and education. One course doesn't make a statistician, but it can help you develop basic understanding of and fluency in the technical jargon of statistics. This is useful, since statistics is the universal language for communicating research findings.

The word "statistics" has several meanings; usually it refers either to a collection of techniques for making decisions based on data or to functions describing data, such as the mean and range. Your study of statistics should help you to (1) read the professional literature in your field, (2) design and analyze simple experiments, and (3) detect statistical fallacies in the mass media and technical reports. In addition, you may learn new, more critical and analytical ways of thinking.

Research questions usually concern characteristics of populations. For example, what do people of voting age think about an issue? Is one instructional technology for fifth graders more effective than another? Do 21-year-old women prefer smaller families than men of the same age? The populations are, respectively, the attitudes of voters on the issue, the achievement scores of fifth graders, and preferred family sizes of 21-year-old women and men. The term "population" originally referred to all the inhabitants of the state. In statistics it refers to the collection of all objects or observations having one or more specified characteristics. Either for practical reasons or because the population is infinite in size, it is rarely possible to observe all the elements of a population. Instead, we conduct research on a sample of elements. A sample can contain a single element or all but one of the population elements. If every sample of a particular size has an equal chance of being selected from the population, the sampling process is said to be random.

Statistics can be applied to data from samples or from populations to obtain a clearer understanding of their characteristics. If we obtained a random sample of 21-year-old women and men, we might find that on the average they prefer, respectively, 2.2 and 2.4 children. In addition, we might learn that the range for women was 0 to 14 and that for men was 0 to 8.

The numbers, 2.2 and 2.4, and the ranges, 0–14 and 0–8, are descriptive statistics; they summarize properties of the two samples. Description is one important application of statistics. A second is the inferring of characteristics of a population by observing a sample. The sample statistics for preferred family size, for example, provide our best guess about the corresponding population values. These two uses of statistics, description and inference, are discussed in the first and second halves of this book.

Once an experimenter has identified the population of interest and the characteristic to be observed, he or she must decide how the characteristic should be measured. Mathematicians and statisticians have historically classified variables as qualitative (ordered or unordered) or quantitative (discrete or continuous). This scheme evolved because different mathematical tools are used in derivations and proofs for different kinds of variables.

Behavioral scientists, on the other hand, developed a classification scheme that reflected their concern with the degree to which numerals are isomorphic with the characteristics they re present. A four-level classification of measurement resulted: nominal, ordinal, interval, and ratio. Today we recognize that there are more than four levels of measurement and that the measurement of many variables in the behavioral sciences and education lies somewhere between the ordinal and interval levels. Much of the controversy surrounding the use of arithmetic operations with data between these levels can be resolved by adhering to two sets of rules. When the tools of statistics are used, the mathematician's and statistician's rules must be followed; when the numbers are interpreted as statements about the real world, the behavioral scientist's measurement rules must be followed.

Modern statistics is the culmination of three historical lines of development: national statistics, probability theory, and experimental statistics. Its origins are in antiquity, yet most of the material in this book is the product of the twentieth century. And we can expect to see an acceleration in the development of new statistical tools and theory—an acceleration made possible, in part, by the advent of the computer with its phenomenal capacity for information processing and storage.

2

Frequency Distributions and Graphs

Section 2.1 Introduction

Variation in the behavior of subjects under the same apparent conditions is inevitable. The problem is more troublesome in the behavioral sciences and education than in the physical sciences. A chemist can be confident that different samples of H_2O will react with another substance the same way under controlled tests. But this kind of uniformity in behavioral research is rare. The variation problem is usually handled by observing many subjects and collecting volumes of data. This in turn calls for procedures for depicting and summarizing data so that they can be more readily comprehended. Two kinds of descriptive tools serve the purpose—graphical and numerical methods. This chapter is devoted to graphical methods; numerical methods will be described in Chapters 3 through 6.

Section 2.2 Frequency Distributions

The first step in summarizing data is to organize the data in some logical fashion. This involves defining equivalence classes and counting the number of observations in each class. *An equivalence class can be a single score value, a collection of score values, or a collection of elements in a qualitative category. A table showing the classes and the frequency of occurrence of their elements is called a **frequency distribution.*** The equivalence classes of a frequency distribution are often called *class intervals.* If each of the class intervals is a single score value, the distribution is said to be *ungrouped.*

Ungrouped Frequency Distribution for Quantitative Variables

Suppose we administered a test of leadership aptitude to all high school football coaches in Punt County, Iowa. Their test scores are shown in Table 2.2-1. If we examine the table carefully, we can see that the smallest score is 30 and the largest is 68 and that most of the scores are in the high 40s or low 50s. The same information can be extracted much more easily from the ungrouped frequency distribution in Table 2.2-2, which associates with each score value, X, the frequency of its occurrence, f. In constructing the frequency distribution we followed the convention of putting the largest score at the upper left of the table. In addition, each number between the largest and the smallest scores is listed in the

Table 2.2-1. Leadership aptitude scores

Coach	Score	Coach	Score	Coach	Score
John Walker	55	William Kleis	39	Frank Vance	45
Harold Wilson	46	David Hays	47	Dave Baker	33
Jack Thornton	50	Thomas Brown	52	Bill Hicks	50
Heyward Helm	51	Bill Adams	54	Howard Clarke	51
John Hansen	48	Alan Ellis	48	Mike Gerbing	54
Mike Needham	50	Norman Jones	46	Ed Black	59
Chuck Carter	30	Douglas Kahn	68	Jack Wyman	49
Stephen Shanon	53	Bill Fluet	44	Don Campbell	42
Jack Patterson	57	Robert Vandorn	49	Chip Baty	56
Sharon Hamner	62	Joseph Schmidt	52	Mac Barker	53

Table 2.2-2. Ungrouped frequency distribution for leadership scores from Table 2.2-1

Score X	Frequency f	Score X	Frequency f	Score X	Frequency f	Score X	Frequency f
68			58	0	48	||	38	0
67	0	57	|	47	|	37	0
66	0	56	|	46	||	36	0
65	0	55	|	45	|	35	0
64	0	54	||	44	|	34	0
63	0	53	||	43	0	33	|
62	|	52	||	42	|	32	0
61	0	51	||	41	0	31	0
60	0	50	|||	40	0	30	|
59	|	49	||	39	|		

distribution so that every possible score can be tallied and the gaps between scores can be easily detected.

The frequency distribution is an effective organizing device, but some information is lost. We cannot tell from Table 2.2-2 which coach made the highest score, which coach made the lowest score, or that one of the coaches is a woman. We must refer to the original data for this information.

Grouped Frequency Distribution for Quantitative Variables

If the spread of scores for a quantitative variable is large, it is often helpful to construct a *grouped frequency distribution* in which the class intervals span two or more score values, as shown in Table 2.2-3. Such class intervals for a quantitative variable have a *nominal lower limit* and a *nominal upper limit*; for the class interval 66–68 they are, respectively, 66 and 68. However, the interval 66–68 actually includes any number *equal to or greater than* 65.5 and *less than* 68.5. The numbers 65.5 and 68.5 are called the *real limits* of the interval.[1] They extend 0.5 below

---

[1] If our measurement were accurate to the nearest tenth, so that we had scores such as 6.7, the class interval nominal limits would be 6.6 and 6.8 and the real limits would be 6.55 and 6.85. These values are obtained by adding and subtracting 0.05 instead of 0.5 from the nominal limits. Similarly, if our measurement were accurate to the nearest hundredth, the nominal limits would be 0.66 and 0.68 and the real limits would be 0.655 and 0.685, which differ from the nominal limits by ± 0.005.

Table 2.2-3. Grouped frequency distribution for leadership aptitude scores from Table 2.2-1

Class interval	Frequency, f
66–68	1
63–65	0
60–62	1
57–59	2
54–56	4
51–53	6
48–50	7
45–47	4
42–44	2
39–41	1
36–38	0
33–35	1
30–32	1
	$n^a = 30$

[a] n denotes the total number of scores in the frequency distribution.

the nominal lower limit and, in actuality, 0.4999 . . . above the nominal upper limit. The nominal limits are used to denote each class interval; the real limits indicate the underlying continuity of the class intervals and are used to compute the class interval size. The size of a class interval, denoted by i, is equal to the real upper limit minus the real lower limit; for example,

$$i = 68.5 - 65.5 = 3.$$

The concepts of real limits and class interval size also apply to the class intervals in ungrouped frequency distributions such as in Table 2.2-2. For the class interval 68, for example, the real limits are 67.5 and 68.5.[2] The class interval size is

$$i = 68.5 - 67.5 = 1.$$

Several conventions are followed in constructing a frequency distribution. They are not inviolate rules, but rather are guidelines for constructing easily interpreted tables.

1. For quantitative variables, the class interval containing the largest score should be at the top of the table. For qualitative variables, the order of class intervals should be determined by convenience and/or logic.
2. The class intervals should be mutually exclusive; that is, class intervals should be chosen so that a score belongs to one and only one interval.
3. For quantitative variables, there should be no gaps between the class intervals. In Table 2.2-3, even though no scores fall in the class intervals 63–65 and 36–38, the intervals are included for completeness.

[2] Some variables do not follow this convention. A common example is age. If a person is 21, this means that the twenty-first birthday has passed, but the twenty-second has not. The real limits for the age 21 are 21.0 and 21.999 . . . .

4. All quantitative class intervals should have the same width or size.[3]
5. There should be 10–20 class intervals unless the number of scores is very small, in which case it may be desirable to use fewer class intervals. For qualitative variables, the number of class intervals is usually dictated by the nature of the variable. For example, if the variable is sex, there may be three class intervals—male, female, and unknown.
6. For quantitative variables, one of the preferred class interval sizes should be used: 1, 2, 3, 5, 10, 15, 20, ...
7. The nominal lower limit of each quantitative class interval should be an integer multiple of the class interval size. For example, the nominal lower limit of the class interval 30–32 is equal to $3 \times 10 = 30$, where $i = 3$ and 10 is the integer multiplier.

Determining the Number and Size of Class Intervals for a
Quantitative Variable

The conventions for constructing a grouped frequency distribution provide general guidelines for the number and size of class intervals. We know that there should be 10–20 class intervals (unless there are only a few scores), and the preferred class interval sizes are 1, 2, 3, 5, 10, 15, 20, ... With these guidelines in mind, we can estimate the number and size of class intervals in a trial-and-error fashion using the formula

$$\frac{\text{Range}}{\text{Preferred } i} = \text{Number of class intervals.}$$

The range is equal to the real upper limit of the largest score minus the real lower limit of the smallest score. A preferred class interval size ($i = 1$ or 2 or 3 or . . .) is selected so that the formula yields between 10 and 20 class intervals. To illustrate, the largest and smallest scores in Table 2.2-1 are 68 and 30. The range is $68.5 - 29.5 = 39$. If a class interval size of 2 is tried in the formula, there will be $\frac{39}{2} \simeq 20$ class intervals. Since there are only 30 scores, a smaller number of class intervals would be preferable. If a class interval size of 3 is selected, the formula yields $\frac{39}{3} = 13$ class intervals, the number used in Table 2.2-3. A class interval size of 5 should not be used because it would give only $\frac{39}{5} \simeq 8$ class intervals. For most sets of data there will be no more than two class interval sizes that give the desired 10–20 class intervals. As a general rule, when the number of scores is small, use fewer than 15 class intervals; when the number is large, use 15–20.

Suppose that we have administered a test of reading readiness to 26 children enrolled in the first grade. The highest and lowest scores on the test are 132 and 73; the range is $132.5 - 72.5 = 60$. How many class intervals of what size should

[3] Sometimes this is not possible or desirable. Suppose that one subject was unable to learn a list of nonsense syllables in the usual number of trials, 6–10, required by most subjects. After the twentieth trial the subject was still unable to meet the learning criterion and gave up. This subject can't be given an exact score; he or she falls into the top class interval "20 or more." This interval is *open*, since one real limit can be specified but not the other. Or, suppose that the class intervals represent family income. It might be desirable to make the bottom and top class intervals open to include the few families with extremely small or extremely large incomes.

the frequency distribution have? By trial and error and the formula

$$\frac{\text{Range}}{\text{Preferred } i} = \text{Number of class intervals,}$$

we see that two grouping schemes are possible: $\frac{60}{3} = 20$ and $\frac{60}{5} = 12$. The one in which $i = 5$ is preferred because there are only 26 scores. The lowest class interval, following convention 7, given earlier, would be 70–74 since 70 is an integer multiple of $i = 5$—that is, $70 = 5 \times 14$. The highest class interval would be 130–134. Thus, this grouping scheme actually results in 13 instead of 12 class intervals because the smallest and largest scores don't fall at the nominal lower and upper limits of their class intervals. If the extreme scores had been 134 and 70, the procedure for estimating the number of class intervals would have given 13.

Suppose that we had tested 221 children instead of 26. In this case, a class interval size of 3 should be used. The lowest and highest class intervals would be 72–74 and 132–134, since 72 and 132 are integer multiples of 3. Even though the use of $i = 3$ results in 21 class intervals, it is preferred to $i = 5$ because of the large number of scores. The purpose of graphical methods is to make data easier to comprehend, and sometimes this can best be done by departing from the conventions.

The Pros and Cons of Grouping Data

Grouping scores into class intervals where $i > 1$ has its disadvantages. First, some information inevitably is lost. For example, we know from Table 2.2-3 that four scores occur in the class interval 54–56, but we don't know their individual values.

A second disadvantage is that the rules used to group data do not always produce unique distributions. For any set of data, only one ungrouped frequency distribution can be constructed, but for grouped frequency distributions, there is often a choice between two or more class interval sizes. In such cases, the shape of the frequency distribution depends on the size of the class interval chosen.

The disadvantages of loss of information and lack of uniqueness must be weighed against the simplicity achieved by grouping. If the spread of scores is large, a grouped frequency distribution is much more easily interpreted. Also, grouping may be ultimately more convenient if graphs are to be used, since graphs usually are constructed from grouped frequency distributions. Years ago grouping was used to minimize the labor required in computing statistics. The advent of hand-held calculators and computers eliminated this use for grouped frequency distributions.

Relative Frequency Distributions

To aid in the interpretation of a frequency distribution, it is often helpful to express each frequency as either a proportion or a percentage of the total number of scores. The formulas for proportionate frequency and percentage frequency are

$$\text{Prop} f = \frac{f}{n} \quad \text{and} \quad \% f = \frac{f}{n} \times 100,$$

where f is the frequency in the class interval and n is the total number of scores. *A distribution giving **Prop f** or **% f** is called a **relative frequency distribution.***

Table 2.2-4. Relative frequency distributions
for leadership aptitude scores

Class interval	f	Prop f	% f
66–68	1	0.03	3
63–65	0	0	0
60–62	1	0.03	3
57–59	2	0.07	7
54–56	4	0.13	13
51–53	6	0.20	20
48–50	7	0.23	23
45–47	4	0.13	13
42–44	2	0.07	7
39–41	1	0.03	3
36–38	0	0	0
33–35	1	0.03	3
30–32	1	0.03	3
	$n = 30$	Sum $= 0.98^a$	Sum $= 98^a$

[a] The sums do not equal 1.00 and 100 due to errors
introduced by rounding numbers.

Usually the frequency associated with each class interval is given along with Prop f
or % f. Both kinds of relative frequencies are shown in Table 2.2-4.

The transformations Prop f and % f convert the total number of scores, n, into
standard numbers—either 1.0 or 100. We see that the sum of Prop f is 1.0 and
the sum of % f is 100. Relative frequencies indicate whether a frequency is "relatively large" rather than whether it is "absolutely large." For example, the class
interval 48–50 in Table 2.2-4 contains only seven scores, but this is a relatively
large proportion (almost one-fourth) of the total number of scores. Also, relative
frequencies are particularly useful in comparing two frequency distributions with
different n's. Consider the history achievement scores shown in Table 2.2-5 for high
school students taught by two methods. Because of the great difference in n's, a
comparison of % f's is more meaningful than a comparison of f's.

Cumulative Frequency Distribution

*A **cumulative frequency distribution** shows the number, proportion, or percentage of
scores that occur below the real upper limit of each class interval.* Such a distribution
is helpful in answering certain kinds of questions. For instance, if Susan's score is
62, how many students did better and how many did worse? Or, what scores
divide the top 10% or the bottom 25% of students from the remainder of the
class?

To construct a cumulative frequency distribution, we begin with a frequency
distribution like the one in the first two columns of Table 2.2-6. A given cumulative
frequency is obtained by adding the frequency in column 2 for the class interval
to the cumulative frequency recorded in column 3 for the class interval below it.
For example, in the class interval 30–32, $f = 1$ and there are no scores below, so
the Cum f for that class interval is $1 + 0 = 1$. For the class interval 33–35, $f = 1$,
which, added to the Cum f below, yields a Cum f of $1 + 1 = 2$. The cumulative
frequency recorded for the top class interval should equal n.

Table 2.2-5.   History achievement scores for classes taught by different methods

Achievement scores	Method A		Method B	
	f	$\% f$	f	$\% f$
150–154	1	1	1	3
145–149	0	0	2	6
140–144	2	3	2	6
135–139	4	5	4	12
130–134	6	8	6	19
125–129	8	11	8	25
120–124	9	12	5	16
115–119	10	14	2	6
110–114	8	11	1	3
105–109	8	11	0	0
100–104	6	8	1	3
95–99	5	7	0	0
90–94	3	4	0	0
85–89	2	3	0	0
80–84	1	1	0	0
	$n = 73$	Sum $= 99^a$	$n = 32$	Sum $= 99^a$

[a] Sum is not equal to 100 due to errors introduced by rounding numbers.

Table 2.2-6.   Cumulative frequency distributions for leadership aptitude scores

(1) Class interval	(2) f	(3) Cum f	(4) Cum prop f	(5) Cum $\% f$
66–68	1	30	1.00	100
63–65	0	29	0.97	97
60–62	1	29	0.97	97
57–59	2	28	0.93	93
54–56	4	26	0.87	87
51–53	6	22	0.73	73
48–50	7	16	0.53	53
45–47	4	9	0.30	30
42–44	2	5	0.17	17
39–41	1	3	0.10	10
36–38	0	2	0.07	7
33–35	1	2	0.07	7
30–32	1	1	0.03	3
	$n = 30$			

Cumulative frequencies can be transformed into Cum prop f and Cum $\% f$ by the formulas Cum prop $f = $ Cum f/n and Cum $\% f = $ Cum $f/n \times 100$. These relative frequencies are shown in columns 4 and 5 of Table 2.2-6.

Frequency Distributions for Qualitative Variables

The construction of frequency distributions for qualitative variables is relatively simple since no decisions about size and number of class intervals have to be made—the equivalence classes of the variable become the class intervals.

Table 2.2-7.   Political affiliation of students at Ohio State University

(1) Political affiliation	(2) f	(3) Prop f	(4) %f
Democrat	92	0.42	42
Independent	33	0.15	15
Republican	85	0.38	38
Unspecified or other	11	0.05	5
	$n = 221$	Sum = 1.00	Sum = 100

Consider the unordered qualitative variable of political party affiliation: Democrat, Independent, Republican, and Unspecified or other. If we obtained a random sample of college students at Ohio State University and determined their political affiliation, we could construct a frequency distribution like the one in columns 1 and 2 of Table 2.2-7. The equivalence classes are ordered alphabetically for lack of a more logical sequence. For ordered qualitative variables, class intervals should preserve the order inherent in the original equivalence classes.

The frequencies in column 2 of Table 2.2-7 are converted to Prop f in column 3 and %f in column 4. Cumulative frequencies are not shown; they are not meaningful since the order of the class intervals was arbitrarily determined.

Review Exercises for Section 2.2

*1. A marriage counselor asked his clients to keep a record of the number of arguments they had during the week. The following data for 23 couples were obtained. Construct an ungrouped frequency distribution for these data.

```
2   5    4   9    6
4   3    3   5   10
5   0   13   4    2
1   7    6   3
4   5    4   4
```

2. Construct an ungrouped frequency distribution for study-abroad candidates' ages at last birthday. The data are as follows.

```
18   20   19   20
20   19   19   19
23   18   20   21
17   20   18   20
```

*3. Assembly-line workers were asked to complete a job-satisfaction questionnaire. Construct an ungrouped frequency distribution for the following scores.

```
 7    8    4   25    9    8    4   15   11    9
 6    9    7    7   10   17    5   10    5    8
 3    7   11    8   13   22    7    8    7    6
10    6    7    9    4    8    6    6    8   11
 7    5    6    6    7    7   12    5    5    6
15   21    5   11    6    9    5   12   10    8
```

*4. List the guidelines for constructing an ungrouped frequency distribution.
*5. For the following nominal class intervals, give the real limits and the class interval size.

    *a. 50–54   *b. 74   c. 16
    *d. 18.0–19.9   e. 60–69   f. 18.00–19.99
    g. 12.0–14.9   h. 0–0.4   i. 1.50–1.74

*6. For each of the following give (a) the number of class intervals, (b) the size of the class interval, and (c) the nominal limits of the lowest class interval.

	Largest Score	Smallest Score	Number of Scores
*a.	68	22	53
*b.	260	106	21
*c.	254	92	91
d.	37	8	106
e.	62	23	273
f.	164	126	29
g.	52	0	22

*7. A test of mechanical aptitude was given to seniors at Middlecenter High School. Construct a grouped frequency distribution for the following data.

$$
\begin{array}{ccccccc}
80 & 73 & 51 & 81 & 46 & 85 & 84 \\
75 & 44 & 84 & 77 & 95 & 48 & 88 \\
50 & 35 & 52 & 93 & 43 & 59 & 63 \\
47 & 66 & 55 & 58 & 62 & 51 & 75 \\
86 & 82 & 89 & 51 & 77 & 73 & 59
\end{array}
$$

8. First-line supervisors were asked to complete the job-satisfaction questionnaire mentioned in Exercise 3. Construct a grouped frequency distribution for the following data.

$$
\begin{array}{ccccc}
25 & 23 & 18 & 24 & 14 \\
21 & 17 & 12 & 19 \\
15 & 6 & 22 & 16 \\
20 & 20 & 21 & 18
\end{array}
$$

*9. In a traffic safety research project, the reaction time in milliseconds of 27 subjects to the onset of a light was measured. For the following data, (a) construct two grouped frequency distributions having different i's, and (b) discuss the relative merits of the two grouping schemes.

$$
\begin{array}{cccccc}
186 & 187 & 211 & 185 & 196 & 193 \\
184 & 185 & 191 & 188 & 192 & 190 \\
188 & 190 & 202 & 199 & 189 \\
193 & 186 & 180 & 205 & 187 \\
189 & 195 & 184 & 198 & 202
\end{array}
$$

10. What are the advantages and disadvantages of grouped and ungrouped frequency distributions?
*11. For the data in Exercise 7, construct a relative frequency distribution using Prop f.
12. For the data in Exercise 8, construct a relative frequency distribution using % f.
13. Construct a relative frequency distribution for comparing the job satisfaction of assembly-line workers in Exercise 3 with that of first-line supervisors in Exercise 8.
14. Under what conditions is a relative frequency distribution more informative than an ordinary frequency distribution?

*15. Thirty-two subjects participated in a paired-associates learning experiment in which they were shown 12 nouns written in hiragana (a Japanese writing system) and asked to learn the corresponding English words. The number of trials required to learn to a criterion of two consecutive correct anticipations is shown below. Construct a cumulative frequency distribution for the data.

10	9	11	12	6	14	10	12
11	10	12	10	9	11	16	8
9	7	8	11	10	8	12	12
13	10	10	9	11	13	7	11

*16. For the data in Exercise 1, construct a cumulative proportionate frequency distribution.

17. For the data in Exercise 7, construct a cumulative frequency distribution.

18. For the data in Exercise 8, construct a cumulative percentage frequency distribution.

*19. A random sample of 29 students from each of the following classifications (freshman, sophomore, junior, senior, and graduate student) were asked whether they believed in ESP. The classifications of students who believed in ESP are listed below. Construct a frequency distribution for these data.

Junior	Senior	Junior	Sophomore	Junior
Freshman	Junior	Freshman	Junior	Sophomore
Sophomore	Senior	Senior	Senior	Junior
Graduate	Sophomore	Junior	Freshman	Senior
Freshman	Junior	Junior	Senior	Sophomore
Junior	Senior	Sophomore	Senior	

20. Students enrolling in Introductory Sociology were randomly assigned to one of three classes: traditional lecture (TL), guided reading (GR), or audiotutorial (AT). Following are the class assignments of the top 30 students on the final examination; construct a frequency distribution for these data.

AT	GR	AT	TL	GR	AT
AT	TL	GR	AT	AT	GR
TL	TL	TL	AT	AT	AT
GR	AT	AT	AT	TL	TL
AT	AT	TL	GR	AT	AT

21. Twenty-five physicians were asked what they felt was the main health threat to male executives. The most common responses were occupational stress (OS), obesity (OB), smoking (S), lack of exercise (LE), and other (O). Construct a frequency distribution for these data.

OB	OS	S	OB	LE
S	OB	OB	OS	O
LE	S	OS	S	OB
O	LE	O	LE	LE
OB	LE	O	O	OB

22. Toss a die 50 times and construct a frequency distribution showing the number of times each die face occurred.

23. Contrast the procedures for constructing frequency distributions for qualitative variables with those for quantitative variables.

24. Under what condition would it be meaningless to construct a cumulative frequency distribution for a qualitative variable?
25. Terms to remember:
    a. Frequency distribution
    b. Nominal limits
    c. Real limits
    d. Relative frequency distribution
    e. Proportionate frequency
    f. Percentage frequency
    g. Cumulative frequency distribution

Section 2.3   *Introduction to Graphs*

Frequency distributions present the main features of data succinctly, but they are still abstract numerical representations and require effort to interpret. Graphs can impart the same information and speak to us more directly. Their ease of interpretation makes them particularly useful when we want to present data to the general public.

There are many ways to graph data. In fact, whole books have been devoted to the subject.[4] Our presentation is limited to the six most common graphs: bar graphs, pie charts, histograms, frequency polygons, cumulative polygons, and stem-and-leaf displays. Qualitative variables are usually represented by bar graphs and pie charts, and quantitative variables, by histograms, frequency polygons, cumulative polygons, and stem-and-leaf displays.

Section 2.4   *Graphs for Qualitative Variables*

Bar Graphs

Once a frequency distribution has been made, most of the work of constructing a bar graph has been done. The only step remaining is to cast the data into two dimensions. This is illustrated in Figure 2.4-1 for the data in Table 2.2-7. Class intervals are represented along the horizontal axis (*abscissa*, or X axis) and frequencies, along the vertical axis (*ordinate*, or Y axis). The zero point or origin of the vertical axis is located at the X and Y intercept—the point where the two axes cross. A vertical bar is erected over each class interval such that its height corresponds to the number of scores in the interval. The bars can be any width, but they should not touch, since their noncontiguity points up the discrete, qualitative character of the class intervals. By convention, the height of the tallest bar should be 66 to 75% of the length of the horizontal axis. This results in a rectangular figure with proportions that are the most esthetically pleasing (according to the ancient Greeks). The X and Y axes of the graph should be labeled and a figure caption provided to help the reader interpret the graph.

The Y axis can be used to represent proportionate frequency or percentage frequency, depending on the questions of interest to the experimenter. We saw in Section 2.2 that these transformations are useful in determining whether a fre-

---

[4] One example is Arken and Colton (1938).

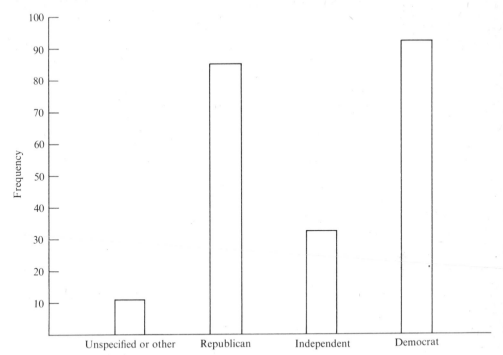

Figure 2.4-1.   Political affiliation of a random sample of $n = 221$ students at Ohio State University. (Data from Table 2.2-7)

quency is relatively large and in comparing graphs with different total numbers of scores.

Pie Charts

*Perhaps the most easily interpreted graph is a **pie chart**, which is merely a circle divided into sectors representing proportionate frequency or percentage frequency of the class intervals.* A pie chart is illustrated in Figure 2.4-2 for the data in Table 2.2-7. To construct a pie chart, think of it as a circle having 60 min like a clock. The sections corresponding to proportionate frequency or percentage frequency are marked off in minutes according to the formulas Prop $f \times 60$ and $\% f/100 \times 60$. For Figure 2.4-2, the minutes corresponding to the four percentage frequencies are

$$42\%/100 \times 60 = 25.2$$

$$15\%/100 \times 60 = 9.0$$

$$38\%/100 \times 60 = 22.8$$

$$5\%/100 \times 60 = 3.0.$$

Thus, 42% corresponds to 25.2 min after 12 o'clock; the next 15% corresponds to $25.2 + 9.0 = 34.2$ min after 12 o'clock; the next 38%, to $25.2 + 9.0 + 22.8 = 57$ min; and so on. The last steps in constructing a pie chart are to label the sections of the pie and provide an appropriate figure caption.

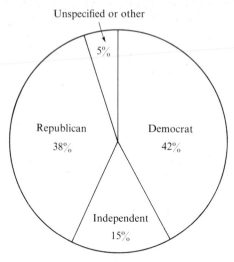

Figure 2.4-2. Political affiliation in percentage frequency of a random sample of $n = 221$ students at Ohio State University. (Data from Table 2.2-7)

Review Exercises for Section 2.4

*26. College students were asked to give their favorite leisure-time activity. The five most commonly mentioned activities were rapping with friends (RF), reading (R), watching television (TV), participating in a sport (PS), and drinking (D). Construct a bar graph for the following data.

RF	PS	D	RF	R	TV	RF	D	PS
RF	RF	R	TV	RF	D	TV	RF	TV
D	TV	RF	RF	D	RF	R	R	RF
R	R	TV	D	TV	D	D	RF	TV
TV	RF	PS	TV	RF	TV	TV	D	
D	D	TV	RF	PS	RF	RF	D	

27. Information from a biographical inventory was used to compute a socioeconomic index for students in a university marching band. Scores above 72 were classified as very high (VH); scores from 61 to 72, as high (H); from 43 to 60, as middle (M); and below 43, as low (L). Construct a bar graph for the following data.

H	H	H	H	M	VH	VH	H	M
M	L	H	M	VH	H	H	H	VH
H	M	M	H	H	VH	H	M	
VH	H	H	M	M	VH	M	L	
H	VH	VH	H	H	M	VH	M	
M	M	VH	L	M	H	H	VH	

*28. A study was conducted in an Arizona nursing school to determine if attitudes toward research would improve after students conducted a project of their own. After completing a required research course, students who expressed positive attitudes toward research were asked to indicate which one of four statements best represented their attitude. Of the 230 student nurses who responded, 31 checked that they would like to be involved in research after graduation. Seventy-three checked that nurses should understand research as a part of their professional responsibility. Sixty checked that they felt confident in their ability to evaluate research in nursing. Sixty-six checked that the required project was responsible for their improved understanding of the research process. Construct a bar graph for these data. (Suggested by Van Bree, Nancee S. Undergraduate research. *Nursing Outlook*, 1981, 39–41.)

29. The value of psychoeducational programs as a means of preventing and relieving problems of daily living is gaining acceptance in the medical community. A health maintenance organization recently surveyed the health needs of its members by means of a questionnaire. The number of respondents who selected one of nine popular programs as the one in which they were most interested is shown in the following table. Construct a bar graph for these data. (Suggested by Burnell, George M., and Taylor, Peter H. Psychoeducational programs for problems in living. *Health and Social Work*, 1982, 7(1), 7–13.)

Program	Number indicating primary interest
Weight reduction	154
Fatigue	101
Marital and sex problems	92
Coping with physical problems	71
Stress	65
Heart disease prevention	61
Assertiveness	60
Stop smoking	48
Headaches	47

30. Construct a bar graph for the data in Exercise 20 (page 33).
31. Construct a bar graph for the data in Exercise 21 (page 33); plot percentage frequency on the *Y* axis.
32. Describe the procedure for constructing a bar graph from a frequency distribution.
*33. Construct a pie chart for the data in Exercise 26.
34. Construct a pie chart for the data in Exercise 27.
35. Construct a pie chart for the data in Exercise 28.
36. Construct a pie chart for the data in Exercise 20 (page 33).
37. Describe the procedure for constructing a pie chart from a frequency distribution.
38. Terms to remember:
     a. Bar graph   b. Abscissa   c. *X* axis
     d. Ordinate   e. *Y* axis   f. Pie chart

Section 2.5   Graphs for Quantitative Variables

Histograms

A **histogram** is similar in appearance and construction to a bar graph, but it is used for quantitative variables rather than for qualitative variables. It is constructed by erecting vertical bars over the **real limits** of each class interval, with the height of each bar corresponding to the number of scores in the interval. The bars are contiguous

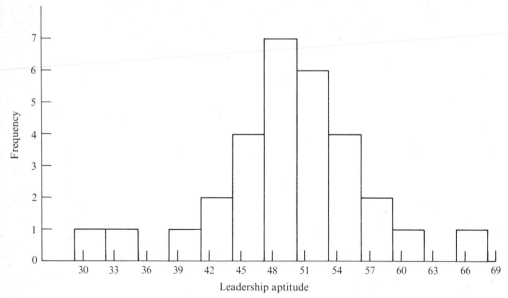

Figure 2.5-1.   Histogram for leadership aptitude scores for $n = 30$ football coaches. (Data from Table 2.2-3)

rather than separated; this emphasizes the continuity of the class intervals. Except for this difference, histograms and bar graphs are constructed in the same manner: (1) the class intervals are represented along the horizontal axis and frequency is represented along the vertical axis, (2) the zero point or origin of each axis is located at the X and Y intercept, (3) the height of the tallest bar is 66 to 75% of the length of the horizontal axis, and (4) the two axes are labeled appropriately and a figure caption is given to help the reader interpret the graph.

A histogram for the data in Table 2.2-3 is shown in Figure 2.5-1. Note that the edges of the bars coincide with the real limits of the class intervals rather than with the nominal limits, for example, 29.5–32.5 and not 30–32. Either frequency or relative frequency can be represented along the vertical axis. (Transformation of frequencies to relative frequencies is discussed in Section 2.2.)

Frequency Polygons

To construct a frequency polygon from a frequency distribution, we begin as though we were making a histogram. The horizontal axis is marked off into class intervals, and the vertical axis, into numbers representing frequencies. However, the frequency of class intervals is not represented by vertical bars but by dots placed at the proper height over the midpoints of the class intervals. Finally, adjacent dots are joined by straight lines. At the extremes of the graph, two additional class intervals containing no scores are identified and lines are dropped to their midpoints so as to anchor the graph to the horizontal axis. The midpoint

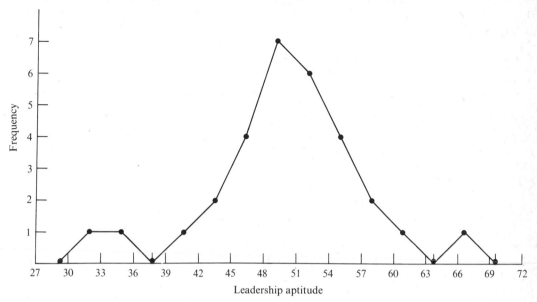

Figure 2.5-2. Frequency polygon for leadership aptitude scores for $n = 30$ football coaches. (Data from Table 2.2-3)

of a class interval is given by

$$\text{Midpoint} = \frac{(\text{Upper limit of class interval} + \text{Lower limit of class interval})}{2}.$$

For the data in Table 2.2-3, the midpoint of the class interval 30–32 is $(32 + 30)/2 = 31$. A frequency polygon for these data is shown in Figure 2.5-2. Frequency polygons and histograms impart the same information; the choice between them is largely a matter of personal preference. The histogram is probably a little easier for the general public to interpret, but the stepwise bars tend to obscure the shape of the distribution. The frequency polygon is preferred when two or more sets of data are represented in the same graph, since superimposed histograms, even markedly different ones, are difficult to interpret.

Cumulative Polygons

We saw in Section 2.2 that a cumulative frequency distribution can be used to show the number, proportion, or percentage of scores that lie below the real upper limit of each class interval. This information can be represented graphically by a cumulative polygon. Instead of placing dots over the midpoints of class intervals, we place them over the real upper limits. The vertical axis can represent Cum f, Cum prop f, or Cum $\% f$. A cumulative percentage polygon for the data in Table 2.2-6 is shown in Figure 2.5-3. As is usually the case in the behavioral sciences and

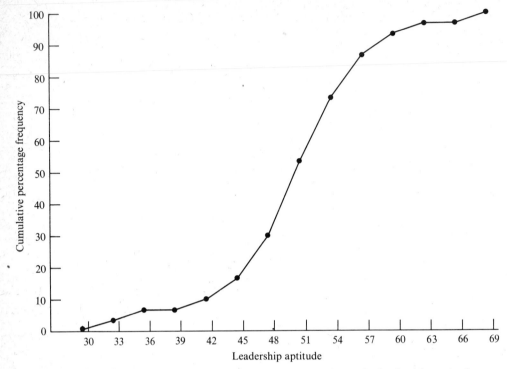

Figure 2.5-3. ·Cumulative percentage frequency polygon for leadership aptitude scores for *n* = 30 football coaches.

education, the cumulative polygon has the characteristic S shape. The S shape occurs whenever there are more scores in the middle of the corresponding frequency distribution than at the extremes. Graphs that are S shaped are called *ogives* (pronounced "oh jives").

Stem-and-Leaf Displays

Another useful graphic procedure is called the *stem-and-leaf display*.[5] It resembles a histogram that has been turned on its side. A stem-and-leaf display is illustrated in Table 2.5-1 for the data in Table 2.2-2. The first step in constructing the display is to specify class intervals following the procedures in Section 2.2. The stems or rows of the display represent class intervals. The leaves consist· of the individual scores in each class interval. A score is represented by its unit digit, for example, 68 = 8, 62 = 2, and 57 = 7; the scores within a stem can be ordered from the smallest to the largest.

A stem-and-leaf display provides all of the information that is in a histogram and, in addition, preserves the values of the individual scores. If desired, the display can be supplemented by a frequency distribution, as in column 3 of Table 2.5-1.

[5] The procedure was popularized by John Tukey (1977).

Table 2.5-1.   Stem-and-leaf display for data in
Table 2.2-2.

(1) Stems (Class intervals)	(2) Leaves (Scores)	(3) Frequency (f)
66–68	8	1
63–65		0
60–62	2	1
57–59	7 9	2
54–56	4 4 5 6	4
51–53	1 1 2 2 3 3	6
48–50	8 8 9 9 0 0 0	7
45–47	5 6 6 7	4
42–44	2 4	2
39–41	9	1
36–38		0
33–35	3	1
30–32	0	1
		$n = 30$

Review Exercises for Section 2.5

*39. The following data represent the number of cigarettes smoked per day by mothers whose first babies were stillborn. Construct a histogram for these data.

27	25	31	22	3	16	15
21	32	29	30	12	14	26
9	27	25	27	30	28	31
30	18	0	23	20	21	
28	16	10	19	13	19	

40. Rats were shown three illuminated symbols; their task was to press the lever below the symbol that differed from the other two. The dependent measure was the number of trials required to learn to the criterion of eight consecutive correct responses. Construct a histogram for these data.

52	34	57	47	54	56	46
60	63	42	20	50	81	41
43	51	36	73	56	77	59
50	42	58	65	42	58	63
66	55	53	63	53	54	61

41. Construct a histogram for the data in Exercise 8 (page 32). Plot percentage frequency on the ordinate.

42. Construct a histogram for the data in Exercise 9 (page 32). Plot proportionate frequency on the ordinate.

43. How does the construction of histograms and bar graphs differ?

*44. Determine the midpoints of the following class intervals.
   *a. 20–24   b. 8–11   *c. 132–133   d. 1.50–1.74
   *e. 15–29   f. 100–104   g. 0–2   h. 60–69

*45. Construct a frequency polygon for the data in Exercise 39.

46. Construct a frequency polygon for the data in Exercise 40.

47. A study was undertaken to determine how well psychological crises resulting from traumatic events are resolved over time. The subjects included 15 female cancer patients who underwent breast surgery for the first time, 15 female patients who underwent less serious surgery (gall bladder removal, hernia repair, and so forth), and 15 physically healthy (nonsurgery) women. Each patient took the Halpern Crisis Scale at intervals of 0, 3, 7, 11, and 15 weeks. The 0 interval represented the night before surgery. The sample of healthy control subjects also took the scale at the same time intervals. The following data, number of women with a Halpern Crisis Scale score over 72, were obtained. A score above 72 is considered a high crisis score. (Suggested by Gottesman, David, and Lewis, Marc S. Differences in crisis reactions among cancer and surgery patients. *Journal of Consulting and Clinical Psychology*, 1982, *50*, 381–388.)

| | *Week number* | | | | |
Group	*0*	*3*	*7*	*11*	*15*
Cancer surgery	11	12	14	12	14
Other surgery	8	11	12	10	8
Nonsurgery	4	5	5	6	5

   Construct a frequency polygon for these data. Plot the data for each group on the same graph; do not anchor the polygon to the horizontal axis.
48. Construct a frequency polygon for the data in Exercise 8 (page 32). Plot percentage frequency on the ordinate.
49. What are the relative merits of histograms and frequency polygons?
*50. (a) Construct a cumulative polygon for the data in Exercise 39; plot Cum $\%f$ on the ordinate. (b) Estimate the score above which 50% of the cases fall.
51. (a) Construct a cumulative polygon for the data in Exercise 40; plot Cum prop f on the ordinate. (b) Estimate the score below which 50% of the cases fall and the score below which 20% of the cases fall.
52. Construct a cumulative polygon for the data in Exercise 9 (page 32).
*53. How can we tell from a frequency distribution whether a cumulative polygon for the data would have an S shape?
*54. Construct a stem-and-leaf display for the data in Exercise 39.
55. Construct a stem-and-leaf display for the data in Exercise 40.
56. Terms to remember:
    a. Histogram   b. Frequency polygon
    c. Class interval midpoint   d. Cumulative polygon
    e. Ogive   f. Stem-and-leaf display

Section 2.6   Shapes of Distributions

Graphs come in many different shapes. Some shapes occur with enough regularity that they have been given special names. These are shown in Figure 2.6-1.

Bell-Shaped Distributions

The frequency polygon in Figure 2.6-1(a) approximates the shape of the normal distribution, which is discussed in Chapter 10. This important distribution is symmetrical—the right half is the mirror image of the left half—and it has a

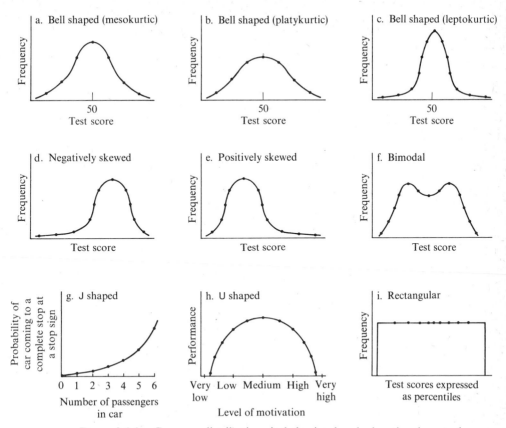

Figure 2.6-1.   Common distributions in behavioral and educational research.

particular degree of peakedness. *The property of being peaked, flat, or in between is referred to as* **kurtosis.** The normal distribution is *mesokurtic;* "meso-" means intermediate. Distributions that are flatter than the normal distribution are called *platykurtic;* "platy-" means flat or broad. Those that are more peaked are designated "lepto-," meaning slender or narrow. Examples of these distributions are shown in Figures 2.6-1(b) and (c). These distributions and (a) all center around the same test score, 50. The point around which a distribution centers is referred to as its *central tendency.* Another important characteristic of a distribution is its *dispersion*—the extent to which scores are spread out around a central point. The scores in Figure 2.6-1(c), for example, have less dispersion or scatter than those in (a) and (b).

Skewed Distributions

Distributions are either symmetrical or asymmetrical. *If the longer tail of an asymmetrical distribution extends toward the X and Y origin, as in Figure 2.6-1(d), the distribution is* **negatively skewed.** *If the longer tail extends away from the intercept, as in Figure 2.6-1(e), the distribution is* **positively skewed.** A negatively skewed distribution results, for example, if a very easy test is given to subjects. Since most of the subjects score high and only a few score low, the longer tail trails off toward the X and Y origin. A positively skewed distribution results if the test is very hard.

Bimodal Distributions

A distribution is **bimodal** *if it has two humps, each having the same maximum frequency.* Bimodal distributions often result when two distinct populations are represented on a single graph. For example, a graph like Figure 2.6-1(f) would result if we plotted the masculinity scores of 50 men and 50 women.

A graph with three or more humps, each having the same maximum frequency, is **multimodal.** Technically, a distribution is bimodal or multimodal only if its humps have the same frequency. Nevertheless, distributions with pronounced but unequal humps are commonly described as bimodal or multimodal.

J, U, and Rectangular Distributions

J *and* U *distributions are so named because their shapes resemble those letters.* A J-shaped curve like the one in Figure 2.6-1(g) is obtained, for example, if the probability of coming to a complete stop at a stop sign is plotted as a function of the number of passengers in the car. A reversed J curve is obtained if the number of people arriving for church is plotted as a function of the number of minutes they are late. Similar results are obtained in most studies of conforming social behavior; most people conform to social conventions and laws, so fewer and fewer people exhibit larger degrees of nonconformity.

A U curve like the one in Figure 2.6-1(h) is obtained, for example, if performance on a difficult task is plotted as a function of the level of motivation of the subjects.

A **rectangular** *or* **uniform** *distribution is one in which each class interval has the same frequency.* A rectangular distribution is produced when test scores are converted to percentiles (see Section 4.2) and the number of scores in the class intervals 0–10th percentile, 10–20th percentile, . . . , 90–100th percentile is graphed. It follows that the resulting graph will be rectangular since each class interval by definition must contain 10% of the scores.

In this section we have described some common distributions, and in the process we have mentioned four characteristics of distributions: (1) central tendency, (2) dispersion, (3) symmetry or lack of symmetry, and (4) kurtosis. In Chapters 3 and 4 we will see how to compute numbers that represent each of these characteristics.

Review Exercises for Section 2.6

*57. Indicate whether the following statements are true or false.
   *a. A normal distribution is symmetrical and mesokurtic.
   *b. If the upper half of a distribution is not the mirror image of the lower half, the distribution is asymmetrical.
   c. A distribution that is flatter than the normal distribution is called mesokurtic.
   d. "Lepto" in leptokurtic means slender or narrow.

e. The tail of a negatively skewed distribution extends away from the X and Y intercept.

f. A distribution with three maximum humps each having the same frequency is bimodal.

*58. Draw the shape of a frequency polygon that would occur in each of the following experiments. Identify each distribution.

 *a. Beauty contestants take a masculinity test.

 *b. An intelligence test is given to a large sample of sixth-grade children.

  c. Students at Juilliard School of Music take a test of musical aptitude.

  d. Students are surprised with a pop quiz immediately after the Easter vacation.

  e. Subjects attempt to solve 20 complex puzzles under five levels of motivation: very low, low, medium, high, and very high.

  f. Crime statistics per 1000 population are collected for five cities; it turns out that the cities have the same crime rate.

  g. Engineering majors' and business majors' scores on a test of mechanical aptitude are plotted.

  h. Strength of grip is measured for males; the sample contains all ages but a preponderance of young boys, men in their early 20s, and men over 65.

  i. Arrival time is recorded for people who are late for a concert.

  j. The number of persons contracting polio in the United States from 1940 to 1970 is determined from hospital records.

59. Terms to remember:

a. Normal distribution	b. Symmetrical distribution
c. Kurtosis	d. Mesokurtic
e. Leptokurtic	f. Platykurtic
g. Central tendency	h. Dispersion
i. Skewness (negative and positive)	j. Bimodal
k. Multimodal	l. J distribution
m. U distribution	n. Rectangular distribution

Section 2.7   Misleading Graphs

Graphs should be constructed so that they accurately portray the essential characteristics of data. Not all graphs do this—some even defy correct interpretation. Graphs of the same data can convey entirely different impressions, as shown in Figures 2.7-1(a) and (b), which report crime statistics for three similar neighborhoods. In neighborhood A, cruising patrol cars were eliminated during a 3-month trial period; B had 5 cruising cars during the period; while C was flooded with 15 cars. Your conclusions about the effects of patrol cars would probably depend on which graph you saw. Figure 2.7-1(a) gives the impression that the presence or absence of patrol cars is associated with a dramatic difference in crime rate. Note, however, that the largest difference—1000 versus 970—is only 3%. Such a small difference could just as well be attributed to chance factors or to differences in crime-reporting procedures. The graph is misleading because it violates the height–width rule and because the Y axis begins with a frequency of 960 crimes instead of 0 crimes.[6] This is inimical to the aim of statistics, which is to help the user make sense out of data.

---

[6] Huff (1954) illustrates many more misleading techniques.

Figure 2.7-1. Number of reported crimes in three similar neighborhoods during a 3-month test period. Note how graph (a) falsely gives the impression of a great difference in crime rate across the three conditions.

A more subtle form of misrepresentation can occur in pictograms. *A pictogram represents quantity by means of pictures of the objects being compared.* Pictograms are often used in the mass media in place of bar graphs and histograms to enliven a presentation. Consider Figure 2.7-2, in which sales for three brands of microcomputers are represented by two types of pictograms. Figure 2.7-2(a) is inherently misleading because our perception of the sales of the three brands is influenced not only by the heights of the pictures but also by their areas, and area

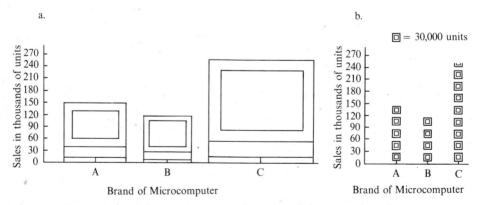

Figure 2.7-2. Pictograms representing sales of three popular microcomputers. Pictogram (a) is misleading because our perception of sales is influenced by the heights of the pictures and by their areas, and area is an irrelevant dimension.

is an irrelevant dimension. For example, sales for brand C are approximately twice those for brand B, but the area of brand C's picture is 4.3 times larger than that of brand B. The pictogram in Figure 2.7-2(b) provides a more realistic representation of sales.

Review Exercises for Section 2.7

*60. Prepare two bar graphs for the following data. Design one to deliberately suggest that government spending has been stable, the other to suggest a dramatic increase in government spending.

Month	Spending	Month	Spending
June	$29,400,000	October	$29,500,000
July	29,200,000	November	29,600,000
August	29,300,000	December	29,800,000
September	29,600,000	January	30,200,000

61. The following data are sales figures for vacuum cleaner salespeople. Prepare graphs that suggest that (a) all the salespeople are producing at a uniformly high level, (b) Chapman should be fired, and (c) they should all be fired.

Chapman	$66,000	Mayes	$68,200
Hays	$67,300	Belew	$71,000
Toland	$69,900	Reynolds	$71,100

62. (a) Collect examples from newspapers and news magazines of graphs that appear to be misleading. (b) Why is each one misleading?

63. In the news media, how do graphs reporting financial statistics differ from those used in advertising? Note whether the X and Y axes of the graphs are clearly labeled and whether they have figure captions.

Section 2.8   Summary

This chapter presented two descriptive devices that make data easily comprehensible: frequency distributions and graphs. As a first and sometimes final step in summarizing data, an experimenter may report the various equivalence classes and the number of observations that fall into each. This results in a table called a frequency distribution. If each equivalence class is a single score value, the distribution is ungrouped; if the classes contain two or more score values, the distribution is grouped. Grouping simplifies the interpretation of data by collapsing them into 10–20 class intervals.

A graph is a pictorial representation of a frequency distribution and hence is much easier to interpret. The most common graphs for qualitative variables are bar diagrams and pie charts. Histograms, frequency polygons, cumulative polygons, and stem-and-leaf displays are commonly used to represent quantitative variables.

A graph should present data accurately and unambiguously in such a way that its main characteristics can be seen at a glance. To achieve this, certain conventions are followed: (1) plotting frequency on the Y axis and equivalence classes on the X axis; (2) placing the zero point (or origin) at the X and Y intercept; (3) making the height of the graph 66 to 75% of its width; (4) labeling the X and Y axes; and (5) providing a figure caption.

†*Section 2.9   Computer Supplement*

There was a time when performing statistical computations was sheer drudgery, but hand-held calculators and digital computers have changed that. This chapter supplement provides a brief survey of these computing aids and software packages, which are the instructions that tell a computer what to do.

2.9-1   Hand-Held Calculators

Hand-held calculators fall into one of two broad categories: nonprogrammable and programmable. A nonprogrammable calculator requires the user to key in the operations that make up a calculation each time the calculation is performed. A programmable calculator can store a sequence of operations as it is entered from the keyboard. This allows the user to perform repeatedly the calculations by simply initiating the sequence.

Nonprogrammable calculators.   The simplest of the nonprogrammable calculators is the four-function $(+, -, \times, \div)$ calculator, which can be purchased for as little as $5. Such calculators usually have special keys for obtaining percents and square roots. For slightly more money (less than $20) an expanded-function calculator can be purchased that provides an eight-digit display, one or more registers for storing intermediate results, a constant facility for using the same value in repeated calculations, and keys for obtaining percents, change in percents, squares, square roots, π, and logarithmic, exponential, and trigonometric functions. Such calculators with four or more fully addressable storage registers are well suited to performing statistical calculations. Special-purpose statistical calculators, available for under $45, provide a ten-digit display, eight or more fully addressable storage registers, and one-key computation of a variety of statistics.

Programmable calculators.   With the development of sophisticated programmable calculators, the calculator–computer distinction has become blurred. Programmable calculators can store data and instructions (programs) and then execute the instructions at high speed without human intervention. And like computers, the more expensive programmable calculators can (1) branch within a program—if a value is, say, greater than 1, perform one set of instructions; if it is not, perform a different set, (2) store programs on an external medium, (3) interact with a variety of peripheral devices such as a printer, and (4) accept instructions written in one of the higher-level programming languages. Although the distinction between programmable calculators and computers has become blurred, computers still have the edge in speed of processing instructions, storing voluminous amounts of data and instructions, and flexibility in interacting with peripheral devices. In the sections that follow we will examine digital computers and software packages.

2.9-2   Computer Hardware

Types of digital computers.   In 1951 the U.S. Bureau of the Census purchased a Univac I (*UNIV*ersal *A*utomatic *C*omputer), the first computer designed for the

---

†This and similarly marked sections can be omitted without loss of continuity.

marketplace. It was a monstrous machine containing thousands of vacuum tubes. By precomputer standards, its computational speed was phenomenal—up to 5000 additions per second. The development of the transistor paved the way for the second generation of computers, introduced in 1958. They could perform 500,000 additions per second and had greatly expanded memories for storing data and instructions. The third generation of computers, using integrated circuits—tiny chips that combine all the capabilities of many transistors and other circuit components—was developed in 1964. Such computers are capable of performing 10 million additions per second.

As computers became more powerful, they also became smaller as a result of technological breakthroughs and advances in microelectronics. Today, digital computers are classified as *large (mainframe) computers*, *minicomputers*, or *microcomputers*. As the names suggest, distinctions are based largely on size. If it requires a truck to carry it, it is a mainframe computer. If two burly men can carry it, it is a minicomputer. If you can put it under your arm, it is a microcomputer.

Components of a computer.   Regardless of size, all general-purpose digital computers have the following components: a central processing unit (CPU), main computer storage, and input–output (I/O) devices. A schematic diagram of a computer is shown in Figure 2.9-1. The central processing unit contains the electronic circuits that enable the computer to carry out the instructions in a program, perform tests, and control input and output devices in response to instructions.

Bits and bytes.   Instructions and data are stored in the computer's memory in the form of a binary code. The basic unit is the *bit* (*bi*nary digi*t*). This unit can be in one of two states: *on* or *off*. A single bit with only two states is obviously not sufficient to code the numbers 0 to 9, letters A to Z, and special characters. One coding scheme, the Extended Binary Coded Decimal Interchange Code (EBCDIC), uses eight bits to accomplish this. The eight bits taken together form a *byte* or computer word. One byte is required to store each decimal digit, letter, or special

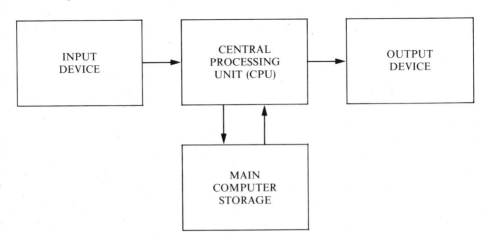

Figure 2.9-1.   Basic components of a general-purpose computer.

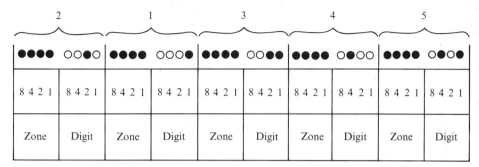

Figure 2.9-2.   EBCDIC coding scheme for the number 21345. Five bytes are required to store the number. Numbers in this scheme are represented with all zone bits "on" and with different combinations of the digit bits. Letters and special characters are represented by different combinations of both zone and digit bits. There are $2^8 = 256$ possible code combinations in an eight-bit byte.

character. The coding scheme for the number 21345 is shown in Figure 2.9-2. Another coding scheme called the American Standard Code for Information Interchange (ASCII) is widely used with microcomputers.

Main computer storage.   Each byte in main computer storage has a unique address. This enables the computer to access the address and retrieve the information stored there. Large computers typically can store between 512,000 (512K) and 16 million (16 mega) bytes. Minicomputers can store between 64K and 1 megabytes, while most microcomputers can store only between 2K and 128K bytes. But the storage capacity of micros is increasing. For example, when the Apple II was introduced in 1977 it had a storage capacity of 16K bytes that could be expanded to 48K. Its successor, the Apple II Plus, has a capacity of 64K bytes; the most recent version, the Apple IIe, has a maximum storage capacity of 128K bytes.

The type of storage we have just described is called *random access memory* (RAM). Information can be both written into and read from each memory location. Another type of memory is called *read only memory* (ROM). The information in ROM is placed there when the computer is manufactured. The information can be read and used but not altered.

Auxiliary storage.   Because main computer memory is expensive, it is customary to supplement it with less expensive, auxiliary storage media. The earliest of these was the punched card, developed in the late 1880s by Dr. Herman Hollerith. The modern version, introduced by IBM in 1928, has 80 columns and 12 rows. Numbers, letters, and special characters are coded by one or more punches in a column. Because reading cards into a computer is a relatively slow process and because the cards are bulky, they have been superseded by other auxiliary storage media—most notably magnetic tape and magnetic disks.

Magnetic tape resembles that used with audio tape recorders. A 10-inch-diameter reel of tape for a minicomputer or mainframe computer is capable of storing up to 6250 bytes per inch; common storage densities are 800 and 1600 bytes per inch. Considering that the tape is 2400 feet long, this is a lot of storage capacity.

The data on a magnetic tape is stored sequentially. Since a finite length of time is required to run the tape past the computer's read head, access time to a particular datum is not instantaneous. In order to reduce access time, magnetic disks resembling a stack of LP phonograph records were developed. The last decade has seen amazing improvements in disk technology. One variant is the $5\frac{1}{4}$-inch-diameter floppy disk or *diskette* that is used with microcomputers. Such disks can store 175K bytes—roughly half of a typical novel. Eight-inch diskettes are available but are less popular. A more recent innovation in disk technology is the $5\frac{1}{4}$- and 8-inch hard or rigid disks. Unlike floppy disks that the user inserts into the computer's disk drive, hard disks are permanently housed in a disk drive. Their advantage is greater storage capacity. For example, $5\frac{1}{4}$-inch hard disks have capacities that range from 5 to 50 megabytes. The smallest hard disk can store approximately 15 paperback novels.

Input–output devices.   Input devices enable the user to communicate with the central processing unit and place information in memory. Input media include the punched card, the typewriter-like keyboard of a cathode ray tube terminal (CRT terminal), magnetic tape, and magnetic disks.

The final step in the data processing cycle is to obtain output from the computer. For many years the primary output media were the printed report and the punched card. The punched card has been largely supplanted by the CRT terminal, which provides a TV-like display.

Batch versus interactive computing.   Thus far we have described the major types of digital computers and their components, often referred to as *computer hardware*. Before turning to computer software, we will distinguish between two modes of processing instructions and data—batch and interactive processing—and describe microcomputers in more detail.

The earliest computers operated in the *batch processing mode*. In this mode the computer executes a complete program without any interaction with the user. In the *interactive mode* the user monitors the progress of the computations at a CRT terminal, responds to queries or prompts from the computer, and provides information or specifications as required. The user also may alter the computational process through the use of special instructions called preempts. As we will see, software packages are designed to operate in one of these two processing modes.

More about microcomputers.   In 1974 a small electronics firm in Albuquerque, New Mexico, named MITS announced that a home personal computer could be purchased in a kit form for about $500. It came with 1K of RAM—hardly enough to store 200 words of English text. Nevertheless, the firm sold thousands of computers in the first year. Thus began the phenomenal growth of the microcomputer industry. At the time of this writing, over a million microcomputers had been sold in the USA alone. One software services firm in California estimates that the number of microcomputers in use is growing at the rate of about 24% a year.

Most microcomputers puchased today come assembled with a CRT as the primary output device and a typewriter-like keyboard as the primary input device. Many also have auxiliary storage, either a cassette tape similar to that used in audio tape recorders or a floppy disk.

The big four in terms of sales are the Apple II, Radio Shack's TRS-80, IBM's Personal Computer, and the Commodore PET and VIC 20. A TRS-80 model III with a printer, cassette magnetic-tape drive, and the software packages for word processing and budget management can be purchased for as little as $1299. Atari, Texas Instruments, and Timex offer microcomputers that use the owner's TV screen as the primary output device. These units have been advertised for under $200, $190, and $100, respectively. The central processing unit for these computers, called a *microprocessor* (MPU), supports an eight-bit word. But 16- and 32-bit microcomputers are available; IBM's Personal Computer and Apple's Lisa are examples, respectively. Such computers are inherently faster than eight-bit computers because they manipulate bigger chunks of data at a time. A 16-bit word, for example, can represent four digits or two characters. Also, 16- and 32-bit computers tend to have more sophisticated instruction repertoires; therefore they can accomplish more with each instruction. With the development of 16- and 32-bit microprocessor chips and hard disks with their greater storage capacity, microcomputers can begin to challenge minicomputers in terms of speed and number-crunching ability.

Microcomputers typically are sold with 16 to 64K of RAM. Those microcomputers with less than 64K usually have provisions for expanding their memory capacity. For example, the Timex Sinclair 1000 comes with 2K of RAM, which can be expanded to 16K by the purchase of a $49 module. As a general principle, the more memory the better. For example, 64K is sufficient to store slightly over 10,000 words of English text. Although 32K is sufficient for many statistical applications, complex procedures and large data sets can require many times more memory. With continuing price declines and increased speed and memory capacity, it is likely that the microcomputer will become the primary computing tool in statistics.

2.9-3   Computer Software

The term *software* refers to programs written for a computer. There are two broad categories of software—systems software and applications software. *Systems software* consists of a series of programs that reside permanently in memory and control the input–output operations of the computer, schedule the resources of the computer system, and communicate with the computer operator. In this chapter supplement we will be concerned only with *applications software*—those programs written for particular applications such as processing data.

Computer languages.   As we saw earlier, computers respond to a binary code. The earliest programs were written in a language called *machine language* that could be interpreted directly by the computer's electronic circuitry. Unfortunately, programming in machine language is a difficult and burdensome task. To simplify the programming process, symbolic programming languages called *assembler languages* were developed. Programs written in assembler languages are very efficient in terms of program execution time and main storage use. However, a given assembler language reflects the particular architecture of the computer for which it was written and generally will not run on a different type of computer.

Higher-level languages, which are described next, are not related to the internal characteristics of a particular computer. As a result, programs written in such languages can run on different computers with few or no alterations. Furthermore, higher-level languages bear a resemblance to the English language and, hence, make programming easier and less error-prone.

Over 200 different programming languages are in use today. The major higher-level languages are FORTRAN, COBOL, PL/I, BASIC, and PASCAL. FORTRAN, BASIC, and PASCAL are widely used in the behavioral sciences and education; COBOL is a ·business-oriented language. BASIC, which stands for *B*eginner's *A*ll-Purpose *S*ymbolic *I*nstruction *C*ode, is commonly used with microcomputers.

Statistical software packages.   One does not have to learn a programming language to use a computer. Software packages written in one of the higher-level languages are available for almost every conceivable application—from balancing your checkbook to circumnavigating the earth. The major statistical software packages are BMDP (Biomedical Computer Programs), IMSL (International Mathematical and Statistical Libraries), LISREL V, MINITAB, OMNITAB II, OSIRIS, SAS (Statistical Analysis System), SCSS Conversational System, SSP/360 (Scientific Subroutine Package), and SPSS (Statistical Package for the Social Sciences). SCSS is an interactive package; MINITAB, OMNITAB II, and OSIRIS can run in either mode. In general, interactive packages are easier to learn to use than batch packages.

The packages just mentioned require a large amount of storage and usually are implemented on large computers or minicomputers. Manufacturers of microcomputers offer their own sets of programs. Programs for microcomputers also are available from independent software suppliers and computer buffs. Advertisements for these programs appear in magazines devoted to microcomputers.

Software packages enable the novice to perform the most complex statistical procedures in a matter of seconds. However, such computing power may encourage the unwary to use procedures that they do not fully understand. This can result in inappropriate analyses and incorrect interpretations of the analyses. To combat this problem, a significant shift in the emphasis of statistics courses is taking place. Computational matters are being given less attention, whereas more attention is being devoted to examining the assumptions underlying statistics, the conditions under which they are appropriate, and their proper interpretation.

To acquaint you with the output of statistical packages, computational examples using various packages will be given in the computer supplement at the end of most chapters. The SCSS and BMDP packages were used to construct a frequency distribution and histogram for the leadership aptitude data in Table 2.2-1. The computer printouts are shown in Tables 2.9-3 and 2.9-4.

Table 2.9-3.   Printout from the SCSS univariate program (Data from Table 2.2-1)

```
LOGFILE OK
EXPLAIN? (Enter YES or NO)
· NO
STYLE OF PROMPTING?
VERBOSE
WORKFILE NAME?  (NONE* or filename)
KIRK.WRK
MASTERFILE NAME:  KIRK.MSF
PROCEDURE?   (Procedure name followed by varlist)
UNIVARIATE
VARIABLE LIST?  (Varlist)
LAS
CELLS?  (ALL* or VALUE,N,TOTPCT,NMPCT,CUMPCT,LABELS or NONE)
VALUE,N,LABELS
STATISTICS? (MEAN*, STDDEV* or list of statistics)
DISPLAY?  (STATS, CELLS, BOTH and/or item list)
CELLS
```

LAS LEADERSHIP APTITUDE SCORE

VALUE	N	LABEL
13	1	66—68
11	1	60—62
10	2	57—59
9	4	54—56
8	6	51—53
7	7	48—50
6	4	45—47
5	2	42—44
4	1	39—41
2	1	33—35
1	1	30—32

TOTAL N = 30   VALID N = 30

```
DISPLAY? (STATS, CELLS, BOTH and/or item list)
/HISTOGRAM NONNORMAL DECREASING
```

LAS LEADERSHIP APTITUDE SCORE

PCT	N	LOW BND	(* = 1 CASES)
3.3	1	13.00	*
.0	0	12.00	
3.3	1	11.00	*
6.7	2	10.00	**
13.3	4	9.00	****
20.0	6	8.00	******
23.3	7	7.00	*******
13.3	4	6.00	****
6.7	2	5.00	**

```
3.3      1      4.00 *
 .0      0      3.00
3.3      1      2.00 *
3.3      1      1.00 *
TOTAL N = 30    VALID N = 30

DISPLAY? (STATS, CELLS, BOTH and/or item list)
/STOP
$
```

Table 2.9-4.   Printout from the BMDP histograms and univariate plots program (P5D)
(Data from Table 2.2-1)

```
PAGE   5      GROUPED FREQUENCY DISTRIBUTION FOR LEADERSHIP APTITUDE
HISTOGRAM OF VARIABLE    1 SCORES
                              SYMBOL   COUNT      MEAN      ST.DEV.
                                 X        30     48.300       7.484
                        EACH SYMBOL REPRESENTS      1 OBSERVATIONS
INTERVAL                                          FREQUENCY PERCENTAGE
NAME           5   10   15   20   25   30   35   40  INT. CUM. INT.  CUM.
      +----+----+----+----+----+----+----+----+
66-68  +X                                            1    1   3.3   3.3
63-65                                                0    1   0.0   3.3
60-62  +X                                            1    2   3.3   6.7
57-59  +XX                                           2    4   6.7  13.3
54-56  +XXXX                                         4    8  13.3  26.7
51-53  +XXXXXX                                       6   14  20.0  46.7
48-50  +XXXXXXX                                      7   21  23.3  70.0
45-47  +XXXX                                         4   25  13.3  83.3
42-44  +XX                                           2   27   6.7  90.0
39-41  +X                                            1   28   3.3  93.3
36-38                                                0   28   0.0  93.3
33-35  +X                                            1   29   3.3  96.7
30-32  +X                                            1   30   3.3 100.0
      +----+----+----+----+----+----+----+----
        5   10   15   20   25   30   35   40
NUMBER OF INTEGER WORDS OF STORAGE USED IN PRECEDING     PROBLEM     416
CPU TIME USED       3.370 SECONDS
```

   The SCSS program (Nie *et al.*, 1980) that was used to construct the frequency distribution and histogram in Table 2.9-3 is called *Univariate*. It is one of nine statistical programs in the package. At first glance, the printout may appear to be somewhat cryptic. As we will see, it is really not difficult to interpret. Of course in order to fully understand a printout, the user should always consult the user's guide for the package.

   Recall that SCSS is an interactive package. The output in Table 2.9-3 shows the sequence of prompts or questions asked by SCSS (statements followed by a question mark, ?); specifications or answers provided by the user; and preempts or commands issued by the user (statements preceded by a slash, /). The program begins by asking the user if explanations for using the program are desired. The user responded with NO. The program provides three styles of prompting—verbose, normal, and terse. The user chose VERBOSE. Following this, work and master files were named and the procedure UNIVARIATE was requested. The user identified LAS (leadership aptitude scores) as the variable list in memory that was to be analyzed. The **cells?** and **statistics?** prompts ask the user to specify the specific items of information that are to be displayed; in this example the user requested VALUE (class interval designations), N (class interval frequencies), and LABEL (class interval labels). Statistics such as the mean were not requested. In response to the first **display?** prompt, the user requested a printout of the CELLS information, which produced the frequency distribution shown in Table 2.9-3. Note that class intervals with zero frequency are not printed.

A histogram was requested in response to the second **display?** prompt. Due to the graphic restrictions imposed by printers and CRT displays, SCSS produces an approximation of a histogram in which the traditional vertical bars are replaced by horizontal lines of asterisks. The first column of the histogram gives the percentage in each class interval, the second gives the frequency, and the third gives the lower boundary, or value, of the class interval designation. As is indicated in the printout, each asterisk represents one case.

The printout in Table 2.9-4 was produced by the the *Histograms and Univariate Plots Program (P5D)* of BMDP (Dixon *et al.*, 1981). This package was originally designed to run in the batch mode, although recent versions provide some interactive capability. Unlike SCSS, it does not provide a running dialogue with the user; virtually all instructions and program options must be specified prior to running the program. The printout in Table 2.9-4 gives a horizontal histogram, frequency distribution, cumulative frequency distribution, percents, and cumulative percents, as well as two descriptive statistics, the mean and standard deviation (ST. DEV.), which are discussed in Chapters 3 and 4 of the text, respectively. Other examples of computer printouts will be found in the computer supplement at the end of most chapters.

Review Exercise for Section 2.9

64. Terms to remember:
    a. Large (mainframe) computer
    b. Minicomputer
    c. Microcomputer
    d. Central processing unit (CPU)
    e. Input/output (I/O) devices
    f. Bit
    g. Byte
    h. Random access memory (RAM)
    i. Read only memory (ROM)
    j. Auxiliary storage
    k. Floppy disk (diskette)
    l. Hard disk
    m. Cathode ray tube (CRT) terminal
    n. Batch and interactive computing
    o. Computer hardware
    p. Computer software
    q. Microprocessor (MPU)
    r. Systems software
    s. Applications software
    t. Machine language
    u. Assembler language
    v. Higher-level language

3

Measures of
Central Tendency

Section 3.1   Introduction

A frequency distribution or a graph summarizes data, but sometimes it is desirable to summarize further by describing certain properties of the data numerically. The most interesting property is usually *central tendency*, the score value around which a distribution tends to center. This value is popularly called the *average*, and it connotes what is typical, usual, representative, normal, or expected. Because of these different connotations, statisticians prefer more precise terms. The *mode*, *mean*, and *median* are three conceptions of central tendency described in this chapter.

Close behind central tendency in importance is *dispersion*—the extent to which scores differ from one another—that is, their scatter or heterogeneity. Several ways of describing dispersion are discussed in Chapter 4. Chapter 4 also covers two other properties of data—*skewness* and *kurtosis*. Measures of skewness tell us whether a distribution is symmetrical or asymmetrical; measures of kurtosis tell us whether it is peaked or flat. These four properties of data—central tendency, dispersion, skewness, and kurtosis—can be measured to yield a relatively complete summary of the information presented by frequency distributions and graphs. In many cases, a knowledge of only two of these—central tendency and dispersion—is sufficient for our purposes.

Section 3.2   The Mode

The simplest of the three conceptions of central tendency is the mode, denoted by *Mo*. The **mode** *is the score or qualitative category that occurs with greatest frequency*. Consider the following scores, which represent the number of times in September that 11 students called their parents long distance: 0, 0, 0, 1, 1, 1, 1, 2, 2, 3, 9. We note that 0 occurs three times; 1, four times; 2, twice; and 3 and 9, once. The mode is 1, since it occurs with the greatest frequency. If data are tabulated in an ungrouped frequency distribution, we can determine the mode at a glance. This can be seen for the distribution of family size of college professors shown in Table 3.2-1. The largest frequency, 10, is associated with a family size of four; hence the mode is 4. This tells us that the most *typical* family size for this sample is four, an easy-to-understand concept. As these examples show, the mode is determined by inspection rather than by computation. The mode can be used to describe the central tendency of both qualitative and quantitative variables, but it is most often

Table 3.2-1.   Frequency distribution of family size of college professors

X	f
11	1
10	0
9	0
8	1
7	1
6	2
5	4
4	10
3	8
2	8
1	5
	$n = 40$

used for qualitative variables. We will see why this is true later, when we compare the three measures of central tendency.

The mode should be computed from an ungrouped frequency distribution if possible. If this can't be done, the midpoint of the class interval with the greatest frequency is designated as the mode. This is, of course, imprecise since the midpoint would be different for different grouping schemes.

As a measure of central tendency, the mode has one particularly serious limitation. As we saw in Section 2.6, a distribution can have two nonadjacent scores or two nonadjacent class intervals with the same maximum frequency. Such distributions are called bimodal and cannot be described by a single mode. It is customary in such cases to cite two modes—the scores associated with the two maximum frequencies—but then the concept of the mode as the most typical score no longer applies.

Review Exercises for Section 3.2

*1. The behavior of members of the university wine-tasting club was rated following their biweekly learn-by-doing meeting. The following scale was used: N = no change in behavior, S = slight change in verbal and/or emotional expressions, M = marked change in verbal and/or emotional expressions, C = clumsiness in locomotion, and G = gross intoxication. (a) Determine the mode for the following data: N, S, S, G, M, N, S, M, M, C, G, N, S, M, C, S, S, M, S, S. (b) What type of variable do the data represent?

2. In a paired-associates learning experiment, data representing the number of trials to reach the criterion of three consecutive errorless trials were 10, 6, 11, 10, 9, 8, 10, 11, 14, 12, 10, 9, 11, 10, 12, 9, 8, 9. (a) Determine the mode. (b) What type of variable do the data represent?

3. The electoral systems of 11 emerging nations were classified as N = noncompetitive, P = partially competitive, and C = competitive. (a) Determine the mode for the following data: N, P, N, C, N, P, P, N, N, C, N. (b) What type of variable do the data represent?

*4. The ruling structures of 11 emerging nations were classified as 1 = premobilized authoritarian, 2 = conservative authoritarian, and 3 = premobilized democratic. (a) Determine the mode for the following data: 1, 3, 1, 1, 2, 3, 1, 3, 3, 1, 3. (b) What type of variable do the data represent?

5. Why should the mode be computed from ungrouped rather than grouped data whenever possible?

Section 3.3   The Mean

The most widely used and familiar measure of central tendency is the **arithmetic mean**—*the sum of scores divided by the number of scores.* The mean[1] is commonly known as the average. The usual symbol for a sample mean, $\bar{X}$, is read "X bar."[2] The letter X identifies the variable that has been measured, and the bar above it denotes the mean. Other letters toward the end of the English alphabet, for example, Y and Z, are also used as symbols for variables, and the corresponding means are denoted by $\bar{Y}$ and $\bar{Z}$. In later chapters on inferential statistics, it is important to distinguish between the mean computed for a sample and the population mean. It is customary to use lowercase Greek letters to represent population characteristics. Following this convention, the mean of a population is denoted by μ, the Greek letter mu, and is read "mew." Different population means can be identified by number or letter subscripts, for example, μ_1 and μ_2 or μ_X and μ_Y. The distinction between samples and populations appears in another way—a descriptive measure for a sample is called a *statistic*; a descriptive measure for a population is called a *parameter*.

Computational Formula Using Summation Notation

The mean of a sample is obtained by dividing the sum of the scores by the number of scores. At this point we will describe a useful notation for the sum of scores. Suppose that we are interested in frequency of movie attendance of college students. We can denote this variable by the capital letter X and individual values of the variable by X and a subscript: $X_1, X_2, \ldots, X_i, \ldots, X_n$. According to this notation, X_1 is the frequency of movie attendance for person 1, X_2 is the frequency for person 2, and X_n denotes the frequency for the nth or last person in the sample. We will let i be a general subscript that designates an unspecified one of the $i = 1, \ldots, n$ persons. The i in X_i can be replaced by any integer between 1 and n inclusive.[3] Suppose that the following values of X_i were obtained: $X_1 = 3$, $X_2 = 1$, $X_3 = 4$, $X_4 = 2$. The mean of these $n = 4$ scores is given by

$$\bar{X} = \frac{X_1 + X_2 + X_3 + X_4}{n} = \frac{3 + 1 + 4 + 2}{4} = \frac{10}{4} = 2.5.$$

This formula is tedious to write, so it is customarily abbreviated by using the *summation* symbol $\sum$, the Greek capital sigma. The symbol $\sum$, like $+$, indicates

---

[1] There are several kinds of means, but only the arithmetic mean is discussed in this book.

[2] Some books in the behavioral sciences and education denote the mean by M; $\bar{X}$ or $\bar{x}$ is preferred by statisticians (Halperin, Hartley, & Hoel, 1965).

[3] The letter i is also used to denote the size of a class interval; this use is discussed in Section 2.2. (Since there are only 26 letters, it is not surprising that they have multiple meanings.)

addition. However, $+$ indicates the addition of only two numbers, while $\sum_{i=1}^{n}$ means to perform addition until all $i = 1, \ldots, n$ numbers have been added.[4] The equation

$$\sum_{i=1}^{n} X_i = X_1 + X_2 + \cdots + X_n$$

says to let the first value of X_i be X_1; add to this the second value, X_2, and so forth until the X_nth value has been added. In the notation $\sum_{i=1}^{n}$, i is called the *index of summation*, $i = 1$ is the initial value of i, and n is its terminal value. Using summation notation, the formula for the mean movie attendance of four students is written $\bar{X} = \sum_{i=1}^{4} X_i/4$, which is equivalent to $\bar{X} = (X_1 + X_2 + X_3 + X_4)/4$. *The general formula for $\bar{X}$ is written*

$$\bar{X} = \sum_{i=1}^{n} X_i/n.$$

When the initial and terminal values for the summation are clearly understood, the formula may be simplified to $\bar{X} = \sum X_i/n$ or $\sum X/n$.

Computing the Mean from an Ungrouped Frequency Distribution

The formula $\bar{X} = \sum_{i=1}^{n} X_i/n$ is appropriate for data in their original unordered state. *If the data have been ordered by an ungrouped frequency distribution, the mean can be computed from* $\bar{X} = \sum_{j=1}^{k} f_j X_j/n$, *where X_j denotes the value of the jth class interval, f_j is the frequency of scores in the jth class interval, $\sum_{j=1}^{k}$ is the sum over the $j = 1, \ldots, k$ class intervals, and n is the number of scores.* The use of this formula is illustrated in Table 3.3-1. The data are scores on the Wakefield Self-Assessment Depression Inventory for a sample of 20 males facing exploratory cancer surgery.
    Two formulas for computing the mean have been described:

$$\bar{X} = \frac{\sum_{i=1}^{n} X_i}{n}, \qquad \text{where } i = 1, \ldots, n \ (n \text{ is the number of scores})$$

and

$$\bar{X} = \frac{\sum_{j=1}^{k} f_j X_j}{n}, \qquad \text{where } j = 1, \ldots, k \ (k \text{ is the number of class intervals}).$$

In the first formula, X_i denotes the value of the ith score. To compute the mean, the scores are summed and then divided by n, the number of scores. In the second formula, X_j denotes the value of the jth class interval, and f_j, the frequency of scores in that class interval. To compute the mean, we first obtain $f_j X_j$ for each class interval. Next we sum these products, and finally we divide the sum by n, the number of scores.

---

[4] Rules of summation are described in the Technical Notes, Section 3.9-1.

Table 3.3-1.   Depression scores for males facing exploratory cancer surgery (A score of 25 or above indicates extremely high depression.)

(i) *Data* (X_j *denotes the value of the jth class interval,* f_j *is the frequency in the jth class interval,* $j = 1, \ldots, k$, *and n is the number of scores*):

X_j	f_j	$f_j X_j$
28	1	$(1)(28) = 28$
27	0	$(0)(27) = 0$
26	1	$(1)(26) = 26$
25	2	$(2)(25) = 50$
24	3	$(3)(24) = 72$
23	4	$(4)(23) = 92$
22	3	$(3)(22) = 66$
21	0	$(0)(21) = 0$
20	1	$(1)(20) = 20$
19	2	$(2)(19) = 38$
18	1	$(1)(18) = 18$
17	0	$(0)(17) = 0$
16	1	$(1)(16) = 16$
15	0	$(0)(15) = 0$
14	1	$(1)(14) = 14$
$n = 20$		$\sum_{j=1}^{k} f_j X_j = 440$

(ii) *Computation of* $\bar{X}$ *from ungrouped frequency distribution:*

$$\bar{X} = \frac{\sum_{j=1}^{k} f_j X_j}{n} = \frac{440}{20} = 22$$

Review Exercises for Section 3.3

*6. Identify:
   *a. X_1 b. $\bar{Y}$ *c. μ_1
   d. $\bar{Z}_3$ e. $\bar{X}$ f. μ_Z
   g. Y_2 *h. X_i i. Y_j
   j. Z_k k. Y_n l. n
   m. f_j n. k *o. X_j

*7. Write out the following, listing individual values of the variable.

   *a. $\displaystyle\sum_{i=1}^{n} X_i$ b. $\displaystyle\sum_{i=1}^{5} Y_i/n$ *c. $\displaystyle\sum_{\substack{i=1 \\ i \neq 3}}^{4} Z_i/n$

   *d. $\displaystyle\sum_{j=1}^{k} f_j X_j/n$ e. $\displaystyle\sum_{j=1}^{6} f_j Y_j/n$ f. $\displaystyle\sum_{\substack{j=1 \\ j \neq 2}}^{4} f_j Z_j$

   g. $\displaystyle\sum_{i=1}^{n} n_i \bar{X}_i/n_i$

*8. The socioeconomic level of white families in a predominantly black neighborhood was rated on the basis of income, educational attainment, physical condition of dwelling, and number of home appliances. Compute the mean using $\sum_{i=1}^{n} X_i/n$ for the following socioeconomic scores.

$$5 \quad 4 \quad 9 \quad 5 \quad 3 \quad 4$$
$$4 \quad 6 \quad 7 \quad 5 \quad 3 \quad 2$$
$$6 \quad 2 \quad 5 \quad 1 \quad 7$$

9. In the same neighborhood described in Exercise 8, socioeconomic ratings were obtained on black families. Compute the mean using $\sum_{i=1}^{n} X_i/n$.

$$
\begin{array}{cccccccccc}
5 & 6 & 4 & 5 & 10 & 6 & 3 & 5 & 7 & 6 \\
3 & 4 & 5 & 8 & 5 & 4 & 7 & 1 & 6 & 7
\end{array}
$$

*10. The following data represent the number of suicides per 10,000 inhabitants in pre-dominantly rural prefectures in Japan. Compute the mean using $\sum_{i=1}^{n} X_i/n$.

$$
\begin{array}{ccccccccc}
22 & 10 & 12 & 2 & 10 & 9 & 16 & 11 & 8 \\
14 & 11 & 8 & 13 & 10 & 9 & 12 & 0 & 10 \\
12 & 8 & 11 & 5 & 7 & 10 & 7 & 9 & 9 \\
9 & 8 & 7 & 8 & 5 & 14 & 3 & 10 & 11
\end{array}
$$

11. Suicide data for predominantly urban prefectures in Japan are as follows. Compute the mean using $\sum_{i=1}^{n} X_i/n$.

$$
\begin{array}{cccccccccc}
23 & 24 & 21 & 19 & 23 & 24 & 25 & 22 & 21 & 27 \\
24 & 23 & 23 & 22 & 20 & 23 & 26 & 25 & 24 & 22 \\
20 & 17 & 26 & 23 & 21 & 25 & 14 & 21 & 23 \\
26 & 24 & 23 & 22 & 25 & 23 & 25 & 28 & 24
\end{array}
$$

*12. For the data in Exercise 8, construct an ungrouped frequency distribution and compute the mean using $\bar{X} = \sum_{j=1}^{k} f_j X_j/n$.

13. For the data in Exercise 9, construct an ungrouped frequency distribution and compute the mean using $\bar{X} = \sum_{j=1}^{k} f_j X_j/n$.

*14. For the data in Exercise 10, construct an ungrouped frequency distribution and compute the mean using $\bar{X} = \sum_{j=1}^{k} f_j X_j/n$.

15. For the data in Exercise 11, construct an ungrouped frequency distribution and compute the mean using $\bar{X} = \sum_{j=1}^{k} f_j X_j/n$.

16. Terms to remember:

    a. Mu   b. Statistic

    c. Parameter   d. Index of summation

Section 3.4   The Median

*The **median** is the point in a distribution that divides the data into two groups having equal frequency.* It is denoted by Mdn. As its name suggests, it is the middle score when scores have been arranged in order of size and n is odd. When n is even, the median is the midway point between the two middle scores. The procedure for determining the median is slightly different depending on whether n, the number of scores, is odd or even and whether a frequency distribution has been constructed for the data. If the number of scores is small, the median can be determined by inspection. Consider the case in which n is odd, and the scores are 5, 3, 8, 9, 11, 12, 2. When the scores are rank ordered from smallest to largest along the number line, as in Figure 3.4-1, it is immediately apparent that 8 is the middle score. Three scores fall above 8 and three below; therefore, 8 is the median. Figure 3.4-1, which represents scores as the distance between real limits on a number line, illustrates that the median does not divide the interval from 2 to 12 in half; rather it divides the ordered scores into two equal size samples. When, as in Figure 3.4-1, the number of scores is odd, the median is the $(n + 1)/2$th score

Figure 3.4-1.   Determination of the median when *n* is odd.

from either end of the number line. For example, $(n + 1)/2 = (7 + 1)/2 = 4$; hence, the median is the fourth score counting from either end.

Figure 3.4-2 illustrates the location of the median along the line when *n* is even and the scores are 3, 5, 8, 9, 11, 12. Any point along the number line larger than 8 and less than 9 would qualify as the median, since there are two scores below 8 and two scores above 9. By convention, the median is taken as the midway point between the $n/2$th score and the $(n/2) + 1$th score. For example, $n/2 = 3$ and $(n/2) + 1 = 4$. The midway point between the third score (8) and the fourth score (9), counting from the left, is $(8 + 9)/2 = 8.5$, which is the median.

Frequencies greater than 1 at the middle score value may present special problems. The median for Figure 3.4-3(a) is obviously 8, but what about Figure 3.4-3(b)? According to our definition, the median should be the $(n + 1)/2 = (7 + 1)/2 = 4$th score from either end. This score is 8, but below 8 there are three scores and above 8, only two scores. The problem is resolved by arbitrarily assigning half the interval 7.5–8.5 to each score. This results in two smaller intervals, 7.5–8 and 8–8.5, as shown in the upper part of Figure 3.4-3(b). Going four scores from the lower end, we reach the score defined by 7.5–8, which has a midpoint at $(7.5 + 8)/2 = 7.75$; similarly, four scores from the upper end is also the score defined by 7.5–8. Thus, the median is 7.75, the midpoint of the score defined by the interval 7.5–8. Now consider the scores in Figure 3.4-4. Each of the three scores in the interval 7.5–8.5 is assumed to occupy one-third of the interval, resulting in the smaller intervals 7.500–7.833, 7.833–8.167, 8.167–8.500, as shown in the upper part of the figure. Since *n* is even, the median is the score value that is midway between the $n/2 = 4$th and the $(n/2) + 1 = 5$th scores. These scores are defined by the intervals 7.500–7.833 and 7.833–8.167, respectively. The median is the midway point $(7.667 + 8.000)/2 = 7.833$.

Computation of the Median in an Ungrouped Frequency Distribution

We determined the median in Figures 3.4-3 and 3.4-4 by interpolating—finding the value within an interval such that the definition was satisfied. For data that

Figure 3.4-2.   Determination of the median when *n* is even.

a.

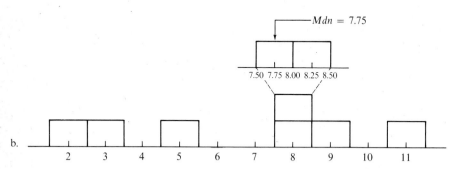

b.

Figure 3.4-3.   Determination of the median when the frequency of the middle score value is greater than 1.

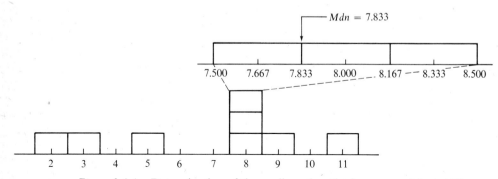

Figure 3.4-4.   Determination of the median when the frequency of the middle score value is greater than 1.

have been ordered in a frequency distribution, the interpolation can be done automatically by a formula. This computation is described in Table 3.4-1 for the data in Figure 3.4-4. The computational procedure works fine as long as the median falls in a class interval with a frequency greater than zero. The formula will not yield an answer if f_i, the number of scores in the class interval containing the median, is equal to zero, because then the fraction $i(n/2 - \sum f_b)/f_i$ is undefined. (See Appendix A, Rule 9d.) In this case, the median is taken to be the midpoint of the class interval containing the median.

Table 3.4-1.   Procedure for computing the median from a frequency distribution

(i) *Data and computational formula:*

X_j	f_j	Cum f^a
11	1	8
10	0	7
9	1	7
8	3	6
7	0	3
6	0	3
5	1	3
4	0	2
3	1	2
2	1	1
	$n = 8$	

$$Mdn = X_{ll} + i\left(\frac{n/2 - \sum f_b}{f_i}\right)$$

$$= 7.5 + 1\left(\frac{8/2 - 3}{3}\right)$$

$$= 7.5 + 1\left(\frac{4 - 3}{3}\right)$$

$$= 7.5 + 0.33 = 7.83$$

(ii) *Definition of terms:*

X_j = value of jth class interval
f_j = frequency of jth class interval
X_{ll} = real lower limit of class interval containing the median
i = class interval size
n = number of scores
$\sum f_b$ = number of scores below X_{ll}
f_i = number of scores in the class interval containing the median

(iii) *Computational sequence:*

1. Compute $n/2 = 8/2 = 4$.
2. Locate the class interval containing the $n/2 = 4$th score in the Cum f column. The median will fall somewhere in this class interval. The fourth score occurs in the class interval 8. This class interval contains the fourth, fifth, and sixth scores; X_{ll} for this class interval is 7.5.
3. Compute i: i = (Real upper limit of class interval − Real lower limit of class interval), for example, $i = 8.5 - 7.5 = 1$.
4. Determine $\sum f_b$.
5. Determine f_i.

^{a} Cumulative frequency is discussed in Section 2.2.

Review Exercises for Section 3.4

17. For a small number of scores, how is the median determined when (a) n is odd and (b) n is even?
*18. Determine the median for the following scores.
   *a. 9, 3, 16, 5, 21   b. 2, 8, 11, 19, 3, 26, 28
   *c. 16, 19, 17, 31   d. 3, 1, 3, 4
   *e. 3, 1, 3, 4, 5   *f. 3, 4, 4, 2, 8
   g. 3, 5, 5, 4, 8   h. 3, 5, 5, 4, 8, 5
*19. For the data in Exercise 10 (page 64), construct an ungrouped frequency distribution and compute the median using $X_{ll} + i\left(\dfrac{n/2 - \sum f_b}{f_i}\right)$.

20. For the data in Exercise 11 (page 64), construct an ungrouped frequency distribution and compute the median using $X_{ll} + i\left(\dfrac{n/2 - \sum f_b}{f_i}\right)$.

*21. The computational procedure for the median illustrated in Table 3.4-1 calculates the median from below—that is, by coming halfway through the scores, starting from the lowest class interval. Alternatively, the median can be computed by coming down halfway from above—from the highest class interval. The computational formula is $X_{ul} - i\left(\dfrac{n/2 - \sum f_a}{f_i}\right)$. By analogy with the definitions in Table 3.4-1, define each of the symbols in the alternative formula.

22. For the data in Table 3.4-1, compute the median using $X_{ul} - i\left(\dfrac{n/2 - \sum f_a}{f_i}\right)$.

23. For the data in Exercise 11 (page 64), construct an ungrouped frequency distribution and compute the median using $X_{ul} - i\left(\dfrac{n/2 - \sum f_a}{f_i}\right)$.

Section 3.5   Relative Merits of the Mean, Median, and Mode

Computation is fairly simple for all three measures of central tendency. Which one should an experimenter use for a given problem? The choice should be based on (1) the shape of the distribution, (2) the intended uses of the statistic, (3) the nature of the variable, and (4) the mathematical properties and merits of the three measures.

Although they all are measures of central tendency, the mean, median, and mode impart somewhat different information. Consider the scores in Figure 3.5-1. By inspection, we see that the mode is 3. The median is the $(n + 1)/2 = $ 3rd score from either end of the number line. This score falls in the interval with real limits 2.5–3.5. When the interval is divided in half, the real limits of the third score are 3–3.5 and its midpoint is 3.25; hence, the median is 3.25. The mean is $\bar{X} = (2 + 3 + \cdots + 8)/5 = 4$. The three values obtained represent different conceptions of the point around which scores cluster. For a unimodal set of data plotted as a histogram,

1. the *mode* is the score value with the largest frequency—the most typical score;
2. the *median* is the score point that divides the ordered scores into two equal-size samples;

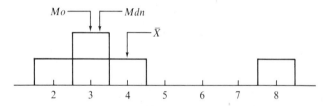

Figure 3.5-1.   Comparison of $\bar{X}$, *Mdn*, and *Mo*. The number line can be thought of as a teeter-totter whose balance point is the mean.

3. the *mean* is the score point at which the distribution balances—its center of gravity.

If a distribution is asymmetrical, as in Figure 3.5-1, the mean and the median are unequal; the value of the mode may or may not differ from those for the mean and the median. If a distribution is symmetrical, the mean and the median are equal; if, in addition, the distribution is unimodal, all three measures are equal.

Mean

The mean has a number of mathematical properties that make it the preferred measure of central tendency for relatively symmetrical distributions and for quantitative variables. One of these properties is its sampling stability.[5] Suppose that from an extremely large population we repeatedly drew random samples of size *n*. If we computed the mean for each sample, we would expect the means to be similar but not identical. Suppose that we also computed the median and the mode for each sample. The variability from sample to sample would be greatest for the mode and least for the mean. The better *sampling stability* of the mean is an important advantage, especially when one uses inferential statistics to draw conclusions about the central tendency of a population by observing a single sample.

Another advantage of the mean is that it is amenable to arithmetic and algebraic manipulations in ways that the other measures are not. Therefore, if further statistical computations are to be performed, the mean is usually the measure of choice. This property accounts for the appearance of the mean in the formulas for many important statistics.

The mean is the only one of the three measures that reflects every score and its value. Obviously, since $\bar{X} = \sum X/n$, the mean is a function of the value of every score in addition to the number of scores. On the other hand, the median is independent of the values of scores (other than the median value itself) as long as the number of scores above and below the median is not altered. If, for example, the score of 8 in Figure 3.5-1 is changed to 5, the values of the median and the mode are unchanged; the value of the mean, however, is changed from 4 to 3.4.

It is no accident that the balance point of the scores in Figure 3.5-1 coincides with the mean. This fulcrum property of the mean follows from the mathematical statement $\sum_{i=1}^{n} (X_i - \bar{X}) = 0$, the sum of the deviations of scores from the mean is zero. In Figure 3.5-1, for example, $\sum_{i=1}^{n} (X_i - \bar{X}) = (2 - 4) + (3 - 4) + \cdots + (8 - 4) = 0$, and this will be true for any distribution. It is also true that the sum of squared deviations of scores from the mean is less than the sum of squared deviations from any other point. This can be stated mathematically: The mean is the point for which $\sum_{i=1}^{n} (X_i - \bar{X})^2 = \text{Minimum}$. If we think of a deviation as a distance, the mean is the point from which (1) the sum of the distances to all the scores is zero and (2) the sum of the squared distances is a minimum. For proofs of these properties, see Sections 3.9-2 and 3.9-3.

There are three situations in which the mean is not the preferred measure of central tendency: when the distribution is very skewed, when the data are

---

[5] See Technical Notes, Section 3.9, for more about these properties.

qualitative in character, and when the distribution is *open-ended,* that is, the values of extreme scores are unknown. We will discuss the first two situations here and the third in the following section on the median.

Suppose that the following data were obtained for the number of minutes required to solve math problems: 10.1, 10.3, 10.5, 10.6, 10.7, 10.9, 56.9. The mean is $120/7 = 17.1$; the median is 10.6. Which number best represents the seven scores? Most would agree that it is 10.6, the median. The mean is unduly affected by the lone extreme score of 56.9. Any time a distribution is extremely asymmetrical, the mean is strongly affected by the extreme scores and, as a result, falls farther away from what would be considered the distribution's central area.

The mean is not the preferred measure of central tendency when the data are qualitative in character. Suppose that the dependent variable is eye color and we collect the following data: blue, brown, brown, gray, blue, brown. There is no meaningful way to represent these data by a mean; we could, however, compute the mode and say that the most typical eye color is brown.

Median

Although the mean is usually the preferred measure of central tendency, there are several situations in which the median is preferred. As was mentioned earlier, the median is not sensitive to the values of the scores above and below it—only to the number of such scores. Unlike the mean, it is not affected by extreme scores, and thus it is a more representative measure of central tendency for very skewed distributions. Also, it can be computed when the values of the extreme scores are unknown. Suppose, for example, that we recorded the number of trials required to learn a list of paired adjectives and Japanese kana (writing) symbols. The data are as follows: 12, 17, 17, 18, 21, 24, >41. After the forty-first trial, the poorest learner was still unable to learn the list and gave up; his score is some number greater than 41. The distribution is open-ended, because one of the extreme values is unknown. The median can be computed for these data and is 18; the mean cannot be computed because the value of the extreme score is unknown.

The median has the added advantage of being easy to compute; when the number of scores is small, it can be determined by inspection.

We saw earlier that the mean is the point from which the sum of the squared deviations is a minimum. We expressed this idea mathematically by saying that $\sum_{i=1}^{n}(X_i - \bar{X})^2 = $ Minimum. We can think of the median as the point from which the sum of the deviations of the scores, disregarding the sign of the deviations, is a minimum. We can state this mathematically: The median is the point for which $\sum_{i=1}^{n}|X_i - Mdn| = $ Minimum; that is, the sum of the deviations of scores from the median, disregarding their signs, is as small as or smaller than from any other point.[6] Note the similarity between this statement and the mathematical statement describing the minimum squared deviation property of the mean. For the data in Figure 3.5-1, $\sum_{i=1}^{n}|X_i - Mdn| = |2 - 3.25| + |3 - 3.25| + \cdots + |8 - 3.25| = 1.25 + 0.25 + \cdots + 4.75 = 7.25$. If we take the deviations from, say, the mean, and

---

[6] The proof of this statement is given by Paul Horst (1931).

disregard their signs, the sum is 8, which is larger than the sum for the median or any other point in the interval 3–3.5 that defines the median.

The principal disadvantages of the median relative to the mean are (1) its poorer sampling stability and (2) its poorer mathematical tractability. For these and other reasons it is not widely used in advanced descriptive and inferential statistical procedures.

Mode

The mode is the only measure of central tendency that can be used with unordered qualitative variables such as eye color, blood type, race, and political party affiliation. For quantitative variables that are inherently discrete, such as family size, it is sometimes a more meaningful measure of central tendency than the mean or the median. Who ever heard of a family with 4.2 members? It makes more sense to say that the most typical family size is 4, the mode. Other than these two applications, the mode has little to recommend it except its ease of estimation.

Let us consider why the mode is called the most typical score. We have seen that the mean is the value for which $\sum_{i=1}^{n} (X_i - \bar{X})^2 = $ Minimum and the median is the value for which $\sum_{i=1}^{n} |X_i - Mdn| = $ Minimum. The mode is the score value that occurs most often. We can state this mathematically: The mode is the value for which the number of $(X_i \neq Mo) = $ Minimum; that is, the number of scores not equal to the mode is a minimum. In Figure 3.5-1, for example, there are three scores that differ from the mode, but four that differ from the mean and five that differ from the median; hence, the mode is the most typical score.

The mode has a number of limitations. Its sampling stability is much poorer than that of the mean and the median, and it is also less mathematically tractable. Therefore, it is rarely used in advanced descriptive and inferential statistics. However, the mode, like the median, can be computed for an open-ended distribution if the distribution is known to be unimodal and if the unknown scores don't have the greatest frequency. However, because of the median's superior mathematical properties, it is preferred for this application.

A mode may not exist for a set of data, as when the distribution is bi- or multimodal. In such cases, it may be useful to report two or more modes. Since most variables in the behavioral sciences are approximately normally distributed, the existence of two modes suggests the presence of two underlying distributions. This would occur if we administered a test of masculinity to a sample containing an equal number of males and females. To report a mean or a median for such data would be misleading without also reporting that the distribution is bimodal and the values of the modes.

Summary of the Properties of the Mean, Median, and Mode

The mean is

1. the balance point of a distribution—the point for which $\sum_{i=1}^{n} (X_i - \bar{X}) = 0$ and the point for which $\sum_{i=1}^{n} (X_i - \bar{X})^2 = $ Minimum;
2. the preferred measure for relatively symmetrical distributions and quantitative variables;

    3. the measure with the best sampling stability;

    4. widely used in advanced statistical procedures;

    5. mathematically tractable;

    6. the only measure whose value is dependent on the value of every score in the distribution;

    7. more sensitive to extreme scores than the median and the mode and, hence, is not recommended for markedly skewed distributions;

    8. not appropriate for qualitative data; and

    9. not appropriate for open-ended distributions.

The median is

    1. the point that divides the ordered scores into two equal-size samples— the point for which $\sum_{i=1}^{n} |X_i - Mdn| = \text{Minimum}$;

    2. second to the mean in usefulness;

    3. widely used for markedly skewed distributions, because it is sensitive only to the number rather than to the values of scores above and below it;

    4. the most stable measure that can be used with open-ended distributions;

    5. more subject to sampling fluctuation than the mean;

    6. less mathematically tractable than the mean; and

    7. less often used in advanced statistical procedures.

The mode is

    1. the score value that occurs most often and, therefore, the most typical value—the value for which the number of $(X_i \neq Mo) = \text{Minimum}$;

    2. the only measure appropriate for unordered qualitative variables;

    3. more appropriate than the mean or the median for quantitative variables that are inherently discrete;

    4. the easiest measure to compute;

    5. much more subject to sampling fluctuation than the mean and the median;

    6. less mathematically tractable than the mean and the median;

    7. not necessarily existent, as when a distribution has two or more points with the same maximum frequency; and

    8. rarely used in advanced statistical procedures.

Review Exercises for Section 3.5

*24. For the following sets of data, what measures of central tendency would you compute? Justify your choices.

    *a. 9, 6, 5, 7, 1, 6, 7, 8, 10, 6, 5, 4, 3, 6, 9, 7, 4, 5, 6, 8, 3, 2

    b. 4, 3, 7, 5, 4, 2, 12, 6, 5, 4, 3, 3, 2, 7, 1, 6, 4, 5, 3, 5

    *c. 6, 5, 9, 6, 7, 5, 6, 8, 3, 4, 5, 7, 5, 4, 8, 5

    d. Eye color: blue, brown, brown, blue, green, brown, gray, brown, blue

    e. 7, 8, 6, 7, 8, 9, 1, 6, 5, 3, 7, 8, 7, 6, 7, 8, 5, 7

    f. Family size: 4, 3, 5, 4, 1, 2, 4, 6, 5

*25. Rank the three measures of central tendency with respect to the following characteristics; let 1 = most or hardest and 3 = least or easiest.

    *a. Sampling stability

   b. Suitability for advanced applications
   c. Mathematical tractability
   d. Sensitiveness to value of each score
   *e. Appropriateness for qualitative variables
   f. Ease of computation
26. Term to remember:
   Sampling stability

Section 3.6   Location of $\bar{X}$, Mdn, and Mo in a Distribution

If a distribution is unimodal and symmetrical, the mean, median, and mode have the same value. If the distribution is unimodal but skewed, the three measures will be arranged in a predictable order. This is illustrated in Figure 3.6-1. In both examples, the mean is on the side of the distribution having the longest tail and the median falls about one-third of the distance from the mean to the mode. To remember the order—mean, median, mode—note that it is alphabetical, starting from the longer tail. This order occurs because the mean is affected by the value of extreme scores; the median is affected by the presence of extreme scores but not by their value; and the mode is not affected by extreme scores unless they happen to have the greatest frequency of occurrence.

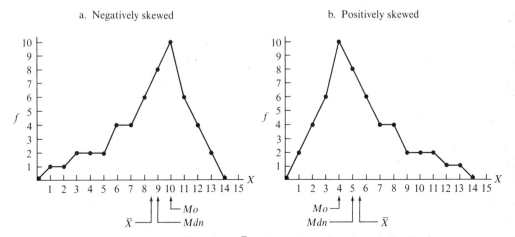

Figure 3.6-1. Location of the $\bar{X}$, Mdn, and Mo for skewed distributions.

The relative position of the measures provides a rough guide as to whether a distribution is positively or negatively skewed. For negatively skewed distributions, it is virtually always true that $Mdn > \bar{X}$; for positively skewed distributions, $\bar{X} > Mdn$. If, for example, we know that the median is 25 and the mean is 20, we would strongly suspect that the distribution is negatively skewed. The greater the discrepancy between the two values, the greater the departure from symmetry.[7]

---

[7] A more sophisticated measure of skewness is described in Section 4.6.

Review Exercises for Section 3.6

27. For each of the following distributions, indicate on the X axis the approximate location of $\bar{X}$, Mdn, and Mo.

a.   b.   c.

d.

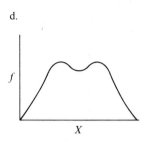

*28. Determine the shape of each distribution from the following measures of central tendency. Assume a unimodal distribution.

   *a. $\bar{X} = 16$, $Mdn = 10$ b. $\bar{X} = 21$, $Mdn = 21$, $Mo = 21$

   c. $Mdn = 109$, $Mo = 116$ d. $\bar{X} = 73$, $Mdn = 84$

   *e. $\bar{X} > Mdn$ f. $Mo > Mdn$

   g. $Mdn > \bar{X}$ h. $\bar{X} = Mdn = Mo$

Section 3.7   Mean of Combined Subgroups

Suppose that two introductory sociology classes obtained the following mean scores on a departmental examination: 80 and 90. What is the mean of the two means? If each class had the same number of students, we could compute the mean of the means by $\bar{X} = (\bar{X}_1 + \bar{X}_2)/2 = (80 + 90)/2 = 85$. If, as is more likely, the classes contain different numbers of students, we must weight the means proportionally to their respective sample sizes. Assume that $\bar{X}_1 = 80$ and $n_1 = 20$ and that $\bar{X}_2 = 90$ and $n_2 = 40$. The weighted mean, $\bar{X}_w$, is given by

$$\bar{X}_w = \frac{n_1\bar{X}_1 + n_2\bar{X}_2 + \cdots + n_n\bar{X}_n}{n_1 + n_2 + \cdots + n_n} = \frac{20(80) + 40(90)}{20 + 40} = \frac{5200}{60} = 86.7.$$

The weighted mean is closer to 90 than to 80; this reflects the larger n_2 associated with $\bar{X}_2 = 90$.

Review Exercises for Section 3.7

*29. For the following data, compute weighted means.
*a. $\bar{X}_1 = 30, n_1 = 10; \bar{X}_2 = 50, n_2 = 20$
b. $\bar{X}_1 = 50, n_1 = 20; \bar{X}_2 = 100, n_2 = 30$
c. $\bar{X}_1 = 8, n_1 = 10; \bar{X}_2 = 12, n_2 = 30; \bar{X}_3 = 18, n_3 = 20$
d. $\bar{X}_1 = 100, n_1 = 20; \bar{X}_2 = 200, n_2 = 20$

Section 3.8   Summary

Three measures of central tendency are described in this chapter—the mean, median, and mode. The different indexes result from different ways of conceptualizing the point around which scores cluster. The mean is the point around which the distribution balances—its center of gravity; the median is the point that divides the ordered scores into two equal-size samples; and the mode is the score value with the greatest frequency—the most typical score.

The mean is the most widely used of the indexes, partly because of its superior sampling stability and partly because many advanced statistical procedures are based on it. The median and the mode, by contrast, are *terminal statistics*; their usefulness in advanced descriptive and inferential procedures is very limited.

There are three situations in which the mean is not the preferred measure of central tendency: when the distribution is markedly skewed, when the variable is qualitative in character, and when the distribution is open-ended. For markedly skewed distributions, the median is preferred because it is not so sensitive as the mean to the presence of extreme scores. For unordered qualitative variables, the only measure that can be computed is the mode. In addition, the mode may be more meaningful for inherently discrete variables such as family size.

Section 3.9   Technical Notes

The following notes are designed to provide additional insights about selected topics in this chapter. In keeping with the rest of the book, no knowledge of mathematics beyond high school math is assumed, and the presentation is appropriately simplified.

3.9-1   Summation Rules

Four simple summation rules are widely used in statistical proofs and derivations. An understanding of these rules will go far toward taking derivations out of the realm of magic.

Rule 3.9-1-1.   The sum of a constant. Let c be a constant; the sum over $i = 1, \ldots, n$ of the constant can be written as the product of the upper limit of the summation, n, and c. That is,

$$\sum_{i=1}^{n} c = \underbrace{c + c + \cdots + c}_{n \text{ terms}} = nc.$$

For example, let $c = 2$ and $i = 1, \ldots, 3$; then

$$\sum_{i=1}^{3} 2 = 2 + 2 + 2 = 3(2) = 6.$$

Rule 3.9-1-2[8]. The sum of a variable. Let V be a variable with values $V_1, V_2, \ldots, V_n$; the sum over $i = 1, \ldots, n$ of the variable is

$$\sum_{i=1}^{n} V_i = V_1 + V_2 + \cdots + V_n.$$

For example, let $V_1 = 2$, $V_2 = 3$, and $V_3 = 4$; then

$$\sum_{i=1}^{3} V_i = 2 + 3 + 4 = 9.$$

Rule 3.9-1-3. The sum of the product of a constant and a variable. The sum $\sum_{i=1}^{n} cV_i$ can be written as the product of the constant and the sum of the variable—that is,

$$\sum_{i=1}^{n} cV_i = c \sum_{i=1}^{n} V_i.$$

For example, let $c = 2$ and $V_1 = 2$, $V_2 = 3$, and $V_3 = 4$; then

$$\sum_{i=1}^{3} cV_i = 2(2) + 2(3) + 2(4)$$

$$= c \sum_{i=1}^{n} V_i$$

$$= 2(2 + 3 + 4) = 18.$$

Rule 3.9-1-4. Distribution of summation. If the only operation to be performed before summation is addition or subtraction, the summation sign can be distributed among the separate terms of the sum. Let V and W be two variables; then

$$\sum_{i=1}^{n} (V_i + W_i) = \sum_{i=1}^{n} V_i + \sum_{i=1}^{n} W_i.$$

For example, let $V_1 = 2$, $V_2 = 3$, $V_3 = 4$, $W_1 = 5$, $W_2 = 6$, and $W_3 = 7$; then

$$\sum_{i=1}^{3} (V_i + W_i) = (2 + 5) + (3 + 6) + (4 + 7)$$

$$= \sum_{i=1}^{3} V_i + \sum_{i=1}^{3} W_i$$

$$= (2 + 3 + 4) + (5 + 6 + 7) = 27.$$

[8] This rule was introduced in Section 3.3.

This rule applies to any number of terms. For example, let V, W, and X be variables and a, b, and c be constants; then according to Rules 3.9-1-1, 3.9-1-2, and 3.9-1-4,

$$\sum_{i=1}^{n} (V_i + W_i + X_i + a + b + c)$$

$$= \sum_{i=1}^{n} V_i + \sum_{i=1}^{n} W_i + \sum_{i=1}^{n} X_i + na + nb + nc.$$

3.9-2   Proof that the Mean Is a Balance Point

In Section 3.5 we said that the mean is the point such that $\sum_{i=1}^{n}(X_i - \bar{X}) = 0$. We can construct a simple proof of this assertion using Rules 3.9-1-1, 3.9-1-2, and 3.9-1-4. In the term $\sum_{i=1}^{n}(X_i - \bar{X})$, X_i is a variable; but for any set of scores, $\bar{X}$ is a constant. Hence,

$$\sum_{i=1}^{n} (X_i - \bar{X}) = \sum_{i=1}^{n} X_i - n\bar{X} \qquad \left(\text{By definition} \quad \bar{X} = \sum_{i=1}^{n} X_i/n, \right.$$

$$= \sum_{i=1}^{n} X_i - \sum_{i=1}^{n} X_i \quad \text{so} \quad \left. n\bar{X} = \sum_{i=1}^{n} X_i. \right)$$

$$= 0.$$

3.9-3   Proof that $\sum (X_i - \bar{X})^2 = Minimum$

Because the mean is the balance point of a distribution, the sum of the squared deviations from the mean, $\sum_{i=1}^{n}(X_i - \bar{X})^2$, is smaller than the sum about any other point. We can show this as follows. Let c be a constant not equal to zero that is added to the mean; then

$$\sum_{i=1}^{n} [X_i - (\bar{X} + c)]^2$$

$$= \sum_{i=1}^{n} [(X_i - \bar{X}) - c]^2 \qquad \text{(According to the associative law)}$$

$$= \sum_{i=1}^{n} [(X_i - \bar{X})^2 - 2c(X_i - \bar{X}) + c^2]$$

$$= \sum_{i=1}^{n} (X_i - \bar{X})^2 - 2c \sum_{i=1}^{n} (X_i - \bar{X}) + nc^2. \quad \text{(Rules 3.9-1-1–3.9-1-4)}$$

But we saw in Section 3.9-2 that $\sum_{i=1}^{n}(X - \bar{X}) = 0$. Therefore,

$$\sum_{i=1}^{n} [X_i - (\bar{X} + c)]^2 = \sum_{i=1}^{n} (X_i - \bar{X})^2 + nc^2.$$

Thus, for $c \neq 0$, nc^2 is greater than 0, and

$$\sum_{i=1}^{n} (X_i - \bar{X})^2 + nc^2 > \sum_{i=1}^{n} (X_i - \bar{X})^2.$$

The sum of squared deviations about the mean appears in many advanced statistics. We will see in Chapter 4 that $S = \sqrt{\sum_{i=1}^{n}(X_i - \bar{X})^2/n}$ is used to describe the dispersion of scores in a distribution; S is called the standard deviation.

Review Exercises for Section 3.9

*30. Write the following expressions as the sum of individual values of the variables, X and Y, or constant, a; for example,

$$\sum_{i=1}^{n} X_i = X_1 + X_2 + \cdots + X_n.$$

*a. $\displaystyle\sum_{i=1}^{3} X_i$ *b. $\displaystyle\sum_{i=1}^{4} Y_i$ c. $\displaystyle\sum_{i=1}^{5} X_i$

*d. $\displaystyle\sum_{j=1}^{3} f_j X_j$ *e. $\displaystyle\sum_{j=1}^{k} f_j X_j$ f. $\displaystyle\sum_{j=1}^{4} f_j Y_j$

*g. $\displaystyle\sum_{i=1}^{3} aX_i$ h. $\displaystyle\sum_{i=1}^{4} aY_i$ *i. $\displaystyle\sum_{i=1}^{n} (X_i + a)$

*31. Let X and Y denote variables and let a and b denote constants. Assume that the values of the variables and the constants are as follows.

$$\begin{array}{lll} X_3 = 4 & Y_4 = 9 & a = 2 \\ X_2 = 3 & Y_3 = 4 & b = 3 \\ X_1 = 2 & Y_2 = 2 & \\ & Y_1 = 1 & \end{array}$$

Determine the values of the following expressions.

* a. $\displaystyle\sum_{i=1}^{3} a$ *b. $\displaystyle\sum_{i=1}^{4} b$ c. $\displaystyle\sum_{i=1}^{2} b$

d. $\displaystyle\sum_{i=1}^{2} a$ *e. $\displaystyle\sum_{i=1}^{n} X_i$ *f. $\displaystyle\sum_{i=1}^{n} Y_i$

* g. $\displaystyle\sum_{i=1}^{2} X_i$ h. $\displaystyle\sum_{i=1}^{3} Y_i$ *i. $\displaystyle\sum_{i=1}^{n} aX_i$

j. $\displaystyle\sum_{i=1}^{n} bY_i$ *k. $\displaystyle\sum_{i=1}^{3} (X_i + a)$ l. $\displaystyle\sum_{i=1}^{4} (Y_i - b)$

*m. $\displaystyle\sum_{i=1}^{4} (Y_i + a - b)$ *n. $\displaystyle\sum_{i=1}^{2} (X_i + a)$ o. $\displaystyle\sum_{i=1}^{3} (X_i + Y_i)$

*32. The following proofs show the effect on the mean of adding a constant to each score or multiplying each score by a constant. For each proof, identify the summation operations and the number of the summation rules that were used from Section 3.9-1.

*a. Let $\bar{X}_{X+c}$ be the mean of a distribution that has been altered by adding a constant c to each score—that is, $X_1 + c, X_2 + c, \ldots, X_n + c$. Then

$$\bar{X}_{X+c} = \frac{\displaystyle\sum_{i=1}^{n} (X_i + c)}{n} = \frac{\displaystyle\sum_{i=1}^{n} X_i + nc}{n} = \frac{\displaystyle\sum_{i=1}^{n} X_i}{n} + c = \bar{X} + c.$$

Thus, the effect of adding a constant c to each score is to change $\bar{X}$, the mean of the original scores, to $\bar{X} + c$. Similarly, it can be shown that the effect of subtracting a constant c from each score is to change $\bar{X}$ to $\bar{X} - c$.

b. Let $\bar{X}_{cX}$ be the mean of a distribution that has been altered by multiplying each score by a constant c—that is, $cX_1, cX_2, \ldots, cX_n$. Then

$$\bar{X}_{cX} = \frac{\sum\limits_{i=1}^{n} (cX_i)}{n} = \frac{c \sum\limits_{i=1}^{n} X_i}{n} = c\bar{X}.$$

Thus, the effect of multiplying each score by a constant c is to change $\bar{X}$, the mean of the original scores, to $c\bar{X}$. Similarly, it can be shown that the effect of dividing each score by a constant is to change $\bar{X}$ to $\bar{X}/c$.

†Section 3.10   Computer Supplement

The Statistical Package for the Social Sciences (SPSS), version M, Release 9.1, (Nie *et al.*, 1975) was used to compute descriptive statistics for the data in Table 3.3-1, which represent depression scores for males facing surgery. SPSS, which is a batch mode package, has two subprograms for computing descriptive statistics: Condescriptive and Frequencies. The Frequencies subprogram was used to obtain the printout in Table 3.10-1.

The Frequencies subprogram provides a frequency distribution with (1) class intervals denoted in the printout by Code, (2) frequencies, denoted by Absolute Freq, (3) relative frequencies (percentage frequencies with missing values included), (4) adjusted frequencies (percentage frequencies with missing values excluded), and (5) cumulative frequencies (percentages). In addition, the user can request a variety of descriptive statistics and a histogram. In this example the mean, median, and mode were requested. Since there were no missing cases (scores), the values of the relative frequencies with missing values included (Relative Freq) and adjusted relative frequencies with missing values excluded (Adjusted Freq) are the same.

Table 3.10-1. Printout from the SPSS subprogram frequencies: One-way frequency distributions with descriptive statistics (Data from Table 3.3-1)

COMPUTATION OF MEAN DEPRESSION SCORES FOR MALES FACING SURGERY

FILE   NONAME   (CREATION DATE = 02/01/83)

SCORES   DEPRESSION SCORES

CATEGORY LABEL	CODE	ABSOLUTE FREQ	RELATIVE FREQ (PCT)	ADJUSTED FREQ (PCT)	CUM FREQ (PCT)
	14.	1	5.0	5.0	5.0
	16.	1	5.0	5.0	10.0
	18.	1	5.0	5.0	15.0
	19.	2	10.0	10.0	25.0
	20.	1	5.0	5.0	30.0
	22.	3	15.0	15.0	45.0
	23.	4	20.0	20.0	65.0
	24.	3	15.0	15.0	80.0
	25.	2	10.0	10.0	90.0
	26.	1	5.0	5.0	95.0
	28.	1	5.0	5.0	100.0
	TOTAL	20	100.0	100.0	

MEAN	22.000	MEDIAN	22.750	MODE	23.000
VALID CASES	20	MISSING CASES	0		

† This section can be omitted without loss of continuity.

The packages described thus far were developed for use with mainframe computers and minicomputers. The package described here was written by Brad Wilson (1984) for Apple II microcomputers having 64K of memory. This package, which contains both statistical programs and tutorial demonstrations, was designed for use with *Elementary Statistics, 2d ed.*

The package is very easy to use. Once the user has loaded the floppy disk (diskette) in the computer, it guides the analysis or demonstration session with a series of prompts. The interactive character of the package is illustrated in Table 3.10-2 for the depression data in Table 3.3-1. The user has 14 programs from which to choose. In this example the user selected 2—the Mean, Median, and Mode Program.

Table 3.10-2.   Printout from the Wilson Stat Package: Mean, median, and mode program (Data from Table 3.3-1)

```
******************************************
*                                        *
*            Brad Wilson's               *
*            Stat Package                *
*                                        *
******************************************
*                                        *
*      Behavioral Statistics Program     *
*           Baylor University            *
*                                        *
*                                        *
******************************************
Press RETURN to continue>>>>>

When you respond to a prompt just
type in your response and press
the RETURN key.

For all yes-no (Y/N) questions
it is not necessary to press the
RETURN key.

Press RETURN to continue>>>>>

Would you like to

  1. Perform statistical analyses

  2. See a demonstration of the
     Central Limit Theorem

  3. End the session?

Select a number: 1

Please wait a few moments.

Here are the available
statistical programs
    1.  Frequency distribution and
        histogram
    2.  Mean, median, and mode
```

(*continued*)

Table 3.10-2. (*continued*)

```
3.   Standard deviation
4.   Index of dispersion
5.   Correlation
6.   Regression
7.   One-sample t test
8.   One-sample chi-square test
9.   Two-sample independent t test
10.  Two-sample dependent t test
11.  Analysis of variance for a
     completely randomized design
12.  Pearson's chi-square for
     testing goodness of fit
13.  Pearson's chi-square for
     testing independence
14.  Pearson's chi-square for
     testing equality of proportions

Indicate the program you want
by selecting a number: 2

Please wait a few moments.

Do you want to enter data from

1. The terminal

2. A file, or

3. Use the data from the
   previous session, or do
   you want to

4. Return to the menu?

Select a number: 1

What is the label for
variable 1? DEP. SCORE

How many values in DEP. SCORE? 20

     variable            # of cases
-----------------------------------------
  1. DEP. SCORE              20

Do you want to
1. Go on to the next step
2. Examine your data specifications
   again
3. Change your variable name(s)
4. Change your # of cases?
Select one: 1
```

(*continued*)

Table 3.10-2. (*continued*)

```
Now you will be prompted for
the individual values:
    DEP. SCORE  1: 28
    DEP. SCORE  2: 14
    DEP. SCORE  3: 19
    DEP. SCORE  4: 18
    DEP. SCORE  5: 16
    DEP. SCORE  6: 20
    DEP. SCORE  7: 19
    DEP. SCORE  8: 22
    DEP. SCORE  9: 22
    DEP. SCORE 10: 23
    DEP. SCORE 11: 23
    DEP. SCORE 12: 24
    DEP. SCORE 13: 25
    DEP. SCORE 14: 26
    DEP. SCORE 15: 25
    DEP. SCORE 16: 22
    DEP. SCORE 17: 23
    DEP. SCORE 18: 24
    DEP. SCORE 19: 24
    DEP. SCORE 20: 23


Press RETURN to continue>>>

Would you like to check the data for
errors? (Y/N) N

If you have a second disk drive, you can
store your data on a separate diskette.
Would you like to store the data? (Y/N) N

Please wait a few moments.

Do you want the results sent to the

 1. Screen

 2. Printer?

Select a number: 1

NOTE: Use CTRL and S simultaneously to
stop and start the scrolling of
the screen.

Press RETURN to continue>>>

The mean of DEP. SCORE is 22.000

The median is 22.750

The mode is 23.000
```

(*continued*)

Table 3.10-2.   (*continued*)

```
Press RETURN to continue>>>

Would you like to

   1. Perform statistical analyses

   2. See a demonstration of the
      Central Limit Theorem

   3. End the session?

Select a number: 3
```

4

Measures of Dispersion, Skewness, and Kurtosis

Section 4.1   Introduction to Measures of Dispersion

Mr. Jacques and Ms. Shapley are taking a well-deserved break in the teachers' lounge. The conversation turns to Ms. Shapley's third-grade class. "I've got a bunch of little monsters this year. I can't seem to keep their interest for more than 10 minutes. I had to discipline Emerson twice this morning for flying paper airplanes during arithmetic, and Waldo is still picking fights. I just can't understand it; this class has the same average IQ as my class last year, and you remember

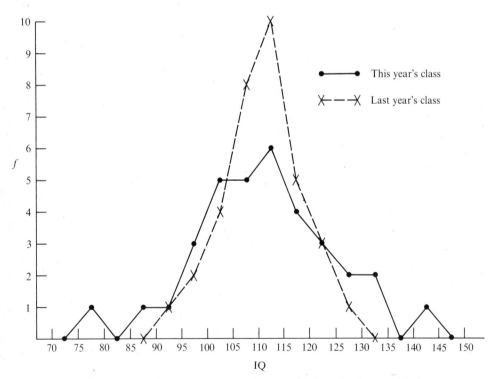

Figure 4.1-1.   Frequency polygons for two third-grade classes with the same central tendency but different dispersions.

how good those kids were." As Ms. Shapley contemplates her options—should I face the class for 7 more months, drop out and become a lottery player, or go back to school in molecular physics—we wonder what makes one class a joy and the other a disaster. The frequency polygon in Figure 4.1-1 illustrates the reason. Although the two classes have almost identical mean IQs, this year's class is much more heterogeneous in learning aptitude. Last year, for example, there were no children with IQs below 90; this year there are two. That's Waldo in the class interval 75–79—moderately retarded. At the other end of the distribution in the 140–144 class interval is our paper-plane thrower—a potential genius. It is small wonder that this year's class, with its wide range of aptitude, is giving Ms. Shapley difficulties.

Information about central tendency is important, but this tells only part of the story; the heterogeneity or dispersion of scores is often just as informative. The measures of central tendency in Chapter 3 represent points around which a distribution tends to center. As we will see, the most widely used dispersion measures summarize the spread or scatter of scores in terms of distance along the X axis. Many measures of dispersion have been proposed. We will examine the four most useful measures in the behavioral sciences and education; they are the range, semi-interquartile range, standard deviation, and index of dispersion.

Section 4.2   Four Measures of Dispersion

Range

*Intuitively, the simplest measure of dispersion is the **range**—the distance between the largest and smallest scores, inclusive. The range is denoted by R and is computed from the formula*

$$R = X_{ul(\text{largest score})} - X_{ll(\text{smallest score})},$$

where X_{ul} is the real upper limit of the largest score and X_{ll} is the real lower limit of the smallest score. (The midpoints of these scores are sometimes used instead of their real limits.) If, for example, Emerson's 144 is the highest IQ and Waldo's 76 is the lowest, the range is $144.5 - 75.5 = 69$. The range of 69 IQ points is a distance that includes 100% of the scores. In general, the larger the range, the greater the spread or scatter of scores.

In spite of its simplicity, the range is not widely used. For one thing, its value is determined by the two most extreme scores, and so its sampling stability is quite poor. It also is mathematically intractable and is not meaningful for qualitative data. These and other disadvantages discussed in Section 4.3 limit its usefulness as a measure of dispersion.

As we will see, each measure of dispersion is typically reported with a particular measure of central tendency. For quantitative data, the range can be reported with the mode, thereby giving a more complete picture of data.[1]

---

[1] The range also can be reported with a rarely used measure of central tendency called the *midrange*. For a set of scores, $X_1, X_2, \ldots, X_n$, arranged in order of magnitude, the midrange is defined as $(X_1 + X_n)/2$.

However, since the mode often is used with qualitative data, a different measure of dispersion is needed. The index of dispersion described later fills this need.

Semi-Interquartile Range

We have seen that the sampling stability of R is poor because it is computed from the two most extreme scores in a distribution. A second measure of dispersion, the semi-interquartile range, is based on two scores closer to the center of the distribution. Hence, it is considerably more stable than R. The ***semi-interquartile range,*** denoted by Q, is defined as one-half of the distance between the first quartile point, Q_1, and the third quartile point, Q_3. These points and the median are shown in Figure 4.2-1. The formula for Q is

$$Q = \frac{Q_3 - Q_1}{2}.$$

The computation of Q_1 and Q_3 is similar to that for the median and is illustrated in Table 4.2-1. The data are IQ scores from Ms. Shapley's current class. The semi-interquartile range for these data is 8.4. The larger the value of Q, the greater the distance between Q_1 and Q_3. In general, the larger Q is, the greater the spread or scatter of scores.

The semi-interquartile range is often reported along with the median to give a more complete description of data. For a symmetrical distribution, the median plus and minus the semi-interquartile range, $Mdn \pm Q$, is a distance that contains 50% of scores. This is evident from Figure 4.2-1. For the data in Table 4.2-1,

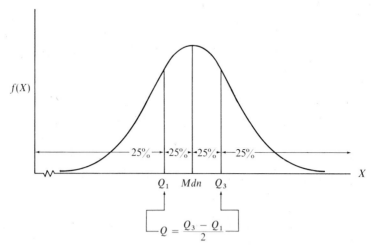

Figure 4.2-1. Q_1 is a point below which 25% of the scores fall and above which 75% fall; Q_3 is a point below which 75% fall and above which 25% fall. The median is sometimes referred to as Q_2, since it is a point that divides the distribution of scores in half. The semi-interquartile range, Q, is one-half the distance from Q_3 to Q_1.

Table 4.2-1.   Computational procedures for Q_1, Q_3, and Q (Data from Figure 4.1-1, This Year's Class)

(i) *Data and computational formulas:*

$^a X_j$	f_j	Cum f
142	1	
134	1	
131	1	
128	1	
125	1	
122	3	
118	1	26
117	3	25
111	6	22
109	2	16
105	3	14
101	5	11
99	2	6
96	1	4
94	1	3
87	1	2
76	1	1
	$n = 34$	

$$Q_1 = X_{ll} + i\left(\frac{n/4 - \sum f_b}{f_i}\right)$$

$$= 100.5 + 1\left(\frac{8.5 - 6}{5}\right)$$

$$= 100.5 + 0.50 = 101.0$$

$$Q_3 = X_{ll} + i\left(\frac{3n/4 - \sum f_b}{f_i}\right)$$

$$= 117.5 + 1\left(\frac{25.5 - 25}{1}\right)$$

$$= 117.5 + 0.5 = 118.0$$

$$Q = \frac{Q_3 - Q_1}{2}$$

$$= \frac{118.0 - 101.0}{2} = 8.4$$

(ii) *Definition of terms:*

X_j = value of jth class interval
f_j = frequency of jth class interval
X_{ll} = real lower limit of class interval containing Q_1 or Q_3
i = class interval size
n = number of classes
$\sum f_b$ = number of scores below X_{ll}
f_i = number of scores in class interval containing Q_1 or Q_3

(iii) *Computational sequence illustrated for Q_1:*
1. Compute $n/4 = \frac{34}{4} = 8.5$.
2. Locate the class interval containing the $n/4 = 8.5$th score in the Cum f column; the 8.5th score occurs in the class interval 101. For this class interval, X_{ll} is 100.5.
3. Compute i: i = Real upper limit of class interval − Real lower limit of class interval = 101.5 − 100.5 = 1.
4. Determine $\sum f_b$.
5. Determine f_i.

^{a} To conserve space, class intervals with $f_j = 0$ have been omitted.

$Mdn \pm Q$ is 110.7 ± 8.4 (or 102.3–119.1). This interval, however, does not contain exactly 50% of the scores, since the distribution is asymmetrical.

   The semi-interquartile range, like the median, is a terminal statistic; its usefulness in advanced descriptive and inferential procedures is very limited. It shares both the advantages and the disadvantages of the median, since it is computed from "medianlike" descriptive statistics, Q_1 and Q_3. We will now digress for a moment to describe another medianlike statistic—the percentile.

Percentile

*A **percentile point**, also called a **percentile** or **centile**, is a score point in a distribution below which a specified percentage of scores falls. The term **percentile rank** denotes the percentage of scores that falls below the percentile point.* Procedures for computing percentile points corresponding to the 25th, 50th, and 75th percentile ranks already have been described, since these points correspond to Q_1, Mdn, and Q_3, respectively. Percentiles corresponding to other percentile ranks can be computed from the formula

$$P_\% = X_{ll} + i\left(\frac{n(\%/100) - \sum f_b}{f_i}\right),$$

where the subscript in $P_\%$ identifies the rank of the percentile point and the $\%$ that appears in the fraction denotes the percentile rank of interest. The remaining symbols are defined in Table 4.2-1; $P_\%$ should be substituted for Q_1 or Q_3 where it appears.

To determine P_{60} for the data in Table 4.2-1, first compute $n(60/100) = 34(60/100) = 20.4$. By following the computational sequence illustrated for Q_1, we obtain

$$P_{60} = 110.5 + 1\left(\frac{20.4 - 16}{6}\right) = 110.5 + 0.7 = 111.2.$$

This tells us that 60% of the scores fall below an IQ of 111.2.

Percentiles are widely used in reporting the performance of individuals on psychological tests. We will return to percentiles in Chapter 10.

Standard Deviation

The standard deviation, denoted by S for a sample and by σ for a population, is the most important and most widely used measure of dispersion. The formulas for S and σ are, respectively,[2]

$$S = \sqrt{\frac{\sum_{i=1}^{n}(X_i - \bar{X})^2}{n}} \quad and \quad \sigma = \sqrt{\frac{\sum_{i=1}^{n}(X_i - \mu)^2}{n}},$$

---

[2] When the population standard deviation, σ, is estimated from sample data, a better estimator,

$$\hat{\sigma} = \sqrt{\frac{\sum_{i=1}^{n}(X_i - \bar{X})^2}{n-1}},$$

is denoted by $\hat{\sigma}$; this statistic is used in Chapters 12 to 16 on inferential statistics.

where $\bar{X}$ and μ denote the sample and population means, respectively. We can develop an intuitive understanding of the standard deviation by examining the formula for S. First note that, unlike R and Q, S is computed from every score in a distribution. Each score is expressed as a deviation from the mean, $(X_i - \bar{X})$, squared, and then summed. What would happen if we didn't square the deviations? We know from Chapter 3 that for any distribution,[3]

$$\sum_{i=1}^{n} (X_i - \bar{X}) = 0,$$

so squaring or some other operation on the deviations is necessary for the sum to equal a value other than zero.[4] The sum of the squared deviations is divided by n, which gives us the mean squared distance by which the scores deviate from the mean. To convert $\sum (X_i - \bar{X})^2/n$ back into deviations expressed in the original unit of measurement, we take its square root.[5]

To summarize, the standard deviation is a number that (1) is based on every score in a distribution and (2) represents the square root of the mean-squared distance of scores from the mean. The larger the value of S, the larger the mean-squared deviation of scores from the mean. Put another way, the larger the value of S, the greater is the spread or scatter of scores. Since the standard deviation is based on every score in the distribution, its sampling stability is much better than that of other measures of dispersion. For this reason and because it is mathematically tractable, the standard deviation is widely used in advanced descriptive and inferential statistics.

As we saw earlier, each measure of dispersion is typically reported with a particular measure of central tendency—R with Mo and Q with Mdn. The standard deviation is reported with the mean. One special type of distribution, called the *normal distribution*, is often approximated by behavioral science data (see Section 2.6 and Chapter 10). For this distribution, the mean plus and minus the standard deviation ($\bar{X} \pm S$) is an interval that contains 68.27% of scores, as Figure 4.2-2 illustrates. The other two dispersion measures are shown in the figure for comparison.

Computation of the standard deviation using the formula

$$S = \sqrt{\frac{\sum_{i=1}^{n} (X_i - \bar{X})^2}{n}}$$

---

[3] For a proof see Section 3.9-2.

[4] An obvious alternative would be to take the absolute value of each deviation, that is, to ignore the sign of $(X_i - \bar{X})$. Years ago, a statistic called the *mean deviation (MD)*, or *average deviation (AD)*, was developed that used this procedure. Its formula is $MD = \sum |X_i - \bar{X}|/n$, where $|X_i - \bar{X}|$ means to treat all differences as positive. The mean deviation has fallen into disuse, because it lacks certain advantages possessed by the standard deviation.

[5] The statistics $\sum (X_i - \bar{X})^2/n$ and $\sum (X_i - \bar{X})^2/(n-1)$ are called *variance* and are denoted, respectively, by S^2 and $\hat{\sigma}^2$; $\hat{\sigma}^2$ is said to be an unbiased estimator of σ^2, because the mean of an indefinite but very large number of $\hat{\sigma}^2$'s computed from random samples is equal to σ^2. This is discussed in Section 10.3. The estimator $\hat{\sigma}^2$ is widely used in inferential statistics. We will return to $\hat{\sigma}^2$ in Chapters 12 and 14 and in Chapter 16, when we discuss analysis of variance.

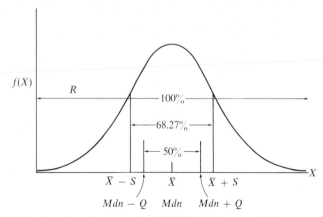

Figure 4.2-2.   A region that contains 68.27% of the area of the normal distribution is marked off by $\bar{X} \pm S$; $Mdn \pm Q$ contains 50% of the area, and the R contains 100% of the area.

is illustrated in Table 4.2-2 (see columns 2 and 3). The data are socioeconomic level ratings of white families in a predominantly black neighborhood. For these data, $\bar{X} = 5$ and $S = 2.3$. If we take $\bar{X} \pm S$, we obtain an interval of 2.7–7.3. It can be shown using the formula in Exercise 6 (page 97) that this interval contains 63.34% of the 12 scores since 2.7 corresponds to the 18.33rd percentile and 7.3 corresponds to the 81.67th percentile. The number 63.34% is reasonably close to the 68.27% we would find for the normal distribution, even though the distribution in the example deviates appreciably from the normal form.

The formula for S just illustrated is called the *deviation formula,* since it requires the subtraction of $\bar{X}$ from each X_i. If $\bar{X}$ is not an integer, this computational procedure can lead to a sizeable rounding error because squaring the n deviations $(X_i - \bar{X})$ magnifies the rounding error in $\bar{X}$. An alternative formula that circumvents this problem and also is more convenient to use is

$$S = \sqrt{\frac{\sum\limits_{i=1}^{n} X_i^2 - \dfrac{\left(\sum\limits_{i=1}^{n} X_i\right)^2}{n}}{n}}.$$

This is called the *raw-score formula;* it is algebraically equivalent to the deviation formula.[6] Use of the raw-score formula is illustrated in columns 1 and 4 of Table 4.2-2. It looks more complicated than the deviation formula, but actually it is simpler to use. The ultimate in simplicity is the hand-held calculator with a standard deviation key. After all of the scores have been entered, these calculators compute the standard deviation with the press of a key.[7]

---

[6] The equivalence is shown in Section 4.8-1.
[7] Most hand-held calculators compute $\hat{\sigma}$, an inferential statistic (see footnote 2). The sample statistic S can be obtained by multiplying $\hat{\sigma}$ by $\sqrt{(n-1)/n}$; that is, $S = \hat{\sigma}\sqrt{(n-1)/n}$.

Table 4.2-2.   Computation of the standard deviation

(i) *Data:*

(1) X_i	(2) $X_i - \bar{X}$	(3) $(X_i - \bar{X})^2$	(4) X_i^2
5	0	0	25
9	4	16	81
2	-3	9	4
8	3	9	64
6	1	1	36
5	0	0	25
4	-1	1	16
7	2	4	49
4	-1	1	16
3	-2	4	9
1	-4	16	1
6	1	1	36

$$\sum_{i=1}^{n} X_i = 60 \qquad\qquad \sum_{i=1}^{n} (X_i - \bar{X})^2 = 62 \qquad\qquad \sum_{i=1}^{n} X_i^2 = 362$$

$$\bar{X} = \frac{\sum_{i=1}^{n} X_i}{n} = 5$$

(ii) *Computational formulas:*

Deviation formula

$$S = \sqrt{\frac{\sum_{i=1}^{n} (X_i - \bar{X})^2}{n}}$$

$$= \sqrt{\frac{62}{12}}$$

$$= 2.3$$

Raw-score formula

$$S = \sqrt{\frac{\sum_{i=1}^{n} X_i^2 - \dfrac{\left(\sum_{i=1}^{n} X_i\right)^2}{n}}{n}}$$

$$= \sqrt{\frac{362 - \dfrac{(60)^2}{12}}{12}}$$

$$= \sqrt{\frac{362 - 300}{12}}^{\,a}$$

$$= 2.3$$

[a] The numerator of this ratio is always positive; a negative value indicates a computational error. One common error is to use the value of $\sum X_i^2$ for $(\sum X_i)^2$ and vice versa.

The standard deviation also can be computed from an ungrouped frequency distribution. In this case, the two formulas are modified as follows:

Deviation formula

$$S = \sqrt{\frac{\sum_{j=1}^{k} f_j(X_j - \bar{X})^2}{n}}$$

and

Raw-score formula

$$S = \sqrt{\frac{\sum_{j=1}^{k} f_j X_j^2 - \dfrac{\left(\sum_{j=1}^{k} f_j X_j\right)^2}{n}}{n}},$$

where X_j is the value of the jth class interval, f_j is the frequency of scores in the jth class interval, and summation is performed over the $j = 1, \ldots, k$ class intervals.

Index of Dispersion

The three measures of dispersion that have been described are distance measures, which of course are appropriate for quantitative data. If data do not contain distance information, as in the case of qualitative variables, how can we describe dispersion? One approach is to think of it as the distinguishability of the observations—that is, as the number of pairs of observations actually distinguishable relative to the maximum possible number.[8] Consider $c = 2$ qualitative categories, A and B, with observations denoted by a_i and b_j, as shown in Figure 4.2-3(a). The two observations in A are indistinguishable; the four in B are indistinguishable. However, among the $n = 6$ observations there are eight distinguishable pairs (DP): $a_1 b_1, a_1 b_2, a_1 b_3, a_1 b_4, a_2 b_1, a_2 b_2, a_2 b_3, a_2 b_4$. The lower limit of DP, which represents minimum dispersion, is zero; this occurs when all the observations are in one category and hence are indistinguishable. The upper limit of DP, denoted by DP_{max}, occurs when the observations are equally divided among the c categories, as in Figure 4.2-3(b). For this example, representing maximum dispersion, there are nine distinguishable pairs (DP_{max}): $a_1 b_1, a_1 b_2, a_1 b_3, a_2 b_1, a_2 b_2, a_2 b_3, a_3 b_1, a_3 b_2,$ $a_3 b_3$. The ratio DP/DP_{max}—*the number of distinguishable pairs to the maximum possible number of distinguishable pairs for c categories*—is called the **index of dispersion** and is denoted by D. For the data in Figure 4.2-3(a), $D = DP/DP_{max} = \frac{8}{9} = 0.89$.

a.   b.

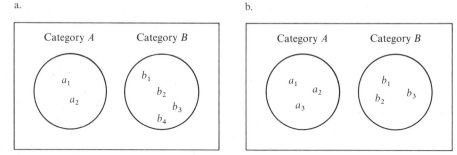

Figure 4.2-3. Observations are assigned to different qualitative categories such that those within a category are indistinguishable with respect to some characteristic.

The minimum value of $D = DP/DP_{max}$ is 0 and occurs when $DP = 0$, which indicates that all the observations are in one category. The maximum value of D is 1 and occurs when $DP = DP_{max}$, which indicates that the observations are equally divided among the c categories. Thus, D ranges over values 0–1; the larger D, the larger the observed number of distinguishable pairs of observations relative to the maximum number for c categories and, hence, the greater the dispersion.

---

[8] Such an index has been suggested by several people, including Hammond, Householder, and Castellan (1970, pp. 111–117) and Mueller and Schuessler (1961, pp. 177–179).

When the number of observations is large, it is tedious to enumerate DP and DP_{max}. A simple alternative formula for D is

$$D = \frac{c\left(n_2 - \sum_{j=1}^{c} n_j^2\right)}{n^2(c-1)},$$

where c is the number of categories, n is the number of observations, and n_j is the number of observations in each of the $j = 1, \ldots, c$ categories.[9] For the data in Figure 4.2-3(a),

$$D = \frac{2[(6)^2 - (2)^2 - (4)^2]}{(6)^2(2-1)} = 0.89,$$

the same value obtained previously.

The value $D \times 100$ can be interpreted as a percentage, since it is the ratio of the number of distinguishable pairs of observations relative to the maximum possible number of pairs for the c categories. For example, if $D = 0.25$, we know that the number of observed distinguishable pairs is $0.25 \times 100 = 25\%$ of the maximum possible number. The index is particularly useful in comparing the dispersions of several distributions based on the same set of c categories. Suppose that we have asked married women with either a high school or a college education to rate their marital happiness. The results of the survey along with the mode and the index of dispersion are shown in Table 4.2-3. We see that the modes are identical;

Table 4.2-3.   Marital happiness ratings of women with either a high school or a college education

(i) *Data:*

Rating	High School Graduates	College Graduates
Very happy	15	12
Moderately happy	28	39
Neutral	16	30
Unhappy	13	12
Very unhappy	8	3
	$n_1 = 80$	$n_2 = 96$
	$Mo_1 =$ Moderately happy	$Mo_2 =$ Moderately happy
	$D_1 = 0.96$	$D_2 = 0.88$

(ii) *Computation of D:*

$$D = \frac{c\left(n^2 - \sum_{j=1}^{c} n_j^2\right)}{n^2(c-1)}$$

$$D_1 = \frac{5[(80)^2 - (15)^2 - (28)^2 - (16)^2 - (13)^2 - (8)^2]}{(80)^2(5-1)} = \frac{24{,}510}{25{,}600} = 0.96$$

$$D_2 = \frac{5[(96)^2 - (12)^2 - (39)^2 - (30)^2 - (12)^2 - (3)^2]}{(96)^2(5-1)} = \frac{32{,}490}{36{,}864} = 0.88$$

[9] See Kirk (1978, pp. 91–93) for the derivation of this formula.

however, the dispersion of the college graduates' distribution is smaller than that for the high school graduates.

When data are qualitative in character, the only appropriate measure of central tendency is the mode; the appropriate measure of dispersion is D. The index of dispersion has two disadvantages: (1) it is a terminal statistic (its usefulness in advanced descriptive and inferential statistics is limited) and (2) it is less familiar than R, Q, and S, which are based on the concept of distance rather than on the number of distinguishable pairs of observations.

Review Exercises for Section 4.2

*1. Compute the range for the following sets of numbers.
  *a. 11, 6, 5, 2, 9, 14, 17, 4   *b. 7, 1, 6, 6, 6, 7, 7, 16
  c. 3, 9, 5, 6, 5, 4, 5   d. 26, 18, 30, 24, 23, 24

2. The ranges in Exercises 1a and 1b are identical, although the first set of numbers appears to be more heterogeneous than the second. Why doesn't the range reflect this difference?

3. For what kind of data can we compute the mode but not the range?

*4. Data representing the length of time required to notice the onset of a warning light during the performance of a simulated driving test are listed in the table. *(a) Compute the median and the semi-interquartile range for these data. *(b) Compute P_{10} and P_{90}. (c) Construct a histogram.

X_j, Time (seconds)	f_j	X_j, Time (seconds)	f_j
32	1	26	3
31	1	25	2
30	2	24	1
29	3	23	0
28	4	22	0
27	6	21	1

5. The emotional stability of a random sample of encounter group participants at Nelase Institute was measured. (a) Compute the median and the semi-interquartile range for the emotional stability scores listed in the table. (b) Compute P_{20}. (c) Construct a histogram.

X_j, Emotional stability	f_j	X_j, Emotional stability	f_j
30	1	19	4
27	1	18	3
25	2	17	3
23	2	16	3
22	2	15	4
21	2	13	2
20	3	12	2
		10	2
		8	1

*6. If we are given a score, its percentile rank (denoted by %) can be computed from

$$\% = \frac{100}{n}\left[\sum f_b + \frac{f_i(P_\% - X_{ll})}{i}\right].$$

This formula can be derived from the formula for computing a percentile point, $P_\%$,

$$P_\% = X_{ll} + i\left(\frac{n(\%/100) - \sum f_b}{f_i}\right).$$

For the data in Exercise 4, compute the percentile rank for $X = 30$ using the formula. For these data, note that $i = 1$ and the real limits of a score, say 27, are 26.5 and 27.5.

7. For the data in Exercise 5, compute the percentile rank for (a) $X = 13$, (b) $X = 19$, and (c) $X = 25$.

8. Describe the nature of the distance represented by the standard deviation.

*9. Preschool children, particularly those who are very intelligent, often create imaginary companions. Compute the mean and the standard deviation using $\sqrt{\sum(X_i - \bar{X})^2/n}$ for the following data, which represent the number of companions per child.

$$
\begin{array}{ccccc}
4 & 2 & 5 & 3 & 1 \\
3 & 2 & 1 & 2 & 3 \\
2 & 4 & 3 & 2 & 0
\end{array}
$$

*10. Compute S for the data in Exercise 9 using $\sqrt{[\sum X_i^2 - (\sum X_i)^2/n]/n}$.

11. Infants are not as passive and undiscriminating about stimulation as we once thought; they show distinct preferences when given an opportunity to control stimuli presented to them. The following data are the number of trials required for infants to learn to control visual stimuli by varying their sucking responses. (a) Compute the mean and the standard deviation for these data. Compute S using both the deviation and the raw-score formulas. (b) Construct a frequency polygon.

$$
\begin{array}{cccccc}
81 & 73 & 75 & 72 & 76 & 74 \\
77 & 72 & 71 & 74 & 72 & 73 \\
73 & 70 & 78 & 73 & 71 & 69 \\
75 & 74 & 68 & 70 & 69 & 73 \\
72 & 70 & 66 & 71 & 75 & 72 \\
76 & 74 & 73 & 77 & &
\end{array}
$$

12. In a concept-learning experiment, chimpanzees were taught to recognize a triangle in different orientations. Compute the mean and the standard deviation for these data. Compute S using both the deviation and the raw score formulas.

X_j, Number of trials	f_j	X_j, Number of trials	f_j
50	1	45	6
49	3	44	4
48	4	43	2
47	6	42	1
46	8	41	1

13. For the data in Exercise 5, (a) compute the mean and the standard deviation. Compute S using both the deviation and the raw score formulas for an ungrouped frequency distribution. (b) Construct a frequency polygon.

*14. The attitudes of a random sample of white female college students toward having a career were surveyed. *(a) For the data in the table, compute the mode and the index of dispersion. (b) Construct a bar graph.

Category	f
Strongly desire career	16
Moderately desire career	23
Undecided about career	19
Don't want career	10

15. The following data for black female high school students were obtained in the survey described in Exercise 14. Compute the mode and the index of dispersion.

Category	f
Strongly desire career	38
Moderately desire career	19
Undecided about career	5
Don't want career	17

16. Approximately a million college-bound high school seniors participated in the College Board's Admissions Testing Program for the 1980–81 year. The responses of males and females to the question "What is the highest level of education you plan to complete beyond high school?" are as follows. (Suggested by *Profiles, College-Bound Seniors, 1981*, New York, N.Y.: College Entrance Examination Board, 1982.)

Category	f_{males}	$f_{females}$
Two-year training program	13,510	15,609
Associate in Arts degree	6,101	15,122
B.A. or B.S. degree	133,795	160,484
M.A. or M.S. degree	116,362	118,046
M.D., Ph.D., other professional degree	83,240	77,559
Undecided	82,805	100,973
	$n = 435,813$	$n = 487,793$

a. Compute the mode and the index of dispersion for the males and the females.

b. Is the magnitude of the dispersion of educational plans for males and females appreciably different?

*17. The following proofs show the effect on the standard deviation of adding a constant to each score or multiplying each score by a constant. For each proof, identify the summation operations and the number of the summation rules that were used from Section 3.9-1.

*a. Let S_{X+c} be the standard deviation of a distribution that has been altered by adding a constant c to each score—that is, $X_1 + c, X_2 + c, \ldots, + X_n + c$. To determine the effect on S of adding a constant, we replace X_i by $(X_i + c)$ and $\bar{X}$ by $\sum_{i=1}^{n} (X_i + c)/n$ in the formula $S = \sqrt{\sum_{i=1}^{n} (X_i - \bar{X})^2/n}$, as follows.

$$S_{X+c} = \sqrt{\frac{\sum_{i=1}^{n}\left[(X_i + c) - \sum_{i=1}^{n}(X_i + c)/n\right]^2}{n}}$$

$$= \sqrt{\frac{\sum_{i=1}^{n}\left(X_i + c - \sum_{i=1}^{n} X_i/n - nc/n\right)^2}{n}}$$

$$= \sqrt{\frac{\sum_{i=1}^{n}(X_i + c - \bar{X} - c)^2}{n}}$$

$$= \sqrt{\frac{\sum_{i=1}^{n}(X_i - \bar{X})^2}{n}}$$

$$= S$$

Thus, adding a constant c to each score doesn't affect the value of the standard deviation. Similarly, it can be shown that subtracting a constant also doesn't affect the value of the standard deviation.

b. Let S_{cX} be the standard deviation of a distribution that has been altered by multiplying each score by a positive constant—that is, $cX_1, cX_2, \ldots, cX_n$. The effect of this alteration can be shown by replacing X_i by cX_i and $\bar{X}$ by $\sum_{i=1}^{n} cX_i/n$ in the formula for S, as follows.

$$S_{cX} = \sqrt{\frac{\sum_{i=1}^{n}\left(cX_i - \sum_{i=1}^{n} cX_i/n\right)^2}{n}}$$

$$= \sqrt{\frac{\sum_{i=1}^{n}\left(cX_i - c\sum_{i=1}^{n} X_i/n\right)^2}{n}}$$

$$= \sqrt{\frac{\sum_{i=1}^{n}(cX_i - c\bar{X})^2}{n}} = \sqrt{\frac{\sum_{i=1}^{n} c^2(X_i - \bar{X})^2}{n}}$$

$$= c\sqrt{\frac{\sum_{i=1}^{n}(X_i - \bar{X})^2}{n}} = cS$$

Thus, the effect of multiplying each score by a positive constant c is to change S, the standard deviation of the original scores, to cS. Similarly, it can be shown that the effect of dividing each score by a positive constant c is to change S to S/c.

If c is a negative constant, $S_{cX} = |c|S$. The use of $|c|$ ensures that $|c|S$ is positive and is consistent with the definition of the standard deviation as the positive square root of $\sum_{i=1}^{n}(X_i - \bar{X})^2/n$.

*18. Interpret the following. *(a) $\bar{X} = 100$, $S = 15$, and the distribution is approximately normal, (b) $Mdn = 50$, $Q = 8$, (c) $Mo = 30$, $R = 5$, (d) $Mo =$ Category of Ford cars, $D = 0.20$.

19. Terms to remember:
   a. Deviation formula for S
   b. Raw-score formula for S

Section 4.3   Relative Merits of the Measures of Dispersion

Standard Deviation

The standard deviation, which is typically reported with the mean, is the most important and most widely used measure of dispersion for quantitative variables whose distributions are relatively symmetrical. Its popularity is due largely to its superior sampling stability and its mathematical tractability. There are two situations, however, in which the standard deviation is not the preferred measure of dispersion: when a distribution is very skewed and when the data are qualitative. Consider the case of a skewed distribution. The value of the standard deviation is determined by squaring the deviation of each score from the mean. The squaring operation gives undue weight to extreme scores in the longer tail and results in a much larger standard deviation than would have been obtained in the absence of extreme scores. This is a disadvantage. Suppose that we wished to compare the dispersion of two distributions that are similar except that one contains several more extreme scores in the longer tail. In spite of their similarity, the standard deviations of the distributions would be quite different and the comparison would be misleading. A few extreme scores exert an influence that is disproportionate to their number.

The standard deviation shouldn't be used with qualitative variables, since it is a distance measure, actually the root-mean-squared distance by which scores deviate from the mean. If data don't contain distance information, as is the case of qualitative variables, the standard deviation is meaningless.

Semi-Interquartile Range

The semi-interquartile range is reported with the median and shares many of its advantages and disadvantages. It is limited to descriptive applications with quantitative variables and is relatively intractable mathematically. Nevertheless, it is preferred over the standard deviation in two situations. We saw in Section 3.5 that the median can be computed for open-ended distributions. So can the semi-interquartile range if the unknown scores lie above Q_3 or below Q_1. The standard deviation can also be computed when there are unknown scores, but none of the procedures for doing so is entirely satisfactory. The semi-interquartile range is also preferred for skewed distributions—it is sensitive to the number but not to the value of scores lying above Q_3 and below Q_1, so it is less influenced by the extreme scores in the longer tail of a distribution than is the standard deviation. However, it is only when distributions are markedly skewed or open-ended that the semi-interquartile range is preferred over the standard deviation.

Range

The range is used for quantitative variables and may be reported with the mode. The great advantage of the range is its simplicity—it is easy to understand and to compute. As a result, it is used widely as a preliminary measure of dispersion. It also is used in deciding how to group data in a frequency distribution, an application that was described in Section 2.2.

The major deficiency of the range is its poor sampling stability. Its value is determined by only two scores (the largest and the smallest), which means that it is not sensitive to most of the score values.

Another deficiency is its dependency on sample size. If scores are randomly sampled from a population, the range will tend to be larger for larger samples because large samples are more likely to include extreme scores. These deficiencies, plus its poor mathematical tractability, limit the range to descriptive applications.

Index of Dispersion

The index of dispersion is the only measure that is appropriate for qualitative variables. Unlike other dispersion measures, it represents not distance but the number of distinguishable pairs of observations relative to the maximum possible number in c categories. It is usually reported with the mode, which is the appropriate measure of central tendency for qualitative variables. However, the index of dispersion is less familiar than the other measures of dispersion and is rarely used in advanced statistical procedures.

Summary of the Properties of the Four Measures of Dispersion

The standard deviation is

1. a distance measure—the root-mean-squared distance by which scores deviate from the mean;
2. the preferred measure for quantitative variables with distributions that are relatively symmetrical;
3. often reported with the mean—for a normal distribution, $\bar{X} \pm S$ is an interval that contains 68.27% of scores;
4. the measure with the best sampling stability;
5. widely used, implicitly or explicitly, in advanced statistics;
6. mathematically tractable;
7. the only measure whose value is affected by the value of every score in the distribution;
8. fairly sensitive to extreme scores, so it is not recommended for markedly skewed distributions; and
9. not appropriate for qualitative variables.

The semi-interquartile range is

1. a distance measure—one-half the distance between the first and the third quartiles;
2. often reported with the median for quantitative variables;
3. closely related to the median, since both are defined in terms of quartile points;
4. sensitive only to the number and not the value of scores above Q_3 and below Q_1; hence it is used widely for markedly skewed distributions;
5. the only relatively stable measure that is appropriate for open-ended distributions;
6. more subject to sampling fluctuation than the standard deviation;
7. less mathematically tractable than the standard deviation; and
8. rarely used in advanced statistical procedures.

The range is

1. a distance measure—the distance between the largest and the smallest scores, inclusive;
2. often reported with the mode for quantitative variables;
3. the simplest measure of dispersion to compute and interpret;
4. used in deciding how to group data in a frequency distribution;
5. much more subject to sampling fluctuation than the other measures of dispersion;
6. dependent on sample size—the larger the sample size, the larger, on the average, the range;
7. less mathematically tractable than the standard deviation; and
8. rarely used in advanced statistical procedures.

The index of dispersion is

1. a measure of the distinguishability of observations—that is, the number of distinguishable pairs of observations relative to the number possible. The index is zero when all observations are in one qualitative category (minimum dispersion), and it has its maximum value of 1 when the observations are evenly distributed over the categories (maximum dispersion);
2. the only measure appropriate for qualitative variables;
3. reported with the mode;
4. rarely used in advanced statistical procedures; and
5. less familiar than R, Q, and S, which are based on the concept of distance.

Review Exercises for Section 4.3

*20. What measure of central tendency and dispersion would you compute for the following data? Defend your choice.

*a.

IQ

b.

Family size

*c.

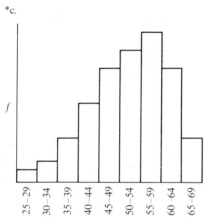

Creativity scores of doctoral candidates
in English

*d.

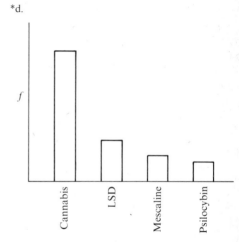

Frequency of drug use among teenagers

e.

Emotional stability scores

f.

REM sleep at five ages

g.

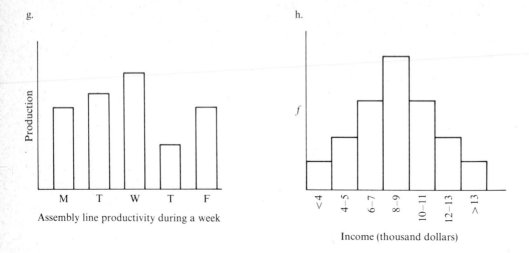

Assembly line productivity during a week

h.

Income (thousand dollars)

Section 4.4   *Dispersion and the Normal Distribution*

The distribution of many variables in the behavioral sciences and education resembles the bell-shaped normal distribution. Because this distribution is so important, its properties have been studied extensively by mathematicians. We saw earlier that 50% of scores fall above and 50% below the mean and that the interval $\bar{X} \pm S$ is a distance that includes 68.27% of scores. Suppose that we are interested in the interval $\bar{X} \pm 2S$ or $\bar{X} \pm 3S$. The percent of scores included within these intervals is shown in Figure 4.4-1. It can be seen that an interval of six standard deviations includes almost all of the scores, 99.73%. Also, $\bar{X} \pm S$ gives the scores that mark the inflection points of the normal distribution—that is, where the curve changes from convex to concave or the reverse.

For a normal distribution, a knowledge of one dispersion measure enables us to determine the other two. For example, $Q = 0.6745S$ and $S = 1.483Q$. Com-

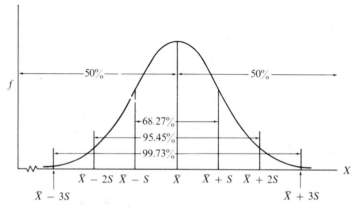

Figure 4.4-1. Percentage of scores contained in selected intervals around the mean for a normal distribution.

puting the range from Q or from S is not as simple, since the range is a function of sample size, as the following approximations illustrate.[10]

$$S \cong \frac{R}{3} \quad \text{and} \quad R \cong 3S \quad \text{when} \quad n = 10$$

$$S \cong \frac{R}{4} \quad \text{and} \quad R \cong 4S \quad \text{when} \quad n = 30$$

$$S \cong \frac{R}{5} \quad \text{and} \quad R \cong 5S \quad \text{when} \quad n = 100$$

$$S \cong \frac{R}{6} \quad \text{and} \quad R \cong 6S \quad \text{when} \quad n = 500$$

For any distribution, the minimum value of S is $R/\sqrt{2n}$ and the maximum is $R/2$.[11] This information can be used to check quickly the value of S for computational errors. For example, if $n = 30$, $R = 60$, and the distribution is approximately normal, the standard deviation should be close to $R/4 = \frac{60}{4} = 15$. If your computed S is 25, you should be suspicious; if it is greater than $R/2 = \frac{60}{2} = 30$, you know that you have made a computational error.

Review Exercises for Section 4.4

*21. For a normal distribution, what percentage of the scores falls *(a) below $\bar{X} + S$? (b) above $\bar{X} - 3S$? (c) above $\bar{X} + 2S$?

*22. Assume a normal distribution, $S = 5$, and $n = 30$. Estimate (a) Q and (b) R.

23. Assume a normal distribution, $S = 10$, and $n = 100$. Estimate (a) Q and (b) R.

*24. If $R = 30$ and $n = 50$, determine the minimum and maximum values for S.

25. If $R = 40$ and $n = 32$, determine the minimum and maximum values for S.

*26. Which of the following values of S are incorrect?
    *a. $S = 10$, $R = 42$, $n = 30$
    b. $S = 15$, $R = 25$, $n = 10$
    c. $S = 18$, $R = 210$, $n = 50$

27. Term to remember:
    Inflection point

[†]Section 4.5   Tchebycheff's Theorem

The percentage of scores contained in an interval bounded by $\bar{X} \pm S$, $\bar{X} \pm 2S$, and $\bar{X} \pm 3S$ is shown in Figure 4.4-1 for a normal distribution. *For nonnormal distributions, what is the minimum percentage of scores contained in these intervals? This question can be answered by* **Tchebycheff's theorem,** *which states that no matter how skewed the distribution,* $\bar{X} \pm k$ *standard deviation units, where* $k \geq 1$, *will include at least* $(1 - 1/k^2) \times 100\%$ *of the scores.* For example, if $k = 2$, $\bar{X} + 2S$ will contain no less than $(1 - 1/2^2) \times 100 = 75\%$ of the scores; the corresponding

---

[10] Adapted from Tippett (1925).
[11] See Kirk (1978, pp. 90–91) for a proof.
[†] This section can be omitted without loss of continuity.

Table 4.5-1. Illustrative values of $1 - 1/k^2$ giving the minimum percentage of scores bounded by $\bar{X} \pm kS$

k	Normal distribution (%)	Nonnormal distribution (%)
1.0	68.27	0
1.5	86.64	55.56
2.0	95.45	75.00
2.5	98.76	84.00
3.0	99.73	88.89

percentage for the normal distribution is 95.45. Table 4.5-1 gives a comparison for selected values of k. Since Tchebycheff's theorem applies to any distribution, it is very conservative. It and the normal distribution can be used to bracket the percentage of scores that would be expected to fall in the interval $\bar{X} + kS$. For example, if $k = 2$, at least 75% will fall within $\bar{X} \pm 2S$; as the distribution approaches the normal form, the percentage approaches 95.45

Review Exercises for Section 4.5

*28. Compute the minimum percentage of scores that can be expected to lie within k standard deviations of the mean when the distribution is not normal and k is *(a) 1.75 (b) 2.25 (c) 4   (d) 4.50.

*29. Suppose that a distribution is believed to be approximately normal. Give the minimum and the probable percentage of scores that can be expected to lie within k standard deviations of the mean when k is equal to *(a) 1   *(b) 2   (c) 2.5   (d) 3.

Section 4.6   Skewness and Kurtosis

To complete our description of a distribution we need two more statistics: indexes of skewness and kurtosis. We saw in Section 2.6 that skewness refers to the asymmetry of a distribution and kurtosis, to its peakedness or flatness.

Skewness

A number of indexes of skewness have been developed; the most widely used one is[12]

$$Sk = \frac{\dfrac{\sum\limits_{i=1}^{n} (X_i - \bar{X})^3}{n}}{\left[\dfrac{\sum\limits_{i=1}^{n} (X_i - \bar{X})^2}{n} \right] \sqrt{\dfrac{\sum\limits_{i=1}^{n} (X_i - \bar{X})^2}{n}}}.$$

If a distribution is symmetrical, $Sk = 0$; if it is positively skewed, $Sk > 0$; and if it is negatively skewed, $Sk < 0$. Computation of the index is illustrated in Table 4.6-1. For these data, $Sk = -0.7$, which indicates that the distribution is negatively skewed, as Figure 4.6-1 shows.

[12] This index, developed by Karl Pearson, is sometimes denoted by $\sqrt{\beta_1}$ and sometimes by g_1.

Table 4.6-1.   Example illustrating computation of measures of skewness and kurtosis

(i) *Data:*

X_i	$(X_i - \bar{X})$	$(X_i - \bar{X})^2$	$(X_i - \bar{X})^3$	$(X_i - \bar{X})^4$
6	2	4	8	16
5	1	1	1	1
5	1	1	1	1
5	1	1	1	1
5	1	1	1	1
4	0	0	0	0
3	−1	1	−1	1
2	−2	4	−8	16
1	−3	9	−27	81

$$\sum_{i=1}^{n} X_i = 36 \quad \sum_{i=1}^{n}(X_i - \bar{X}) = 0 \quad \sum_{i=1}^{n}(X_i - \bar{X})^2 = 22 \quad \sum_{i=1}^{n}(X_i - \bar{X})^3 = -24 \quad \sum_{i=1}^{n}(X_i - \bar{X})^4 = 118$$

$$\bar{X} = 4$$

(ii) *Computation of Sk:*

$$Sk = \frac{\dfrac{\sum_{i=1}^{n}(X_i - \bar{X})^3}{n}}{\left[\dfrac{\sum_{i=1}^{n}(X_i - \bar{X})^2}{n}\right]\sqrt{\dfrac{\sum_{i=1}^{n}(X_i - \bar{X})^2}{n}}} = \frac{\dfrac{-24}{9}}{\left(\dfrac{22}{9}\right)\sqrt{\dfrac{22}{9}}} = \frac{-2.67}{3.82} = -0.7$$

(iii) *Computation of Kur:*

$$Kur = \frac{\dfrac{\sum_{i=1}^{n}(X_i - \bar{X})^4}{n}}{\left[\dfrac{\sum_{i=1}^{n}(X_i - \bar{X})^2}{n}\right]^2} - 3 = \frac{\dfrac{118}{9}}{\left(\dfrac{22}{9}\right)^2} - 3 = \frac{13.11}{5.98} - 3 = -0.8$$

The value of *Sk* can be used to compare the type and the degree of skewness of two distributions independent of any differences in central tendency and dispersion. However, in practice *Sk* is rarely computed, since asymmetry can easily be detected by looking at a frequency distribution or a graph.

Kurtosis

The most common index of kurtosis is[13]

$$Kur = \frac{\dfrac{\sum_{i=1}^{n}(X_i - \bar{X})^4}{n}}{\left[\dfrac{\sum_{i=1}^{n}(X_i - \bar{X})^2}{n}\right]^2} - 3.$$

a.

b.

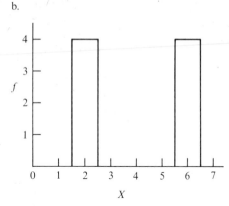

Figure 4.6-1.   (a) Frequency polygon for data in Table 4.6-1; $Sk = -0.7$ and $Kur = -0.8$. (b) Histogram for a perfectly symmetrical bimodal distribution; $Sk = 0$ and $Kur = -2$.

If a distribution is flatter (has a broader hump) than the normal distribution, it is called *platykurtic*, and $Kur < 0$. If its peakedness is the same as that of the normal distribution, it is *mesokurtic*, and $Kur = 0$. If it is more peaked (has a narrower hump) than the normal distribution, it is *leptokurtic*, and $Kur > 0$. Computation of the index is illustrated in Table 4.6-1.

A graph for these data and one for a perfectly symmetrical bimodal distribution are given in Figure 4.6-1. In Figure 4.6-1(a) $Kur = -0.8$, and in Figure 4.6-1(b) $Kur = -2$. It should be apparent that the interpretation of Kur is not as straightforward as that of Sk. It turns out that the value of Kur is dependent not only on the central peak of a distribution, but also on the fullness of its tails as well. Therefore, for distributions that deviate appreciably from the normal form, like those in Figure 4.6-1, the interpretation of Kur is ambiguous.[14] It is doubtful whether any single statistic can adequately measure the quality of peakedness.

Review Exercises for Section 4.6

*30. Age at onset of Parkinson's disease, a degenerative brain disorder, was determined for a sample of adults between 60 and 70 years old. (a) Determine the type and degree of skewness for these data. (b) Construct a histogram. (c) Does it support your decision based on Sk?

$$
\begin{array}{cccccc}
67 & 68 & 60 & 64 & 68 & 63 \\
68 & 70 & 63 & 70 & 68 & 69 \\
70 & 69 & 69 & 69 & 69 & 68 \\
62 & 70 & 70 & 64 & 66 & 66 \\
66 & 69 & 67 & 67 & 70 & 67
\end{array}
$$

31. The reading readiness of preschool children in two neighborhoods was measured. (a) Determine the type and the degree of skewness for these data. (b) Which set of

---

[13] This index is also denoted by g_2. As originally developed by Karl Pearson, the index was equal to $Kur + 3$ and was denoted by β_2.

[14] This point is discussed in detail by Chissom (1970) and Darlington (1970).

data has the greatest skewness? (c) Construct histograms. (d) Do the histograms support your decision based on *Sk*?

Neighborhood A						Neighborhood B				
30	33	32	31	35	33	29	32	28	29	29
32	29	33	30	32	28	30	31	26	30	28
31	31	29	31	26	30	28	29	29	34	30
32	30	33	32	27	32	29	27	30	31	35

*32. One theory predicts that the distribution of reaction times in a paired-associates learning task will be leptokurtic. Do the following learning data support the prediction?

28	28	27	29	29
29	31	32	28	30
27	28	30	31	25
24	27	28	25	27
28	24	32	27	28
29	28	27	28	29

*33. Determine the type and the degree of kurtosis for the data in Exercise 30.

34. Determine which set of data in Exercise 31 deviates most from the normal distribution in terms of kurtosis.

35. Why is *Kur* not an entirely satisfactory measure of peakedness?

36. Terms to remember:
   a. Positively and negatively skewed
   b. Platykurtic
   c. Mesokurtic
   d. Leptokurtic

Section 4.7   Summary

Measures of dispersion summarize the extent to which scores differ from each other, either quantitatively in terms of the spread or scatter of scores or qualitatively in terms of their distinguishability.

Of the four measures of dispersion discussed in this chapter, three are based on the concept of distance and are appropriate for quantitative variables. They are the range, the semi-interquartile range, and the standard deviation. The most important and widely used of the three is the standard deviation, which is typically reported with the mean.

The index of dispersion, which is reported with the mode, describes the distinguishability of observations. Specifically, it indicates the number of distinguishable pairs of observations relative to the maximum possible number of distinguishable pairs in *c* categories. The lower bound of the index, 0, occurs when all observations are in one category; its upper bound, 1, occurs when the observations are evenly distributed over the categories.

Dispersion and central tendency are generally the most important characteristics of a distribution, and they completely describe a normal distribution, which is by definition symmetrical and mesokurtic. For nonnormal distributions, *Sk* and *Kur* provide interesting but somewhat less important information about, respectively, skewness (symmetry) and kurtosis (peakedness).

---

†*Section 4.8   Technical Note*

4.8-1   *Proof of the Equivalence of Raw-Score and Deviation Formulas for S*

The algebraic equivalence of

$$\sqrt{\frac{\sum\limits_{i=1}^{n}(X_i - \bar{X})^2}{n}} \quad \text{and} \quad \sqrt{\frac{\sum\limits_{i=1}^{n}X_i^2 - \dfrac{\left(\sum\limits_{i=1}^{n}X_i\right)^2}{n}}{n}}$$

can be shown by deriving the formula on the right from the formula on the left as follows.

$$S = \sqrt{\frac{\sum\limits_{i=1}^{n}(X_i - \bar{X})^2}{n}}$$

$$= \sqrt{\frac{\sum\limits_{i=1}^{n}(X_i^2 - 2X_i\bar{X} + \bar{X}^2)}{n}}$$

$$= \sqrt{\frac{\sum\limits_{i=1}^{n}X_i^2 - 2\bar{X}\sum\limits_{i=1}^{n}X_i + n\bar{X}^2}{n}} \qquad \text{Rules 3.9-1-1–3.9-1-4}$$

$$= \sqrt{\frac{\sum\limits_{i=1}^{n}X_i^2 - \dfrac{2\sum\limits_{i=1}^{n}X_i}{n}\sum\limits_{i=1}^{n}X_i + \dfrac{n\left(\sum\limits_{i=1}^{n}X_i\right)^2}{n^2}}{n}} \qquad \begin{array}{l}\text{By definition}\\[4pt]\bar{X} = \sum\limits_{i=1}^{n}X_i/n.\end{array}$$

$$= \sqrt{\frac{\sum\limits_{i=1}^{n}X_i^2 - \dfrac{2\left(\sum\limits_{i=1}^{n}X_i\right)^2}{n} + \dfrac{\left(\sum\limits_{i=1}^{n}X_i\right)^2}{n}}{n}}$$

$$= \sqrt{\frac{\sum\limits_{i=1}^{n}X_i^2 - \dfrac{\left(\sum\limits_{i=1}^{n}X_i\right)^2}{n}}{n}}$$

---

† This note, which is not essential to the text, can be omitted without loss of continuity.

†Section 4.9   Computer Supplement

The BMDP package was used to compute descriptive statistics for the data in Table 4.2-2. The printout in Table 4.9-1 was produced by the Detailed Data Description, Including Frequenices Program (P2D). As the program title suggests, the printout provides a rich description of the data.

To aid the reader in interpreting the printout, circled numbers have been added to the table. The following program notes are keyed to these circled numbers. For a complete description of the printout, refer to the user's guide for the program (Dixon *et al.*, 1981).

1. One variable, X, was analyzed.
2. Twelve cases were read and all 12 values were used in the data analysis. Since all values fell within the acceptable range for the variable X, none were excluded from the analysis. There were nine distinct values of X; these values are listed in the ⑨ portion of the table.
3. *MAXIMUM* is the value of the largest score.
   MINIMUM is the value of the smallest score.
   $RANGE = MAXIMUM - MINIMUM$.
   $VARIANCE = \sum (X_i - \bar{X})^2/(n - 1)$.
   ST. DEV. (standard deviation) $= \sqrt{\sum (X_i - \bar{X})^2/(n - 1)}$. The sample standard deviation can be obtained from this estimate of the population standard deviation (see Section 4.2) by *ST. DEV.* $\sqrt{(n - 1)/n} = 2.3741026\sqrt{(12 - 1)/12} = 2.3$.
   $(Q_3 - Q_1)/2$ is the semi-interquartile range.
   MX. ST. SC. (maximum standard score). Standard scores are discussed in Section 10.1 of the text.
   MN. ST. SC. (minimum standard score).
4. The distribution does not have a mode since three scores, 4, 5, and 6, have the same maximum frequency of 2. The *ST. ERROR* (standard error) is discussed in Section 10.3 of the text.
5. This histogram uses the letter H to form the bars for each class interval. $L = 0$ and $U = 10$ denote the lower and the upper class intervals, respectively, that define the horizontal axis of the histogram.

6. $$SKEWNESS = \frac{\dfrac{\sum (X - \bar{X})^3}{n}}{\left[\dfrac{\sum (X_i - \bar{X})^2}{n - 1}\right]\sqrt{\dfrac{\sum (X_i - \bar{X})^2}{n - 1}}}$$

$$KURTOSIS = \frac{\dfrac{\sum (X_i - \bar{X})^4}{n}}{\left[\dfrac{\sum (X_i - \bar{X})^2}{n - 1}\right]^2} - 3$$

---

† This section can be omitted without loss of continuity.

Table 4.9-1. Printout from the BMDP detailed data description, including frequencies program (P2D) (Data from Table 4.2-2)

THREE LOCATION ESTIMATES ARE PRINTED & PLOTTED FOR COMPARISON.

WHEN ANY ESTIMATES ARE TOO CLOSE TOGETHER FOR BOTH TO APPEAR ON THE
PLOT, THE PRIORITY OF PLOTTING IS THE ORDER IN WHICH THE ESTIMATES
ARE LISTED.

Q1 AND Q3 ARE THE 1ST & 3RD QUARTILES. S— AND S+ REPRESENT THE

QUANTITIES (MEAN-ST. DEV.) AND (MEAN+ST. DEV.).

BASED ON INPUT FORMAT SUPPLIED   1 RECORDS READ PER CASE.

NUMBER OF CASES READ.   12

```
*   X1  *
************

①  VARIABLE NUMBER . . . . . .      1
②  NUMBER OF DISTINCT VALUES .      9
   NUMBER OF VALUES COUNTED. .     12
   NUMBER OF VALUES NOT COUNTED     0

LOCATION ESTIMATES
 ④  MEAN          5.0000000
    MEDIAN        5.0000000
    MODE         NOT UNIQUE
```

```
③  MAXIMUM         9.0000000
   MINIMUM         1.0000000
   RANGE           8.0000000
   VARIANCE        5.6363630
   ST. DEV.        2.3741026
   (Q3-Q1)/2       1.5000000
   MX. ST. SC.     1.68
   MN. ST. SC.    -1.68

            ST. ERROR
            0.6853444
            0.8660259
```

```
                  HHH
 ⑤  HHHHHHHHH
    L---------U

EACH '-' ABOVE =     1.0000        EACH 'H'
                L =  0.0000        REPRESENTS
                U = 10.0000            1
                                  COUNT(S)

CASE NO. OF MIN. VAL. =  11
CASE NO. OF MAX. VAL. =   2
```

112

⑦ Q1 = 3.5000000
Q3 = 6.5000000
S— = 2.6258974
S+ = 7.3741026

	VALUE	VALUE/S.E.
⑥ SKEWNESS	0.00	0.00
KURTOSIS	-1.14	-0.80

EACH '.' BELOW = 0.0750

```
        S        Q        M        Q        S        M
        -        1        E        3        +        A
                          A                          X
                          N
⑧ M I N ......................................................
```

⑨ VALUE	COUNT	PERCENTS CELL	CUM
1.	1	8.3	8.3
2.	1	8.3	16.7
3.	1	8.3	25.0

VALUE	COUNT	PERCENTS CELL	CUM
4.	2	16.7	41.7
5.	2	16.7	58.3
6.	2	16.7	75.0

VALUE	COUNT	PERCENTS CELL	CUM
7.	1	8.3	83.3
8.	1	8.3	91.7
9.	1	8.3	100.0

CPU TIME USED   1.850 SECONDS

These formulas for computing skewness and kurtosis differ slightly from those given in Section 4.6. The column labeled *VALUE/S.E.* contains the ratios of skewness to its standard error and kurtosis to its standard error. Standard errors are discussed in Section 10.3 of the text.

7. Q_1 is the first quartile.

   Q_3 is the third quartile.

   $S-$ denotes the mean minus the standard deviation.

   $S+$ denotes the mean plus the standard deviation.

8. This graph shows the location on a line of the minimum score (*MIN*), mean minus the standard deviation (*S*), first quartile (Q_1), mean, third quartile (Q_3), mean plus the standard deviation (S_+), and the maximum score (*MAX*).

9. This frequency distribution lists each score value, its frequency (denoted by *COUNT*), the percent of scores having this value (denoted by *CELL*), and the cumulative percent of scores less than or equal to this value (denoted by *CUM*).

The BMDP program P2D provides the reader with an amazing amount of descriptive information concerning the data in Table 4.2-3: over twenty statistics, two graphs, and one frequency distribution. And according to the note at the bottom of the table, the computer required only 1.85 seconds to compute this information.

5

Correlation

Section 5.1 Introduction

Correlation and Regression Distinguished

Correlation and regression, which are described in this chapter and the next, are procedures for describing the relationship between two sets of paired scores. Because the procedures have much in common, perhaps the simplest way to distinguish between them is by means of examples. Suppose we perform an experiment in which different dosages of amphetamine, the independent variable, denoted by X, are administered to children suffering from hyperkinesis, a behavioral disorder characterized by restlessness, inattention, and disruptive behavior. Children are assigned randomly to the dosage conditions. Following administration of the drug, changes in frequency of hyperkinetic behavior, the dependent variable, denoted by Y, are recorded. For each child the experimenter has paired X and Y scores representing, respectively, dosage and behavior change. The experimenter is interested in knowing whether the two variables are related and if so, in predicting Y from a knowledge of X. This experiment illustrates the key features of a problem in *regression analysis*. First, there is a clearly defined independent variable—amount of amphetamine—whose dosage levels were determined in advance by the experimenter. Second, the value of the dependent variable for a given dosage is free to take on different values. This is in contrast to the independent variable whose values were fixed in advance. Finally, the experimenter is interested in predicting Y from a knowledge of X.

Contrast this experiment with one in which tests of reading readiness and intelligence are administered to a sample of children, yielding paired X and Y scores, respectively, for each child. The experimenter is interested in knowing whether variables X and Y are related, and if so, the strength of their association. In addition we might want to predict either variable from a knowledge of the other. This experiment illustrates a problem in *correlation analysis*. How does it differ from the regression problem? First, there is no obvious independent variable. Second, since the experimenter did not preselect the values of either X or Y, both X and Y are free to take on different values. Finally, the experimenter is interested in assessing the strength of association between X and Y and possibly in predicting either variable from a knowledge of the other.

To summarize, both correlation and regression procedures are concerned with assessing the relationship between sets of paired data. They differ in the way in which the samples are drawn, the status of one variable as the independent or predictor variable, and to some extent the kinds of conclusions drawn.

A Bit of History

The concepts of correlation and regression were developed by Sir Francis Galton during his investigations of the genetic transmission of natural characteristics. He was intrigued by the question "How is it possible for a whole population to remain alike in its features during many successive generations, if the average produce of each couple resembles the parents?" Data from one of his studies on the inheritance of stature are reproduced in Table 5.1-1. Parents' height is plotted on the horizontal or X axis and offspring's height on the vertical or Y axis. This representation of the joint frequency of two variables is called a *bivariate frequency distribution* or *scatter diagram (scattergram)*. Consider the entry in the cell at the intersection of column 68.5 and row 69.2; the frequency is 48. This means that for parents whose height was 68–69, there were 48 offspring whose height was 68.7–69.7 in. The circles in Table 5.1-1 denote the class intervals containing the median of each column. We see, as did Galton, that the relationship between height of offspring and height of parents is approximately *linear*; that is, the set of circles approximates a straight line. Galton developed a procedure for finding the "straight line of best fit," thereby laying the foundation for correlation and regression.

Table 5.1-1. Scatter diagram of midparent height and height of adult offspring[a] (female heights multiplied by 1.08)

Height of adult offspring	Midparent height (inches)[b]										
	<64	64.5	65.5	66.5	67.5	68.5	69.5	70.5	71.5	72.5	≥73
≥73.7							5	3	2	4	
73.2						3	4	3	2	2	③
72.2			1		4	4	11	4	9	⑦	1
71.2			2		11	18	20	7	4	2	
70.2			5	4	19	21	25	14	⑩	1	
69.2	1	2	7	13	38	48	㉝	⑱	5	2	
68.2	1		7	14	28	㉞	20	12	3	1	
67.2	2	5	⑪	⑰	㊳	31	27	3	4		
66.2	2	⑤	11	17	36	25	17	1	3		
65.2	①	1	7	2	15	16	4	1	1		
64.2	4	4	5	5	14	11	16				
63.2	2	4	9	3	5	7	1	1			
62.2		1		3	3						
<61.7	1	1	1			1		1			

[a] Galton (1889, p. 208). I am grateful to Edward W. Minium for bringing these data to my attention.
[b] Circles denote the class intervals containing the median of each column.

A straight line provides a reasonably good fit for many relationships found in behavioral and educational research. Even those that are nonlinear are often approximately linear over some portion of their range. But let us return to the question that sparked Galton's interest. How is it that a population remains alike? The answer is in the trend represented by the circles in Table 5.1-1. We note that on the average, short parents have slightly taller offspring, while tall parents have slightly shorter offspring. This is shown more clearly in Figure 5.1-1. Galton referred to this tendency as *regression* or *reversion* toward the mean and called the best-fitting straight line the *regression* or *reversion line.*

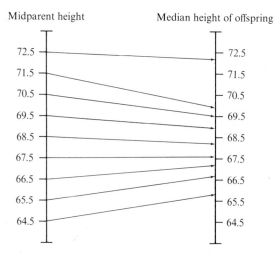

Figure 5.1-1. Arrows relate height (in inches) of parents to median height of offspring. Tall parents tend to have slightly shorter offspring, and short parents, slightly taller offspring. Galton referred to this as reversion toward the mean.

In the discussion that follows, we'll focus our attention on variables like those in Table 5.1-1 that appear to be linearly related.[1] We find that many relationships of interest in the behavioral sciences and education fall into this category.

Review Exercises for Section 5.1

*1. A speech therapist interested in the relationship between two tests of articulation disorders administered the tests to 26 children. (a) Construct a scatter diagram like Table 5.1-1 for the data in the following table. (b) Does the relationship appear to be linear or nonlinear?

---

[1] The nonlinear case is treated in advanced texts such as Glass and Stanley (1970, pp. 150–152) and Kirk (1982, p. 162).

Subject	Test A	Test B	Subject	Test A	Test B
1	26	36	14	33	39
2	28	35	15	22	33
3	25	34	16	24	32
4	21	32	17	27	35
5	25	33	18	29	36
6	26	32	19	32	39
7	26	34	20	28	36
8	31	37	21	25	36
9	27	34	22	24	34
10	20	30	23	25	34
11	23	32	24	27	36
12	30	38	25	26	34
13	29	37	26	26	35

2. A job-satisfaction questionnaire was administered to a random sample of 36 males between the ages of 29 and 34. The investigator was interested in the relationship between years of formal education completed and job satisfaction. (a) Construct a scatter diagram for the data in the following table. (b) Does the relationship appear to be linear or nonlinear?

Subject	Years of education	Job satisfaction	Subject	Years of education	Job satisfaction
1	14	36	19	12	43
2	11	38	20	11	46
3	10	36	21	18	53
4	15	51	22	8	30
5	7	30	23	9	35
6	8	37	24	12	40
7	12	40	25	13	40
8	13	43	26	13	41
9	16	47	27	10	32
10	12	44	28	14	50
11	12	37	29	12	33
12	11	40	30	14	47
13	9	32	31	10	38
14	12	42	32	11	37
15	13	45	33	12	40
16	11	38	34	14	50
17	12	42	35	13	42
18	11	37	36	13	45

*3. Discuss the meaning of the term "regression toward the mean."
4. Terms to remember:
   a. Bivariate frequency distribution or scatter diagram
   b. Linear relationship
   c. Regression line

Section 5.2   A Numerical Index of Correlation

The degree of association or strength of relationship between two variables is represented by a number called a **correlation coefficient.** A correlation coefficient for linearly related sample data is denoted by r_{XY}, or simply r; for a population,

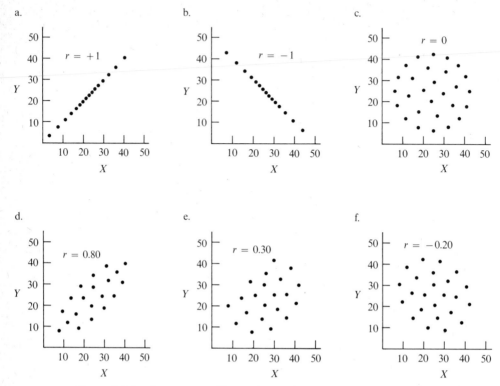

Figure 5.2-1. Scattergrams illustrating various degrees of correlation.

it is denoted by the Greek letter ρ *(rho)*.[2] If the relationship is perfect, the value of r is either $+1$ or -1. A value of $+1$ denotes a perfect positive relationship; this is depicted in the scatter diagram in Figure 5.2-1(a). For this case, all the data points fall on a straight line such that high scores on one variable are paired with high scores on the other, while low scores are paired with low scores. A coefficient of -1 denotes a perfect negative or inverse relationship. For this case, the data points also fall on a straight line, but high scores on one variable are paired with low scores on the other and vice versa, resulting in a line that slopes down instead of up. This is shown in Figure 5.2-1(b). If there is no association between variables, $r = 0$. In this case, the data points tend to fall in a circle, as shown in Figure 5.2-1(c). Intermediate degrees of association are represented by coefficients between -1 and 0 ($-1 < r < 0$) or by coefficients between 0 and 1 ($0 < r < 1$). Some examples of intermediate degrees of association for normally distributed X and Y variables are depicted in Figures 5.2-1(d) through (f). As shown in the figures, the data points for intermediate values of r tend to form an ellipse; the lower the degree of association, the more the ellipse resembles a circle.

[2] Galton originally used the letter r from the word *reversion* to denote the slope of the straight line of best fit.

Review Exercises for Section 5.2

5. Distinguish between r and ρ.

*6. Match the r values 1, -1, 0, 0.4, and -0.9 with the scatter diagrams shown here.

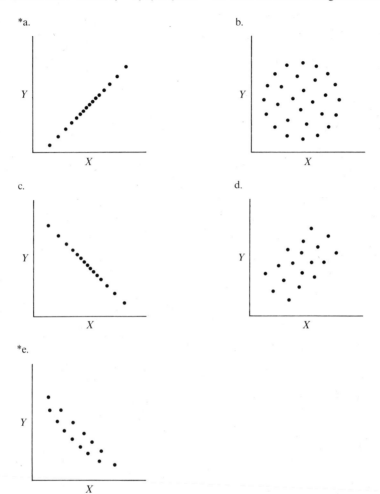

*7. Would you expect the correlation between the following to be positive, negative, or essentially zero?
   *a. Masculinity of fathers and sons
   *b. Reaction time and number of choice alternatives
   c. Mechanical aptitude and birth order
   d. Verbal intelligence and number of trials to learn a list of nonsense syllables
   e. Grades in college and annual income 10 years after graduation
   f. Number of letters in last name and musical aptitude

8. Terms to remember:
   a. Correlation coefficient
   b. Positive relationship
   c. Negative relationship

Section 5.3   Pearson Product-Moment Correlation Coefficient

The most widely used index of correlation is called the **Pearson product-moment correlation coefficient**, after Karl Pearson (1859–1936), who contributed so much to its development. The coefficient is appropriate for describing the relationship between two quantitative variables that are linearly related. The definitional formula for Pearson's r is

$$r = \frac{\dfrac{\sum\limits_{i=1}^{n} (X_i - \bar{X})(Y_i - \bar{Y})}{n}}{\sqrt{\left[\dfrac{\sum\limits_{i=1}^{n} (X_i - \bar{X})^2}{n}\right]\left[\dfrac{\sum\limits_{i=1}^{n} (Y_i - \bar{Y})^2}{n}\right]}}.$$

Table 5.3-1.   Computation of r for father and son authoritarianism scores

(i) *Data:*

Family	Father's score, X	Son's score, Y	XY	X^2	Y^2
1	25	28	700	625	784
2	32	31	992	1,024	961
3	40	41	1,640	1,600	1,681
4	29	33	957	841	1,089
5	31	25	775	961	625
6	16	18	288	256	324
7	28	26	728	784	676
8	36	38	1,368	1,296	1,444
9	33	34	1,122	1,089	1,156
10	29	36	1,044	841	1,296
11	23	20	460	529	400
12	27	28	756	729	784
13	37	30	1,110	1,369	900
14	30	26	780	900	676
15	27	22	594	729	484
16	20	23	460	400	529
17	28	29	812	784	841
18	38	36	1,368	1,444	1,296
19	35	32	1,120	1,225	1,024
20	19	19	361	361	361
	$\sum X_i = 583$	$\sum Y_i = 575$	$\sum X_i Y_i = 17{,}435$	$\sum X_i^2 = 17{,}787$	$\sum Y_i^2 = 17{,}331$

(ii) *Computational procedure:*

$$r = \frac{\sum X_i Y_i - \dfrac{(\sum X_i)(\sum Y_i)}{n}}{\sqrt{\left[\sum X_i^2 - \dfrac{(\sum X_i)^2}{n}\right]\left[\sum Y_i^2 - \dfrac{(\sum Y_i)^2}{n}\right]}} = \frac{17{,}435 - \dfrac{(583)(575)}{20}}{\sqrt{\left[17{,}787 - \dfrac{(583)^2}{20}\right]\left[17{,}331 - \dfrac{(575)^2}{20}\right]}}$$

$$= \frac{673.7500}{(28.1523)(28.2799)} = 0.85$$

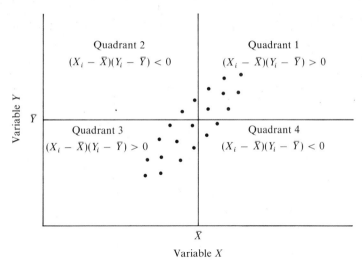

Figure 5.3-1. All the cross products in quadrants 2 and 4 are negative; those in quadrants 1 and 3 are positive.

A more convenient computational formula, called the raw-score formula,[3] is

$$r = \frac{\sum\limits_{i=1}^{n} X_i Y_i - \dfrac{\sum\limits_{i=1}^{n} X_i \sum\limits_{i=1}^{n} Y_i}{n}}{\sqrt{\left[\sum\limits_{i=1}^{n} X_i^2 - \dfrac{\left(\sum\limits_{i=1}^{n} X_i\right)^2}{n}\right]\left[\sum\limits_{i=1}^{n} Y_i^2 - \dfrac{\sum\limits_{i=1}^{n} Y_i}{n}\right]}}.$$

Calculation of r using the raw-score formula is illustrated in Table 5.3-1. The data are 20 paired father–son scores on a test of authoritarianism, which measures rigidity, dependency, and ethnocentrism. The coefficient is equal to 0.85. This tells us two things about the relationship: its strength, represented by the extent to which the value of r differs from zero, and the direction of the relationship, represented by the sign of r. In the following discussion, we will see why r reflects this information. The interpretation of r is discussed in Section 5.4.

Information Contained in the Cross Product

The standing of a person with respect to the mean of variables X and Y can be expressed in terms of the size and the algebraic sign of $(X_i - \bar{X})(Y_i - \bar{Y})$, which is the product of the deviation of the X_i and Y_i scores from their respective means. If a person is above the mean on both variables, the algebraic sign of $(X_i - \bar{X})(Y_i - \bar{Y})$, which is called a *cross product*, is positive, and the associated data point falls in quadrant 1 of Figure 5.3-1. If a person is below the mean

---

[3] The algebraic equivalence of this formula and the definitional formula is shown in Section 5.10-1.

on both variables, $(X_i - \bar{X})(Y_i - \bar{Y})$ is also positive since it is the product of two negative numbers, but the corresponding data point falls in quadrant 3 of Figure 5.3-1. If a person is above the mean on one variable but below the mean on the other, the sign of $(X_i - \bar{X})(Y_i - \bar{Y})$ is negative, and the data point falls in either quadrant 2 or 4.

In Figure 5.3-1 most of the data points are in quadrants 1 and 3; hence the algebraic sign of the sum $\sum_{i=1}^{n}(X_i - \bar{X})(Y_i - \bar{Y})$ is positive. When this sum is positive, the two variables are said to be positively related; that is, an increase in one variable is accompanied by an increase in the other. If an inverse relationship exists between X and Y, most of the data points fall in quadrants 2 and 4, and the sign of $\sum_{i=1}^{n}(X_i - \bar{X})(Y_i - \bar{Y})$ is negative. From the foregoing discussion, it is apparent that the algebraic sign of $\sum_{i=1}^{n}(X_i - \bar{X})(Y_i - \bar{Y})$ in the numerator of

$$ r = \frac{\displaystyle\sum_{i=1}^{n}(X_i - \bar{X})(Y_i - \bar{Y})}{n} \Bigg/ \sqrt{\left[\frac{\displaystyle\sum_{i=1}^{n}(X_i - \bar{X})^2}{n}\right]\left[\frac{\displaystyle\sum_{i=1}^{n}(Y_i - \bar{Y})^2}{n}\right]} $$

indicates whether X and Y are positively or inversely related.

The greater the strength of the relationship between X and Y, the larger the absolute value of $\sum_{i=1}^{n}(X_i - \bar{X})(Y_i - \bar{Y})$. A large positive value of $\sum_{i=1}^{n}(X_i - \bar{X})(Y_i - \bar{Y})$ occurs when the largest $(X_i - \bar{X})$ is paired with the largest $(Y_i - \bar{Y})$, the second largest $(X_i - \bar{X})$ with the second largest $(Y_i - \bar{Y})$, and so on. A large negative value of $\sum_{i=1}^{n}(X_i - \bar{X})(Y_i - \bar{Y})$ occurs when the largest $(X_i - \bar{X})$ is paired with the smallest $(Y_i - \bar{Y})$, and so on. If the correlation is positive but less than 1, the cross-product sum may be composed of some large positive values of $(X_i - \bar{X})$ multiplied by some moderate-sized $(Y_i - \bar{Y})$'s or even some negative $(Y_i - \bar{Y})$'s. The resulting value of $\sum_{i=1}^{n}(X_i - \bar{X})(Y_i - \bar{Y})$ is smaller than if the largest $(X_i - \bar{X})$ were paired with the largest $(Y_i - \bar{Y})$, and so on. If $r = 0$, positive $(X_i - \bar{X})$'s are as likely to be paired with negative $(Y_i - \bar{Y})$'s as with positive $(Y_i - \bar{Y})$'s, resulting in a sum of $\sum_{i=1}^{n}(X_i - \bar{X})(Y_i - \bar{Y})$ that is equal to 0.

It should be apparent after some reflection that the absolute value of the sum of the cross product reflects the strength of the association between X and Y, but it also reflects, or is affected by, the number of X and Y scores and the size of the standard deviations of X and Y. To obtain a measure of the strength of the association that is independent of the number of pairs of scores, we compute the mean of the cross-product sum:

$$ S_{XY} = \frac{\displaystyle\sum_{i=1}^{n}(X_i - \bar{X})(Y_i - \bar{Y})}{n}, $$

where n is the number of paired X and Y scores. This mean is called the *covariance* of X and Y and is denoted by S_{XY}. If we divide the covariance by the standard deviations of X and of Y, S_X and S_Y, we obtain a measure of strength of association that is independent of the dispersions of the two variables as well. The resulting statistic,

$$r = \frac{\dfrac{\sum\limits_{i=1}^{n} (X_i - \bar{X})(Y_i - \bar{Y})}{n}}{\sqrt{\left[\dfrac{\sum\limits_{i=1}^{n} (X_i - \bar{X})^2}{n}\right]\left[\dfrac{\sum\limits_{i=1}^{n} (Y_i - \bar{Y})^2}{n}\right]}} = \frac{S_{XY}}{S_X S_Y},$$

was defined earlier as the definitional formula for the Pearson product-moment correlation coefficient.

To summarize, the heart of the correlation formula is the cross-product term $\sum_{i=1}^{n}(X_i - \bar{X})(Y_i - \bar{Y})$. This term reflects both the nature of the relationship between X and Y (positive versus inverse) and the magnitude of the relationship. The cross-product term is divided by n to free it of dependence on the number of paired X and Y scores; it is divided by $S_X S_Y$ to free it of dependence on the size of the dispersions of the X and Y variables. Because of these operations and because X and Y are expressed as deviations from their respective means, the r statistic is a dimensionless index of linear relationship. This means that the value of r does not depend on the unit of measurement of either the X or Y variables or on the value that is designated as the zero or point of origin of either measuring scale. To put it another way, multiplying X or Y by a positive constant and/or adding a constant (a positive linear transformation) does not affect the value of r. As stated earlier, r ranges over the interval -1 to $+1$. In the following section we will discuss ways to interpret r.

Review Exercises for Section 5.3

*9. A reading readiness test and an intelligence test were administered to a random sample of first-grade children. Compute r for the data in the table.

Child	Reading readiness score	IQ score	Child	Reading readiness score	IQ score
1	45	102	11	43	104
2	40	100	12	50	108
3	48	106	13	42	96
4	45	101	14	40	99
5	38	98	15	41	96
6	43	100	16	48	102
7	36	92	17	47	104
8	41	102	18	37	94
9	42	102	19	42	98
10	50	110	20	45	100

10. The Alcohol Dependence Scale was developed to assist the World Health Organization in the classification of alcoholism. Fifteen alcoholics seeking counseling for alcohol-related disabilities took this scale and the Michigan Alcoholism Screening Test, which

yields an index of problems related to drinking. The following data were obtained. (Suggested by Skinner, Harvey A., and Allen, Barbara A. Alcohol dependence syndrome: Measurement and validation. *Journal of Abnormal Psychology*, 1982, *91*, 199–209.)

Counselee	Alcohol Dependence Scale	Michigan Alcoholism Screening Test
1	89	78
2	48	57
3	74	65
4	97	86
5	59	58
6	65	75
7	46	57
8	84	95
9	78	69
10	77	86
11	67	78
12	36	47
13	83	74
14	68	77
15	96	87

a. Construct a scatter diagram for these data and decide if the data appear to be linearly related.

b. Compute r for these data.

11. Researchers have reported that lonely people often describe themselves as shy. To investigate the strength of the relationship between the two variables, a modified version of the Stanford Shyness Survey and the UCLA Loneliness Scale were given to 20 male and 20 female college students. The order of administration of the instruments was randomized independently for each student. The following data for the male students were obtained. (Experiment suggested by Maroldo, Georgette K. Shyness and loneliness among college men and women. *Psychological Reports*, 1981, *48*, 885–886.)

Student	Stanford Shyness Survey	UCLA Loneliness Scale	Student	Stanford Shyness Survey	UCLA Loneliness Scale
1	36	51	11	30	29
2	39	52	12	30	40
3	30	33	13	33	45
4	23	35	14	32	30
5	28	55	15	28	42
6	41	52	16	34	45
7	29	32	17	21	35
8	27	38	18	41	35
9	28	40	19	23	30
10	28	33	20	39	51

a. Construct a scatter diagram for these data and decide if the data appear to be linearly related.

b. Compute r for these data.

*12. Studies have shown that music can affect mood, emotion, task performance, and cognition. It was hypothesized that the tempo of country–western music played in bars was related to the consumption of alcohol. Observers visited three bars featuring recorded country–western music on three Friday nights. They obtained permission to tape record the music and to make observations of patrons at selected tables. When the music began, the rate of sipping an alcoholic beverage was recorded for each patron. The music tapes were analyzed for the tempo (beats per minute) of each song; the mean number of sips during each song was also determined. The following data were obtained. (Suggested by Bach, Paul J., and Schaefer, James M. The tempo of country music and the rate of drinking in bars. *Journal of Studies on Alcohol*, 1979, *40*, 1058–1059.)

Tempo	Mean number of sips	Tempo	Mean number of sips
35	1.150	80	0.900
38	1.150	85	0.725
44	0.400	91	0.725
48	1.075	93	0.875
51	0.950	100	0.525
64	0.975	102	0.800
68	0.950	108	0.775
68	0.925	112	0.750
72	0.875	118	0.625

   a. Construct a scatter diagram for these data and decide if the data appear to be linearly related.

   b. Compute r for these data.

   c. What does the r tell you about the relationship between tempo and sips per minute?

*13. Calculate r for the data in Exercise 1 (pages 118–119)

14. Calculate r for the data in Exercise 2 (page 119)

*15. Calculate $\sum (X_i - \bar{X})(Y_i - \bar{Y})$ for the following data points. In which quadrants of Figure 5.3-1 would the data points fall? Are the variables related, and if so, is the relationship positive or negative?

*a. X	Y	*b. X	Y	c. X	Y	d. X	Y
9	14	9	14	9	17	9	17
11	17	11	14	11	14	11	17
13	17	11	16	13	10	13	13
7	12	9	16	7	19	7	13

16. What does the covariance S_{XY} tell us about the relationship between X and Y? In computing r, why is S_{XY} divided by $S_X S_Y$?

17. For a set of data with $S_X = 4$ and $S_Y = 5$, what is the largest possible value that S_{XY} can be? (*Hint:* The maximum value of $r = +1$ and $r = S_{XY}/S_X S_Y$.)

*18. If $n = 2$, what are the possible values of r? Make a scatter diagram that supports your answer.

19. Terms to remember:

   a. Cross product

   b. Covariance

Section 5.4   Interpretation of r: Explained and Unexplained Variation

As we have seen, a correlation coefficient reflects the nature and the strength of the linear association between two variables. However, two other statistics, both functions of r, are more useful for getting an intuitive notion for the strength of association represented by r. These statistics are the *coefficient of determination*, r^2, which is equal to the square of the correlation coefficient, and the *coefficient of nondetermination*, k^2, which is equal to $1 - r^2$.

If we examine the authoritarianism scores in Table 5.3-1, we see that there is variation among the X scores and among the Y scores. How can we account for this intersubject variability? One reason why subjects differ on Y is that they differ on X. Since X and Y are correlated, $r = 0.85$, a person who is high on X is also likely to be high on Y. Because of the linear relationship between X and Y, some of the variation among the Y scores can be accounted for or explained by variation among the X scores. However, not all the variation can be explained in this way, since some subjects who have the same X score have different Y scores. For a given linear relationship between X and Y we would like to know how much of the Y score variance is accounted for by the X score variance and how much is not accounted for. We will see that this information is given by, respectively, r^2 and k^2. Let us denote the total variance of X and of Y by S_X^2 and S_Y^2, respectively. (Recall from the discussion of the standard deviation, page 91, that variance is a measure of dispersion: the mean squared distance by which scores deviate from the mean.) If we divide S_X^2 by itself and S_Y^2 by itself, we change both variances into proportions with values of 1, each of which can be partitioned into two components, as follows.[4]

$$\frac{S_X^2}{S_X^2} \quad = \quad r^2 \quad + \quad k^2$$

$$\begin{pmatrix} \text{Total } X \text{ variance} \\ \text{expressed as a} \\ \text{proportion} \end{pmatrix} = \begin{pmatrix} \text{Proportion of} \\ X \text{ variance explained} \\ \text{by } Y \text{ variance} \end{pmatrix} + \begin{pmatrix} \text{Proportion of} \\ X \text{ variance not explained} \\ \text{by } Y \text{ variance} \end{pmatrix}$$

$$\frac{S_Y^2}{S_Y^2} \quad = \quad r^2 \quad + \quad k^2$$

$$\begin{pmatrix} \text{Total } Y \text{ variance} \\ \text{expressed as a} \\ \text{proportion} \end{pmatrix} = \begin{pmatrix} \text{Proportion of} \\ Y \text{ variance explained} \\ \text{by } X \text{ variance} \end{pmatrix} + \begin{pmatrix} \text{Proportion of} \\ Y \text{ variance not explained} \\ \text{by } X \text{ variance} \end{pmatrix}$$

Thus, the total variance expressed as a proportion is equal to the coefficient of determination, r^2, plus the coefficient of nondetermination, k^2. To compute r^2 we square the correlation coefficient; k^2 is computed from $k^2 = 1 - r^2$.

---

[4] The derivation for

$$\frac{S_Y^2}{S_Y^2} = r^2 + k^2$$

is given by Kirk (1978, pp. 144–145).

For the authoritarianism data in Table 5.3-1, $r^2 = (0.85)^2 = 0.72$ and $k^2 = 1 - 0.72 = 0.28$. This means that 0.72 of the variance of the Y scores can be explained by the linear relationship between this variable and the corresponding X scores, but 0.28 of the Y variance is not explained by the X variance. To put it more simply, the linear relationship between the fathers' and sons' scores enables us to account for much of the variance in the sons' authoritarianism scores (0.72); however, 0.28 of the variance cannot be accounted for. In all likelihood we could find other variables, such as the sons' levels of education, that would enable us to reduce the proportion of unaccounted-for variance. The index k^2 is a measure of how much of the variance remains to be accounted for.

A visual representation of the proportion of explained and unexplained variance is presented in Figure 5.4-1(a), where the proportions $S_Y^2/S_Y^2 = 1$ and $S_X^2/S_X^2 = 1$ are represented by the areas of circles. The area in which the circles overlap corresponds to r^2; the nonoverlap areas correspond to k^2.

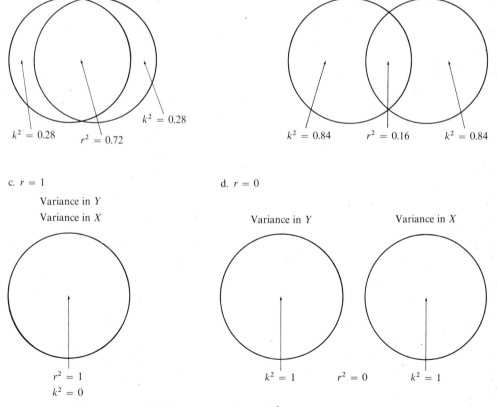

a. $r = 0.85$

Variance in Y Variance in X

$k^2 = 0.28$

$k^2 = 0.28$ $r^2 = 0.72$

b. $r = 0.40$

Variance in Y Variance in X

$k^2 = 0.84$ $r^2 = 0.16$ $k^2 = 0.84$

c. $r = 1$

Variance in Y
Variance in X

$r^2 = 1$
$k^2 = 0$

d. $r = 0$

Variance in Y Variance in X

$k^2 = 1$ $r^2 = 0$ $k^2 = 1$

Figure 5.4-1. Visual representation of r^2, the proportion of variance of one variable that is explained by the variance of the other variable, and k^2, the proportion that is not explained by the variance of the other variable.

If $r = +1$ or -1, the circles completely overlap, and all the variance of one variable is explained by that of the other variable. If $r = 0$, the circles do not overlap, and none of the variance of either variable is explained by that of the other variable.

Most variables of interest to behavioral scientists and educators are affected by a multiplicity of factors. School performance, for example, is affected by academic aptitude, scholastic motivation, health, and parental support for achievement, to name only a few. The correlation between performance and academic aptitude is 0.30, which tells us that we have accounted for $(0.30)^2 = 0.09$ of performance variance and that we have to look to other variables to account for the remaining $1 - 0.09 = 0.91$ of the variance. Note that since $r^2 \leq |r|$, values of $|r|$ close to 1 are required to account for an appreciable proportion of variance. Not until $r = 0.71$ does $r^2 = 0.50$.

Review Exercises for Section 5.4

20. What do r^2 and k^2 tell us about the relationship between X and Y?
*21. For the following experiments, compute r^2 and k^2 and interpret them verbally and by means of diagrams like those in Figure 5.4-1.
    *a. The correlation between freshman English grades and grades in a physical education bowling class was 0.22.
    b. The correlation between the number of hours that rats had been deprived of food and the time to traverse a maze with sunflower seeds in the goal box was 0.80.
    c. The correlation between the last two digits of students' social security numbers and the number of trials to learn nonsense syllables was 0.02.
22. Terms to remember:
    a. Coefficient of determination, r^2
    b. Coefficient of nondetermination, k^2

Section 5.5   Some Common Errors in Interpreting r

Error: Interpreting r in Direct Proportion to Its Size

Correlation coefficients are often incorrectly interpreted. A common error is to interpret the size of r as directly reflecting the percentage of association. For example, it is incorrect to say there is a 60% association between two variables when $r = 0.60$. Such a statement is meaningless. Does it mean that 60% of the elements are associated? An r does not indicate the percentage of association but rather is a measure of strength of association on a scale of -1 to $+1$. A related misconception is saying, for example, that $r = 0.80$ represents twice the relationship indicated by $r = 0.40$ or that an increase in correlation from 0.10 to 0.20 is equivalent to an increase in correlation from 0.60 to 0.70.

Error: Interpreting r in Terms of Arbitrary Descriptive Labels

Various schemes have been suggested to help students interpret correlation coefficients. A common but misleading scheme is the classification of certain r values as "very high" (for example, $r \geq 0.90$), "high" ($r = 0.70$–0.89), "medium" ($r =$

0.30–0.69), or "low" ($r < 0.30$). The problem with such classifications is that what constitutes a high or low correlation depends on what is being correlated with what, and what use is to be made of r once it has been computed. This will be illustrated for the concepts of *reliability* and *validity*—two desirable character-istics of psychological tests. One type of reliability, called *test–retest reliability*, is determined by administering a test to a group of subjects, waiting, and then readministering the test to the same subjects. The test's reliability, or consistency of measurement, is the correlation between the two sets of scores. Reliability coef-ficients of 0.90 or higher are common for tests of intellectual aptitude. A test–retest reliability coefficient below 0.80 would raise serious questions about the reliability of an intelligence test; however, the system described above would classify $r = 0.80$ as high. Equally misleading designations result when these schemes are used to interpret validity coefficients. The validity of a test is the degree to which it measures what it is supposed to measure. To assess the validity of, say, a college aptitude test, students' aptitude scores can be correlated with their grade point averages. The best aptitude tests rarely have validity coefficients above 0.60. It is misleading to label a validity coefficient of 0.60 as medium when higher coefficients are seldom obtained. An $r = 0.60$ is an extremely high validity coefficient, but a very, very low reliability coefficient. It should be apparent from these examples that no single classification scheme for interpreting r is applicable to all situations.

Error: Inferring Causation from Correlation

Another common error in interpreting a correlation coefficient is to infer that because two variables are correlated, one causes the other. *A nonzero correlation coefficient simply means there is a* **concomitant relationship** *between X and Y; that is, variation in one variable is associated in some way with variation in the other.* It is true that if X causes Y, there must be a correlation between the variables. How-ever, the converse of this statement is not true. A concomitant relationship is necessary but not sufficient for inferring causality. A concomitant relationship often exists because both variables are caused by a third variable. For example, it does not necessarily follow from the positive correlation between Sunday school attendance and honesty that attending Sunday school causes honesty. In all likeli-hood both variables are caused by a third variable—certain early childhood training practices in the home.

It is easy to fall into the trap of inferring causality from correlation, especially when one variable occurs before the other. Consider the well-publicized positive correlation between years of formal education and income. Does this mean that going to college causes one to earn more money? Before giving an affirmative answer we would have to know how much college graduates would have earned if they hadn't gone to college. A causal relationship may in fact exist, but this can't be ascertained from the correlation. Some or all of the correlation between education and income might be explained in terms of other causal variables. For example, colleges attract two kinds of students—the bright and the rich. We know that bright individuals tend to rise to better-paying jobs whether or not they have gone to college and that few children of rich parents end up poor.

Review Exercises for Section 5.5

*23. Which of the following are incorrect interpretations of a correlation coefficient, and why?

*a. The strength of association between test forms L and M is 0.96.

b. The correlation between height and weight at age 6 is 0.40; this correlation is twice as high as that at age 16, when $r = 0.20$.

*c. There is a medium correlation, $r = 0.67$, between the age at which babies can roll over and the age at which they can sit up alone.

d. Since the correlation between reaction time and number of automobile accidents is 0.20, 96% of the variance in frequency of accidents is unaccounted for.

e. We can conclude from the high correlation between level of motivation and number of elective offices sought that office-seeking behavior is caused at least in part by motivation.

24. What is wrong with interpreting r

a. in direct proportion to its size?

b. in terms of arbitrary descriptive labels?

c. as indicating causality?

*25. In an attempt to help children with low IQs improve their school performance, a special perceptual awareness program was instituted. Suppose that the program was completely ineffective. The group's mean IQ before the program was 72. Would you expect it to change after the special program, and if so, in what direction? (If you don't see the issue, reread Section 5.1.)

26. Employees with the worst accident rates were required to complete a safety course. Following the course, the employees had fewer accidents. Can we conclude that the course was effective? What controls could be used in the experiment to make the outcome easier to interpret?

27. Terms to remember:

a. Reliability

b. Validity

c. Concomitant relationship

Section 5.6   Factors that Affect the Size of r

Nature of the Relationship between X and Y

There are many ways in which two variables can be related. It is sufficient for our purposes to classify them as linear or nonlinear. Three examples showing the straight or curved lines of best fit for bivariate data are presented in Figure 5.6-1. In general, the more closely data points hug the line of best fit, whether it is a straight or a curved line, the higher the correlation. We saw in Section 5.2 that when $r = +1$ or -1, the data points fall on a straight line. If X and Y are normally distributed and have equal variances, as the absolute value of r decreases, the points form fatter and fatter ellipses until finally, when $r = 0$, they tend to form a circle. The Pearson product-moment correlation always fits data points by a straight line. This works fine if the relationship is linear but not so well if the relationship is nonlinear, as in Figure 5.6-1(c). If a nonlinear relationship is fitted by a straight line, the data points will not hug the line as closely as they would an appropriate curved line; consequently, r underestimates the strength of association. In fact, an $r = 0$ can be obtained even though X and Y are highly correlated.

a.   b.   c.

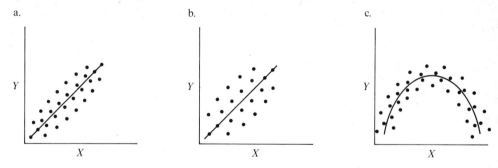

Figure 5.6-1.   Parts a and b illustrate linear relationships and part c illustrates a nonlinear relationship. The higher the correlation, the better the data points hug the line of best fit.

A different correlation measure called the **correlation ratio** *or* **eta squared (η^2)** *has been developed for determining the strength of association between nonlinearly related variables.* Eta squared fits data points by whatever line is appropriate. If the relationship is linear, a straight line is used and $\eta^2 = r^2$. For nonlinear relationships in which the correlation is not equal to zero, η^2 fits the points by a curved line, and its value is always larger than that for r^2. A discussion of the correlation ratio can be found in more advanced texts.[5]

How can we determine whether the relationship between X and Y is linear or nonlinear and hence, whether to use r or η^2? Statistical tests can be used;[6] however, the simplest method is to examine the scatter diagram for evidence of nonlinearity—the so-called eyeball test. Usually, visual inspection is adequate to detect cases in which r would underestimate strength of association.

In summary, r is a measure of the linear relationship between two quantitative variables. If the relationship is not linear, r underestimates the strength of association.

Truncated Range

The size of the Pearson product-moment correlation coefficient is affected by the range of the X and Y variables. If the range of either variable is restricted, the size of r will be reduced. Suppose that we have administered an aptitude test to assembly-line job applicants at a new factory. Because of the large number of jobs to be filled, all the applicants were hired regardless of their scores. Six months later we construct a scatter diagram like that in Figure 5.6-2, compute the correlation between aptitude scores and employee productivity, and find that $r = 0.55$. This is a very respectable validity coefficient. In the future if we had a surplus of applicants, we could improve productivity by hiring only those with high aptitude scores. Suppose that instead of hiring all the applicants when the plant opened, we had artificially restricted the range of aptitude scores by hiring only applicants with scores of 60 or above. For this case, the correlation between aptitude and productivity would have been 0.06 instead of 0.55. In this case, we would have incorrectly concluded that the test is of little value in selecting

[5] See Glass and Stanley (1970, pp. 150–152) and Kirk (1982, p. 162).
[6] See, for example, Hays (1981, pp. 510–514).

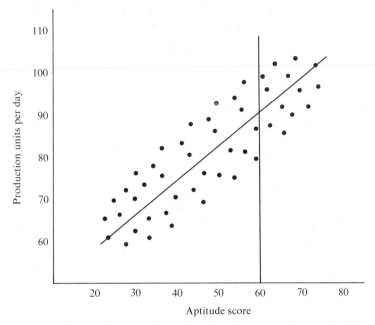

Figure 5.6-2. Scatter diagram illustrating the effect on *r* of restricting the range of *X* to scores of 60 or above. The *r* for the unrestricted range is 0.55; that for the restricted range is 0.06.

employees. The reason why the restriction or *truncation* of the range of the *X* variable results in a misleadingly low correlation coefficient can be seen from Figure 5.6-2. The effect would have been the same had the range of the *Y* variable been truncated.

The truncated range problem is common in behavioral and educational research, since much of it is conducted with college students who have been carefully screened for intelligence and related variables and, consequently, constitute a relatively homogeneous population. It's not surprising that college aptitude scores do not correlate very highly with grades, since admission offices truncate the range by admitting only students with high aptitude scores.

Spurious Effects Due to Subgroups with Different Means or Standard Deviations

A substantial correlation between *X* and *Y* may occur because the sample of subjects contains two or more subgroups with means that differ for both variables.[7] Suppose that we are interested in the correlation between school achievement (*Y*) and anxiety level (*X*) as measured by the Taylor Manifest Anxiety Scale and that we obtain random samples of students from lower- and middle-class families. The correlation coefficient computed for the combined samples will be much higher than that for either sample taken alone. This occurs because the means for the

[7] This problem has been examined in detail by Sockloff (1975).

two subgroups differ with respect to both X and Y. Those subjects from middle-class families tend to perform better in school and to be somewhat more anxious than children from lower-class families. When the subgroups are combined, the correlation between achievement and anxiety is misleadingly high because of the differing means. The reason for this can be seen in Figure 5.6-3(a), where the letters L and M denote data points for children from lower- and middle-class families, respectively. Figure 5.6-3(b) illustrates a situation in which the means of two subgroups, A and B, differ only on X. The correlation coefficient computed from the combined samples is lower than that for either sample taken alone.

A spurious correlation may occur when the standard deviations of the subgroups but not their means differ for one or both variables. This situation is depicted in Figure 5.6-3(c) and (d), where the letters A and B denote the subgroups. Figures 5.6-3(e) and (f) depict other ways in which subgroups can produce spurious correlations.

From the foregoing discussion, it is apparent that the inclusion of subgroups with different means or standard deviations on X and Y can affect the size and the sign of r. Unfortunately, we are not always aware that our sample contains distinct subgroups. Our first clue may come when we construct a scatter diagram and note in retrospect that the scores that cluster together tend to come from subjects who have some common distinguishing attribute.

Sometimes an experimenter intentionally conducts research with *extreme groups*—groups at opposite ends of a continuum. The use of introverts and extroverts, high and low achievers, or normals and neurotics enhances the likelihood of detecting other variables on which the groups differ. This is a useful research strategy, but it may lead to spuriously high correlation coefficients. Frequently, the means of the groups differ on both X and Y, and the data points have the shape illustrated in Figure 5.6-4. The data were selected from Table 5.3-1 so as to contain two extreme groups: the four fathers with the highest authoritarianism scores and the four with the lowest scores. The correlation for all 20 father–son pairs in Table 5.3-1 is 0.85; the correlation based on the two extreme groups is 0.94.

Extreme groups constitute one type of *discontinuous distribution*. A discontinuous distribution results any time we restrict our samples to extreme groups or to a relatively small number of points along a continuum. It should be apparent from this discussion that correlation coefficients involving discontinuous distributions should be viewed with suspicion.

Normality and Heterogeneity of Array Variances

If the distributions of X and Y are markedly skewed, the value of r will be less than if the variables are approximately normally distributed. The reason for this can be seen in Figure 5.6-5, which shows various combinations of skewed X and Y distributions.

The presence of skewed X and Y distributions is often accompanied by unequal dispersion of the Y scores for different values of X and similarly unequal dispersion of the X scores for different values of Y. This condition is called **heterogeneity of array**

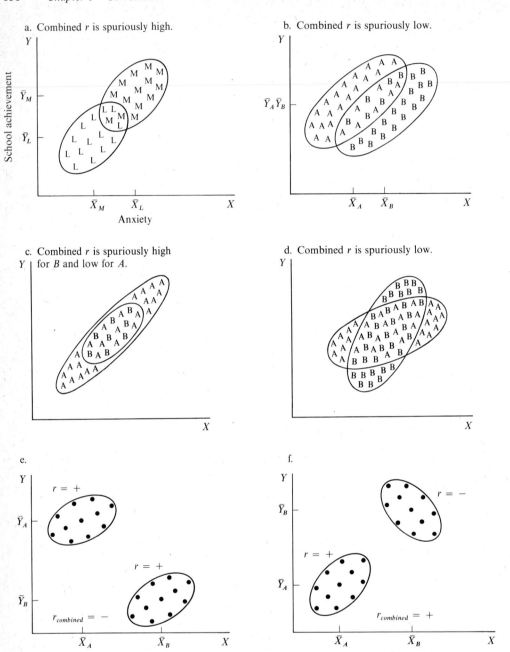

a. Combined *r* is spuriously high.

b. Combined *r* is spuriously low.

School achievement

Anxiety

c. Combined *r* is spuriously high for *B* and low for *A*.

d. Combined *r* is spuriously low.

e.

f.

Figure 5.6-3. Scatter diagrams illustrating the effects on *r* of subsamples with means that differ on both variables (parts a, e, and f) or on only one variable (b). Parts (c) and (d) illustrate the effects of heterogeneous standard deviations. Parts (e) and (f) show that the sign of the coefficient for the combined samples may differ from that for one or both of the subsamples.

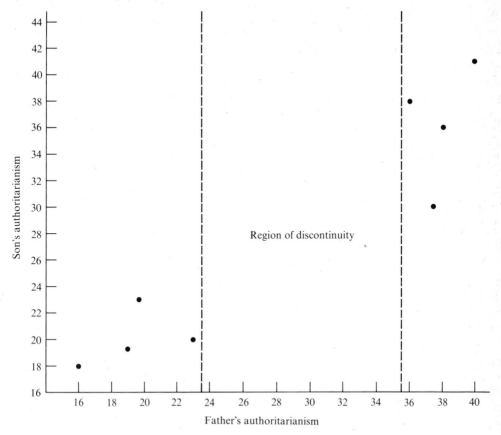

Figure 5.6-4.   Scatter diagram illustrating the effects on *r* of using extreme groups. The data are taken from Table 5.3-1, with the eight data points representing the four highest and the four lowest authoritarianism scores based on the fathers' data.

(*row and column*) **variances** *or* **heteroscedasticity.** *This is illustrated in Figure 5.6-6.* We saw earlier that *r* reflects the average degree to which scores hug the line of best fit. If the dispersion around the line differs for different values of, say, *X* and *Y*, the correlation coefficient will not have the same meaning as for homogeneous array variances. For example, in Figure 5.6-6(a) and (b) the correlation coefficient will underestimate the magnitude of association for low *X* scores and overestimate it for high *X* scores.

   The use of *r* as a descriptive measure of association doesn't require any assumptions regarding the shape of the *X* and *Y* distributions. As we have seen, however, if *X* and *Y* are markedly skewed, the value of *r* will be closer to 0 than if the distributions are approximately normal. Furthermore, under these conditions the interpretation of *r* is altered, since it no longer describes the average degree to which the data points hug the line of best fit. Finally, the presence of skewed

Figure 5.6-5.   Effects of markedly skewed *X* and *Y* distributions on the distribution of data points in a scatter diagram.

X and *Y* distributions is often accompanied by a nonlinear relationship between the variables. This condition calls for the computation of η^2 instead of *r*.

It is apparent from this discussion that the interpretation of *r* as a descriptive measure is simplified if *X* and *Y* are approximately normally distributed. It should be emphasized, however, that normality is not required for purely descriptive purposes, since regardless of the shapes of the *X* and *Y* distributions, *r* is a measure of how closely data points hug a straight line of best fit. We will see in Chapter 12 that normality is required when the sample correlation is used to draw inferences about the population correlation.

The factors that affect the size of *r* are summarized in Table 5.6-1.

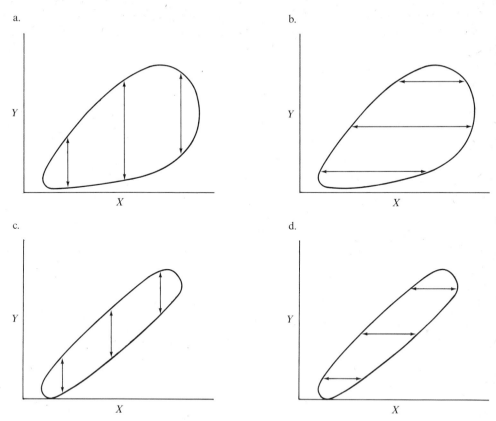

Figure 5.6-6. Parts a and b illustrate heterogeneity of column and row dispersion, respectively; parts c and d illustrate homogeneity of dispersion.

Table 5.6-1. Factors that affect the size of r

r underestimates magnitude of relationship when:	r overestimates magnitude of relationship when:
1. The relationship between X and Y is nonlinear. 2. The range of either X or Y is truncated. 3. The distributions of X and Y are skewed.	1. The sample contains subgroups with means that differ for both variables. 2. The sample is composed of extreme groups.

Review Exercises for Section 5.6

*28. What effects do the following factors have on r as a measure of strength of association? Draw figures like Figures 5.6-1 through 5.6-5 to represent the data.

 *a. The relationship between X and Y looks like an inverted U.

  b. The range of X is reduced by deleting subjects with scores below $\bar{X}$.

  c. The sample contains subgroups a and b with equal standard deviations and means $\bar{X}_a = 16$, $\bar{X}_b = 22$, $\bar{Y}_a = 42$, and $\bar{Y}_b = 31$.

*d. The sample contains subgroups a and b with equal standard deviations and means $\bar{X}_a = 16$, $\bar{X}_b = 22$, $\bar{Y}_a = 31$, and $\bar{Y}_b = 37$.

*e. The sample contains subgroups a and b with equal means and standard deviations $S_{X_a} = 13$, $S_{X_b} = 22$, $S_{Y_a} = 22$, and $S_{Y_b} = 13$.

  f. The sample contains subgroups a and b with equal means and standard deviations $S_{X_a} = 18$, $S_{X_b} = 18$, $S_{Y_a} = 26$, and $S_{Y_b} = 34$.

*g. The distribution of the X variable is negatively skewed; that for the Y variable is positively skewed.

  h. The distributions of the X and Y variables are negatively skewed.

29. How can we detect cases in which η^2 should be used instead of r?

30. What are the potential advantages and disadvantages of using extreme groups in research?

31. The correlation between IQ and grade point average (GPA) for high school seniors was 0.63. For seniors who went on to college the correlation between IQ and college GPA was 0.51. Explain why this correlation is lower.

32. The correlation between IQ and ratings of the creativity of 50 highly creative individuals was 0.18. Can we conclude that IQ is a relatively unimportant factor in creativity? Discuss.

33. Terms to remember:
  a. Correlation ratio
  b. Truncated range
  c. Extreme groups
  d. Discontinuous distribution
  e. Heterogeneity of array variance (heteroscedasticity)

Section 5.7   Spearman Rank Correlation

The **Spearman rank correlation coefficient**, denoted by r_s, is used to describe the degree of agreement between paired data that are in the form of ranks.[8] Such data may occur as a result of ranking scores, as when grade point averages are converted to rank in graduating class, or because rank data are obtained in the original instance, as when freshman English themes are ranked on the basis of creativity. Ranking can often be done when it is difficult or impossible to apply more refined measuring procedures.

The formula for r_s is

$$r_s = 1 - \frac{6 \sum_{i=1}^{n} (R_{X_i} - R_{Y_i})^2}{n(n^2 - 1)},$$

where $R_{X_i} - R_{Y_i}$ is the difference between the ith person's rank on X and rank on Y, and n is the number of pairs of ranks. The computation of r_s is illustrated in Table 5.7-1, where 14 graduate school applicants have been ranked by tenured faculty (R_{X_i}) and nontenured faculty (R_{Y_i}).

The index r_s is a measure of the agreement between two sets of ranks and is interpreted in much the same way as the Pearson product-moment coefficient. The range of r_s is from -1 to $+1$. Values of r_s greater than 0 indicate that large R_X's

---

[8] This coefficient was first used by Sir Francis Galton, but was named for the British psychologist Charles Spearman, who made more extensive use of it.

Table 5.7-1. Computation of r_S for ranks assigned to applicants by tenured faculty (R_{X_i}) and nontenured faculty (R_{Y_i})

(i) *Data:*

Applicant	Rank, R_{X_i}	Rank, R_{Y_i}	$R_{X_i} - R_{Y_i}$	$(R_{X_i} - R_{Y_i})^2$
1	6	8	-2	4
2	3	2	1	1
3	4	5	-1	1
4	12	11	1	1
5	10	9	1	1
6	1	1	0	0
7	5	4	1	1
8	7	7	0	0
9	14	14	0	0
10	2	3	-1	1
11	8	10	-2	4
12	11	12	-1	1
13	9	6	3	9
14	13	13	0	0
			$\sum(R_{X_i} - R_{Y_i})^2 =$	$\overline{24}$

(ii) *Computational procedure:*

$$r_s = 1 - \frac{6\sum_{i=1}^{n}(R_{X_i} - R_{Y_i})^2}{n(n^2 - 1)} = 1 - \frac{6(24)}{14[(14)^2 - 1]} = 1 - \frac{144}{2730} = 0.95$$

tend to be paired with large R_Y's; values less than 0 indicate that large R_X's are paired with small R_Y's, and so on; $r_s = 1$ if and only if each person's X and Y ranks are equal. It can be shown[9] that the formula for r_s is equivalent to that for r when two sets of consecutive untied ranks $1, \ldots, n$ are substituted for X_i and Y_i in the Pearson formula. However, the use of ranks instead of scores alters the meaning of the correlation coefficient.

We saw earlier that r is a measure of the linear relationship between two quantitative variables; r_s is a measure of the *monotonic relationship* between two sets of ranks. *A function $Y = f(X)$ is said to be* **strictly monotonic increasing** *if an increase in the value of X is always accompanied by an increase in Y.*[10] *A strictly monotonic decreasing function is one in which an increase in X is accompanied by a decrease in Y.* Monotonic functions include linear functions ($Y = a + bX$) as well as a number of other functions that are nonlinear ($Y = X^3$; $Y = \log X$). Thus, Spearman's rank correlation coefficient does not necessarily reflect the linear relationship between the variables represented by ranks. It does reflect the strength of the monotonic relationship—a more general relationship. If $r_s = 0$, either the variables represented by ranks are not related or the form of the relationship is nonmonotonic.

---

[9] See Kirk (1978, pp. 122–124) for the derivation.
[10] If for all $X_n < X_{n+1}$, $f(X_n) \leq f(X_{n+1})$, the function $f(X)$ is said to be monotonic increasing. If for all $X_n < X_{n+1}$, $f(X_n) < f(X_{n+1})$, the function is strictly monotonic increasing.

The Problem of Tied Ranks

Occasionally, two or more objects or individuals are tied with respect to the variable being ranked. The usual practice is to give them the mean of the ranks they would have received collectively if they had been distinguishable. For example, if Jack, Debbie, and Bill are considered equally gregarious, each is given the mean of the ranks they would have occupied, say, $(8 + 9 + 10)/3 = 9$. Unfortunately, the presence of tied ranks violates the assumptions underlying the derivation of the computational formula for r_s. A correction for ties can be incorporated in the formula, but the computation is tedious. The most desirable solution is to force those making ratings to discern differences among the objects or individuals, thereby eliminating tied ranks. If this is done, the uncorrected formula can be used. If raters persist in assigning tied ranks, the next best solution is to treat the sets of ranks as though they were scores and to compute a Pearson product-moment correlation coefficient. The result can be regarded as a Spearman rank correlation coefficient that has been corrected for ties.

Review Exercises for Section 5.7

34. List the similarities and differences between r and r_s.

*35. A random sample of freshman psychology majors rank ordered various fields of psychology according to vocational attractiveness. The students again ranked the fields when they were seniors. Compute the correlation between their freshman and senior rankings.

Field	Freshman rank	Senior rank
Social	5	2
Experimental	7	6
Human factors	6	7
Clinical	1	1
Statistics and measurement	8	8
Industrial	3	3
Educational	4	4
Counseling	2	5

36. A psychiatric social worker and an occupational therapist ranked 11 Veterans' Administration patients with respect to extent of recovery following 3 months of therapy. Compute the correlation between the two sets of rankings.

Patient	Social worker	Occupational therapist
1	7	7
2	2	1
3	1	2
4	3	5
5	8	9
6	10	10
7	4	3
8	9	8
9	11	11
10	6	6
11	5	4

37. Subjects rated the attractiveness of one set of geometric shapes before smoking marijuana and a similar set after smoking the marijuana. One shape in the two sets was the same. The following data are the ratings for that shape. A rating of 1 means very attractive; a rating of 20, very unattractive. Transform the ratings to ranks, and compute the correlation between the two sets of ranks.

Subject	Before smoking	After smoking
1	6	3
2	8	7
3	14	16
4	7	2
5	10	12
6	9	15
7	5	1
8	15	20
9	12	17

38. Suppose that for the data in Exercise 37, subject 6 had assigned a rating of 12 instead of 15 to the geometric shape after smoking marijuana. This rating results in tied ranks. How would this affect the computational procedure for the correlation coefficient?

*39. The debate format can be a useful adjunct to traditional teaching methodologies for presenting complex issues. Graduate student nurses were exposed to a debate on the issue of third-party reimbursement. A questionnaire was used to evaluate pre- and postdebate knowledge of 13 affirmative and negative arguments concerning the issue. The results are listed in the following table; a rank of 1 was assigned to the argument known by the most student nurses. Compute the correlation between the two sets of ranks. (Suggested by Archold, Patricia G., and Hoeffer, Beverly. Reframing the issue: A debate on third-party reimbursement. *Nursing Outlook*, 1981, 423–427.)

Argument	Pretest rank	Posttest rank
Legitimize role and service of nurses	1	7
Increase health care cost	2	7
Increase access of consumer to nursing services	3	4.5
Nursing services are undefined and dependent on physicians	4	12.5
Decrease health care cost	6	4.5
Provide equal opportunity in a free-market system	6	11
Supports health-care delivery system not based on need	6	7
Cumbersome process for individual nurses	8	12.5
Increase accountability of nurses for their services	9	9.5
Increase power and autonomy of nursing to influence health care delivery system	11.5	2
Supports inequitable/discriminatory health-care delivery system	11.5	1
Elitist/divisive to nursing	11.5	3
No increase in accessibility	11.5	9.5

*40. Which of the following are strictly monotonic functions?
　　 *a. $Y = 1 + 2X$　　 *b. $Y = X^2$
　　　c. $Y = X^3$　　　　 d. $Y = 1/X$
 41. Terms to remember:
　　　Strictly monotonic increasing and decreasing functions

---

†Section 5.8   Other Kinds of Correlation Coefficients

Three correlation coefficients have been mentioned thus far: r, η^2, and r_s. A fourth coefficient, Cramer's $\hat{V}$ or $\hat{\phi}'$, appropriate for unordered qualitative variables, is discussed in Section 17.3. Other coefficients are also available, but they are beyond the scope of this book. A summary of the more widely used coefficients is given in Table 5.8-1.

Table 5.8-1.   Summary of the major correlation coefficients

Coefficient	Symbol	Characteristics	Reference[a] and pages
1. Pearson product moment	r	X and Y quantitative, linear relationship	(a) 109–116 (b) 446–449 (c) 176–178
2. Eta squared (correlation ratio)	η^2	X and Y quantitative, curvilinear relationship	(a) 150–152
3. Spearman rank coefficient	r_s	X and Y ranked, monotonic relationship	(a) 172–176 (b) 596–598 (c) 221–222
4. Kendall's tau	τ	X and Y ranked, monotonic relationship	(b) 598–605 (c) 223–225
5. Point biserial	r_{pb}	One variable quantitative, the other dichotomous[b]	(a) 163–164 (c) 212–214
6. Biserial	r_b	X and Y quantitative, but one variable forced into a dichotomy	(a) 168–172 (c) 216–217
7. Tetrachoric	r_t	X and Y quantitative, but both forced into dichotomies	(a) 165–167
8. Phi or fourfold	ϕ	X and Y both dichotomous	(a) 158–162 (c) 217–219

[a] (a) Glass and Stanley (1970), (b) Hays (1981), (c) Howell (1982).
[b] Dichotomous classifications assign elements to one of two categories, for example, pass–fail, male–female, married–not married, and IQ > 100–IQ ≤ 100.

Section 5.9   Summary

The term correlation denotes the concomitance between two or more quantitative or qualitative variables. A correlation coefficient is a measure of the degree of concomitance. The two most widely used coefficients in the behavioral sciences and education are the Pearson product-moment correlation coefficient, r, and the Spearman rank correlation coefficient, r_s. Pearson's r reflects the strength and the nature of the linear relationship between two quantitative variables. It is a number

† This section can be omitted without loss of continuity.

that varies between -1 and 1, with 0 indicating the absence of a linear relationship. Negative values indicate an inverse relationship between the variables; positive values indicate a positive or direct relationship. Spearman's r_s measures the strength and the nature of the monotonic relationship between two ordered qualitative variables—that is, ranked data. It, like r, varies between -1 and 1, with 0 indicating the absence of a monotonic relationship.

Two statistics, both functions of r, are useful in interpreting a particular r value—the coefficient of determination, r^2, and the coefficient of nondetermination, $k^2 = 1 - r^2$. For a given linear relationship between X and Y, r^2 reflects the proportion of the X score variance that can be explained by the Y score variance and vice versa; k^2 reflects the proportion that can't be explained. If, for example, $r = 0.50$, we know that based on the linear relationship between the variables, 25% of the variance of one variable can be explained by the variance of the other variable, while 75% remains to be accounted for.

The Pearson product-moment correlation coefficient is appropriate for linearly related quantitative variables. For descriptive purposes, no other assumptions regarding the variables are required. However, in interpreting r, keep in mind that the size of r can be affected by such factors as the shape of the X and Y distributions, the presence of a truncated X or Y range, and the presence of subgroups with standard deviations or means that differ on both variables.

†*Section 5.10   Technical Note*

5.10-1   *Proof of Algebraic Equivalence of the Definitional Formula and Computational Formula for r*

The definitional formula for r given in Section 5.3 is

$$r = \frac{\dfrac{\sum\limits_{i=1}^{n} (X_i - \bar{X})(Y_i - \bar{Y})}{n}}{\sqrt{\left[\dfrac{\sum\limits_{i=1}^{n} (X_i - \bar{X})^2}{n}\right]\left[\dfrac{\sum\limits_{i=1}^{n} (Y_i - \bar{Y})^2}{n}\right]}}$$

$$= \frac{\dfrac{1}{n}\left[\sum\limits_{i=1}^{n} (X_iY_i - \bar{Y}X_i - \bar{X}Y_i + \bar{X}\bar{Y})\right]}{\dfrac{1}{n}\sqrt{\left[\sum\limits_{i=1}^{n} (X_i^2 - 2\bar{X}X_i + \bar{X}^2)\right]\left[\sum\limits_{i=1}^{n} (Y_i^2 - 2\bar{Y}Y_i + \bar{Y}^2)\right]}}$$

$$= \frac{\sum\limits_{i=1}^{n} X_iY_i - \bar{Y}\sum\limits_{i=1}^{n} X_i - \bar{X}\sum\limits_{i=1}^{n} Y_i + n\bar{X}\bar{Y}}{\sqrt{\left(\sum\limits_{i=1}^{n} X_i^2 - 2\bar{X}\sum\limits_{i=1}^{n} X_i + n\bar{X}^2\right)\left(\sum\limits_{i=1}^{n} Y_i^2 - 2\bar{Y}\sum\limits_{i=1}^{n} Y_i + n\bar{Y}^2\right)}},$$

---

† This note, which is not essential to the text, can be omitted without loss of continuity.

but $\bar{X} = \sum X_i/n$, $\bar{X}^2 = (\sum X_i)^2/n^2$, $\bar{Y} = \sum Y_i/n$, and $\bar{Y}^2 = (\sum Y_i)^2/n^2$. Substituting,

$$r = \frac{\sum X_i Y_i - \dfrac{\sum Y_i}{n} \sum X_i - \dfrac{\sum X_i}{n} \sum Y_i + n \dfrac{\sum X_i}{n} \dfrac{\sum Y_i}{n}}{\sqrt{\left[\sum X_i^2 - \dfrac{2\sum X_i}{n} \sum X_i + \dfrac{n(\sum X_i)^2}{n^2}\right]\left[\sum Y_i^2 - \dfrac{2\sum Y_i}{n} \sum Y_i + \dfrac{n(\sum Y_i^2)}{n^2}\right]}}$$

$$= \frac{\sum X_i Y_i - \dfrac{2\sum X_i \sum Y_i}{n} + \dfrac{\sum X_i \sum Y_i}{n}}{\sqrt{\left[\sum X_i^2 - \dfrac{2(\sum X_i)^2}{n} + \dfrac{(\sum X_i)^2}{n}\right]\left[\sum Y_i^2 - \dfrac{2(\sum Y_i)^2}{n} + \dfrac{(\sum Y_i)^2}{n}\right]}}.$$

Finally, adding like terms gives

$$r = \frac{\sum X_i Y_i - \dfrac{(\sum X_i)(\sum Y_i)}{n}}{\sqrt{\left[\sum X_i^2 - \dfrac{(\sum X_i)^2}{n}\right]\left[\sum Y_i^2 - \dfrac{(\sum Y_i)^2}{n}\right]}}.$$

†Section 5.11   Computer Supplement

A microcomputer package developed by Steinmetz, Romano, and Patterson (1981) was used to compute the correlation between the father and son authoritarianism scores in Table 5.3-1. This interactive package was written in APPLESOFT BASIC for Apple II microcomputers with 48K of memory. It contains 12 programs.

The package begins by displaying a list of available programs—the menu. Once a program is selected, it prompts the user to enter information as required. The sequence of prompts and responses for computing a correlation coefficient is shown in Table 15.11-1. The user has the option of entering data that have been previously stored on, say, a disk or of entering the data using the computer's keyboard. Once data have been entered, the user can edit the data and make corrections as required.

Table 5.11-1.   Printout from the Steinmetz, Romano, and Patterson Correlation Program (Data from Table 5.3-1)

```
[ CORRELATION PROGRAMS ]

THIS PROGRAM CALCULATES EITHER A
PEARSON'S 'R' OR SPEARMAN'S RHO
CORRELATION COEFFICIENT.

AVAILABLE OPTIONS:

     (1) DATA ENTRY CORRECTION ROUTINE
     (2) INPUT DATA FROM DISK
     (3) SAVE DATA TO DISK
PRESS <RETURN> TO CONTINUE 1
WHICH CORRELATION COEFFICIENT DO YOU
WANT TO CALCULATE ?

     (1) PEARSON'S R
     (2) SPEARMAN'S RHO

(ENTER A 1 OR 2) 1
HOW MANY PAIRS OF SCORES ARE INVOLVED  IN THE CORRELATION ? 20
WOULD YOU LIKE TO USE DATA THAT IS
STORED ON A DISK (Y/N) ? N
ENTER THE 20 PAIRS OF SCORES:

-----------------------------------
SCORE X          SCORE Y
-------          -------
 25
                  28
 32
                  31
```

(continued)

---

† This section can be omitted without loss of continuity.

Table 5.11-1. *(continued)*

```
40
                41
29
                33
31
                25
16
                18
28
                26
36
                38
33
                34
29
                36
23
                20
27
                28
37
                30
30
                26
27
                22
20
                23
28
                29
38
                36
35
                32
19
                19
WOULD YOU LIKE THE DATA
DISPLAYED (Y/N) ? N
ANY CHANGES (Y/N) ? N
THE CALCULATIONS HAVE BEEN COMPLETED.
HOW SHOULD THE RESULTS BE REPORTED ?

    (1) VIDEO DISPLAY ONLY
    (2) PRINTER AND VIDEO

(ENTER A 1 OR 2): 2

VARIABLE X LABEL: FATHER'S SCORE
VARIABLE Y LABEL: SON'S SCORE
```

(continued)

Table 5.11-1.   (continued)

ENTER PRINTER SLOT# 1
PEARSON'S R COEFFICIENT

X = FATHER'S SCORE

Y = SON'S SCORE

R = .846268777

****

6

Regression

Section 6.1 Introduction

Jean's score on the Law School Aptitude Test (LSAT) is 69. What grade point average can she expect to make in law school? Bertha is on a 750-calorie diet. How many pounds should she be able to lose in a month? Since the variables are correlated, we can predict GPA from an LSAT score and weight loss from calorie intake with better than chance accuracy. The higher the correlation, the more accurate the prediction. For $r = \pm 1$, the dependent variable, denoted by Y, can be predicted from the independent variable, X, with perfect accuracy; but if $r = 0$, a knowledge of X is useless in predicting Y. Although a correlation coefficient is indicative of our ability to predict, the actual prediction is made using regression analysis, the subject of this chapter. Strictly speaking, *regression analysis* applies to paired data (X_i, Y_i), where X is the independent variable with values X_i that are selected in advance, and Y is the dependent variable with values Y_i that are free to vary. However, regression procedures also are applicable when both X and Y are free to vary, as they are in correlation.

An Overview of Prediction

George, who is taking statistics, copies down the grades from last semester's class and constructs the scatter diagram shown in Figure 6.1-1. He finds that the correlation between the midterm and the final exam was 0.80. His midterm grade was 82, and he wonders how he'll do on the final. According to the scatter diagram, two students made 82; the mean of their grades—and hence George's predicted grade—is $(74 + 84)/2 = 79$.

While this prediction method works, it has a serious disadvantage. The prediction is based on only the two Y scores corresponding to $X_i = 82$; the other ten paired scores are ignored. Predictions based on such small samples tend to be unstable—that is, to vary markedly from sample to sample. Prediction can be improved by utilizing all the data rather than a small subset. George notes that the relationship between the midterm and final grades appears to be linear, so he determines the best-fitting linear regression line. It is shown as a dashed line in Figure 6.1-1. To predict his final grade George draws a vertical line from $X_i = 82$ to the regression line and then a horizontal line to the Y axis. His predicted grade is 78.

Figure 6.1-1.   Scatter diagram for paired midterm and final exam grades.

Predictions based on the regression line take into account all the sample data and hence are more stable than those based on only the mean of the Y scores corresponding to a given X score. Both procedures presuppose that the population represented by the current sample (George's statistics class) doesn't differ from that represented by the earlier sample (last year's class). Obviously, if this assumption isn't tenable, we can have little faith in the prediction. The regression approach also presupposes that the data points have been fitted by the correct regression equation—in our example, the equation for a straight line. Fortunately, the tenability of this assumption is easily checked by looking at the scatter diagram.

Section 6.2   Criterion for the Line of Best Fit

Predicting Y from X

We have frequently referred to the line of best fit without ever defining it—what *is* the line of best fit for data points? Best fit can be defined in a number of ways. It seems reasonable that we should want a line that minimizes some function of the error in predicting Y_i from X_i. A **prediction error,** e_i, *is defined as the difference between the ith person's actual score,* Y_i, *and the score predicted for that person,* Y_i'—*that is,* $e_i = Y_i - Y_i'$. Prediction errors are illustrated in Figure 6.2-1. One def-

Figure 6.2-1.   A prediction error, e_i, is the discrepancy between Y_i, the actual observed score for person i, and Y_i', the predicted score based on the regression line; for example, $e_7 = Y_7 - Y_7'$.

inition of best fit widely used by mathematicians is based on the *principle of least squares* and may be stated as follows. *The line of best fit is the one that minimizes the sum of the squared prediction errors—that is, the line for which $\sum_{i=1}^{n} e_i^2 = \sum_{i=1}^{n}(Y_i - Y_i')^2$ is a minimum.* We will limit our discussion to linearly related data. For this case the predicted values fall on a straight line, called the *regression line*. The equation for a straight line is $Y_i' = a_{Y \cdot X} + b_{Y \cdot X}X_i$, where

> Y_i' is the predicted value
> $a_{Y \cdot X}$ is the point at which the line crosses the Y axis
> $b_{Y \cdot X}$ is the slope of the line
> X_i is a value of the independent variable.

The subscript $Y \cdot X$ is read "Y given X" and indicates that we are predicting Y from X. According to the least squares criterion, we want values of the constants $a_{Y \cdot X}$ and $b_{Y \cdot X}$ such that

$$\sum_{i=1}^{n} e_i^2 = \sum_{i=1}^{n}(Y_i - Y_i')^2 = \sum_{i=1}^{n}[Y_i - (a_{Y \cdot X} + b_{Y \cdot X}X_i)]^2$$

is as small as it possibly can be.

The method of finding numerical values for $a_{Y \cdot X}$ and $b_{Y \cdot X}$ that minimizes $\sum e_i^2$ utilizes the differential calculus and hence is beyond the scope of this text.[1]

---

[1] McNemar (1969, pp. 133–136) describes the method in a text written for behavioral science students.

We shall simply report here the formulas for computing their values:

$$a_{Y \cdot X} = \bar{Y} - b_{Y \cdot X} \bar{X} \quad \text{and} \quad b_{Y \cdot X} = \frac{\sum\limits_{i=1}^{n} (X_i - \bar{X})(Y_i - \bar{Y})}{\sum\limits_{i=1}^{n} (X_i - \bar{X})^2}.$$

A computational example is given in Table 6.2-1 for the data in Figure 6.1-1. The values of the constants from part ii of the table are $a_{Y \cdot X} = 12.1868$ and $b_{Y \cdot X} = 0.8026$. Hence, the linear equation that minimizes the sum of the squared prediction errors is $Y_i' = 12.1868 + 0.8026 X_i$. According to the equation, the line crosses the Y axis at 12.1868. In other words, when $X = 0$, $Y' = 12.1868$. The slope of the line is 0.8026, which means that as X increases 1 unit, Y increases 0.8026 unit. To determine the predicted value for, say, $X_i = 82$, we enter the value of X_i in the equation $Y_i' = 12.1868 + 0.8026(82)$ and solve for Y_i'. The predicted value is 78.

Alternatively, we can determine predicted values by graphic means, as we did in Figure 6.1-1. The first step is to draw the line of best fit. Since a straight line is defined by two points, we can begin by solving for Y_i' when $X_i = 72$ and 94 (the smallest and largest X scores, respectively). The corresponding Y_i' values are, respectively, 69.97 and 87.63. Once a line connecting the (X_i, Y_i) points (72, 69.97) and (94, 87.63) has been drawn, it can be used to obtain Y_i' for other values of X_i.

Predicting X from Y

If the value of Y_i is known, X_i can be predicted from the equation

$$X_i' = a_{X \cdot Y} + b_{X \cdot Y} Y_i.$$

The subscript $X \cdot Y$ indicates that X is predicted from Y. Note that $a_{X \cdot Y}$ is different from $a_{Y \cdot X}$ and $b_{X \cdot Y}$ from $b_{Y \cdot X}$, since they apply to different regression lines. The constants of the linear equation for predicting X from Y are given by

$$a_{X \cdot Y} = \bar{X} - b_{X \cdot Y} \bar{Y} \quad \text{and} \quad b_{X \cdot Y} = \frac{\sum\limits_{i=1}^{n} (X_i - \bar{X})(Y_i - \bar{Y})}{\sum\limits_{i=1}^{n} (Y_i - \bar{Y})^2}.$$

The formulas for $a_{X \cdot Y}$ and $b_{X \cdot Y}$ were derived so as to minimize the sum of the squared prediction errors defined by $\sum_{i=1}^{n} e_i^2 = \sum_{i=1}^{n} (X_i - X_i')^2$. These prediction errors are illustrated in Figure 6.2-2. The computation of $a_{X \cdot Y}$ and $b_{X \cdot Y}$ is illustrated in Table 6.2-1. The regression equation is $X_i' = 19.1320 + 0.8060 Y_i$.

For any set of paired data points we can compute two regression lines—the regression of Y on X, given by $Y_i' = a_{Y \cdot X} + b_{Y \cdot X} X_i$, and the regression of X on Y, given by $X_i' = a_{X \cdot Y} + b_{X \cdot Y} Y_i$. The two lines are shown in Figure 6.2-3 for the data in Table 6.2-1. There are two lines because in predicting Y from X we want to minimize one error, $\sum (Y_i - Y_i')^2$, but in predicting X from Y we minimize a different error, $\sum (X_i - X_i')^2$.

Relationship between r and the Slopes of the Regression Lines

There are a number of interesting relationships between r and the two regression coefficients $b_{Y \cdot X}$ and $b_{X \cdot Y}$. It is a simple matter to show that $r(S_Y/S_X) = b_{Y \cdot X}$ and

Table 6.2-1. Computation of least squares values for constants in a linear equation (Data from Figure 6.1-1)

(i) *Data:*

X_i	Y_i	$(X_i - \bar{X})$	$(Y_i - \bar{Y})$	$(X_i - \bar{X})(Y_i - \bar{Y})$	$(X_i - \bar{X})^2$	$(Y_i - \bar{Y})^2$
72	70	−10	−8	80	100	64
75	74	−7	−4	28	49	16
76	70	−6	−8	48	36	64
78	72	−4	−6	24	16	36
78	78	−4	0	0	16	0
82	74	0	−4	0	0	16
82	84	0	6	0	0	36
84	76	2	−2	−4	4	4
86	86	4	8	32	16	64
88	78	6	0	0	36	0
89	88	7	10	70	49	100
94	86	12	8	96	144	64

$\sum X_i = 984 \quad \sum Y_i = 936$
$\bar{X} = 82 \qquad \bar{Y} = 78$

$\sum(X_i - \bar{X})(Y_i - \bar{Y}) = 374 \qquad \sum(X_i - \bar{X})^2 = 466 \qquad \sum(Y_i - \bar{Y})^2 = 464$

(ii) *Computation of $a_{Y\cdot X}$ and $b_{Y\cdot X}$:*

$$b_{Y\cdot X} = \frac{\sum_{i=1}^n (X_i - \bar{X})(Y_i - \bar{Y})}{\sum_{i=1}^n (X_i - \bar{X})^2} = \frac{374}{466} = 0.8026$$

$$a_{Y\cdot X} = \bar{Y} - b_{Y\cdot X}\bar{X} = 78 - 0.8026(82) = 12.1868$$

(iii) *Computation of $a_{X\cdot Y}$ and $b_{X\cdot Y}$:*

$$b_{X\cdot Y} = \frac{\sum_{i=1}^n (X_i - \bar{X})(Y_i - \bar{Y})}{\sum_{i=1}^n (Y_i - \bar{Y})^2} = \frac{374}{464} = 0.8060$$

$$a_{X\cdot Y} = \bar{X} - b_{X\cdot Y}\bar{Y} = 82 - 0.8060(78) = 19.1320$$

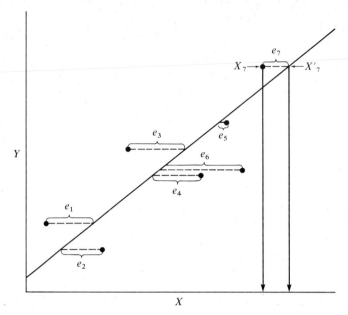

Figure 6.2-2. The error in predicting X_i from Y_i is the discrepancy between X_i, the actual observed value for person i, and X_i', the predicted value based on the regression line; for example, $e_7 = X_7 - X_7'$.

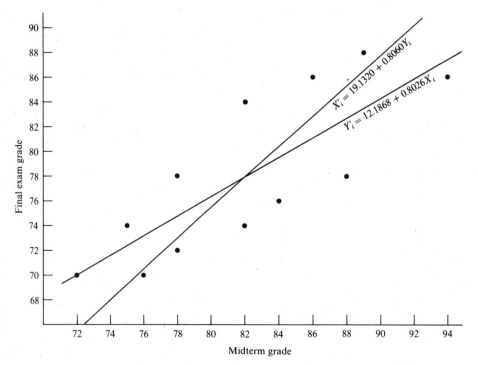

Figure 6.2-3. Regression lines for predicting Y_i from X_i and X_i from Y_i (data from Table 6.2-1).

$r(S_X/S_Y) = b_{X \cdot Y}$ and that $\pm\sqrt{b_{Y \cdot X}b_{X \cdot Y}} = \pm r$.[2] According to the latter relationship, r is the *geometric mean* of the two regression coefficients. *A **geometric mean** is a special kind of central tendency measure often used for averaging rates of change. It is defined as the nth root of the product of n positive numbers.*

A little algebra is all that is necessary to prove these relationships. One of the formulas for r given in Chapter 5 was

$$r = \frac{\sum(X_i - \bar{X})(Y_i - \bar{Y})/n}{\sqrt{[\sum(X_i - \bar{X})^2/n][\sum(Y_i - \bar{Y})^2/n]}}.$$

The standard deviations of X and Y are $S_X = \sqrt{\sum(X_i - \bar{X})^2/n}$ and $S_Y = \sqrt{\sum(Y_i - \bar{Y})^2/n}$. Therefore,

$$r\frac{S_Y}{S_X} = \frac{\sum(X_i - \bar{X})(Y_i - \bar{Y})/n}{\sqrt{[\sum(X_i - \bar{X})^2/n][\sum(Y_i - \bar{Y})^2/n]}}\frac{\sqrt{\sum(Y_i - \bar{Y})^2/n}}{\sqrt{\sum(X_i - \bar{X})^2/n}}$$

$$= \frac{\sum(X_i - \bar{X})(Y_i - \bar{Y})}{\sum(X_i - \bar{X})^2} = b_{Y \cdot X}.$$

$$r\frac{S_X}{S_Y} = \frac{\sum(X_i - \bar{X})(Y_i - \bar{Y})/n}{\sqrt{[\sum(X_i - \bar{X})^2/n][\sum(Y_i - \bar{Y})^2/n]}}\frac{\sqrt{\sum(X_i - \bar{X})^2/n}}{\sqrt{\sum(Y_i - \bar{Y})^2/n}}$$

$$= \frac{\sum(X_i - \bar{X})(Y_i - \bar{Y})}{\sum(Y_i - \bar{Y})^2} = b_{X \cdot Y}.$$

Similarly, we can show that $\sqrt{b_{Y \cdot X}b_{X \cdot Y}} = r$:

$$\sqrt{b_{Y \cdot X}b_{X \cdot Y}} = \sqrt{\left[\frac{\sum(X_i - \bar{X})(Y_i - \bar{Y})}{\sum(X_i - \bar{X})^2}\right]\left[\frac{\sum(X_i - \bar{X})(Y_i - \bar{Y})}{\sum(Y_i - \bar{Y})^2}\right]}$$

$$= \frac{\sqrt{[\sum(X_i - \bar{X})(Y_i - \bar{Y})]^2}}{\sqrt{[\sum(X_i - \bar{X})^2][\sum(Y_i - \bar{Y})^2]}}$$

$$= \frac{\sum(X_i - \bar{X})(Y_i - \bar{Y})}{\sqrt{[\sum(X_i - \bar{X})^2][\sum(Y_i - \bar{Y})^2]}} = r.$$

The last formula for r can be written in the form given earlier by multiplying both the numerator and the denominator by $1/n$.

The linear equation for predicting Y_i from X_i can be expressed in terms of $r(S_Y/S_X)$ instead of $b_{Y \cdot X}$:

$$Y_i' = \overbrace{\bar{Y} - r\frac{S_Y}{S_X}\bar{X}}^{a_{Y \cdot X}} + \overbrace{r\frac{S_Y}{S_X}}^{b_{Y \cdot X}}X_i = \bar{Y} + r\frac{S_Y}{S_X}(X_i - \bar{X}).$$

In this form we can see what happens when $r = 0$; we obtain

$$Y_i' = \bar{Y} + 0\frac{S_Y}{S_X}(X_i - \bar{X})$$

$$= \bar{Y}.$$

---

[2] The sign of r is positive when both regression coefficients are positive and negative when both coefficients are negative; $b_{Y \cdot X}$ and $b_{X \cdot Y}$ always have the same sign.

This means that when the correlation coefficient equals 0, the predicted value of Y is the mean of the Y scores regardless of the X value used to predict Y. In other words, knowing X_i is no help if $r = 0$, since in every case the predicted Y value is $\bar{Y}$.

Review Exercises for Sections 6.1 and 6.2

*1. If Y increases 2 units for every 4-unit increase in X, what is the slope of the regression line of Y on X?

2. If Y decreases 5 units for every 2-unit increase in X, what is the slope of the regression line of Y on X?

*3. In an experiment on sex-typed behavior, a random sample of boys ages 5 to 8 was given choices among such toys as a football, a doll carriage, a dump truck, and dishes. The number of sex-appropriate choices for boys at each age is listed in the table.

Age, X	Number of appropriate choices, Y	Age, X	Number of appropriate choices, Y
7.5	18	7.5	15
7.0	13	5.0	7
5.5	11	5.5	8
8.0	20	6.0	12
6.5	13	8.0	17
6.0	14	7.0	14
5.0	9	6.5	12
8.0	18	5.5	10
6.5	14	5.0	8
6.0	10	7.0	16
7.5	19		

   a. Construct a scatter diagram and decide whether the data appear to be linearly related.

   b. Compute the values of $a_{Y \cdot X}$ and $b_{Y \cdot X}$ for the line of best fit, and draw the line in the scatter diagram. Compute r using the relationship $r = b_{Y \cdot X}(S_X/S_Y)$.

   c. Estimate Y for a 6-year-old boy using both the regression equation and the line of best fit in the scatter diagram.

4. For the experiment described in Exercise 3, data also were obtained for a random sample of girls, as listed in the table.

Age, X	Number of appropriate choices, Y	Age, X	Number of appropriate choices, Y
7.5	10	8.0	14
6.0	11	7.0	11
5.5	10	7.5	13
8.0	15	6.5	9
7.5	14	6.5	11
5.0	6	6.0	10
6.0	8	5.5	8
7.0	12	7.0	10
8.0	12	6.5	13
5.0	7	5.0	9
5.5	9		

a. Construct a scatter diagram and decide whether the data appear to be linearly related.
b. Compute the values of $a_{Y \cdot X}$, $b_{Y \cdot X}$, $a_{X \cdot Y}$, and $b_{X \cdot Y}$ for the two regression equations, and draw the lines of best fit in the scatter diagram.
c. Compute r using the relationship $r = \pm\sqrt{b_{Y \cdot X} b_{X \cdot Y}}$.
d. Estimate Y for a 6-year-old girl and X for a girl who made 11 appropriate choices, using the lines of best fit in the scatter diagram.

*5. In what sense are the regression lines in Exercises 3 and 4 best-fitting lines?
6. What characteristics of the line of best fit do $a_{Y \cdot X}$ and $b_{Y \cdot X}$ describe?
7. Distinguish between $b_{Y \cdot X}$ and $b_{X \cdot Y}$.
*8. The formula for $b_{Y \cdot X}$ in Table 6.2-1 is not the most convenient one for computational purposes. Show that it is algebraically equivalent to

$$\left[\sum X_i Y_i - \frac{(\sum X_i)(\sum Y_i)}{n}\right] \Big/ \left[\sum X_i^2 - \frac{(\sum X_i)^2}{n}\right],$$

which is simpler to compute.
9. If $r = 0$, the predicted Y score for all subjects is the mean of Y. Draw a scatter diagram that illustrates this.
*10. a. If $Y_i' = a + bX_i$ for all i and $a = \bar{Y} - b\bar{X}$, prove that $\sum Y_i' = \sum Y_i$.
   b. If $Y_i' = Y_i$ for all i, what do you know about r?
11. Terms to remember:
   a. Regression analysis   b. Line of best fit
   c. Prediction error   d. Principle of least squares

Section 6.3   Another Measure of Ability to Predict: The Standard Error of Estimate

We have seen that our ability to predict Y from X is a function of the degree of correlation between the two variables. The higher the correlation, the more closely the data points cluster around the regression line and the smaller our prediction error. A measure of the prediction error is given by the **standard error of estimate,** *$S_{Y \cdot X}$, which is a kind of standard deviation. For comparison purposes the formulas for the standard error of estimate and standard deviation are given below.*[3]

$$S_{Y \cdot X} = \sqrt{\frac{\sum(Y_i - Y_i')^2}{n}} \qquad S_Y = \sqrt{\frac{\sum(Y_i - \bar{Y})^2}{n}}$$

In computing $S_{Y \cdot X}$, the deviation $(Y_i - Y_i')$ is from the predicted value or regression line, whereas for S_Y the deviation is from the mean of Y. The two deviations are illustrated in Figure 6.3-1.

Let's look at $S_{Y \cdot X}$ more closely. The regression line denoted by Y' can be thought of as a kind of mean—a "running mean," giving the predicted value of Y for a particular value of X. Whereas $\bar{Y}$ is the mean of all Y's, Y' estimates the mean of Y for a particular value of X. The deviation of a Y score from a running

---

[3] When the population standard error of estimate is estimated from sample data, a better estimator is

$$\hat{\sigma}_{Y \cdot X} = \sqrt{\frac{\sum(Y_i - Y_i')^2}{n - 2}}.$$

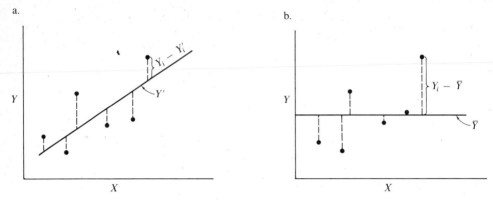

Figure 6.3-1. Comparison of the deviation $Y_i - Y_i'$ used to compute the standard error of estimate (part a) and the deviation $Y_i - \bar{Y}$ used to compute the standard deviation (part b).

mean is a prediction error ($e_i = Y_i - Y_i'$). Viewed in this light, $S_{Y \cdot X}$ (like S_Y) is computed from the sum of squared deviations from means and hence is a standard deviation. However, $S_{Y \cdot X}$ is the standard deviation of predicted errors,[4] while S_Y is the standard deviation of scores; $S_{Y \cdot X}$ can be interpreted in much the same way as a conventional standard deviation.

Interpretation of $S_{Y \cdot X}$

The larger $S_{Y \cdot X}$, the greater the dispersion of Y scores around the regression line and hence the larger the average prediction error. If the distribution of Y scores at every X score is approximately normal and if all the Y-score distributions have the same dispersion, 68.3% of the Y scores will fall within the limits of $Y' \pm S_{Y \cdot X}$. This information is illustrated in Figure 6.3-2. Similarly, 95.4% of the Y scores will fall within the limits $Y' \pm 2S_{Y \cdot X}$, and 99.7% within the limits $Y' \pm 3S_{Y \cdot X}$. This follows from the normal distribution; see Figure 4.4-1.

Section 6.4   *Alternative Formula for $S_{Y \cdot X}$*

The formula for the standard error of estimate given in the last section isn't convenient for computing $S_{Y \cdot X}$. An equivalent formula[5] that is much easier to use is

$$S_{Y \cdot X} = S_Y \sqrt{1 - r^2}.$$

The maximum and minimum possible values of $S_{Y \cdot X}$ can be ascertained readily from this formula. The maximum value occurs when $r = 0$, in which case $S_{Y \cdot X} = S_Y$. In other words, if $r = 0$, the dispersion of Y scores around the regression line is as large as the standard deviation of Y. Knowing the X score doesn't reduce

---

[4] See Kirk (1978, pp. 142–143) for a mathematical development of this idea.

[5] See Kirk (1978, pp. 143–144) for the derivation of the formula. When the population standard error of estimate is estimated from sample data, a better estimator is

$$S_{Y \cdot X} = S_Y \sqrt{\frac{n}{n-2}(1 - r^2)}.$$

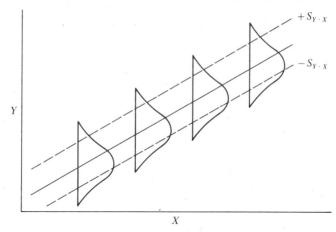

Figure 6.3-2.   Illustration of the standard error of estimate. Approximately 68% of the Y scores fall within the limits $Y' \pm S_{Y \cdot X}$ if the distribution of Y scores at every X score is approximately normally distributed and all the Y score distributions have the same dispersion.

the error of prediction. The minimum value occurs when $r = 1$. In this case $S_{Y \cdot X} = 0$, and there is no dispersion around the regression line and no error in predicting Y from X. To summarize, the upper limit for $S_{Y \cdot X}$ is S_Y and occurs when $r = 0$; the lower limit is 0 and occurs when $r = 1$.

Coefficient of Alienation, k

The relationship between $S_{Y \cdot X}$ and r just described leads to a useful relative index of prediction error called the *coefficient of alienation,* denoted by k. As we have seen, when $r = 0$, we have the worst possible prediction situation—the magnitude of our prediction error is as large as it can be so that $S_{Y \cdot X} = S_Y$. As r approaches ± 1, the prediction error decreases until when $r = \pm 1$, $S_{Y \cdot X} = 0$. The coefficient of alienation compares $S_{Y \cdot X}$, which reflects the magnitude of prediction error, with S_Y, which reflects the error that would be obtained in the worst possible prediction situation, and this occurs when $S_{Y \cdot X} = S_Y$. The coefficient is simply the ratio $S_{Y \cdot X}/S_Y$.

$$k = \left(\begin{array}{l} \text{Comparison of prediction} \\ \text{error for a set of data, with the} \\ \text{prediction error in the worst} \\ \text{possible prediction situation} \end{array} \right) = \frac{S_{Y \cdot X}}{S_Y} = \frac{S_Y \sqrt{1 - r^2}}{S_Y} = \sqrt{1 - r^2}$$

For example, if $r = 0.80$, then $k = \sqrt{1 - (0.80)^2} = 0.60$. This tells us that the standard error of estimate, a measure of our prediction error, is 60% of the size it would be if the correlation were 0. To state it another way, when $r = 0.80$, we have reduced the magnitude of the prediction error by $100\% - 60\% = 40\%$ by taking X into account. *The reduction in prediction error is called the **index of forecasting efficiency,** E,*[6] *and is computed from the formula* $E = 1 - k$.

---

[6] E is also called the *coefficient of prediction efficiency.*

The values of k and E for selected values of r are given in Table 6.4-1. We note from the table that relative prediction error decreases very slowly as r increases from 0 to 1. For $r = 0.85$ the standard error of estimate is 53% of the size it would be if the correlation were 0. We see also that for high correlations a small increase in r produces a greater reduction in relative prediction error than for low correlations. For example, a 0.05 change in r from 0.85 to 0.90 reduces k by 9%, whereas a 0.05 change in r from 0.20 to 0.25 reduces k by only 1%.

Table 6.4-1.   Values of k and E for selected values of r

r	k	E	r	k	E
1.00	0.00	1.00	0.45	0.89	0.11
0.95	0.31	0.69	0.40	0.92	0.08
0.90	0.44	0.56	0.35	0.94	0.06
0.85	0.53	0.47	0.30	0.95	0.05
0.80	0.60	0.40	0.25	0.97	0.03
0.75	0.66	0.34	0.20	0.98	0.02
0.70	0.71	0.29	0.15	0.989	0.011
0.65	0.76	0.24	0.10	0.995	0.005
0.60	0.80	0.20	0.05	0.999	0.001
0.55	0.84	0.16	0.00	1.00	0.00
0.50	0.87	0.13			

Descriptive Application of $S_{Y \cdot X}$

Although the standard error of estimate is most often used in inferential statistics, a descriptive application that was briefly mentioned in Section 6.3 is illustrative. Suppose an experiment was conducted to determine the relationship between length of time (measured in hundredths of a second) necessary to reach a decision, Y, and the number of alternative choices presented, X. The following data were obtained: $S_X = 1.5$, $S_Y = 12.5$, $\bar{X} = 4.5$, $\bar{Y} = 45$, $r = 0.78$, and $n = 100$. Assume that the distribution of Y scores for every X score is approximately normal and that all the Y score distributions have the same dispersion. The predicted reaction time for a person presented with a choice from among, say, three alternatives is given according to the regression equation $Y' = \bar{Y} + r(S_Y/S_X)(X_i - \bar{X})$:

$$Y' = 45 + 0.78 \frac{12.5}{1.5} (3 - 4.5)$$

$$= 45 + 6.5(-1.5)$$

$$= 35.25.$$

The standard error of estimate is

$$S_{Y \cdot X} = S_Y \sqrt{1 - r^2}$$

$$= 12.5 \sqrt{1 - (0.78)^2}$$

$$= 12.5(0.6258)$$

$$= 7.82.$$

We can conclude that approximately 68% of the subjects in the three-choice condition had reaction times between 43.07 and 27.43, as we see from

$$Y' \pm S_{Y \cdot X} = 35.25 \pm 7.82 = 43.07 \text{ and } 27.43.$$

Similarly, approximately 95% had reaction times between

$$Y' \pm 2S_{Y \cdot X} = 35.25 \pm 2(7.82) = 50.89 \text{ and } 19.61.$$

Section 6.5   *Assumptions Associated with Estimate of Prediction Error*

When we make predictions using the regression equation $Y'_i = a + bX_i$, we assume only that the relationship between X and Y is linear. If the assumption is tenable, the principle of least squares ensures that $a + bX_i$ provides the best possible fit for the data points. For prediction purposes we don't have to make any assumptions regarding the shape of the X and Y distributions.

The use of the standard error of estimate and k and E involves more stringent assumptions. In addition to the linearity assumption, we must also assume that (1) for any value of X the associated Y scores are approximately normally distributed and (2) the standard deviation of Y scores, $S_{Y \cdot X}$, is the same for all the X scores. The latter assumption is referred to as the *homoscedasticity* assumption. The converse situation, heteroscedasticity, in which the dispersions of the Y scores for different values of X are unequal, was discussed in Section 5.5.

In making predictions from Y to X the same assumptions are required, but they must be rephrased to reflect the revised roles of X and Y.

Review Exercises for Sections 6.3, 6.4, and 6.5

*12. Chimpanzees were exposed to white noise 8 hours a day for 3 months to determine if it affected their hearing. Ten animals were randomly assigned to the following noise levels: 75 dBA,[7] 85 dBA, 95 dBA, 105 dBA, and 115 dBA.

Animal	Noise level (dBA), X	Hearing loss (dB at 1000 Hz), Y	Animal	Noise level (dBA), X	Hearing loss (dB at 1000 Hz), Y
1	105	11	6	85	9
2	85	6	7	105	13
3	95	10	8	115	11
4	115	15	9	75	5
5	75	7	10	95	8

a. Compute $S_{Y \cdot X}$ using the formula $S_Y \sqrt{1 - r^2}$.

b. Assuming a large sample in which the distribution of Y scores for every X score is approximately normal and all the distributions have the same dispersion, compute the limits of the interval that will contain 68% of the scores for a noise level of 115 dBA.

---

[7] The abbreviation dBA stands for the sound pressure level in decibels as measured on the A weighting network of a sound-level meter.

c. Compute the value of $S_{Y \cdot X}$ for $r = 0$ and $r = 1$.

d. Compute k and E and interpret.

13. The relationship between birth order and participation in dangerous sports such as football, soccer, and boxing was investigated. College records were screened to obtain four males who were first-born, four who were second-born, and so on. The data in the table were obtained.

Subject	Birth order, X	Number of dangerous sports, Y	Subject	Birth order, X	Number of dangerous sports, Y
1	4	1	11	1	0
2	3	1	12	2	0
3	2	0	13	3	2
4	4	2	14	5	1
5	1	0	15	3	1
6	5	2	16	2	1
7	1	0	17	5	2
8	4	1	18	1	1
9	2	1	19	3	1
10	5	3	20	4	2

a. Compute $S_{Y \cdot X}$ using the formula $S_Y \sqrt{1 - r^2}$.

b. Assuming a large sample in which the distribution of Y scores for every X score is approximately normal and all the distributions have the same dispersion, compute the limits that will contain 68% of the scores for fourth-born males.

c. Compute the value of $S_{Y \cdot X}$ for $r = 0$ and $r = 1$.

d. Compute k and E and interpret.

14. In what sense is Y' a mean?

15. How is $S_{Y \cdot X}$ related to the magnitude of prediction error?

16. How is k related to the magnitude of prediction error?

17. Describe the effect of changes in r on the value of k.

18. Compare the assumptions associated with predictions using r, Y', and $S_{Y \cdot X}$.

19. Terms to remember:

a. Standard error of estimate

b. Coefficient of alienation

c. Index of forecasting efficiency (coefficient of predictive efficiency)

d. Homoscedasticity

Section 6.6   Summary

The term "regression" was used first by Sir Francis Galton to refer to the tendency for short parents to have slightly taller offspring and tall parents to have slightly shorter offspring. We still speak of regression toward the mean, but today the term has a broader meaning. It refers to any analysis of paired data (X_1, Y_1), $(X_2, Y_2), \ldots, (X_n, Y_n)$, where X is the independent variable and Y, the dependent variable.

In regression analysis the line of best fit, called the regression line, is used to predict Y from a knowledge of X. The line of best fit according to the least squares principle is the one for which the sum of the squared prediction errors, the discrepancy between the observed value of Y_i and the predicted value, is as small as it can possibly be.

If $r = 1$, the value of Y_i can be predicted perfectly from the equation $Y_i' = a + bX_i = \bar{Y} + r(S_Y/S_X)(X_i - \bar{X})$. If $|r| < 1$, there is likely to be some discrepancy between the observed value Y_i and the predicted value Y_i'. A measure of the prediction error is given by the standard error of estimate $S_{Y \cdot X}$, which is a kind of standard deviation of scores around the regression line. The maximum value of $S_{Y \cdot X}$ is the standard deviation of Y, S_Y, and occurs when $r = 0$. The minimum value is 0 and occurs when $r = 1$.

The coefficient of alienation, k, is a relative index of prediction error. It is the ratio of $S_{Y \cdot X}$, which reflects the magnitude of prediction error, to S_Y, which reflects the magnitude of prediction error in the worst possible prediction situation—when $S_{Y \cdot X} = S_Y$. The coefficient of alienation is conveniently computed from $k = S_{Y \cdot X}/S_Y = \sqrt{1 - r^2}$. If $r = 0.60$, then $k = 0.80$, which means that $S_{Y \cdot X}$ is 80% of the size it would be if the correlation between X and Y were 0. A related statistic, the index of forecasting efficiency, $E = 1 - k$, measures the reduction in magnitude of the prediction error over what it would have been had the value of X been ignored in predicting Y.

In predicting Y from X we assume only that the relationship between the variables is linear. Interpretations involving $S_{Y \cdot X}$, k, or E, also assume that the distribution of Y scores at every X score is approximately normal and that all the Y score distributions have the same dispersion. When prediction involves different samples, as when the performance of one group of students is predicted from that of another, we also must assume that the populations represented by the two samples are identical with respect to the relevant characteristics.

†*Section 6.7   Computer Supplement*

An interactive microcomputer package developed by Brad Wilson (1984) was described in Section 3.10. The regression program of this package was used to analyze the midterm–final exam data in Table 6.2-1. The printout is shown in Table 6.7-1.

Table 6.7-1.   Printout from the Wilson Stat Package:
Regression program (Data from Table 6.2-1)

```
***********************************************
*                                             *
*             Brad Wilson's                   *
*             Stat Package                    *
***********************************************
*                                             *
*      Behavioral Statistics Program          *
*            Baylor University                *
***********************************************
Press RETURN to continue>>>>>

When you respond to a prompt just
type in your response and press
the RETURN key.

For all yes-no (Y/N) questions
it is not necessary to press the
RETURN key.

Press RETURN to continue>>>>>

Would you like to

 1. Perform statistical analyses

 2. See a demonstration of the
    Central Limit Theorem

 3. End the session?

Select a number: 1

Please wait a few moments.

Here are the available
statistical programs
   1.   Frequency distribution and
        histogram
   2.   Mean, median, and mode
   3.   Standard deviation
   4.   Index of dispersion
   5.   Correlation
   6.   Regression
   7.   One-sample t test
   8.   One-sample chi-square test
```

(continued)

---

† This section can be omitted without loss of continuity.

Table 6.7-1.   (continued)

```
 9.   Two-sample independent t test
10.   Two-sample dependent t test
11.   Analysis of variance for a
      completely randomized design
12.   Pearson's chi-square for
      testing goodness of fit
13.   Pearson's chi-square for
      testing independence
14.   Pearson's chi-square for
      testing equality of proportions
```

Indicate the program you want
by selecting a number: 6

Please wait a few moments.

Do you want to enter data from

1. The terminal

2. A file, or

3. Use the data from the
   previous session, or do
   you want to

4. Return to the menu?

Select a number: 1

What is the label for
variable 1? MIDTERM

What is the label for
variable 2? FINAL

How many pairs of values do you have? 12

variable	# of cases
1. MIDTERM	12
2. FINAL	12

Do you want to

1. Go on to the next step

2. Examine your data specifications
   again

3. Change your variable name(s)

4. Change your # of cases?

Select one: 1

(continued)

Table 6.7-1.   (continued)

```
Now you will be prompted for
the individual values

MIDTERM   1:  72
MIDTERM   2:  75
MIDTERM   3:  76
MIDTERM   4:  78
MIDTERM   5:  78
MIDTERM   6:  82
MIDTERM   7:  82
MIDTERM   8:  84
MIDTERM   9:  86
MIDTERM  10:  88
MIDTERM  11:  89
MIDTERM  12:  94
FINAL     1:  70
FINAL     2:  74
FINAL     3:  70
FINAL     4:  72
FINAL     5:  78
FINAL     6:  74
FINAL     7:  84
FINAL     8:  76
FINAL     9:  86
FINAL    10:  78
FINAL    11:  88
FINAL    12:  86

Press RETURN to continue>>>
Would you like to check the data for
errors? (Y/N) N

If you have a second disk drive, you can
store your data on a separate diskette.
Would you like to store the data? (Y/N) N

Please wait a few moments.

Do you want the results sent to the

 1. Screen

 2. Printer?

Select a number: 1

NOTE: Use CTRL and S simultaneously to
stop and start the scrolling of
the screen.

Press RETURN to continue>>>
```

(continued)

Table 6.7-1.   (continued)

```
                            STANDARD
               N    MEAN    DEVIATION
**********************************************************************
MIDTERM    12  82.000       6.509
FINAL      12  78.000       6.495

Press RETURN to continue>>>
```

```
Scatterplot of MIDTERM
 vs. FINAL

Variable Key:
 Y = FINAL
 X = MIDTERM
The maximum value of X = 94.000
The minimum value of X = 72.000
The maximum value of Y = 88.000
The minimum value of Y = 70.000

Symbol table:   * = 1 case
                2 = 2 cases
                3 = 3 or more cases

Y
+                                              *
+
+                                     *                    *
+
+                        *
+
+
+
+
+
+                  *                      *
+
+                           *
+
+        *              *
+
+           *
+
+ *        *
+ + + + + + + + + + + + + + + + + + + + + + + + + + + + + +X

The correlation (r) between
MIDTERM    and FINAL     is 0.804

Press RETURN to continue>>>
```

(continued)

Table 6.7-1.   (continued)

The coefficient of alienation (k) = 0.594

The regression equation for
predicting FINAL   from MIDTERM
is:
FINAL   = 12.189 + 0.803(MIDTERM   )

The standard error of estimate = 3.695

Press RETURN to continue⟩⟩⟩

The regression equation for
predicting MIDTERM   from FINAL
is:

MIDTERM   = 19.129 + 0.806(FINAL   )

The standard error of estimate = 3.703

   Press RETURN to continue⟩⟩⟩

   Would you like to

   1. Perform statistical analyses

   2. See a demonstration of the
      Central Limit Theorem

   3. End the session?

   Select a number: 3

   Just turn off the computer, remove your
   diskettes, and have a nice day!!

7

Probability

Section 7.1   Introduction to Probability

Everyone has some intuitive notion of what probability is. However, its definition is a topic for continuing debate among mathematicians. Three views of probability will be described: (1) the subjective–personalistic view, (2) the classical or logical view, and (3) the empirical relative-frequency view. Fortunately, the different views are not incompatible.

The Subjective–Personalistic View

According to the **subjective view,** *probability is a measure of the strength of one's expectation that an event will occur.* For example, one might assert, "Chances are I'll pass statistics," or "I think I'll go home this weekend." Such assertions express a degree of belief concerning an event whose outcome is at the moment uncertain. Subjective probabilities affect our lives since they enter into our decision-making process. For most people the subjective probability of being struck by a car while crossing the street is low, so they proceed as if the event won't happen. But if a person's subjective probability of, say, being invited to a New Year's party is high enough, all suitable preparations for the event's occurrence will be made.

Although our behavior is influenced by subjective probability, there are difficulties in incorporating it into a formal decision-making process. Equally knowledgeable individuals often disagree on the probability that should be assigned to an event. We find that some people's subjective probabilities follow closely the rules of probability described later, but other people's don't. Hence, a subjective probability can't be considered apart from the person holding it. The measurement of subjective probability poses another problem, although behavioral scientists are beginning to find solutions to this problem. Despite the problems, a formal approach to decision making that utilizes subjective probability has been developed. It is popular in economics and business management and is beginning to find acceptance in behavioral research. This approach, called *Bayesian inference,*[1] enables an experimenter to make decisions about some true state of affairs using not only sample data but also any prior information that is available, either from previous samples or simply in the form of informed opinions or beliefs.

---

[1] Bayesian inference is named for the Reverend Thomas Bayes (1702–1761), whose theorem laid the groundwork for the approach. For a fuller discussion see Hays (1981, pp. 39–41), McGee (1971, chaps. 10 and 11), and Novick and Jackson (1974).

The approach represents an extension of classical inferential methods. Bayesian methods, however, emphasize the steady accumulation and utilization of information from many sources. Furthermore, they lead to different ways of interpreting data, but not necessarily different conclusions about data. The body of classical statistical methods introduced in Chapters 11 through 15 is pretty well mapped out; this isn't true for Bayesian methods. The next decade will see many new developments.

The Classical or Logical View

Suppose that we want to know the probability of rolling a 2 with a fair die. We reason that since a fair die is symmetrical and dynamically balanced, all six faces are equally likely. Of the six possible events, only one is a 2, and therefore the probability of rolling a 2, $p(2)$, is $\frac{1}{6}$. *According to the* **classical view,** *the probability of an event, say A, is given by the number of events favoring A, n_A, divided by the total number of equally likely events, n_S.*[2] *Thus, $p(A) = n_A/n_S$.* The value of $p(A)$ is always a number between 0 and 1 inclusive, because the number of events favoring A can never exceed the total number of events—that is, $n_A \leq n_S$.

The classical view of probability is based on logical analysis. We reason, for example, that when a fair coin is tossed, there are two possible outcomes—a head or a tail—and that the outcomes are equally likely. It follows that the probability of a head is $p(H) = n_H/n_S = \frac{1}{2}$. The probabilities $\frac{1}{2}$ for a head and $\frac{1}{6}$ for a 2 were arrived at by logical analysis of dynamic, event-generating processes. In effect, we developed a mathematical model based on a postulate and logic. We postulated that certain events are equally likely and deduced the consequences. If our logic is correct, the deductions $p(H) = \frac{1}{2}$ and $p(2) = \frac{1}{6}$ are formally correct, although they may not correspond to empirical results since for any particular coin or die the postulate may be incorrect. Experience has demonstrated that the classical view generates probability estimates that closely approximate empirical probabilities for certain fairly simple event-generating processes such as coin tossing. Consequently, this approach is useful for practical problems.

The Empirical Relative-Frequency View

A third view of probability can be adopted for event-generating processes that can be repeated without changing their characteristics. Probability according to this view is estimated from experience—by performing an experiment and determining the ratio of the number of events of interest to the total number of events. This leads to our final definition of probability. *According to the* **empirical relative-frequency view,** *the probability of event A, $p(A)$, is a number approached by the ratio n_A/n as the total number of observations, n, approaches infinity.* For example, in a simple experiment such as tossing a coin, the probability of a head can be estimated by making many tosses of the coin and recording the outcomes. If a head is obtained 12 times in 20 tosses, the best estimate of the probability of heads is $n_A/n = \frac{12}{20} = 0.6$. If a head is obtained 120 times in 200 tosses, our

---

[2] The letter S, denoting a sample space, will be defined in Section 7.2.

confidence in the estimate $120/200 = 0.6$ is even greater. We assume that the sample estimate n_A/n moves closer and closer to some "true probability" as n approaches infinity, and thus we have greater confidence in larger samples.

The empirical view of probability is useful and intuitively simple, but it too has certain difficulties. It is meaningful to speak of the probability of rain tomorrow or the probability of getting an A on Tuesday's quiz; however, there is only one tomorrow and only one such quiz. The interpretation of probability as the number approached by n_A/n as the number of tomorrows approaches infinity is unconvincing.

We must conclude that none of the views of probability is completely adequate. Since they are all useful and they are not incompatible, they coexist amicably in the mathematician's bag of conceptual tools. In the discussion that follows we will rely most on the classical and empirical views.

Review Exercises for Section 7.1

1. Why is subjective probability difficult to incorporate into a formal decision-making process?
2. What distinguishes Bayesian inference from traditional classical methods?
*3. (a) According to the classical view, what is the probability of observing an odd number on the toss of a die? (b) What assumptions were required to arrive at the answer?
4. (a) What is the probability of drawing the king of hearts from a well-shuffled deck of 52 cards? (b) What assumptions were required to arrive at the answer?
*5. (a) According to the relative-frequency view, what is the probability that a head will occur on the next toss of a fair coin if a head appeared on 52 of the last 100 tosses? (b) According to the logical view, what is the probability that a head will occur?

Section 7.2 Basic Concepts

For behavioral scientists and educators, probability theory is a means to an end. It is the vehicle for making inferences about the characteristics of populations by observing samples drawn from the populations. Sample data are obtained by observing events in nature or experimenting in the laboratory. We will denote either procedure by the term *experiment*. In particular, we are interested in experiments with outcomes that can't be predicted with certainty. For example, will desensitization therapy result in more symptom relief than symbolic modeling therapy? Will a rat turn right or left in a T maze?

Simple and Compound Events

One of the simplest experiments we can perform is tossing a die and observing the number that appears on the upper face. Some of the possible outcomes are the following.

1. Event A—observe an odd number.
2. Event B—observe an even number.
3. Event C—observe a number less than 4.
4. Event E_1—observe a 1.

5. Event E_2—observe a 2.
6. Event E_3—observe a 3.
7. Event E_4—observe a 4.
8. Event E_5—observe a 5.
9. Event E_6—observe a 6.

An *event* is an observable happening. Events *A*, *B*, and *C* are called *compound events* because they can be decomposed into simpler events. For example, event *A* is the occurrence of one of the simpler events E_1, E_3, or E_5. Events $E_1, \ldots, E_6$ are called *simple events* because they can't be decomposed. An experiment in which two coins are tossed has four possible simple events.

	Coin 1	Coin 2
Event E_1	Head	Head
Event E_2	Head	Tail
Event E_3	Tail	Head
Event E_4	Tail	Tail

The event—observe two like coins—is a compound event because it can be decomposed into the simpler events E_1 and E_4. A single trial of an experiment will result in one and only one simple event. A list of simple events provides a breakdown of all possible outcomes of the experiment.

Graphing Simple Events

It is convenient to represent the simple events in an experiment by a graph called a Euler diagram.[3] To do this we assign to each simple event a point called a *sample point*. The symbol E_i is used to denote both the *i*th simple event and its sample point. *The set of all sample points is called the* **sample space** *and is denoted by the letter S.* A Euler diagram representing the sample space for the die-tossing experiment is shown in Figure 7.2-1.

Compound events are represented in the diagram by encircling the sample points in that event. For example, earlier we defined event *A* as observing an odd

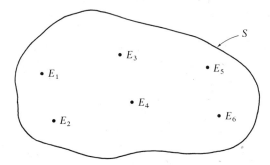

Figure 7.2-1. Euler diagram for the die-tossing experiment. The set of all sample points $E_1, \ldots, E_6$ defines the sample space *S* of the experiment.

---

[3] The diagram was developed by Leonhard Euler (1707–1783), a Swiss mathematician.

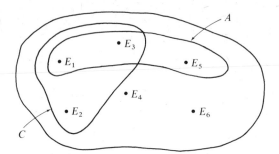

Figure 7.2-2.   Euler diagram for event A, observing an odd number, and event C, observing a number less than 4.

number on the toss of a die and event C as a number less than 4. The two events are represented in Figure 7.2-2 by two subsets of the sample points.

The probability of event A according to the classical view is $p(A) = n_A/n_S = \frac{3}{6}$; the probability of event C is $p(C) = n_C/n_S = \frac{3}{6}$. By examining the sample space, we also can determine the probability for combined events. What is the probability that when a die is tossed the outcome will be an odd number *and* less than 4? We could observe an odd number and a number less than 4, events A and C, if either E_1 or E_3 occurred. Hence, the probability that the outcome will represent both events A and C is $\frac{2}{6} = \frac{1}{3}$. We could observe an odd number *or* a number less than 4, event A or C or both A and C, in four ways: if $E_1, E_2, E_3,$ or E_5 occurred. Hence, the probability of A or C or both A and C is $\frac{4}{6} = \frac{2}{3}$. Formal rules for computing the probabilities of combined events are presented in Section 7.3, but first we will describe three properties of probabilities.

Formal Properties of Probability

Probability theory can be thought of as a system of definitions and operations pertaining to a sample space. According to the classical view described in Section 7.1, the probability of event A is the ratio of the number of sample points that are examples of A to the total number of sample points, provided all sample points are equally likely. For the die-tossing experiment represented in Figure 7.2-1, $p(E_1) = p(E_2) = \cdots = p(E_6) = n_{E_i}/n_S = \frac{1}{6}$. This follows from the assumption that all six faces are equally likely—$n_{E_i} = 1$ for the $i = 1, \ldots, 6$ sample points—and the fact that there are six sample points in the sample space—$n_S = 6$. To each point in the sample space we can assign a number called the probability of E_i such that (1) $0 \le p(E_i) \le 1$ for all i, (2) $\sum_{i=1}^{n_S} p(E_i) = 1$, and (3) $p(S) = 1$. In other words, (1) the probability assigned is a number greater than or equal to 0 and less than or equal to 1, (2) the sum of the probabilities over the sample space equals 1, and (3) the probability of the sure event, S, is always 1.

Review Exercises for Section 7.2

*6. An experiment consists of tossing three fair coins. (a) Represent the sample space by a Euler diagram, and encircle the sample points corresponding to observing two heads, event A, and observing at least one head, event B. (b) What is the probability of event A? (c) What is the probability of event B?

7. A class contains six psychology (P) majors, one sociology (S) major, and three history (H) majors. Assume that no students have double majors. (a) Represent the sample space by a Euler diagram. (b) If a student is selected at random, what is the probability that he or she will be a psychology major? (c) What is the probability that the student selected will be a psychology or a sociology major?

8. An experiment consists of tossing two dice, one green and one red, and recording the outcome. (a) Represent the sample space by a Euler diagram, and encircle the sample points corresponding to observing a 7 as the sum of the dice. (b) What is the probability that the sum of two dice is 7? (c) What is the probability that the sum of two dice is less than 5?

*9. Determine the probability (a) that a man chosen randomly from a group of ten men is a psychologist if the group contains three psychologists and (b) that you will win a car if you bought six raffle tickets and 10,000 tickets were sold.

10. A fair die is rolled once. You win $5 if the outcome is even, event A, or if it is divisible by 3, event B. (a) Represent the sample space by a Euler diagram and encircle events A and B. (b) What is the probability of winning the $5?

11. The following are properties of probabilities. In your own words state what each property means. (a) $0 \le p(E_i) \le 1$, for all i. (b) $\sum_{i=1}^{ns} p(E_i) = 1$. (c) $p(S) = 1$.

12. Terms to remember:
   a. Experiment
   b. Simple and compound events
   c. Euler diagram
   d. Sample point
   e. Sample space

Section 7.3   Probability of Combined Events

This section describes rules for determining the probabilities of combined events. For example, we might want to know the probability that the outcome of an experiment will be event A or event B or both A and B. We denote this probability by $p(A \text{ or } B)$.[4] Alternatively, we might want to know the probability that the outcome will be both A and B. We denote this probability by $p(A \text{ and } B)$.[5]

The *union* of two events A and B is the set of elements that belong to A or to B or to both. The probability of the union of two events, $p(A \text{ or } B)$, is computed by using the addition rule of probability. The *intersection* of two events A and B is the set of elements that belong to both A and B. Its probability, $p(A \text{ and } B)$, is computed by the multiplication rule.

Addition Rule of Probability

The **addition rule** *states that the probability of the union A or B is equal to* $p(A \text{ or } B) = p(A) + p(B) - p(A \text{ and } B)$. For example, let event A be an even number when a die is tossed and event B, a number less than 5. The events are represented in Figure 7.3-1. Their probabilities are determined by counting sample points; $p(A) = n_A/n_S = \frac{3}{6} = \frac{1}{2}$ and $p(B) = n_B/n_S = \frac{4}{6} = \frac{2}{3}$. The probability of event A

---

[4] Some texts use the Boolean algebraic symbol $\cup$ in place of *or*.
[5] Some texts use the Boolean algebraic symbol $\cap$ in place of *and*.

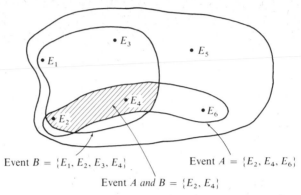

Event $B = \{E_1, E_2, E_3, E_4\}$ Event $A = \{E_2, E_4, E_6\}$

Event A *and* $B = \{E_2, E_4\}$

Figure 7.3-1. Euler diagram for event A, observing an even number, and event B, observing a number less than 5. The intersection of A and B is the shaded area.

and B is the ratio of the number of sample points that are examples of both A and B to the total number of sample points. In symbols, $p(A \text{ and } B) = n_{A \text{ and } B}/n_S = \frac{2}{6} = \frac{1}{3}$, since there are two simple events in both A and B and six in the sample space. Given this information, $p(A \text{ or } B) = \frac{1}{2} + \frac{2}{3} - \frac{1}{3} = \frac{5}{6}$. The probability of observing an even number or a number less than 5 is $\frac{5}{6}$. The value $p(A \text{ and } B) = \frac{1}{3}$ is subtracted from $p(A) + p(B)$ to avoid counting the simple events E_2 and E_4 twice, since they are contained in event A and event B.

The information contained in Figure 7.3-1 can be presented in a table, as shown in Table 7.3-1. This mode of presentation is easier to interpret, especially when the number of events exceeds two.

Table 7.3-1. Tabular presentation of information in Figure 7.3-1

Event

		A	*Not A*	
Event	B	A and $B = \{E_2, E_4\}$	*Not A* and $B = \{E_1, E_3\}$	$B = \{E_1, E_2, E_3, E_4\}$
	Not B	A and *Not B* $= \{E_6\}$	*Not A* and *Not B* $= \{E_5\}$	*Not B* $= \{E_5, E_6\}$
		$A = \{E_2, E_4, E_6\}$	*Not A* $= \{E_1, E_3, E_5\}$	

Addition Rule for Mutually Exclusive Events

Two events may contain no sample points in common, in which case the events are said to be *mutually exclusive*. For example, consider the events observing an even number on the toss of a die, event A, and observing an odd number, event B. As shown in Figure 7.3-2, the intersection A *and* B contains no sample points, and hence A and B are mutually exclusive.

For mutually exclusive events the addition rule is simplified because for this case $p(A \text{ and } B) = 0$. The rule is
$$p(A \text{ or } B) = p(A) + p(B).$$

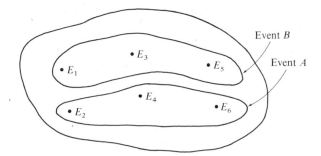

Figure 7.3-2.   Euler diagram for event *A*, observing an even number, and event *B*, observing an odd number. Since the intersection *A and B* contains no sample points, the events are mutually exclusive.

The probability of observing an even number or odd number is $p(A \text{ or } B) = \frac{3}{6} + \frac{3}{6} = 1$. Since the probability is 1, we know that when a die is tossed, one of the events must occur. *Events for which the probability of their union equals 1 are called* **collectively exhaustive** *or simply* **exhaustive.**

Addition Rule for Three Events

The addition rule can be extended to three or more events. For three events *A*, *B*, and *C*, the probability of *A or B or C* is given by

$$p(A \text{ or } B \text{ or } C) = p(A) + p(B) + p(C) - p(A \text{ and } B) - p(A \text{ and } C)$$

$$- p(B \text{ and } C) + p(A \text{ and } B \text{ and } C).$$

The rationale for the formula is apparent from the Euler diagram in Figure 7.3-3. If the events are mutually exclusive, the formula simplifies to

$$p(A \text{ or } B \text{ or } C) = p(A) + p(B) + p(C).$$

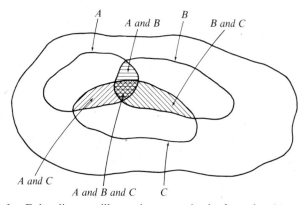

Figure 7.3-3.   Euler diagram illustrating terms in the formula $p(A \text{ or } B \text{ or } C) = p(A) + p(B) + p(C) - p(A \text{ and } B) - p(A \text{ and } C) - p(B \text{ and } C) + p(A \text{ and } B \text{ and } C).$

Multiplication Rule of Probability

The multiplication rule is used to compute the probability of the joint occurrence, or intersection, of two or more events. For example, suppose that 100 psychology majors have been classified according to sex and class level. The number of students in each category is given in Table 7.3-2. If a student is selected by lottery, what is the probability that the student will be both a female and a lowerclassperson? As we will see, the multiplication rule lets us determine the probability that the student selected will be in the intersection *female and lowerclassperson—* that is, both a female and a lowerclassperson. This information is different from that given by the addition rule, which tells us the probability that the student selected will be a female or a lowerclassperson or a female-lowerclassperson.

Table 7.3-2. Number of psychology majors by sex and class level

	Lowerclassperson, L	Upperclassperson, U	Marginal total
Women, W	10 $p(W \text{ and } L) = n_{W \text{ and } L}/n_S$ $= \frac{10}{100}$ $= 0.10$	20	30 $p(W) = n_W/n_S$ $= \frac{30}{100}$ $= 0.30$
Men, M	40	30	70 $p(M) = n_M/n_S$ $= \frac{70}{100}$ $= 0.70$
Marginal total	50 $p(L) = n_L/n_S$ $= \frac{50}{100}$ $= 0.50$	50 $p(U) = n_U/n_S$ $= \frac{50}{100}$ $= 0.50$	100

Before presenting the multiplication rule, we need to discuss the concept of conditional probability. Two events often are related so that the probability of one event depends upon whether the other has or has not occurred. Consider these events: Your roommate reports that he feels bad, event *A*, and his temperature is 103, event *B*. The two events are obviously related since the probability of an elevated temperature, $p(B)$, is much higher if a person feels bad than if he feels good. We refer to this type of relationship as a *conditional probability*. The **conditional probability** of *B*, given that *A* has occurred, is denoted by $p(B|A)$ and is equal to

$$p(B|A) = p(A \text{ and } B)/p(A).$$

Similarly, the conditional probability of *A*, given that *B* has occurred, is

$$p(A|B) = p(A \text{ and } B)/p(B).$$

The vertical line | is read "given," or "given that."

The calculation of conditional probability will be illustrated for the data in Table 7.3-2. The probability that a student selected by lottery is a woman, given that you know the student is a lowerclassperson, is

$$p(W\,|\,L) = p(W \text{ and } L)/p(L) = \left(\frac{10}{100}\right) \bigg/ \left(\frac{50}{100}\right) = 0.20.$$

We see that the condition of being a lowerclassperson limits the outcome to the left column of Table 7.3-2, which is a smaller sample space of size 50. The probability of selecting a woman is a subset of this smaller sample space, namely, 10 events out of 50. The probability of selecting a woman in the absence of information about class level is $p(W) = \frac{30}{100} = 0.30$. The events W and L are related, since a knowledge of one event affects the probability of the other.

The multiplication rule can be stated now. *Given two events A and B, the probability of obtaining both A and B jointly is the product of the probability of obtaining one event times the conditional probability of the other event. Stated symbolically, the joint probability p(A and B) is given by*

$$p(A \text{ and } B) = p(A)p(B\,|\,A)$$
$$= p(B)p(A\,|\,B).$$

For the events defined in Table 7.3-2, the probability of selecting a student who is both a female and a lowerclassperson is

$$p(W \text{ and } L) = p(W)p(L\,|\,W) = p(W)[\,p(W \text{ and } L)/p(W)\,]$$
$$= 0.30(0.10/0.30) = 0.10$$
$$= p(L)p(W\,|\,L) \;\; = p(L)[\,p(W \text{ and } L)/p(L)\,]$$
$$= 0.50(0.10/0.50) = 0.10.$$

The multiplication rule may seem unnecessarily complicated since if the intersection $(A \text{ and } B)$ is known, the rule is not needed. However, sometimes only a *marginal probability*, $p(A)$ or $p(B)$, and a conditional probability, $p(A\,|\,B)$ or $p(B\,|\,A)$, are known. For example, suppose that we want to know the probability of drawing two aces from a 52-card deck. On the first draw, $p(ace \text{ on } first \text{ } draw) = \frac{4}{52}$. If an ace is drawn on the first draw and is not replaced in the deck, the conditional probability of drawing an ace on the second draw is $p(ace \text{ on } second \text{ } draw\,|\,ace \text{ on } first \text{ } draw) = \frac{3}{51}$. The probability of drawing two aces on two draws is

$$p(two \text{ } aces) = p(ace \text{ on } first \text{ } draw)p(ace \text{ on } second \text{ } draw\,|\,ace \text{ on } first \text{ } draw)$$
$$= \left(\frac{4}{52}\right)\left(\frac{3}{51}\right) \simeq 0.0045.$$

Multiplication Rule for Statistically Independent Events

Two events A and B are **statistically independent** *if the probability of one event's occurring is unaffected by the occurrence of the other. Stated symbolically, A and B are statistically independent if* $p(A\,|\,B) = p(A)$. If this equality holds, it must also be true that

$$p(B|A) = p(B).$$

The events $p(W)$ and $p(L)$ in Table 7.3-2 are not statistically independent since $p(W|L) = 0.20$ is not equal to $p(W) = 0.30$. We can easily construct an example in which the events are independent. Consider an experiment in which a fair coin and a die are tossed. Since the coin can land in one of two ways, H or T, and the die, in one of six ways, $1, \ldots, 6$, the possible outcomes are $H1$, $T1$, $H2$, $T2$, $\ldots$, $H6$, $T6$. The sample space for the experiment is shown in Figure 7.3-4. Let event A be a head and B be a 5. The probabilities required to demonstrate independence of A and B are $p(A) = n_A/n_S = \frac{6}{12} = \frac{1}{2}$ and $p(A|B) = (n_{A \text{ and } B}/n_S)/(n_B/n_S) = n_{A \text{ and } B}/n_B = \frac{1}{2}$. Since $p(A) = p(A|B)$, the events are statistically independent; this agrees with our intuition that what happens on the toss of a die can in no way affect the outcome of tossing a coin.

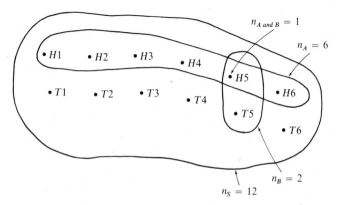

Figure 7.3-4. Euler diagram for event A, observing a head, and event B, observing a 5, when a coin and die are tossed.

For statistically independent events the multiplication rule is simplified because

$$p(A|B) = p(A)$$

and

$$p(B|A) = p(B).$$

The rule

$$p(A \text{ and } B) = p(A)p(B|A)$$
$$= p(B)p(A|B)$$

becomes

$$p(A \text{ and } B) = p(A)p(B).$$

As we just saw, the probability of observing a head and a 5 are independent; hence, the probability of their joint occurrence is $p(A \text{ and } B) = (\frac{1}{2})(\frac{1}{6}) = \frac{1}{12}$.

Multiplication Rules for Three Events

If we want to find the joint probability of three events, the multiplication rule is

$$p(A \text{ and } B \text{ and } C) = p(A)p(B|A)p(C|A \text{ and } B).$$

To illustrate, suppose that we have a box containing five marbles, each a different color: blue, green, red, yellow, and orange. Three marbles are selected from the box at random, one at a time. Once drawn, a marble is not replaced. This procedure is called *sampling without replacement*. Consider the following events.

$$A\text{—first marble is blue}$$

$$B\text{—second marble is green}$$

$$C\text{—third marble is red}$$

According to the classical view, $p(A) = n_A/n_S = \frac{1}{5}$. If a blue marble is selected first, then four remain—one of which is green. The probability of selecting a green marble given that the first is blue is $p(B|A) = \frac{1}{4}$. Similarly, the probability of selecting a red marble given that a blue one and a green one have been selected is $p(C|A \text{ and } B) = \frac{1}{3}$. The joint probability for A, B, and C is $p(A \text{ and } B \text{ and } C) = \frac{1}{5} \times \frac{1}{4} \times \frac{1}{3} = \frac{1}{60}$.

If the marbles sampled from the box are replaced (*sampling with replacement*), the multiplication rule for three events is $p(A \text{ and } B \text{ and } C) = p(A)p(B)p(C)$. This follows, since the selection of the first marble does not decrease the number of sample points in the sample space, and hence $p(B|A) = p(B)$ and $p(C|A \text{ and } B) = p(C)$. If we had replaced each marble after it was drawn from the box, the joint probability for events A, B, and C would be $p(A \text{ and } B \text{ and } C) = \frac{1}{5} \times \frac{1}{5} \times \frac{1}{5} = \frac{1}{125}$.

Common Errors in Applying the Rules of Probability

The probability rules described in this section often are used incorrectly. Some of the more common errors are the following.

1. Using the addition rule $p(A \text{ or } B) = p(A) + p(B)$ for events that are not mutually exclusive. For example, let event A be the classification "psychology major" and event B, "biology major." If $p(A) = 0.20$ and $p(B) = 0.15$, one might conclude that the probability of a student's being either a psychology or a biology major is $p(A \text{ or } B) = (0.20) + (0.15) = 0.35$. This is incorrect because some students have a double major and these students have been counted twice—once in computing $p(A)$ and once in computing $p(B)$. If $p(A \text{ and } B) = 0.03$, the correct probability is given by $p(A \text{ or } B) = p(A) + p(B) - p(A \text{ and } B) = 0.20 + 0.15 - 0.03 = 0.32$.

2. Using the addition rule when the multiplication rule should be used, and vice versa. For example, the probability of observing a 3, event A, or a 5, event B, on the toss of a die is given by $p(A \text{ or } B) = p(A) + p(B) = \frac{1}{6} + \frac{1}{6} = \frac{2}{6}$ and not by $p(A \text{ and } B) = p(A)p(B) = (\frac{1}{6})(\frac{1}{6}) = \frac{1}{36}$.

3. Using the multiplication rule $p(A \text{ and } B) = p(A)p(B)$ when the events are not statistically independent. Suppose the probability of seeing an advertisement for a product, event A, is 0.40 and the probability of buying the product, event B, is 0.30. If the dependency between A and B is ignored, the probability of both seeing an advertisement and buying the product is $p(A \text{ and } B) = (0.40)(0.30) = 0.12$. The correct probability takes into account the conditional probability of buying the product given that the ad has been seen, $p(B|A) = 0.50$, so that $p(A \text{ and } B) = p(A)p(B|A) = (0.40)(0.50) = 0.20$.

Review Exercises for Section 7.3

*13. A standard deck contains 52 cards: 10 number cards of each suit (counting the ace as a 1) and three face cards of each suit. If a card is drawn from the deck at random, what is the probability that it will be (a) an ace? (b) a heart? (c) an ace or a heart or both? (d) a heart or a spade? (e) a face card? (f) a card less than 5?

*14. Events A, B, and C are mutually exclusive and exhaustive, each having a probability of $\frac{1}{3}$. Determine (a) $p(A$ or $B)$, (b) $p(A$ or B or $C)$, *(c) $p(Not\ A)$, and (d) $p[Not(A$ or $B)]$.

*15. Events A and B are independent; $p(A) = 0.6$ and $p(B) = 0.8$. What is the probability that (a) both will occur? (b) neither will occur? (c) one or the other or both will occur?

*16. Recent highway accident statistics show that 10% of all automobile accidents and half of all fatal automobile accidents are caused by drunken drivers. Four in 1000 reported accidents are fatal. (a) Fill in the table with the appropriate probabilities. (b) What is the joint probability that a fatal accident is caused by a drunken driver?

	Fatal, F	Nonfatal, Not F	
Drunken driver, D	$p(D$ and $F) =$	$p(D$ and $Not\ F) =$	$p(D) =$
Other cause, O	$p(O$ and $F) =$	$p(O$ and $Not\ F) =$	$p(O) =$
	$p(F) =$	$p(Not\ F) =$	

*17. You ask your roommate to mail a letter. The probability that she will mail it is 0.98. The probability that the post office will fail to deliver it, given that it was mailed, is 0.15. What is the probability that the letter will be mailed and the post office will fail to deliver it?

18. For the data in the table determine if the events "attend college" and "male" are statistically independent.

	Attend college		
	Yes	No	
Male	0.30	0.20	0.50
Female	0.10	0.40	0.50
	0.40	0.60	1.00

19. Data were obtained on the incidence of rheumatic disease and the presence of grimacing in schizophrenic patients. In a sample of 1942 patients, 6% had a known history of rheumatic disease, 21.8% grimaced, and 1.8% had a history of both rheumatic disease and grimacing. (a) Fill in the table with the appropriate probabilities. (b) What is the probability of grimacing, given a history of rheumatic disease? (c) Are grimacing and rheumatic disease statistically independent?

	Grimacing	No grimacing
History of rheumatic disease		
No history of rheumatic disease		

20. A smoker has ten pipes, three of which are meerschaums. Of his six curved-stem pipes, two are meerschaums. He asks his son to bring him a curved-stem meerschaum. Since the boy doesn't know a meerschaum from other pipes, he picks up a curved-stem pipe at random. (a) Fill in the table with the appropriate probabilities. (b) What is the probability that the son picked the right pipe?

	Meerschaum, M	Other kind of pipe, O	
Curved Stem, C	$p(C$ and $M) =$	$p(C$ and $O) =$	$p(C) =$
Straight Stem, S	$p(S$ and $M) =$	$p(S$ and $O) =$	$p(S) =$
	$p(M) =$	$p(O) =$	

*21. One hundred students are enrolled in a university course. Fifty are males (M) and 50 are females (F). Of the 100 students, 60 are undergraduates (U) and 40 are graduate students (G). Of the 100, 20 are both males and undergraduates. For a student selected at random from the class, compute the following probabilities. *(a) $p(F$ and $U)$, (b) $p(F$ and $G)$, (c) $p(M$ and $G)$, (d) $p(M|U)$, (e) $p(F|G)$. [*Hint:* $p(U) = p(M$ and $U) + p(F$ and $U)$.]

22. A family of three goes to a photography studio to have a group picture made. They consider a picture good only if everyone in it looks good. Suppose the probability that any one person looks good is 0.5. Assume independence. (a) What is the probability that the picture will be good? (b) If two pictures are made, what is the probability that at least one of them will be good?

23. A survey of 100 students taking courses in algebra (A), history (H), and psychology (P) revealed the following enrollments: $A = 65$, $H = 37$, $P = 59$, A and $H = 17$, A and $P = 44$, H and $P = 14$. (a) Represent the sample space by a Euler diagram. If a student is selected at random, what is the probability that he or she is taking (b) all three subjects? (c) algebra but not psychology? (d) history but not algebra? (e) psychology but not algebra or history?

24. A drawer contains six intelligence tests (I), four personality tests (P), and five aptitude tests (A). If three tests are drawn randomly from the drawer, what is the probability that the tests are drawn in the order $I-P-A$ if each test (a) is replaced in the drawer after it is drawn? (b) is not replaced after it is drawn?

25. Terms to remember:
    a. Union
    b. Intersection
    c. Addition rule
    d. Mutually exclusive events
    e. Exhaustive events
    f. Multiplication rule
    g. Conditional probability
    h. Marginal probability
    i. Statistical independence
    j. Sampling with or without replacement

Section 7.4   Summary

Probability is an abstract mathematical concept that can be defined in a number of ways. The three most useful views of probability are the subjective–personalistic view, the classical or logical view, and the empirical relative-frequency view.

Our interest in probability is pragmatic; we want to make statements about the likelihood of observing various outcomes in experiments. An experiment is any well-defined act or process that leads to an outcome. An outcome is either a compound event that can be decomposed into simpler events, such as observing an even number on the toss of a die, or a simple event that can't be decomposed. If we assign to each simple event a point called a sample point, the possible outcomes of an experiment can be represented by a Euler diagram. The set of all sample points is called the sample space S.

Regardless of one's view, probability is based on a system of definitions and operations pertaining to a sample space. If S is the sample space for an experiment and n_S is the number of sample points in S, we can associate with each event E_i a real number called the probability of E_i or $p(E_i)$, satisfying the following properties.

1. $0 \leq p(E_i) \leq 1$, for all i

2. $\displaystyle\sum_{i=1}^{n_S} p(E_i) = 1$

3. $p(S) = 1$

These properties describe probabilities, but they don't tell us how to compute them. If we adopt the classical view, the probability of an event A is computed from the formula $p(A) = n_A/n_S$, where n_A is the number of events favoring A and n_S is the total number of equally likely events in the sample space S. This view of probability is based on logical analysis. We reason that an experiment has n_S possible outcomes, that the outcomes are equally likely, and that n_A of the outcomes favor A. If our reasoning is correct, the value we compute for $p(A)$ will agree closely with that based on the relative-frequency view.

According to the relative-frequency view, the probability of event A is the number approached by n_A/n as the total number of observations, n, approaches infinity. The estimate n_A/n is based on experience, since it is computed for a sample from the population of possible experiments. On the average, the larger the sample, the closer the estimate is to the true probability.

Probabilities for combined events can be computed by the addition rule and the multiplication rule. The addition rule states that the probability that an event will be A or B or both is $p(A \text{ or } B) = p(A) + p(B) - p(A \text{ and } B)$. For mutually exclusive events, $p(A \text{ and } B) = 0$, and the rule simplifies to $p(A \text{ or } B) = p(A) + p(B)$. The multiplication rule states that the probability that an event will be both A and B is $p(A \text{ and } B) = p(A)p(B|A) = p(B)p(A|B)$. For statistically independent events, $p(B|A) = p(B)$ and $p(A|B) = p(A)$, and the rule simplifies to $p(A \text{ and } B) = p(A)p(B)$.

8

More about Probability

Section 8.1   *Counting Simple Events*

Listing all the simple events in an experiment can be tedious. Even a small experiment, such as recording the outcome of tossing three dice, has a large sample space—in this case $6 \times 6 \times 6 = 216$ sample points. Fortunately, it is not necessary to enumerate simple events to compute probabilities. The required information can be determined using the counting rules discussed in this section.

Fundamental Counting Rule[1]

Suppose that an event can occur in n_1 ways and a second, in n_2 ways and that each of the first event's n_1 ways can be followed by any of the second's n_2 ways. Then, according to the **fundamental counting rule,** *event 1 followed by event 2 can occur in $n_1 n_2$ ways.* To illustrate, suppose that we toss a coin and then a die. The number of possible outcomes of the experiment is

$$n_1 n_2 = (2)(6) = 12,$$

since a coin can land heads or tails ($n_1 = 2$) and a die has six faces ($n_2 = 6$). The simple events are enumerated on the tree diagram of Figure 8.1-1.

The fundamental counting rule can be extended to $k > 2$ events. If there are k events (event 1 having n_1 outcomes, followed by event 2 having n_2 outcomes, and so forth), the outcome can occur in $n_1 n_2 \cdots n_k$ ways. For example, the number of possible outcomes of tossing three dice and a coin is $6 \times 6 \times 6 \times 2 = 432$.

Permutation of n Objects Taken n at a Time, $_nP_n$

Suppose that we have n different objects and that we want to find the number of different ordered sequences in which the objects can be arranged. In how many ordered sequences can the letters A, B, and C be arranged? The answer is six: *ABC, ACB, BAC, BCA, CAB, CBA.* Arranging n objects is equivalent to putting them into a big box with n ordered compartments. The first compartment can be filled in any of n ways,

1	2	3	$\cdots$	n
n ways	$n-1$ ways	$n-2$ ways		1 way

[1] Also called the *multiplication principle.*

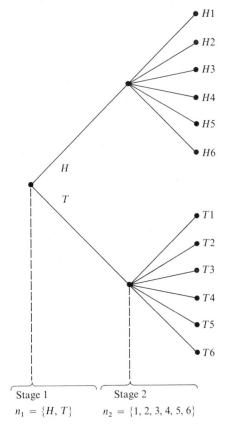

Figure 8.1-1.   Tree diagram of possible outcomes of tossing a coin and then a die.

which uses up one of the objects; the second compartment can be filled in any of $n - 1$ ways, ..., and the last compartment, in only one way. Applying the fundamental counting rule, the number of ordered arrangements of n objects is the product $n(n - 1)(n - 2) \cdots 1$. The quantity $n(n - 1)(n - 2) \cdots 1$ is denoted by the symbol $n!$, which is read "*n factorial.*"

   *An ordered sequence of n objects taken all together is called a **permutation** of the objects. The total number of such permutations, denoted by $_nP_n$, is given by*

$$_nP_n = n! = n(n - 1)(n - 2) \cdots 1.$$

The symbol $_nP_n$ is read "the permutation of n objects taken n at a time."

   Suppose that we are doing a taste preference experiment in which a panel of ten experts rates five nondairy coffee creamers. In how many ordered sequences can coffee prepared with the creamers be presented? The answer is $5! = 5(4)(3)(2)(1) = 120$. Finding ten experts willing to sit through the 120 sequences is probably impossible. However, to control for sequence effects the experiment can be performed by presenting the coffee in 12 of the sequences to one expert, 12 different sequences to another expert, and so on.

Permutation of n Objects Taken r at a Time, $_nP_r$

The number of permutations of n different objects taken r at a time, where $r \leq n$, is denoted[2] by $_nP_r$ and is equal to $n(n - 1)(n - 2) \cdots (n - r + 1)$. For example, the number of ordered sequences of five letters, A, B, C, D, E, taken three at a time is $_5P_3 = 5(5 - 1)(5 - 3 + 1) = 5(4)(3) = 60$. The rationale behind the formula is as follows. Consider the box

1	2	3
n ways	$n - 1$ ways	$n - (r - 1)$ ways

with $r = 3$ ordered compartments. The first compartment can be filled in any of $n = 5$ ways and the second, in $n - 1 = 4$ ways. When we come to the $r = 3$rd compartment, we have used $r - 1 = 2$ of the n letters so that $n - (r - 1) = n - r + 1 = 3$ letters are left to fill the last compartment. According to the fundamental counting rule, the number of ordered sequences is the product of $n(n - 1) \cdots (n - r + 1)$. Therefore, the number of ordered sequences of the five letters taken three at a time is $5(4)(3) = 60$.

The right side of the formula $_nP_r = n(n - 1)(n - 2) \cdots (n - r + 1)$ is often written $n!/(n - r)!$.[3] To see that the expressions are equivalent, note that the next term in $n(n - 1)(n - 2) \cdots (n - r + 1)$ is $(n - r)$, followed by $(n - r - 1) \cdots (3)(2)(1)$. The terms $(n - r)(n - r - 1) \cdots (3)(2)(1)$ can be denoted by $(n - r)!$. If we multiply $n(n - 1) \cdots (n - r + 1)$ by $(n - r)!/(n - r)!$, we obtain the more familiar formula $_nP_r = n!/(n - r)!$ since $[n(n - 1) \cdots (n - r + 1)](n - r)! = n!$.

The taste-preference experiment described earlier could be performed using the method of paired comparisons. In this method an expert sips first one and then a second cup of coffee prepared with two of the creamers and indicates a preference. This procedure is repeated until each creamer has been compared twice with every other creamer, once in the first cup sipped and once in the second cup. In how many ordered sequences can five creamers be presented two at a time? The answer is given by

$$_5P_2 = 5(5 - 2 + 1) = 5(4) = 20$$

or

$$_5P_2 = \frac{5!}{(5 - 2)!} = \frac{(5)(4)(3)(2)(1)}{(3)(2)(1)} = 20.$$

The method of paired comparisons would require each expert to make 20 judgments to compare each creamer with every other creamer.

Permutation of n Objects When Some of the Objects Are Alike, $_nP_{r_1, r_2, \ldots, r_k}$

Suppose that we have n objects that can be divided into k subsets such that each subset contains like objects, but the objects in any given subset are different from those in the other subsets. Let r_1 be the number of like objects of one kind and

---

[2] Also denoted by P_r^n, $P(n, r)$, and $(n)_r$.
[3] In computations involving $(n - r)!$, remember that $1! = 1$ and that by definition $0! = 1$.

r_2 be the number of like objects of another kind, and so on for the k subsets, with $r_1 + r_2 + \cdots + r_k = n$. *The number of permutations of n objects in which $r_1, r_2, \ldots, r_k$ are alike is denoted by $_nP_{r_1, r_2, \ldots, r_k}$ and is equal to $n!/(r_1!r_2! \cdots r_k!)$.* For example, the number of ordered sequences of the ten letters in *PSYCHOLOGY* is

$$_{10}P_{1,1,2,1,1,2,1,1} = \frac{10!}{1!1!2!1!1!2!1!1!} = 907,200.$$

Combination of n Objects Taken r at a Time, $_nC_r$

*Sometimes we are not interested in the number of ordered sequences or permutations of n objects taken r at a time, but instead in the number of different sets of r objects that can be selected from n when order is ignored. This is referred to as the **combination** of n objects taken r at a time and is denoted[4] by $_nC_r$. The formula for $_nC_r$ is*

$$_nC_r = \frac{n!}{r!(n-r)!}.$$

The rationale for the formula is as follows. Consider four letters A, B, C, and D taken two at a time. The number of ordered sequences is $_4P_2 = 4!/(4-2)! = 12$. But suppose that we don't want to distinguish AB from BA, CB from BC, and so on. We note that any sequence of $r = 2$ objects can be permuted in $r! = 2(1) = 2$ ways. If we want to ignore the order of the r objects in $_nP_r$, we can divide $_nP_r$ by $r!$, which gives

$$_nC_r = \frac{_nP_r}{r!} = \frac{\dfrac{n!}{(n-r)!}}{r!} = \frac{n!}{r!(n-r)!}.$$

The number of different sets of $r = 2$ letters that can be selected from $n = 4$ letters, A, B, C, D, is

$$_4C_2 = \frac{4!}{2!(4-2)!} = \frac{4(3)(2)(1)}{2(1)[2(1)]} = 6.$$

The six sets are as follows: AB, AC, AD, BC, BD, CD. Because the order of letters in a pair is of no interest, the six could just as well have been written BA, CA, AD, CB, BD, CD.

The combination of n objects taken r at a time will be used in Chapter 9 to develop the binomial distribution, which describes the possible outcomes of a particular kind of experiment.

Review Exercises for Section 8.1

*1. Determine the number of possible outcomes for the following. (a) Three coins are tossed. (b) Four dice are tossed. (c) A coin and a die are tossed.

2. If there are three candidates for governor and five for mayor, in how many ways can the two offices be filled?

---

[4] Also denoted by C_r^n, $C(n, r)$, and $\binom{n}{r}$

3. The Greasy Spoon menu offers a choice of five appetizers, four salads, eight entrees, seven vegetables, and nine desserts. If a meal consists of one each, in how many ways can you select a dinner?

*4. The four Russian novels *War and Peace, Anna Karenina, Crime and Punishment,* and *The Brothers Karamazov* are to be placed on a shelf. In how many ordered sequences can the books be arranged?

5. Two different psychology books, four different statistics books, and three different sociology books are to be arranged on a shelf. (a) In how many ordered sequences can the books be arranged? (b) If the books in each subject must be kept together, how many ordered sequences are possible?

*6. In how many different ways can ten people be seated four at a time on a bench with only four seats?

7. How many ordered arrangements can be made for the letters in the word *statistics?*

8. If on a statistics examination consisting of 12 questions a student may omit five, in how many ways can the student select the problems to answer?

9. (a) In how many ways can six people be seated in a row at the head banquet table? (b) Suppose the people are to be seated in pairs at separate tables; in how many ways can they be seated if we consider the arrangement *AB* to be different from *BA*? In how many ways if we consider *AB* to be equivalent to *BA*?

*10. Given nine areas from which to choose, in how many ways can a student select (a) a major-minor area? (b) a major and first and second minors? (c) a major and two minors if it is not necessary to designate the order of the minors?

11. How many different committees of three men and four women can be formed from eight men and six women?

12. Terms to remember:
    a. Permutation of *n* objects
    b. Combination of *n* objects

Section 8.2   Summary

This chapter has presented rules developed to simplify the determination of the number of simple events in an experiment. The key rules are as follows.

1. Fundamental counting rule. If there are k events, event 1 followed by event 2, ..., followed by the kth event, the outcome can occur in $n_1 n_2 \cdots n_k$ ways.

2. Permutation of n objects taken n at a time. The number of ordered sequences of n objects taken all together is $_nP_n = n! = n(n-1)(n-2) \cdots 1$.

3. Permutation of n objects taken r at a time. The number of ordered sequences of n objects taken r at a time is $_nP_r = n!/(n-r)!$.

4. Permutation of n objects when some of the objects are alike. The number of ordered sequences of n objects in which $r_1, r_2, \ldots, r_k$ are alike is $_nP_{r_1, r_2, \ldots, r_k} = n!/(r_1! r_2! \cdots r_k!)$.

5. Combination of n objects taken r at a time. The number of sequences, ignoring order, of n objects taken r at a time is $_nC_r = n!/[r!(n-r)!]$.

9

Random Variables and Probability Distributions

Section 9.1   Random Sampling

Inferential statistics are used in reasoning from a sample to the population. Some samples provide a sound basis for this process, while others don't. The difference lies in the method by which the samples are selected. *The method of drawing samples from a population so that every possible sample of a particular size has the same probability of being selected is called* **random sampling** *and the resulting sample, a* **random sample.** As the definition indicates, randomness is a property of the procedure rather than of the particular sample obtained. By the term "random sample" we mean simply a sample produced by a random sampling procedure. Any sampling method based on haphazard or purposeless choices such as utilizing volunteers, students enrolled in a psychology course, or every tenth name in an alphabetical listing is called *nonrandom sampling*. The resulting samples, unlike random samples, don't provide a sound basis for deducing the properties of populations.

As we will see, the inferential procedures described in subsequent chapters assume either random sampling or random assignment of the various conditions of an experiment to subjects.[1] If random sampling is used, there is no guarantee that a particular random sample will resemble the population, but in the long run, random samples are more likely to do so than nonrandom samples. Random assignment ensures that systematic bias isn't introduced, as it would be, for example, if the best subjects were unwittingly assigned to the experimental conditions that are expected to be superior.

Defining the Population

The first step in drawing a random sample is to identify the population. *A* **population** *is the collection of all objects or observations having one or more specified characteristics.* The population is identified when we specify the common characteristics, for example, this year's freshmen at Oregon State University or the outcomes of tossing a die for eternity. *A single object or observation is called an* **element** *of the population.* The elements of the population can be *finite*[2] (or limited) in

---

[1] Random assignment is discussed in Section 13.3.

[2] The probability of drawing a particular sample from a finite population is given by $_rC_r/_nC_r$ (see Section 8.1), where r denotes the sample size and n, the population size. For example, the probability for $r = 2$ and $n = 100$ is $\{2!/[2!(2 - 2)!]\}/\{100!/[2!(100 - 2)!]\} = 1/19,800$.

number, as this year's freshmen at Oregon State, or *infinite* in number, as the outcomes of tossing a die for eternity.

In practice, it is difficult to obtain a random sample from large populations like residents of a city or students at a university. There are two obstacles—obtaining an accurate list of the population elements and securing their participation once they have been selected. Some cities have lists of their residents, but unfortunately the information isn't updated frequently. Telephone directories are more current but exclude certain segments of society more often than others. Both introduce systematic bias into the experiment. An investigator faced with the choice may prefer to redefine the population to fit the more current list. Instead of all city residents, the population is defined as all families in the telephone directory.

Sampling with or without Replacement

After identifying the population, one must decide whether to sample with replacement or without replacement. In sampling **with replacement,** *a sampled element is returned to the population so that it is available to be drawn again; in sampling* **without replacement,** *the element is not replaced and hence can be drawn only once.*[3] Sampling with replacement is rarely appropriate for the kinds of problems investigated in the behavioral sciences and education, since the sampled elements may be significantly and permanently altered by participating in the experiment. For example, once a child has learned an arithmetic unit, that child is no longer a naive learner with respect to the unit; once tissue has been surgically removed, it can't be removed again should the organism happen to be sampled a second time.

Random Sampling Procedures

A variety of procedures can be used to draw a random sample. If the population is finite, each element can be identified on a slip of paper, the slips placed in a container, thoroughly mixed, and then drawn blindly from the container. If sampling with replacement is used, the identity of an element selected is noted and the slip is returned to the container; it is then available to be drawn again. The blind drawing-of-slips procedure seems simple enough, but it isn't always random—as witness the December 1969 draft lottery. More slips containing birth dates in the last 6 months of the year were drawn early than were dates in the first 6 months, much to the dismay of Virgos, Libras, Scorpios, and Sagittarians. The result was attributed to placing the slips in the bowl in chronological order and failing to shake them up thoroughly. Slips for the later months were the last ones in the bowl and the first ones drawn.

Another technique for drawing a random sample is to flip a coin or spin a roulette wheel, with the outcome of the random device determining whether an element is or isn't included in the sample. This procedure is practical for selecting a small sample, but becomes tedious for larger ones.

Most researchers prefer to use a table of random numbers to draw their samples. Random number tables like Appendix Table D.1 were prepared so that

---

[3] The number of samples of size r that can be drawn without replacement from a population of size n is $_nC_r$. The number of samples with replacement is $n_1 n_2 \cdots n_r$.

integers from 0 to 9 occur with about equal frequency and appear in the table in a random fashion. The digits are usually in groups of two or more to make them easier to read, but the grouping has no other significance.

Suppose that we want a random sample of 30 speech therapy majors. A printout listing 273 majors constituting the population is obtained from the computer center, and the students are numbered serially from 001 to 273. We turn to Table D.1 and note that it has two pages with 50 rows and 25 columns each. To decide where to begin in the table we close our eyes and drop our pencil on the table. Suppose it lands on the second page with the point closest to the first number in row 21 and column 13. The numbers reading from left to right are 22 00 20 35 55 . . . . We let the first number, 2, identify the table page on which we begin; the next two digits, 20, the row; and the next two digits, 02, the column. Having previously decided to read the numbers from left to right, although any sequence can be used, we proceed to draw our sample. Numbers are read in groups of three, until we obtain 30 numbers between 001 and 273, inclusive. To sample without replacement, we ignore numbers after their first appearance. The students corresponding to the 30 numbers compose the sample.

In sampling from a list with many pages, such as a telephone directory or a student directory, it isn't necessary to number each population element if the number of names on each page is about the same. Instead, each page and each position on the page is numbered. To select an element, two numbers are drawn from a random number table; the first identifies the directory page and the second, the position of the element on the page. Another procedure, called *systematic sampling*, is sometimes used for sampling from a list. It involves sampling every nth element, say, every 20th person, in the list. Despite the simplicity of this procedure, it can't be recommended, because it doesn't satisfy the definition of random sampling.

Review Exercises for Section 9.1

*1. List the steps involved in drawing a random sample.

 2. What advantages do random samples have over nonrandom samples?

*3. *(a) How many samples of size 5 can be drawn without replacement from a population of size 50? (b) How many samples of size 5 can be drawn with replacement from a population of size 50?

*4. A sample of four supermarkets is to be selected from a total of eight in a small town. (a) How many different random samples without replacement can be drawn? (b) What is the probability that a given sample will be selected? (c) How many random samples with replacement can be drawn?

 5. A sample of five students is to be selected from a class of ten students. (a) How many different random samples without replacement can be drawn? (b) What is the probability that a given sample will be selected? (c) How many random samples with replacement can be drawn?

 6. Use a random number table to draw two random samples of ten students from the following population. For one sample use sampling with replacement and for the other, sampling without replacement. (b) Describe in detail how you used the table.

Debbie	Ed	Bob	Mac	Joe	Jim
Jack	Joan	Roger	Les	Jackie	Mike
Bill	Larry	Helen	Bud	John	Dave

7. Terms to remember:
   a. Random sampling
   b. Sampling with or without replacement
   c. Systematic sampling

Section 9.2   *Random Variables and Their Distributions*

Random Variables

In rolling a pair of dice we can observe the total number of dots; in tossing a coin three times, the total number of heads; in observing a naive rat in a three-choice T maze, the total number of incorrect turns. The variable, number of dots or number of heads or number of incorrect turns, is called a *random variable* because the value of the variable for a particular experiment is determined by chance. In the dice example, the random variable, number of dots, can assume values of 2, . . . , 12; in the T maze example, the random variable can assume values of 0, 1, 2, or 3 errors. Random variables usually are denoted by capital letters toward the end of the alphabet, such as X, Y, or Z. It helps to think of a random variable as the name for the number associated with the outcome of a random experiment before the experiment is performed. Performing the experiment converts the random variable into some specific number.

You may be wondering, why the fancy name? How does a random variable differ from just a plain old variable? We may contrast the two kinds of variables as follows.

1. The variable X is a name for any one of a set of permissible values.
2. The random variable X is a name for any one of a set of permissible numerical values of a random experiment.

Let's pursue the meaning of a random variable a bit further. In Section 7.2 we saw that all the possible outcomes of a random experiment can be represented by points in a sample space. A random variable associates one and only one numerical value with each point; hence, in the language of the mathematician, a random variable is a *function*. To understand this, recall from algebra that a function consists of two sets of elements and a rule that assigns to each element in the first set one and only one element in the second set. The definition of a function is quite general; $\{(a, 1), (b, 5), (c, 6)\}$ is a function, as are $\{(Jack, tall), (Bill, medium), (Debbie, medium)\}$ and $\{(No errors, 0), (One error, 1), (Two errors, 2), (Three errors, 3)\}$. More simply stated, a function is a set of ordered pairs of elements, no two of which have the same first element. If the second element of a pair is a number, the function is said to be numerically valued. A random variable associates one and only one number with each point in a sample space; thus, it is a numerically valued function defined over a sample space. Most readers will find the following definition easier to remember. A random variable is a numerical quantity whose value is determined by the outcome of a random experiment.

Random variables are classified according to the nature of the numbers they can assume. *A random variable is **discrete** if its range can assume only a finite number of values or an infinite number of values that is countable*; for example, family size, number of dates per week, or scores on a test. *A random variable is **continuous** if its range is uncountably infinite*; for example, temperature in Monterey, California, duration of a kiss, or height. We must distinguish between the values the random variable can assume and those yielded by our measuring instruments. A thermometer is usually calibrated in $1°$ steps, a stop watch in 0.1 sec, and a ruler in $\frac{1}{16}$ in. Consequently, our measurement of continuous random variables is always approximate.

Distribution of a Discrete Random Variable

We learned in Chapter 2 that a frequency distribution associates a frequency with each value or class interval of a variable. *A similar representation that associates a probability with each value of a random variable is called a **probability distribution.*** A probability distribution for an experiment of tossing a die is shown in Table 9.2-1, and a graph in which probability is represented by the height of rectangles is shown in Figure 9.2-1. The distribution is said to be uniform, since each value of the random variable has the same probability. Notice that the probabilities sum to 1, because the events $X = 1, \ldots, 6$ are mutually exclusive and collectively exhaustive.

Table 9.2-1.   Probability distribution for outcome of tossing a die

Possible values, r, of the random variable X	$p(X = r)$
1	1/6
2	1/6
3	1/6
4	1/6
5	1/6
6	1/6

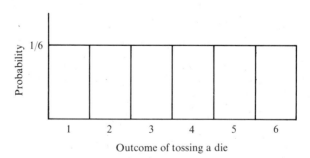

Figure 9.2-1.   Histogram for probability distribution in Table 9.2-1.

Consider next the three-choice T maze experiment mentioned earlier. Suppose that the correct series of turns is *right, left, right* (R, L, R). We know from the fundamental counting rule in Section 8.1 that a rat can traverse the maze in $2 \times 2 \times 2 = 8$ ways, since three right–left choices must be made. The eight ways and the number of errors associated with each are listed in the table. The probability of

	Number of errors
R, L, R	0
R, R, R	1
R, L, L	1
L, L, R	1
R, R, L	2
L, R, R	2
L, L, L	2
L, R, L	3

making 0, 1, 2, or 3 errors—the random variable—can be computed by $p(X = r) = n_r/n_s$, where n_r is the number of events (maze routes) favoring r errors and n_s is the number of events. For example, the probability of one error is $\frac{3}{8} = 0.375$. The probability distribution is given in Table 9.2-2 and a graph of the distribution, in Figure 9.2-2. The table and figure can be used to answer any question about the probability associated with the random variable; for instance, the probability that X is odd: $p(X = 1, 3) = 0.375 + 0.125 = 0.5$, or the probability that X is less than 3: $p(X < 3) = 0.125 + 0.375 + 0.375 = 0.875$.

Table 9.2-2. Probability
distribution for number of errors
in a three-choice T maze

Possible values, r, of random variable X	p(X = r)
0	0.125
1	0.375
2	0.375
3	0.125

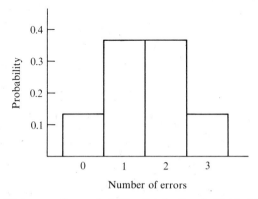

Figure 9.2-2. Histogram for probability distribution in Table 9.2-2.

A probability distribution is similar to a frequency distribution. The probability distribution associates a probability with each value of a variable, while the frequency distribution associates a frequency with each value of the variable.

We saw in Chapter 3 that the arithmetic mean often is used to describe the central tendency of a frequency distribution. A similar index of the central tendency of a probability distribution is called an *expected value.*[4] We now turn to the subject of how to compute expected values.

Expected Value of a Discrete Random Variable

If an extremely large number of naive rats were to run the three-choice T maze, how many errors on the average would we expect them to make? Stated more formally, what is the *expected value* of that random variable? *If X is a discrete random variable that assumes values X_1, $X_2, \ldots, X_n$ with probabilities $p(X_1)$, $p(X_2), \ldots, p(X_n)$, then the expected value of X denoted by $E(X)$ is defined as*[5]

$$E(X) = p(X_1)X_1 + p(X_2)X_2 + \cdots + p(X_n)X_n = \sum_{i=1}^{n} p(X_i)X_i,$$

where $p(X_1) + p(X_2) + \cdots + p(X_n) = 1$. For the T maze example, $E(X)$ for the values in Table 9.2-2 is

$$E(X) = 0.125(0) + 0.375(1) + 0.375(2) + 0.125(3) = 1.5.$$

Thus, we would expect a rat to make on the average 1.5 errors in the maze. Note the similarity between the formula for $E(X)$ and that for the mean of an ungrouped frequency distribution:

$$\bar{X} = \frac{f_1}{n} X_1 + \frac{f_2}{n} X_2 + \cdots + \frac{f_k}{n} X_k.$$

Here, X is a discrete variable that assumes values $X_1, X_2, \ldots, X_k$ with frequencies $f_1, f_2, \ldots, f_k$, where $f_1 + f_2 + \cdots + f_k = n$; $\bar{X}$ and $E(X)$ differ in that $\bar{X}$ is the mean of a sample defined by its frequency distribution; $E(X)$ is the mean of a theoretical population defined by its probability distribution. The latter mean is also denoted by μ.

Originally, the expected value concept was used in games of chance to tell a player what the long-run average loss or gain per play would be. Consider the popular casino game of roulette. A player places a bet, the wheel is spun, and the ball is set in motion. The ball can drop into one of 38 slots; 36 slots are numbered from 1 to 36, with half red and half black, and two green slots are numbered 0 and 00. Suppose a player places $1 on number 7. If the ball drops into the 7 slot, the player receives a $35 payoff; otherwise, the bet is lost. We can calculate the player's expected winnings as shown in Table 9.2-3. According to the table, if the player keeps making $1 bets indefinitely, an average loss of 5.3¢ per bet will be incurred. On any given gamble, the player stands to either win $35 or lose only $1. What the player may choose to ignore is that on the average $35 is won in only 1 out of 38 gambles, while $1 is lost in 37 out of 38.

The term *expected value* is misleading in one sense because $E(X)$ is often not one of the possible outcomes of an experiment. In the T maze example, $E(X) = 1.5$,

---

[4] The terms *expected value* and *expectation* are synonymous.
[5] This definition of $E(X)$ doesn't commit us to a particular view of probability, since the $p(X_i)$'s can be subjective, logical, or empirical.

Table 9.2-3.   Expected value of a bet

Possible winnings, X_i	$p(X_i)$	$p(X_i)X_i$
+ \$35	$\frac{1}{38}$	$\frac{1}{38}(\$35) = \$\frac{35}{38}$
− \$1	$\frac{37}{38}$	$\frac{37}{38}(-\$1) = -\$\frac{37}{38}$

$$E(X) = \sum_{i=1}^{n} p(X_i)X_i = -\$\tfrac{2}{38} = -0.053$$

but the possible values of the random variable are 0, 1, 2, or 3 errors. Similarly, the gambler can win \$35 or lose \$1 on any given play, although $E(X) = -5.3\cent$. In both examples $E(X)$ is an average result, and in this respect it is like a sample mean, $\bar{X}$.

Expected Value of a Continuous Random Variable

Computing the expected value of a discrete random variable is fairly simple because we need only multiply probabilities by random variable values and sum the products—that is, $E(X) = \sum_{i=1}^{n} p(X_i)X_i$. The continuous random variable case is more complicated because the variable can assume an infinite number of values. The probability that a continuous random variable X has a particular value is zero.[6] Consequently, instead of referring to the probability that X has a particular value, we refer to the probability that X lies in an interval between two values of the random variable. This notion is illustrated in Figure 9.2-3. The expected value of a continuous random variable X is the sum of the products formed by multiplying each value that X can assume by the height of the probability distribution curve above that value of X. Since X can assume an infinite number of values, its expected value is not computed by actually physically multiplying each X by the

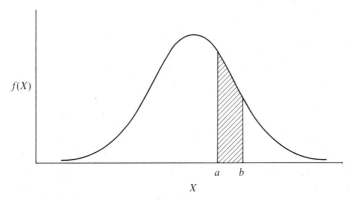

Figure 9.2-3.   The probability that X will assume a value between a and b is equal to the area under the curve between those two points. For many random variables, tables are available that simplify determining the area between two points (see Section 10.1).

---

[6] This may not be obvious. For a discussion of this point, see Hays (1981, pp. 93–96).

height of the curve at X but instead by means of the integral calculus.[7] As we will see, tables for most random variables of interest have been prepared; these tables simplify the calculation of the probability that X lies in an interval.

Standard Deviation of a Discrete Random Variable

In Chapter 4 we learned that the standard deviation is a useful measure of dispersion. For a sample defined by an ungrouped frequency distribution, the standard deviation is given by $S = \sqrt{\sum f_j(X_j - \bar{X})^2/n}$. A similar measure of dispersion for a theoretical population defined by the probability distribution of a discrete random variable is given by

$$\sigma = \sqrt{E\{[X - E(X)]^2\}} = \sqrt{\sum p(X_i)[X_i - E(X)]^2}.$$

We note from the formula on the left that σ is the square root of the expected value of a squared deviation, $[X - E(X)]^2$. We compute this expected value in the same way as we did for $E(X)$, where we multiplied each value of X_i by its probability. To compute $E\{[X - E(X)]^2\}$, we multiply each $[X_i - E(X)]^2$ by its probability $p(X_i)$ and sum the products. The computation of σ is illustrated for the T maze data in Table 9.2-2: $E(X) = 1.5$, and the standard deviation is

$$\sigma = \sqrt{0.125(0 - 1.5)^2 + 0.375(1 - 1.5)^2 + 0.375(2 - 1.5)^2 + 0.125(3 - 1.5)^2}$$

$$= 0.866.$$

The symbol σ is used instead of S, since this standard deviation is a population parameter. The value $\sigma = 0.866$ together with $E(X) = 1.5$ provides a useful summary of the theoretical population.

Review Exercises for Section 9.2

8. Distinguish between a variable and a random variable.
*9. *(a) Construct a probability distribution for a four-choice T maze. Assume that the correct series of turns is *right, right, left, right*. (b) Graph the probability distribution.
*10. Let the random variable X be the number of cars per household. Suppose that in Waco, Texas, X has the probability distribution listed in the table. For a household selected

X	0	1	2	3	4	5
$f(X)$	0.16	0.54	0.23	0.05	0.01	0.01

at random, compute the following.
   a. $p(X \le 2)$ b. $p(X \ge 3)$ c. $p(1 \le X \le 2)$
   d. $E(X)$ e. σ
11. Let the random variable X be the number of children in a family. Suppose that X has the probability distribution listed in the table. For a family selected at random, compute

X	0	1	2	3	4	5	6	7
$f(X)$	0.40	0.18	0.15	0.11	0.09	0.05	0.01	0.01

---

[7] For those familiar with the calculus, the expected value is $E(X) = \int_{X_{min}}^{X_{max}} xf(x)\, dx$, where X_{min} is the smallest value of X and X_{max} is the largest value.

the following.

a. $p(X = 0)$ b. $p(X \geq 4)$ c. $p(X < 3)$

d. $p(2 \leq X \leq 5)$ e. $E(X)$ f. σ

12. How does an expected value differ from the mean of a frequency distribution?

*13. What is the maximum you should be willing to pay to enter a game in which you can win $30 with probability 0.6 and $10 with probability 0.4? (*Hint:* Compute $E(X)$.)

14. If it rains, a fortuneteller loses $12 per day; if it is fair, she earns $110 per day. Assume that the probability of rain is 0.3. What are her expected earnings per day?

15. The random variable X has the probability distribution listed in the table.

X	0	1	2	3	4
$f(X)$	0	$\frac{2}{5}$	$\frac{1}{5}$	$\frac{1}{5}$	$\frac{1}{5}$

(a) Compute $E(X)$. (b) Compute σ.

*16. Suppose that a fraternal organization plans to sell 1000 lottery tickets for $1 each. The prize is a $750 video tape recorder. (a) If you purchase a ticket, what is the probability that you will win? (b) What is your expected gain? Remember to subtract the cost of the ticket from the value of the prize. (c) Does it make economic sense to purchase a ticket? (d) What is the maximum that you should be willing to pay for a ticket? (*Hint:* The maximum you should be willing to pay for a ticket is that amount for which $E(X) = 0$, that is, the amount for which there is no gain or loss over the long run. This amount, denoted by T, can be determined from

$$p(\text{win})(\text{gain value}) + p(\text{lose})(\text{loss value}) = 0,$$

where the gain value is equal to $[750 + (-T)]$ and the loss value is equal to $-T$.)

17. Terms to remember:

a. Discrete random variable   b. Continuous random variable

c. Probability distribution   d. Expected value

Section 9.3   Binomial Distribution

Bernoulli Trial

Many experiments have only two possible outcomes: a new drug is effective or it isn't, an animal takes the correct turn or the wrong turn, a job is given to an applicant or it isn't. These experiments have much in common with the simpler coin-tossing experiment. In each, the random variable is discrete and can assume only two values, often denoted "success" and "failure." Flipping a coin once and noting whether it landed heads or tails or randomly sampling one person from a population of former students and noting whether he or she graduated is called a *Bernoulli trial* or *Bernoulli experiment*.[8] Our interest is usually in several Bernoulli trials. We toss a coin n times and note the number of heads, or we randomly sample n persons and note the number of graduates. The probability that on any given trial we will observe a success is denoted by p and the probability of a failure, by q. Since the two outcomes are mutually exclusive and exhaustive, $p + q = 1$. When there are n Bernoulli trials, the random variable of interest is the number of successes; its value can range from 0 to n.

---

[8] After James Bernoulli (1654–1705), who discussed such trials in his *Ars Conjectandi* (1713).

The characteristics of a Bernoulli trial are as follows.

1. A trial can result in one of two outcomes.
2. The probability of a success remains constant from trial to trial.
3. The outcomes of successive trials are independent.

Very few real-life situations will perfectly satisfy the requirements. Strictly speaking, the last two are satisfied only when sampling is done with replacement or from an infinite population. In most research, sampling is done without replacement from a finite population. This practical departure from the ideal is of little consequence as long as the population is large relative to the sample size.

Below we describe a binomial distribution in which the random variable is a sum—the number of successes observed on $n \geq 2$ Bernoulli trials. *A probability distribution where the random variable is a statistic based on the results of $n \geq 2$ trials is given a special name—sampling distribution.* For convenience, we will examine a relatively simple binomial distribution here and defer discussion of the special properties of sampling distributions to Chapter 10.

The binomial distribution will be encountered repeatedly in subsequent chapters. It is the theoretical model for a variety of statistics, as we will see in Sections 12.4, 14.2, 15.4, 17.2, 17.3, and 17.4.

Binomial Distribution

*The number of successes observed on $n \geq 2$ identical Bernoulli trials is called a **binomial random variable,** and its probability distribution is called a **binomial distribution.**[9] Suppose we toss a fair coin five times. The probability of observing exactly r heads (successes) in n tosses is given by the function rule*

$$p(X = r) = {}_nC_r p^r q^{n-r},$$

where $p(X = r)$ is the probability that the random variable X equals r heads, ${}_nC_r$ is the combination of n objects taken r at a time,[10] p is the probability of success (a head), and $q = 1 - p$. For example, the probability that the random variable X equals four heads is

$$p(X = 4) = {}_5C_4 \left(\frac{1}{2}\right)^4 \left(\frac{1}{2}\right)^{5-4} = \frac{5!}{4!(5-4)!}\left(\frac{1}{2}\right)^4\left(\frac{1}{2}\right)^1 = \frac{5}{32}.$$

The complete probability distribution is given in Table 9.3-1 and a graph of the distribution, in Figure 9.3-1. The probability that X equals or exceeds some value or that it lies in a given interval can be obtained by combining probabilities from

---

[9] So named because the probabilities associated with the distribution can be obtained by raising a binomial (an algebraic expression containing two terms) to the nth power. For example,

$$(p + q)^n = p^n + np^{n-1}q + \frac{n(n-1)}{2(1)}p^{n-2}q^2 + \cdots + q^n,$$

where p is the probability of success, $q = 1 - p$, and n is the number of Bernoulli trials. The first term, p^n, gives the probability of n successes; the second term, $np^{n-1}q$, the probability of $n - 1$ successes, and so on.

[10] Discussed in Section 8.1.

Table 9.3-1. Binomial distribution for
$n = 5$ and $p = \frac{1}{2}$

Number of heads, r	0	1	2	3	4	5
$p(X = r)$	$\frac{1}{32}$	$\frac{5}{32}$	$\frac{10}{32}$	$\frac{10}{32}$	$\frac{5}{32}$	$\frac{1}{32}$

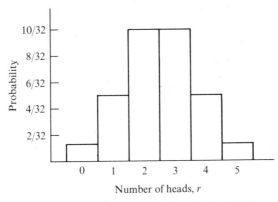

Figure 9.3-1. Histogram for the binomial distribution in Table 9.3-1.

the table or figure. For example, the probability of obtaining four or more heads in five tosses of a fair coin is

$$p(X \geq 4) = p(X = 4) + p(X = 5) = \frac{5}{32} + \frac{1}{32} = \frac{6}{32}.$$

The probability distribution of a binomial random variable is completely specified by n, the number of trials, and the parameter p, the probability of success. When $p < 0.5$, a graph of the probability distribution is positively skewed; for $p = 0.5$ it is symmetrical and for $p > 0.5$ it is negatively skewed. As n increases, the shape of the distribution approaches more and more closely that of the normal bell-shaped distribution. It should be apparent that the binomial distribution is actually a family of distributions, one for each set of p and n values. The thread that binds the distributions into a family is their common function rule, $p(X = r) = {}_nC_r p^r q^{n-r}$.[11] The following example illustrates another member of the binomial family.

Suppose that we are interested in the probability that more than half of a random sample of six patients will show improvement following treatment. Let the probability of improvement for any patient equal 0.7. The probability of observing exactly six successes in six patients is given by

$$p(X = 6) = {}_6C_6(0.7)^6(0.3)^0 = \frac{6!}{6!(6 - 6)!}(0.70)^6(0.30)^0 = 0.118.$$

[11] Other examples of families of discrete probability distributions are the uniform distribution, multinomial distribution, hypergeometric distribution, Poisson distribution, and negative binomial distribution. The first three distributions are discussed in this text. For a discussion of the latter two distributions see Hays (1981, chap. 3), Stilson (1966, chap. 10), and Walpole (1968, chap. 6).

Table 9.3-2. Distribution showing probability of improvement following treatment

Number improved, r	0	1	2	3	4	5	6
$p(X = r)$	0.001	0.008	0.059	0.185	0.324	0.302	0.118

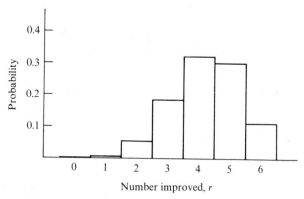

Figure 9.3-2. Histogram for probability that patients will show improvement following treatment.

The complete probability distribution is given in Table 9.3-2 and graphed in Figure 9.3-2. The probability that in a random sample of six patients more than half will show improvement is given by

$$p(X \geq 4) = p(X = 4) + p(X = 5) + p(X = 6) = 0.324 + 0.302 + 0.118 = 0.744.$$

Expected Value and Standard Deviation of Binomial Distribution

The expected value of a discrete random variable always can be computed from $E(X) = \sum_{i=1}^{n} p(X_i)X_i$. For a binomial random variable there is a simpler formula for computing the expected value of X (number of successes):[12]

$$E(X) = np,$$

where n is the number of trials and p is the probability of a success on any trial. For the probability distribution in Table 9.3-2, the expected number of patients showing improvement is $E(X) = 6(0.7) = 4.2$. The same result is obtained using the longer formula $E(X) = \sum_{i=0}^{6} p(X_i)X_i = 0.001(0) + 0.008(1) + \cdots + 0.118(6) = 4.2$.

The standard deviation of a binomial distribution is given by $\sigma = \sqrt{npq}$. For the probability distribution in Table 9.3-2 the standard deviation is

$$\sigma = \sqrt{6(0.7)(0.3)} = 1.12.$$

[12] A derivation of the formula is given by Hays (1981, pp. 154–155).

Multinomial and Hypergeometric Distributions

As we have seen, the binomial distribution is the appropriate model when (1) randomly sampling from a population with elements that belong to one of two classes and (2) the probability of obtaining an element in a class remains constant from trial to trial, as when sampling with replacement or from an infinite population. In subsequent chapters we will refer to two other models that apply when one or both of these conditions aren't satisfied. These models—the multinomial and the hypergeometric distributions—are described next.

Consider first an experiment in which a trial results in an outcome from one of $k \geq 2$ classes and the probabilities associated with the classes remain constant. Such an experiment is called a *multinomial experiment,* and its associated probability distribution is called a *multinomial distribution.*[13] *If n observations are made independently and at random from $k \geq 2$ mutually exclusive and exhaustive classes with probabilities $p_1, p_2, \ldots, p_k$, then the probability that exactly n_1 will be of kind 1, n_2 will be of kind 2, ..., and n_k will be of kind k, where $n_1 + n_2 + \cdots + n_k = n$, is given by the* **multinomial functional rule:**

$$p(X_1 = n_1 \text{ and } X_2 = n_2 \text{ and } \cdots \text{ and } X_k = n_k) = \frac{n!}{n_1! n_2! \cdots n_k!} (p_1)^{n_1}(p_2)^{n_2} \cdots (p_k)^{n_k}.$$

When $k = 2$, this rule is the same as the binomial function rule because then $n_1 = r$, $n_2 = n - r$, $p_1 = p$, and $p_2 = 1 - p_1 = q$. A multinomial experiment is simply an extension of a binomial experiment and applies when there are two or more classes.

The computation of probabilities using the multinomial rule is not too difficult when n and the number of categories is small. Suppose that we want to know the probability of drawing 2 red, 1 white, and 0 blue marbles from a box that contains 4 red, 2 white, and 2 blue marbles. If we randomly draw one marble at a time and replace it before drawing the next, the probability is given by

$$p(X_R = 2 \text{ and } X_W = 1 \text{ and } X_B = 0) = \frac{n!}{n_R! n_W! n_B!} (p_R)^{n_R}(p_W)^{n_W}(p_B)^{n_B}$$

$$= \frac{3!}{2!1!0!} (0.50)^2(0.25)^1(0.25)^0$$

$$= 0.188.$$

Another computational example is given in Section 17.6-2. When n and the number of categories is large, the multinomial function rule requires a prohibitive amount of computation. As we will see in Chapter 17, this problem can be circumvented through the use of the chi-square approximation of the multinomial.

Suppose that instead of replacing marbles after they are drawn from the box, we sample without replacement. What is the probability of drawing 2 red, 1 white, and 0 blue marbles? Now the probabilities associated with drawing the different

---

[13] So named because the probabilities associated with the distribution can be obtained by raising a multinomial (an algebraic expression containing three or more terms) to the nth power.

colored marbles will change for each draw. This experiment is called a *hypergeometric experiment*, and its associated probability distribution is called a *hypergeometric distribution*. If n observations are drawn at random and without replacement from a finite population containing a total of t elements that are divided into k mutually exclusive and exhaustive classes with t_1 in class 1, t_2 in class 2, ..., t_k in class k, then the probability that exactly n_1 will be of kind 1, n_2 will be of kind 2, ..., and n_k will be of kind k, where $n_1 + n_2 + \cdots + n_k = n$ and $t_1 + t_2 + \cdots + t_k = t$, is given by the **hypergeometric function rule**:

$$p(X_1 = n_1 \text{ and } X_2 = n_2 \text{ and } \cdots \text{ and } X_k = n_k) = \frac{(_{t_1}C_{n_1})(_{t_2}C_{n_2}) \cdots (_{t_k}C_{n_k})}{_tC_n}.$$

The probability for our marble experiment is given by

$$p(X_R = 2 \text{ and } X_W = 1 \text{ and } X_B = 0) = \frac{(_4C_2)(_2C_1)(_2C_0)}{_8C_3}$$

$$= \frac{\{4!/[2!(4-2)!]\}\{2!/[1!(2-1)!]\}\{2!/[0!(2-0)!]\}}{\{8!/[3!(8-3)!]\}}$$

$$= 0.214.$$

This probability is different from the 0.188 given by the multinomial rule. The two rules give similar results when the population is extremely large.

Review Exercises for Section 9.3

18. Compare a Bernoulli random variable with a binomial random variable.

19. Interpret the statement $p(X = 3) = 0.2$.

*20. Let the random variable X be the number of males in a random sample of size two taken from a population that contains 60% males and 40% females. (a) Determine the probability of the sample's containing 0, 1, or 2 males. (b) Graph the probability distribution. (c) Compute $E(X)$ and σ.

21. Suppose that 20% of eligible voters in a given city voted in the last election. A random sample of ten eligible voters is obtained to investigate reasons for the poor turnout. (a) If X is the number of people who didn't vote, determine the probability distribution for X. (b) Compute $E(X)$ and σ.

*22. Thirty percent of elementary students in a school system have a reading ability below the national standard for their grade level. (a) If ten children are selected at random, what is the probability that no more than one will be functioning below grade level? (b) Compute $E(X)$ and σ.

23. Ten percent of patients fail to improve after being placed on medication. (a) If five patients are selected at random, what is the probability that two or more will not show improvement? (b) Compute $E(X)$ and σ.

*24. Of 800 families with five children each, how many would you expect to have (a) three girls? (b) five boys? (c) either two or three girls? Assume equal probabilities for girls and boys.

25. What is the probability of guessing correctly at least 6 of 10 answers on a true–false examination?

*26. Suppose that a box contains 50% nickels, 30% dimes, and 20% quarters. If ten coins are drawn at random and with replacement, what is the probability of drawing six nickels, three dimes, and one quarter?

27. Suppose that in Exercise 26 the box contained 20 coins and that ten are drawn at random and without replacement. What is the probability of drawing six nickels, three dimes, and one quarter?

28. A graduate admissions committee has 20 doctoral applicants, but it can admit only ten. What is the probability that the ten selected will include the five most qualified applicants? (*Hint:* $t = 20$, $t_1 = 5$, and $t_2 = 15$.)

29. Terms to remember:
   a. Bernoulli trial
   c. Multinomial experiment
   b. Binomial random variable
   d. Hypergeometric experiment

Section 9.4   Summary

Some kind of random procedure should be a part of all research with samples; most often it takes the form of random sampling from a population or random assignment of experimental conditions to subjects. Randomness is a property of a procedure rather than of a sample. Any procedure for drawing samples from a population so that every possible sample of a particular size has the same probability of being selected is called random sampling, and the resulting sample is called a random sample.

A random variable is a numerical quantity with values determined by the outcomes of a random experiment. A table showing the possible values of a random variable and the associated probabilities is called a probability distribution. Probability distributions and the frequency distributions discussed in Chapter 2 are similar; each associates a number with each of the possible values of a variable. However, for a frequency distribution, the number is a frequency; for a probability distribution, it is a probability. This reflects a fundamental difference between them. A frequency distribution describes a set of data that actually has been observed; it is empirical. A probability distribution describes data that might be observed under certain well-specified conditions; hence, it is hypothetical or theoretical. Probability distributions are used in inferential statistics as models of how random variables are expected to behave. If empirical data deviate appreciably from the predictions of a model, doubt is cast on the correctness of the model and/or its assumptions. For example, if five fair coins are tossed, according to the binomial model we should observe five heads on the average once in every 32 tosses. If instead of observing five heads once we observe them in ten of the first 32 tosses, we would begin to question the assumption that the coin is fair.

The central tendency of a theoretical population defined by a probability distribution can be described in the same way as the central tendency of a sample—by a mean. The mean of random variable values for a theoretical population is called an *expected value* and is given by $E(X) = \sum p(X_i)X_i$.

An experiment is called a Bernoulli trial if (1) its random variable has only two possible outcomes, denoted "success" and "failure," (2) the probability of a success remains constant from trial to trial, and (3) the outcomes of successive trials are independent. The probability distribution of a Bernoulli random variable could hardly be simpler, since it represents the possible outcomes of a single trial. In a series of Bernoulli trials the number of successes in n identical Bernoulli trials is a discrete random variable that can assume integer values from 0 to n. Its distribution is called a binomial distribution. The binomial distribution is one of the more useful models of how a discrete random variable should behave. Two other useful models are the multinomial and the hypergeometric distributions. The multinomial distribution is an extension of the binomial model for the case in which a trial results in an outcome from one of $k \geq 2$ classes. If elements are sampled at random but without replacement from a finite population with $k \geq 2$ classes, the results are described by the hypergeometric distribution.

10

Normal Distribution and Sampling Distributions

Section 10.1 Normal Distribution

The normal distribution is the most important probability distribution in statistics. It is important partly because so many variables in science and nature have probability distributions that closely resemble it. For example, people's heights and weights are approximately normally distributed, as are intelligence, mechanical aptitude, introversion, and most other psychological attributes. The normal distribution is important also because it is a convenient model for estimating other theoretical distribution probabilities. We will see later that it provides an excellent approximation to the binomial distribution when the number of trials is large. It will become apparent that the normal distribution is a useful model, but this hardly accounts for its eminent position in statistical theory. To understand this we must consider the distribution of a sample statistic such as the mean. Suppose that from a population we drew 100 random samples of size n (where n is fairly large), computed the mean of each sample, and constructed a histogram for the sample means. We would find that the resulting graph closely resembled the normal distribution. This might not surprise us if the sampled population were normally distributed, but the striking thing is that if n is sufficiently large the resemblance holds regardless of the population's shape. The tendency for the distribution of a sample statistic to approximate the normal form as n increases plays a key role in inferential statistics; more will be said about this when we discuss the central limit theorem in Section 10.3.

Serendipity has produced many breakthroughs in science, and one of them is the normal distribution. Abraham de Moivre (1667–1754), a mathematics tutor, was searching for a shortcut method of computing probabilities for binomial random variables and in the process derived the function rule for the ubiquitous normal distribution. If we toss ten coins, it doesn't take too much effort to compute the probability of observing 0 heads, 1 head, and so on. But suppose we toss 100 coins. The amount of work necessary to calculate the probabilities associated with 100 heads is prohibitive.

Consider the graph in Figure 10.1-1 of the probability distribution for 16 fair coins. If we superimpose the graph of a normal distribution on the histogram, it provides a fairly good fit. The fit would be even better if we had graphed the distribution for 50 coins. If the number of coins were increased indefinitely, the num-

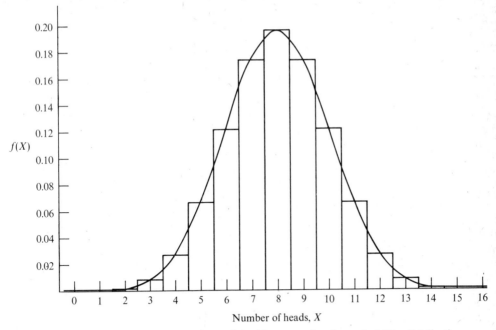

Figure 10.1-1.   Comparison of the histogram for the probability distribution of 16 fair coins and the normal curve.

ber of bars in the histogram would increase, and their outline would eventually coincide with that of the normal curve. De Moivre derived the function rule giving the height of the curve for any value of a random variable.

Characteristics of the Normal Distribution

A random variable X is said to be normally distributed if its probability distribution is given by the normal function rule. *The function rule for the* **normal distribution** *is*

$$f(X) = \frac{1}{\sigma\sqrt{2\pi}}\, e^{-(X-\mu)^2/(2\sigma^2)},$$

where $f(X)$ is the height of the curve at X, π is approximately 3.142, e (the base of the system of natural logarithms) is approximately 2.718, and μ and σ identify a particular normal distribution in the family of normal distributions. Fortunately, we don't have to use the rule to determine areas under the curve between various values of X; as we will see, they can be determined from Appendix Table D.2.

The normal curve is shaped like a bell. Since it is unimodal and symmetrical, its mean, median, and mode have the same value, and that value corresponds to the highest point on the curve. The mean plus and minus the standard deviation, as shown in Figure 10.1-2, defines the inflection points of the curve—that is, the points at which the curve changes from being concave to convex or vice versa.

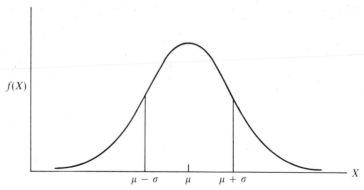

Figure 10.1-2.   Graph of the normal curve. The inflection points occur at $\mu + \sigma$ and $\mu - \sigma$.

Although not shown in the figure, the tails of the curve extend indefinitely in both directions, never quite touching the horizontal axis. The total area under the curve is equal to 1.

Converting Scores to Standard Scores

There are as many normal curves as there are possible values of μ and σ, the parameters that identify a particular curve. To avoid having to develop an infinite number of tables, statisticians have made one particular curve the standard. It has a mean equal to 0 and a standard deviation equal to 1 and is called the *standard normal curve*. This is the curve whose areas are tabulated in Table D.2. Random variable values for this curve are called *standard scores* and are denoted by z.

If a random variable doesn't happen to have a mean of 0 and standard deviation of 1, the variable must be transformed so that $\mu = 0$ and $\sigma = 1$ in order to use the standard normal curve table (Appendix Table D.2). This transformation is accomplished by

$$z = \frac{X - \mu}{\sigma},$$

where z is a standard score, X is a random variable value, μ is the population mean, and σ is the population standard deviation. A comparable formula for a sample is

$$z = \frac{X - \bar{X}}{S},$$

where X is a random variable value, $\bar{X}$ is the sample mean, and S is the sample standard deviation. In the z formula for a sample we first compute $X - \bar{X}$ and then divide the resulting deviation by S. If we apply this transformation to each X in a distribution, will the resulting standard scores have a mean of 0 and a standard deviation of 1? The answer is yes, and the reason is as follows. It can be shown (see Exercise 32, Chapter 3) that if a constant c is subtracted from each score in a distribution, the mean of the new distribution, $\bar{X}_{X-c}$, is reduced by that constant. Hence if we subtract $c = \bar{X}$ from each score, the new mean of the scores will equal $\bar{X} - c = 0$, where $\bar{X}$ is the mean of the original scores. It can also be shown (see Exercise 17, Chapter 4) that if each score is divided by a constant c, the standard deviation of the resulting scores, $S_{X/c}$, is equal to S/c. Hence if we divide each $(X - \bar{X})$ by $c = S$, the new standard deviation of the $(X - \bar{X})$'s will equal $S/c = 1$, where S is the standard deviation of the original scores. Thus applying the transformation $z = (X - \bar{X})/S$ to each X score results in a new variable called a standard score, whose mean is 0 and whose standard deviation is 1.

Suppose that the mean of a random variable is 100 and its standard deviation is 15. The z score corresponding to $X = 130$ is $z = (130 - 100)/15 = 2$. A z score transformation alters the mean and standard deviation of a random variable, but not the relative location of scores in the distribution. For example, $X = 130$ is two standard deviations above the mean of 100, since $130 = \bar{X} + 2(S) = 100 + 2(15)$. Similarly, the transformed value $z = 2$ is also two standard deviations above the mean of 0, since $2 = 0 + 2(1) = 2$. If we were to graph the distribution of the random variable and the distribution of the transformed variable, we would find that they are identical in shape although they differ in central tendency and dispersion. Transforming scores to standard scores doesn't change the shape of the distribution or the relative position of scores—only the mean and the standard deviation. As we will see in Section 10.2, standard scores are particularly useful for comparing the performance of individuals on tests having different means and/or standard deviations.

For approximately normal distributions most z scores are between -3 and $+3$. As we learned in Chapter 4, 99.73% of the area under the normal curve lies within ± 3 standard deviations of the mean.

Finding Areas under the Normal Curve

If a random variable is approximately normally distributed, the standard normal distribution (see Table D.2) can be used to find the proportion of the total area falling between any two scores. The areas tabulated in Table D.2 are shown in Figures 10.1-3(a) and (b).

1. *Area between μ and a score above it (area A)*
   Suppose that the distribution of college student's IQs is approximately normal with $\mu = 115$ and $\sigma = 15$ and that you want to know the proportion of students with IQs between the mean and 130. The first step is to convert $X = 130$ into a standard score: $z = (X - \mu)/\sigma = (130 - 115)/15 = 1$. According to Table D.2, the proportion of the area from μ to $z = 1$ is 0.3413; thus, approximately 34% of students have IQs between $\mu = 115$ and $X = 130$. This area is shown as area A in Figure 10.1-3(a).
2. *Smaller area in the tail (area B)*
   The proportion of students with IQs above 130, which corresponds to a standard score of 1, is shown as area B in Figure 10.1-3(b). It is equal to 0.1587. Since the normal distribution is symmetrical, area A + area B = 0.5. Standard scores are sometimes denoted by z and a subscript that indicates the proportion of the normal distribution that falls beyond the score. For example, the standard score of 1 is denoted by $z_{.1587}$. The symbol z_α denotes the standard score beyond which α proportion of the normal distribution falls.

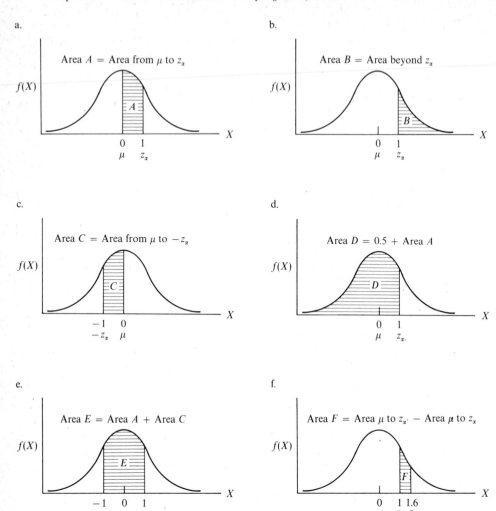

Figure 10.1-3. Illustrations of the areas of the standard normal distribution. Areas *A* and *B* are given in Table D.2. A standard score is denoted by z_α, where α indicates the proportion of the standard normal distribution that falls beyond the score. In considering area *D*, recall that the mean divides the total area in half, so that 0.5 falls above the mean and 0.5, below the mean.

3. *Area between μ and a score below it (area C)*

To determine the proportion of the total area from μ to a score below the mean, say 100, we first convert the score to a z score: $z = (100 - 115)/15 = -1$. Table D.2 gives only areas for positive z scores, but since the distribution is symmetrical, the area from μ to $z = -1$ is the same as that from μ to $z = +1$. Thus, area *C* is obtained by ignoring the negative sign and looking up the z score in area *A*. The area from μ to $z = -1$ is 0.3413.

4. *Larger area in the tail (area D)*

   To find area D for a score of, say, 130 (with a z score of 1), we find area A and add 0.5 to it. For example, area $D = 0.5 + 0.3413 = 0.8413$. A student with an IQ of 130 scores above approximately 84% of college students.

5. *Area between scores on opposite sides of the mean (area E)*

   To find the proportion of the total area between two scores on opposite sides of the mean, add areas A and C. For example, if the scores are 130 and 100, the z scores are 1 and -1. The sum of areas A and C is $0.3413 + 0.3413 = 0.6826$.

6. *Area between scores on the same side of the mean (area F)*

   Suppose that we want to determine the proportion of the total area between scores of 130 and 139. We first transform the scores to z scores: $(130 - 115)/15 = 1$ and $(139 - 115)/15 = 1.6$. Area A for $z = 1$ is 0.3413 and for $z = 1.6$ is 0.4452. Area F for these z values is given by (area μ to $z = 1.6$) $-$ (area μ to $z = 1$) $= 0.4452 - 0.3413 = 0.1039$.

Finding Scores when the Area Is Known

A different kind of problem arises when we have a percentile rank in mind or know the relative size of the area above or below a point in a distribution and we want to determine the untransformed score corresponding to that rank or point. If we know the size of the area, we can determine from Table D.2 the z score that marks the boundary of the area. In the previous examples we knew X, μ, and σ, and solved for z using the formula $z = (X - \mu)/\sigma$. If we know z, μ, and σ, it is a simple matter to solve for X. A little algebra is all that is needed to express the formula in the desired form.

$$z = \frac{(X - \mu)}{\sigma}$$

$$\sigma z = X - \mu$$

$$X = \mu + \sigma z$$

Suppose that we want to know the IQ score corresponding to the 80th percentile rank. We know that 0.80 of the area under the normal curve falls below the score and that $0.80 - 0.50 = 0.30$ of the area falls between μ and the z score. To find the z score, we look in column 2 of Table D.2 until we locate 0.30. The corresponding z score is approximately 0.84. Knowing that $\mu = 115$ and that $\sigma = 15$, we have all the information necessary to solve for X in the formula $X = \mu + \sigma z$. Substituting in the formula, we obtain $115 + 15(0.84) = 127.6$. Thus, a score of 127.6 corresponds to the 80th percentile rank.

To take one more example, suppose that we want to know the IQ score corresponding to the 40th percentile rank. We know that the score is below the mean and that 0.40 of the area lies beyond (below) the score. To find the z score corresponding to the area beyond 0.40, we look in column 3 of Table D.2 until we locate 0.40 and find that $-z \simeq -0.25$. Remember that the sign of z scores below the mean is negative. Substituting in the formula $X = \mu + \sigma z$ gives $X = 115 + 15(-0.25) = 111.25$, the score corresponding to the 40th percentile rank.

Normal Approximation to the Binomial Distribution

The normal distribution function rule was originally derived by de Moivre to estimate binomial distribution probabilities when the number of trials, n, is large. As we will see, the approximation is excellent even when n is small.

Consider an experiment in which a fair coin is tossed five times. The random variable of interest is the number of heads. A graph of the distribution for $n = 5$ and $p = 0.5$ is given in Figure 10.1-4. A normal distribution has been superimposed on the graph. The probability of observing four or more heads can be computed from Table 9.3-1 (page 205): $p(X \geq 4) = \frac{5}{32} + \frac{1}{32} = \frac{6}{32} = 0.1875$.

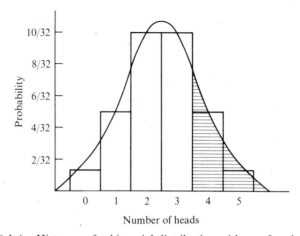

Figure 10.1-4.   Histogram for binomial distribution with $n = 5$ and $p = 0.5$. A normal distribution is superimposed over the histogram. The normal distribution area corresponding to the probability of observing four or more heads is represented by the shaded area.

The probability can be estimated using the normal distribution table by finding the area including and to the right of four heads. Since the normal distribution is continuous, we must think of four heads as occupying a class interval from 3.5 to 4.5; the lower limit of the class interval is 3.5. To find the area above the lower limit of four heads, we first convert 3.5 to a z score. Recall from Section 9.3 that the mean and standard deviation of a binomial distribution are given by, respectively, $E(X) = np$ and $\sigma = \sqrt{npq}$. For our example, $E(X) = 5(0.5) = 2.5$ and $\sigma = \sqrt{5(0.5)(0.5)} = 1.118$. The z score is

$$z = \frac{X - E(X)}{\sigma} = \frac{3.5 - 2.5}{1.118} = 0.894.$$

According to Table D.2, the area beyond $z = 0.894$ is 0.1857, which is close to the exact value of 0.1875 computed from the binomial distribution. So we see that although the normal distribution approximation was not intended to be used for such a small n, it yielded a value quite close to the exact probability.

Review Exercises for Section 10.1

1. a. Why is the normal distribution so important in statistics?

   b. How does a standard normal curve differ from other normal curves?

*2. Which of the following variables do you think approximate the normal distribution? For those that you don't think are normal, sketch the form of the distribution you would expect. (a) Amount of coffee per cup dispensed by a vending machine, (b) Extroversion scores of college students, (c) Incomes of families in the United States, (d) Time spent looking at a painting in a museum, (e) Ages of residents in Normal, Ohio, and (f) The time at which students arrive for class.

*3. A set of scores has a mean of 20 and a standard deviation of 5. Transform the following to z scores.

   *a. 30   *b. 12   *c. 15
   d. 0   e. 40   f. 20
   g. 25   h. 10   i. 18

*4. If z is a normally distributed random variable with $\mu = 0$ and $\sigma = 1$, determine the percentage of the area under the standard normal curve for the following.

   *a. Above $z = 1.5$ *b. Below $z = -2$
   *c. From μ to $z = 3$ d. Between $z = 1$ and $z = -2$
   *e. Between $z = 1$ and $z = 2$ f. Between $z = -2$ and $z = -3$
   g. From μ to $z = -1$ h. Between $z = -1$ and $z = 1.5$

*5. Determine the percentage of the area of the standard normal distribution that falls between $\mu - k\sigma$ and $\mu + k\sigma$, where k is equal to the following.

   *a. 1.0   *b. 1.64   *c. 1.96
   d. 2.58   e. 0.5   f. 3.30
   g. 2.33   h. 0.25   i. 0.67

*6. Compute the score corresponding to each of the following z scores. Assume that the original distribution had a mean of 150 and a standard deviation of 20.

   *a. 2.0   *b. -1.5   *c. 3.1
   d. 0   e. 1   f. -2.1
   g. 2.5   h. -1   i. 1.8

*7. Find the z score such that at least the following proportion of the area under the standard normal distribution falls above it.

   *a. 0.50   *b. 0.05   *c. 0.40
   d. 0.70   e. 0.95   f. 0.16
   g. 0.025   h. 0.84   i. 0.01

8. In the general population, Stanford–Binet IQs are nearly normally distributed, with a mean of 100 and a standard deviation of 16. (a) What is the probability that a randomly selected person will have an IQ between 100 and 124? (b) What percentage of the population will have IQs above 132?

*9. Junior college grade point averages (GPAs) have $\mu = 2.8$ and $\sigma = 0.24$. A university is considering raising its minimum entrance score from 2.2 to 2.5. If GPA is normally distributed, how will the proposed change affect the percentage of students eligible to enter the university from junior colleges?

10. Grading on the curve means assigning grades according to the normal distribution. The mean of a test is 50, with a standard deviation of 10. If 10% of the class receives A, what is the lowest score that receives an A?

11. The time from conception to birth in humans is approximately normally distributed, with a mean of 280.5 days and a standard deviation of 8.4 days. In a paternity case it was proved that the time from the alleged conception to the birth of a $6\frac{1}{2}$-lb baby was at least 306 days. (a) Compute the percentage of women having this or a longer gestation time. (b) Discuss the significance of the evidence.

*12. Use the normal approximation to the binomial distribution to determine the probability of guessing correctly (a) at least 12 of 20 answers on a true–false examination and (b) at least 24 of 40 answers.

*13. Suppose that 10% of physicians' diagnoses at a clinic are incorrect. Use the normal approximation to the binomial distribution to determine the probability that of 400 diagnoses (a) at most 30 will be incorrect, (b) between 30 and 50 will be incorrect, (c) more than 50 will be incorrect.

14. Terms to remember:
   a. Inflection point
   b. Standard normal curve
   c. Standard (z) score

Section 10.2   Interpreting Scores in Terms of z Scores and Percentile Ranks

Your roommate announces that she got a 62 on the midterm. Not knowing whether to rejoice with her or to sympathize, you ask, "What was the class average?" Forty-one, she replies. You press further: "What was the range?" The lowest score was 22 and the highest was 62. A celebration is in order.

This example illustrates a problem in interpreting scores. A score by itself is uninterpretable; a frame of reference is needed to know whether a score is good or bad. The frame of reference in the example was provided by the central tendency of the distribution and its dispersion. The score became interpretable when it was related to the performance of other students.

Standard Score

It would be convenient to have one number that provides all the information necessary to interpret a score instead of having to relate it to the mean and the standard deviation or perhaps the range. We have already discussed two such numbers that can be used to interpret a score: standard score and percentile. A **standard score** *is a single number that expresses the value of a score relative to the mean and the standard deviation of its distribution.* For example, if $\bar{X} = 50$ and $S = 10$, a score of 70 corresponds to a z score of

$$z = \frac{X - \bar{X}}{S} = \frac{70 - 50}{10} = \frac{20}{10} = 2,$$

which is two standard deviations above the mean. A standard score tells us the location in standard deviation units of a score relative to the mean.

Percentile Rank

The second kind of number that can be used to interpret a score is percentile rank, which is discussed in Chapter 4. *The **percentile rank** of a score indicates the percentage of the scores of the distribution that fall below that score. For example, if a score has a percentile rank of 80, we know that 80% of scores fall below it and 20% fall above.* The range of the transformed scale is from the 0th percentile rank to the 100th percentile rank, and the median is the 50th percentile rank. Transforming a score to a percentile rank locates it on a scale from 0 to 100 and indicates the percentage of scores below. Thus, as in the case of a standard score, a single number, the percentile rank, is sufficient to interpret a score.

Because percentile ranks are familiar to most people, they are used widely in presenting psychological test scores. Standard scores, on the other hand, are less familiar but, as we will see, possess a number of advantages over percentiles.

Relative Advantages of z Scores and Percentile Ranks

Consider the distribution of IQ scores in Figure 10.2-1(a); it is slightly negatively skewed. A graph of the percentile ranks corresponding to scores in Figure 10.2-1(a) is shown in part (b) of the figure. The percentile rank graph has a rectangular shape. We can see from the figure that the transformation of scores to percentile ranks has altered four characteristics of the distribution: (1) central tendency, (2) dispersion, (3) skewness, and (4) kurtosis. The only characteristic that isn't changed by the transformation is the rank order of scores within the distribution. In addition, we see that the 10-point difference between, for example, the 50th and 60th percentiles corresponds to a relatively small difference between IQ scores, but a 10-point difference between the 80th and 90th percentiles corresponds to a large difference between IQs. To put it another way, there is a greater difference in intellectual functioning between two individuals at the 80th and 90th percentiles than between individuals at the 50th and 60th percentiles. Thus, the

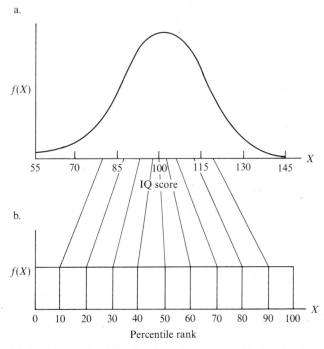

Figure 10.2-1.   (a) Graph of distribution of IQ scores. (b) Graph of distribution of percentile ranks. The transformation of scores to percentile ranks alters the shape of the distribution.

interpretation of a 10-point difference between percentile ranks depends on where the difference is on the 0–100 scale. This problem doesn't occur with standard scores.

As we have seen, transforming scores to percentile ranks alters four characteristics of the distribution; a standard score transformation alters only two characteristics—central tendency and dispersion. Standard scores have the added advantage that they can be manipulated arithmetically. For these reasons, people who use or develop psychological tests prefer standard scores over percentile ranks even though they are less familiar to the average person.

Other Kinds of Standard Scores

The standard scores we have described range approximately from -3 to $+3$ and have a mean of 0 and a standard deviation of 1. It is a minor inconvenience to have to deal with negative numbers, and fortunately this inconvenience can be avoided. If a sufficiently large constant is added to each z score, all the z scores will be positive, with a new mean equal to the constant. Similarly, if each z score is multiplied by a constant, the standard deviation is changed from 1 to the value of the constant. The formula

$$z' = \frac{(X - \bar{X})}{S} S' + \bar{X}'$$

is used to change the mean and standard deviation of z scores to any desired values, where z' is the transformed standard score, S' is the value of the desired standard deviation, and $\bar{X}'$ is the value of the desired mean.

A surprising number of test scores are actually transformed z scores. Most IQ scores are really z scores that have been multiplied by 15 and then had 100 added to the product. The resulting z' scores have $\bar{X} = 100$ and $S = 15$. Other examples of common transformed standard scores are shown in Figure 10.2-2.

Comparing Performance on Different Tests

Randy got a raw score of 68 in arithmetic and a 42 in English. He seems to be doing better in arithmetic than in English, but is he really? His teacher converts the class's arithmetic and English scores to standard scores with a mean of 50 and a standard deviation of 10. A different picture of Randy's performance emerges; his arithmetic z score is 40, 1 standard deviation below the class mean, but his English score is 65, 1.5 standard deviations above the mean. So Randy is much better in English than in arithmetic, relative to others in his class. As this example illustrates, z scores are useful for determining an individual's strengths and weaknesses—that is, for making intra-individual comparisons. They permit us to compare performance on different tasks that are measured on different scales, as were Randy's arithmetic and English tests. For the comparisons to be meaningful it is necessary only that for both variables the z scores be based on the same or equivalent reference groups. Reference groups are equivalent with respect to a variable if their distributions have essentially the same mean, standard deviation, and shape. It wouldn't have been possible to compare Randy's z scores if he had been in an accelerated arithmetic class and a remedial English class.

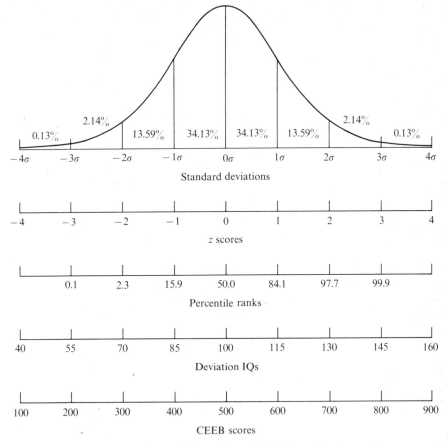

Figure 10.2-2.   Comparison of percentiles and widely used systems of standard scores.

Review Exercises for Section 10.2

15. The statement "Jane got a 29 on the quiz" is uninterpretable. Discuss.

16. Compare the relative merits of standard scores and percentile ranks for interpreting scores.

*17. Suppose that three tests were given in your statistics course. The class means and standard deviations were as listed in the table. Your scores on tests 1, 2, and 3 were, respectively, 72, 61, and 63.

Test	μ	σ
1	60	11
2	44	17
3	53	8

On which test did you do your best and on which did you do your worst?

18. On a mechanical aptitude test, Bill scored 110 and Debbie scored 85. The national mean for men is 104, with a standard deviation of 20. The comparable norms for women are 70 and 30. Which of the two did better considering the norms for their sexes?

*19. Your statistics professor returned the midterm exam and said that the mean was 82 and the standard deviation was 14. The top 15% of test scores received an A. Your score was 99. Did you get an A?

*20. Suppose that the mean of a test was 22 and the standard deviation was 5. Transform a score of 18 to standard scores with the following means and standard deviations.

  *a. $\bar{X} = 100, S = 15$ *b. $\bar{X} = 50, S = 10$
  *c. $\bar{X} = 10, S = 2$ d. $\bar{X} = 100, S = 16$
  e. $\bar{X} = 500, S = 100$ f. $\bar{X} = 80, S = 10$

Section 10.3   Sampling Distributions

Looking Ahead

So far we have covered descriptive statistics, probability, and probability distributions. These topics provide the necessary background for moving on to inferential statistics, the subject of the second half of this book. Inferential statistics are procedures for using sample data to make inferences about one or more population parameters. Two kinds of procedures are categorized under inferential statistics—estimation and hypothesis testing.

The term *estimation* is used in statistics in much the same way it is used in everyday language. A student might estimate that the mean grade point average of members of a sailing club is 2.9 or that it is between 2.7 and 3.1. We call the first type of estimate a *point estimate* because the one number representing the estimate may be associated with a point on a line. The second type, involving two numbers, is called an *interval estimate* because the two numbers and associated points define an interval on a line. *An **estimator** is a rule, usually in the form of a formula such as $\sum X_i/n$, that tells us how to calculate an estimate of a population parameter using sample information. The estimate is the numerical value that results from applying the rule to a sample.* The value of a point estimate will vary from one random sample to the next; hence the value for a particular sample is likely to differ from the population parameter. Interval estimation is used in conjunction with point estimation to specify an interval that has a high likelihood of containing the parameter of interest. These procedures are discussed in Chapter 15.

The other approach to statistical inference, *hypothesis testing*, is similar in many respects to the scientific method. The scientist observes nature, formulates a hypothesis, and then proceeds to test the hypothesis by comparing its predictions with data. Similarly, hypothesis testing begins with a question about nature that leads to a hypothesis regarding the value of one or more population parameters. The experimenter obtains a sample from the population and compares the sample value to the hypothesized value. If the sample value is inconsistent with the hypothesis, the hypothesis is rejected; otherwise it isn't rejected. These procedures are discussed in Chapters 11 through 14.

In summary, estimation is concerned with getting a reasonable idea of the value of a parameter. Hypothesis testing is concerned with deciding whether a hypothesis is or isn't true. In estimation, the result is a number or an interval bounded by two numbers. In hypothesis testing the result is a decision about a hypothesis.

Before turning to hypothesis testing, which is the subject of Chapter 11, we will lay a little more groundwork for statistical inference.

Introduction to Sampling Distributions

As we have learned, inferential statistics is concerned with reasoning from a sample to the population—from the particular to the general. Such reasoning is based on a knowledge of the sample-to-sample variability of a statistic—that is, on its sampling behavior. Before data have been collected we can speak of a sample statistic such as $\bar{X}$ in terms of probability. Its value is yet to be determined and will depend on which score values happen to be randomly selected from the population. Thus at this stage of an investigation a sample statistic is a random variable, since it is computed from score values obtained by random sampling. *Like any random variable, a sample statistic has a probability distribution that gives the probability associated with each value of the statistic over all possible samples of the same size that could be drawn from the population. The distribution of a statistic is called a **sampling distribution** to distinguish it from a probability distribution for, say, a score value.* Sampling distributions play a key role in statistical inference, since they describe the sample-to-sample variability of statistics computed from random samples. In subsequent chapters we will use sampling distributions to (1) determine the tenability of the hypothesis that a population parameter is equal to a particular value and (2) specify a range of values that has a high likelihood of including the parameter.

Sampling Distribution of the Mean

Some of the important characteristics of a sampling distribution will be introduced by an example that, though obviously unrealistic, has the virtue of allowing a concrete approach to the topic. Our discussion will focus on the sampling distribution of a mean, but the ideas we will develop apply to any sampling distribution. Suppose that we have a discrete uniform (rectangular) population consisting of N equal to four scores 1, 2, 3, and 4. A graph of the population is shown in Figure 10.3-1. The mean of the population is $\mu = \sum_{i=1}^{N} X_i/N = (1 + 2 + 3 + 4)/4 = 2.5$, and its standard deviation is

$$\sigma = \sqrt{\sum_{i=1}^{N} (X_i - \mu)^2/N}$$

$$= \sqrt{[(1 - 2.5)^2 + (2 - 2.5)^2 + (3 - 2.5)^2 + (4 - 2.5)^2]/4}$$

$$= 1.118.$$

If we draw all possible samples of size $n = 2$ with replacement, $k = 16$ different samples can be drawn (see Table 10.3-1). This follows, since the first element can be drawn in any one of four ways and the second, in any one of four ways, making a total of $4 \times 4 = 16$ samples. The probability of drawing a particular sample is,

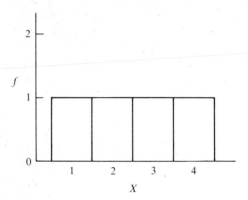

Figure 10.3-1. Histogram of a discrete uniform population.

Table 10.3-1. All possible samples of size two from the population in Figure 10.3-1

(i) *Data (The sample mean for each of the $j = 1, \ldots, k$ samples is given by $\bar{X}_j = \sum_{i=1}^{n} X_i/n$, where $k = 16$ and $n = 2$):*

Sample number	Sample values	$\bar{X}_j$	Sample number	Sample values	$\bar{X}_j$
1	1, 1	1.0	9	2, 3	2.5
2	1, 2	1.5	10	3, 2	2.5
3	2, 1	1.5	11	2, 4	3.0
4	1, 3	2.0	12	4, 2	3.0
5	3, 1	2.0	13	3, 3	3.0
6	1, 4	2.5	14	3, 4	3.5
7	4, 1	2.5	15	4, 3	3.5
8	2, 2	2.0	16	4, 4	4.0

(ii) *Mean and standard deviation of mean:*

$$\mu_{\bar{X}} = \frac{\sum_{j=1}^{k} \bar{X}_j}{k} = \frac{1.0 + 1.5 + \cdots + 4.0}{16} = \frac{40}{16} = 2.5$$

$$\sigma_{\bar{X}} = \sqrt{\frac{\sum_{j=1}^{k} (\bar{X}_j - \mu)^2}{k}} = \sqrt{\frac{(1.0 - 2.5)^2 + (1.5 - 2.5)^2 + \cdots + (4.0 - 2.5)^2}{16}} = 0.791$$

according to the multiplication rule, $(\frac{1}{4})(\frac{1}{4}) = \frac{1}{16}$. The 16 equally likely samples and their means are given in Table 10.3-1. As shown in the table, the mean of the 16 means, denoted by $\mu_{\bar{X}}$, is equal to 2.5; the standard deviation of the means, denoted by $\sigma_{\bar{X}}$, is equal to 0.791. A chart depicting the sampling procedure along with a graph of the sampling distribution of the mean is presented in Figure 10.3-2.

Three characteristics of the sampling distribution are especially significant.

1. The distribution of the sample means doesn't resemble the original population (which was rectangular in appearance), but instead resembles the normal

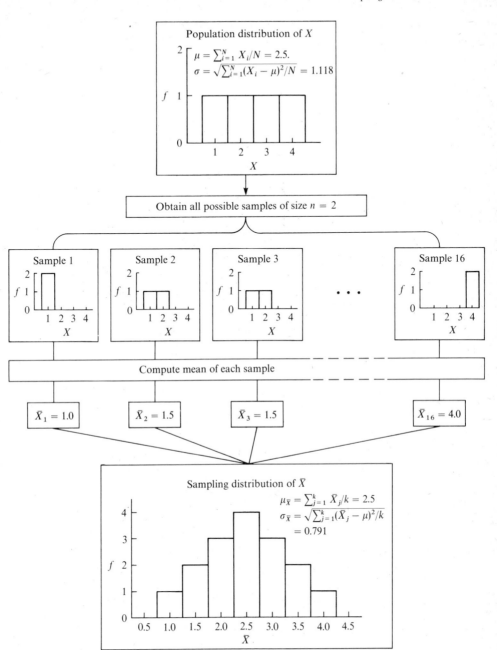

Figure 10.3-2. Graph of the sampling procedure used to construct a sampling distribution for samples of size $n = 2$ from a discrete, uniform population. Note that $\mu_{\bar{X}} = \mu = 2.5$ and that $\sigma_{\bar{X}} = \sqrt{\sum_{j=1}^{k}(\bar{X}_j - \mu)^2/k} = 0.791$, which is equal to $\sigma/\sqrt{n} = 1.118/\sqrt{2} = 0.791$.

distribution. We could show that if the sample size were increased from $n = 2$ to $n = 3, 4, \ldots$, the number of possible $\bar{X}_j$ values would increase and the distribution of $\bar{X}_j$'s would resemble more and more closely the normal distribution.

2. The mean of the 16 sample means, $\mu_{\bar{X}} = 2.5$, equals the mean of the four score values in the population, $\mu = 2.5$.

3. The standard deviation of the 16 sample means, $\sigma_{\bar{X}} = 0.791$, is equal to the standard deviation of the four scores in the population divided by the square root of the sample size—that is, $\sigma_{\bar{X}} = \sigma/\sqrt{n} = 1.118/\sqrt{2} = 0.791$.

Several implications of the third point are easily overlooked. It says in effect that we can compute the standard deviation of sample means in two ways—from $\sqrt{\sum(\bar{X}_j - \mu)^2/k}$ or $\sigma/\sqrt{n}$. Since in practical situations the distribution of sample means isn't available, we rely on a knowledge of, or an estimate of, the population standard deviation and the formula $\sigma/\sqrt{n}$ in computing $\sigma_{\bar{X}}$. The formula $\sigma_{\bar{X}} = \sigma/\sqrt{n}$ also gives us a reason for having greater confidence in large samples. We know that the standard deviation of the population (σ) is a constant. Therefore, it follows from $\sigma_{\bar{X}} = \sigma/\sqrt{n}$ that as n (the sample size) increases, $\sigma_{\bar{X}}$ (the dispersion of sample means) decreases, and hence the closer a randomly selected mean is likely to be to μ. In general, the larger the sample size, the more probable it is that the sample mean comes arbitrarily close to the population mean. This fact, often called the *law of large numbers*, is one justification for using random samples to learn about populations. If the sample is large enough, the sample information is likely to be very accurate.

Central Limit Theorem

The characteristics of the sampling distribution of a mean are succinctly stated in the *central limit theorem*, one of the most important theorems in statistics. In one form the theorem states that *if random samples are selected from a population with mean μ and finite standard deviation σ, as the sample size n increases, the distribution of $\bar{X}$ approaches a normal distribution with mean μ and standard deviation $\sigma/\sqrt{n}$.* Probably the most significant point is that regardless of the shape of the sampled population, the means of sufficiently large samples will be nearly normally distributed. Just how large is sufficiently large? This depends on the shape of the sampled population; the more a population departs from the normal form, the larger n must be. For most populations encountered in the behavioral sciences and education, a sample size of 100 is sufficient to produce a nearly normal sampling distribution of $\bar{X}$. The tendency for the sampling distributions of statistics to approach the normal distribution as n increases helps to explain why the normal distribution is so important in statistics.

Standard Error of a Statistic

We have used the term *standard deviation* to refer to a measure of dispersion, both for scores in a frequency distribution and statistics in a sampling distribution. To avoid confusion in the future we will use the term *standard error* to denote the latter measure. The symbol for a standard error always includes a subscript indicating the statistic to which it applies, for example, $\sigma_{\bar{X}}$, σ_S, and σ_r. There are as many standard errors as there are sample statistics, but they are all interpreted analogously to a standard deviation.

In the future whenever you encounter a standard error, think of it simply as a measure of the sample-to-sample variability of a statistic computed from a large number of random samples. It reflects the magnitude of the dispersion of a statistic over many samples; a standard deviation reflects the magnitude of the dispersion of scores.

Two Properties of Good Estimators

As we have discussed, the value of sample statistics varies from one random sample to the next. Consequently, it is unlikely that a given statistic will equal the population parameter it is used to estimate. This is a little frustrating, but it is something we have to live with. However, we can require that the mean of the distribution of estimates yielded by an estimator must equal the parameter it estimates and that the estimates must vary from one random sample to the next as little as possible. Statistics that satisfy these two intuitively reasonable requirements are said to be unbiased estimators and minimum variance estimators. *More formally, an estimator $\hat{\theta}$ is an **unbiased estimator** for the parameter θ if $E(\hat{\theta}) = \theta$. An estimator $\hat{\theta}$ is a **minimum variance estimator** for the parameter θ if the variance of $\hat{\theta}$, denoted by* Var $(\hat{\theta})$, *is smaller than that for any other possible unbiased estimator of θ.* The sample mean is a good estimator for μ because it satisfies both of these requirements; $E(\bar{X}) = \mu$ and Var$(\bar{X})$ is a minimum. It can be shown that the sample median is also an unbiased estimator for μ for normally distributed populations, but it is not a minimum variance estimator. This can be seen by comparing the variance error of the median with that for the mean: Var$(Mdn) = 1.57\sigma^2/n > $ Var$(\bar{X}) = \sigma^2/n$. This confirms what we stated in Section 3.5, namely, that the sample mean is more stable than the sample median. The sample variance S^2 isn't a good estimator for σ^2, since $E(S^2) \neq \sigma^2$. Instead, we use $\hat{\sigma}^2 = \sum(X_i - \bar{X})^2/(n-1)$ to estimate σ^2, because $E(\hat{\sigma}^2) = \sigma^2$.[1]

Test Statistics

The statistics we have presented thus far—$\bar{X}$, Mdn, S, and so on—are useful for describing samples. If they are computed from a random sample, they can be used also to estimate population parameters, although, as we have seen, some are better for this purpose than others. In subsequent chapters we will introduce a new kind of statistic that is used for testing hypotheses about the values of population parameters. These statistics are called *test statistics.* Consider a test statistic used to test a hypothesis about a population mean μ. Its formula is

$$z = \frac{\bar{X} - \mu_0}{\sigma_{\bar{X}}},$$

---

[1] A demonstration of this is given in Technical Note 10.5-1.

where $\bar{X}$ is the mean of a random sample, μ_0 is the hypothesized value of the population mean, and $\sigma_{\bar{X}}$ is the standard error of the mean. If the sampled population is normal or the sample size is sufficiently large, it is possible to specify the sampling distribution of the z test statistic—it is the standard normal distribution. The similarity in appearance between this z test statistic and a z score $(z = (X - \mu)/\sigma)$ is obvious. In either case, z is obtained by subtracting the mean (μ or the hypothesized value of the mean, μ_0) from the statistic and dividing this difference by the standard deviation of the statistic (σ in the case of X, and $\sigma_{\bar{X}}$ in the case of $\bar{X}$). Other test statistics that will be introduced in later chapters include $t = (\bar{X} - \mu_0)/\hat{\sigma}_{\bar{X}}$ (also used to test a hypothesis about μ), $\chi^2 = (n - 1)\hat{\sigma}^2/\sigma_0^2$ (used to test a hypothesis about σ^2), and $F = \hat{\sigma}_1^2/\hat{\sigma}_2^2$ (used to test the hypothesis that $\sigma_1^2 = \sigma_2^2$).

Review Exercises for Section 10.3

21. Distinguish a sampling distribution from a sample (frequency) distribution.
*22. A population consists of four scores: 0, 1, 2, and 3. (a) List the $(4)(4) = 16$ samples of size two that can be drawn with replacement from the population. (b) Compute the mean and standard error of the mean using the formulas $\mu = \sum X_i/N$ and $\sigma_{\bar{X}} = \sigma/\sqrt{n}$, where $\sigma = \sqrt{[\sum(X_i - \mu)^2]/N}$. (c) Compute the mean and standard error of the mean using the formulas $\mu_{\bar{X}} = \sum \bar{X}_j/k$ and $\sigma_{\bar{X}} = \sqrt{[\sum(\bar{X}_j - \mu)^2]/k}$.
23. For the population in Exercise 22, (a) list the $_4C_2 = 6$ samples of size two that can be drawn without replacement. (b) Compute the mean and standard error of the mean using the formulas $\mu = \sum X_i/N$ and $\sigma_{\bar{X}} = \sigma/\sqrt{n}$, where $\sigma = \sqrt{[\sum(X_i - \mu)^2]/N}$. (c) Compute the mean and standard error of the mean using the formulas $\mu_{\bar{X}} = \sum \bar{X}_j/k$ and $\sigma_{\bar{X}} = \sqrt{[\sum(\bar{X}_j - \mu)^2]/k}$. The means computed from the two formulas should be equal, but the standard error computed from $\sigma/\sqrt{n}$ overestimates the true value, since it assumes an infinite population and/or sampling with replacement. A correction for a finite population or when one is sampling without replacement can be made: $\sigma_{\bar{X}} = (\sigma/\sqrt{n})\sqrt{(N - n)/(N - 1)}$, where N is the number of scores in the population and n is the number in the sample. (d) Apply the correction to $\sigma/\sqrt{n}$. (e) One rule of thumb states that the finite population correction can be ignored when $n/N \leq .05$. Use an example to show why this rule is reasonable.
24. Distinguish a standard error from a standard deviation.
25. How is the dispersion of the sampling distribution of $\bar{X}$ related to σ and n?
*26. A sample of size n is to be drawn from a population with a mean of 100 and a standard deviation of 10. Complete the table.

	n	$\sigma_{\bar{X}}$		n	$\sigma_{\bar{X}}$
*a.	2		e.	32	
*b.	4		f.	64	
c.	8		g.	128	
d.	16		h.	256	

*27. The registrar claims that the mean IQ of students at a university (μ_0) is 120, with a standard deviation (σ) of 10. You obtain a random sample of 25 students and find that their mean ($\bar{X}$) is 115. What is the probability of obtaining a mean of 115 or lower if the true mean is 120? (*Hint:* Transform $\bar{X}$ to a z score and use the standard normal distribution to find the area below 115.)

*28. An elevator has a maximum safe load of 1638 lb. If mens' weights are approximately normally distributed with a mean of 165 lb and a standard deviation of 15 lb, what is the probability that nine men (whose weights can be assumed to be independent) will overload the elevator?

29. Terms to remember:
   a. Point estimate
   c. Estimator
   e. Law of large numbers
   g. Standard error
   i. Minimum variance estimator

   b. Interval estimate
   d. Sampling distribution
   f. Central limit theorem
   h. Unbiased estimator
   j. Test statistic

Section 10.4   Summary

Two theoretical distributions provide a bridge between descriptive and inferential statistics—a probability distribution and its close relative, a sampling distribution. A probability distribution associates a probability with each value of a random variable. If the random variable is some function of two or more population elements, say, a mean or a standard deviation, a probability distribution is called a *sampling distribution.*

The normal distribution is the most widely applicable theoretical model in statistics. It provides an excellent approximation to the binomial distribution and to other theoretical distributions with probabilities that are laborious to work out when n is large. In addition, it serves as a model for the many variables in science and nature that are approximately normally distributed. But its most important use is as a model for the sampling distribution of statistics based on large n's. According to the central limit theorem, as the sample size n increases, the distribution of $\bar{X}$'s from random samples approaches a normal distribution, with mean μ and standard deviation $\sigma/\sqrt{n}$, regardless of the shape of the original population.

The normal distribution is actually a family of distributions, one for each possible combination of μ and σ. The distribution with $\mu = 0$ and $\sigma = 1$ is called the *standard normal curve* and has the areas under the curve given in Table D.2. To use the standard normal table a score is transformed into a standard (z) score by the formula $z = (X - \mu)/\sigma$. The transformation doesn't affect the shape of the original distribution but does change its mean and standard deviation to 0 and 1, respectively. Standard scores are widely used for reporting psychological test scores, since they contain all the information necessary to interpret a score.

An important new measure of dispersion was introduced in this chapter—the standard error, which is the standard deviation of a statistic. It is a random variable that has been computed from two or more population elements. The standard error describes the dispersion of a statistic over all possible samples of the same size. It is denoted by σ, with a subscript identifying the statistic; for example, $\sigma_{\bar{X}}$ denotes the standard error of a mean. In the following chapters we will see how the elements—standard error, sampling distribution, and test statistic—are used in inferential statistics.

†Section 10.5   Technical Note

10.5-1   Demonstration Showing that $\hat{\sigma}^2$ and $\hat{\sigma}^2_{est}$ Are Unbiased Estimators, but S^2 Is a Biased Estimator

In Section 10.3 we drew all possible samples of size two with replacement from a finite population in order to show that the standard error of the mean, $\sigma_{\bar{X}}$, is equal to the standard deviation of the population divided by the square root of the sample size, $\sigma/\sqrt{n}$. In this technical note we will use the same sampling procedure and data to show that $E(\hat{\sigma})^2 = \sigma^2$ and $E(S^2) \neq \sigma^2$, which means that $\hat{\sigma}^2$ is an unbiased estimator for the parameter σ^2 but that S^2 is a biased estimator. The values of $\hat{\sigma}_j{}^2$ and $S_j{}^2$ are shown in Table 10.5-1 and are based on the population in Figure 10.3-1 and the random samples in Table 10.3-1. For the moment we will ignore $\hat{\sigma}^2_{est\,j}$ in column 6 of Table 10.5-1. Since $\hat{\sigma}^2$ is a discrete random variable that assumes values $\hat{\sigma}_1{}^2, \hat{\sigma}_2{}^2, \ldots, \hat{\sigma}_k{}^2$ with probabilities $p(\hat{\sigma}_1{}^2), p(\hat{\sigma}_2{}^2), \ldots, p(\hat{\sigma}_k{}^2)$, the expected value of $\hat{\sigma}^2$ is given by $E(\hat{\sigma}^2) = \sum_{j=1}^{k} p(\hat{\sigma}_j{}^2)\hat{\sigma}_j{}^2$ (see pp. 200–201). Similarly, the expected value of S^2 is given by $E(S^2) = \sum_{j=1}^{k} p(S_j{}^2)S_j{}^2$. It can be seen from the computations in Table 10.5-1 (part ii) that $\hat{\sigma}^2$ is an unbiased estimator for σ^2, since $E(\hat{\sigma}^2) = 1.25 = \sigma^2$; however, $E(S^2) = 0.625 < \sigma^2$, which means that S^2 is a biased estimator for σ^2. Thus, dividing $\sum_{i=1}^{n}(X_i - \bar{X})^2$ by $n-1$ instead of by n provides an unbiased estimator of the population variance.

A second unbiased estimator for the population variance is $\hat{\sigma}^2_{est}$, where $\sum_{i=1}^{n}(X_i - \mu)^2$ is divided by n. According to Table 10.5-1 (part ii), $E(\hat{\sigma}^2_{est}) = 1.25 = \sigma^2$, which means that $\hat{\sigma}^2_{est}$, like $\hat{\sigma}^2$, is an unbiased estimator. It is also a better estimator for σ^2 than is $\hat{\sigma}^2$, since $\hat{\sigma}^2_{est}$ varies less from sample to sample. This is shown in part iii of the table: $\text{Var}(\hat{\sigma}^2_{est}) = 0.5000 < \text{Var}(\hat{\sigma}^2) = 2.0625$. This means that σ^2_{est} is a minimum variance estimator.

To compute $\hat{\sigma}^2_{est}$ we need to know μ, one of the parameters of the population. Since μ is rarely known in real-life situations, we rely instead on $\hat{\sigma}^2$. In computing $\hat{\sigma}^2$, we use $\bar{X}$ in place of μ; in effect, $\bar{X}$ is used to estimate the unknown population parameter. As a consequence, $\sum(X_i - \bar{X})^2$ must be divided by $n-1$ (1 is the number of parameters estimated in the computation) instead of by n to obtain an unbiased estimator of σ^2.

In summary, we have just demonstrated that $\hat{\sigma}^2$ and $\hat{\sigma}^2_{est}$ are unbiased estimators for the population variance and that S^2 is a biased estimator. Furthermore, $\hat{\sigma}^2_{est}$ is a minimum variance estimator. Unfortunately, $\hat{\sigma}^2_{est}$ cannot be used in practice, since it requires a knowledge of the population mean, μ. Consequently, we use $\hat{\sigma}^2$ to estimate σ^2. As we have just seen, this estimator has the desirable property of being unbiased, although it is not a minimum variance estimator.

---

† This technical note can be skipped without loss of continuity.

Table 10.5-1.   Computation of $\hat{\sigma}^2$, S, and $\hat{\sigma}^2_{est}$ for all possible samples of size two from the population in Figure 10.3-1 (for this population, $\mu = 2.5$ and $\sigma^2 = 1.25$)

(i) *Data (The three variance estimators, $\hat{\sigma}_j^2$, S_j^2, and $\hat{\sigma}^2_{est\,j}$, are each computed from $i = 1, \ldots, n$ scores, where $n = 2$. There are $j = 1, \ldots, k$ estimates, where $k = 16$):*

(1) Sample number	(2) Sample values	(3) $\bar{X}_j$	(4) $\hat{\sigma}_j^{\,2} = \dfrac{\sum\limits_{i=1}^{n}(X_i - \bar{X})^2}{n-1}$	(5) $S_j^{\,2} = \dfrac{\sum\limits_{i=1}^{n}(X_i - \bar{X})^2}{n}$	(6) $\hat{\sigma}^2_{est\,j} = \dfrac{\sum\limits_{i=1}^{n}(X_i - \mu)^2}{n}$
1	1, 1	1.0	0.0	0.00	2.25
2	1, 2	1.5	0.5	0.25	1.25
3	2, 1	1.5	0.5	0.25	1.25
4	1, 3	2.0	2.0	1.00	1.25
5	3, 1	2.0	2.0	1.00	1.25
6	1, 4	2.5	4.5	2.25	2.25
7	4, 1	2.5	4.5	2.25	2.25
8	2, 2	2.0	0.0	0.00	0.25
9	2, 3	2.5	0.5	0.25	0.25
10	3, 2	2.5	0.5	0.25	0.25
11	2, 4	3.0	2.0	1.00	1.25
12	4, 2	3.0	2.0	1.00	1.25
13	3, 3	3.0	0.0	0.00	0.25
14	3, 4	3.5	0.5	0.25	1.25
15	4, 3	3.5	0.5	0.25	1.25
16	4, 4	4.0	0.0	0.00	2.25

(ii) *Computation of expected value:*

$$p(\hat{\sigma}_j^2) = p(S_j^2) = p(\sigma^2_{est\,j}) = \tfrac{1}{16} = 0.0625$$

$$E(\hat{\sigma}^2) = p(\hat{\sigma}_1^2)\hat{\sigma}_1^2 + p(\hat{\sigma}_2^2)\hat{\sigma}_2^2 + \cdots + p(\hat{\sigma}_k^2)\hat{\sigma}_k^2$$

$$= 0.0625(0) + 0.0625(0.5) + \cdots + 0.0625(0) = 1.25$$

$$E(S^2) = p(S_1^2)S_1^2 + p(S_2^2)S_2^2 + \cdots + p(S_k^2)S_k^2$$

$$= 0.0625(0) + 0.0625(0.25) + \cdots + 0.0625(0) = 0.625$$

$$E(\hat{\sigma}^2_{est}) = p(\hat{\sigma}^2_{est\,1})\hat{\sigma}^2_{est\,1} + p(\hat{\sigma}^2_{est\,2})\hat{\sigma}^2_{est\,2} + \cdots + p(\hat{\sigma}^2_{est\,k})\hat{\sigma}^2_{est\,k}$$

$$= 0.0625(2.25) + 0.0625(1.25) + \cdots + 0.0625(2.25) = 1.25$$

(iii) *Computation of variance:*

$$\text{Var}(\hat{\sigma}^2) = \sum_{j=1}^{k} p(\hat{\sigma}_j^2)[\hat{\sigma}_j^2 - E(\hat{\sigma}^2)]^2$$

$$= 0.0625(0 - 1.25)^2 + 0.0625(0.5 - 1.25)^2 + \cdots + 0.0625(0 - 1.25)^2$$

$$= 2.0625$$

$$\text{Var}(\hat{\sigma}^2_{est}) = \sum_{j=1}^{k} p(\hat{\sigma}^2_{est\,j})[\hat{\sigma}^2_{est\,j} - E(\hat{\sigma}^2_{est})]^2$$

$$= 0.0625(2.25 - 1.25)^2 + 0.0625(1.25 - 1.25)^2 + \cdots + 0.0625(2.25 - 1.25)^2$$

$$= 0.5000$$

11

Statistical Inference: One Sample

Section 11.1   Introduction to Hypothesis Testing

Evaluating the effectiveness of a new teaching technology or assessing consumer acceptance of a seaweed substitute for meat involves making a decision on the basis of incomplete information. The experimenter's information is usually incomplete because it is impossible or impractical to observe all the people in the population of interest. Fortunately, there are procedures for making rational decisions about nature that use a sample containing only a small portion of the elements in the population. These procedures, called *statistical inference*, are the subject of this and subsequent chapters.

Several approaches to decision making use information from a sample, but we will limit our discussion to classical statistical inference,[1] which evolved from the work of R. A. Fisher and, more directly, Jerzy Neyman and Egon Pearson.[2] Two complementary topics are categorized under classical statistical inference—hypothesis testing and interval estimation. We will describe hypothesis testing procedures in Chapters 11 through 14 and take up interval estimation in Chapter 15.

Scientific Hypotheses

People are by nature inquisitive. We ask questions, develop hunches, and sometimes put our hunches to the test. Over the years a formalized procedure for testing hunches has evolved—the scientific method. It involves (1) observing nature, (2) asking questions, (3) formulating hypotheses, (4) conducting experiments, and (5) developing theories and laws. Let's examine in detail the third characteristic, formulating hypotheses.

A **scientific hypothesis** *is a testable supposition that is tentatively adopted to account for certain facts and to guide in the investigation of others. It is a statement about nature that requires verification.* Some examples of scientific hypotheses are the following: The child-rearing practices of parents affect the personalities of their offspring. Students involved in student government have higher IQs than average college students. Cigarette smoking is associated with high blood pressure.

---

[1] Other less widely used approaches are the likelihood function and Bayesian decision theory. For a discussion of these approaches see Kirk (1972a, chap. 10).

[2] The major contributions of Fisher and of Neyman and Pearson to statistical theory are summarized by Dudycha and Dudycha (1972, pp. 21–24).

Children who feel insecure engage in overt aggression more frequently than children who feel secure. These hypotheses have three characteristics in common with all scientific hypotheses: (1) They are intelligent, informed guesses about phenomena of interest. (2) They can be stated in the *if–then* form of an implication, for example, "*if* John smokes, *then* he will show signs of high blood pressure." (3) Their truth or falsity can be determined by observation and experimentation. Many interesting hypotheses don't quality as scientific hypotheses because they aren't testable by recourse to experience. Such questions as "Can three or more angels sit on the head of a pin?" and "Does life exist in more than one galaxy in the universe?" can't be investigated because no procedures presently exist for observing angels or other galaxies. This doesn't mean that the question concerning the existence of life in other galaxies can never be investigated. Indeed, with continuing advances in space science it is likely that this question eventually will be answered.

Why Statistical Inference?

We have said that statistical inference is a form of reasoning whereby rational decisions about states of nature can be made on the basis of incomplete information. Rational decisions often can be made without resorting to statistical inference, as when a scientific hypothesis concerns some limited phenomenon that is directly observable, for example, "this rat is running." The truth or falsity of the hypothesis can be determined by observing the rat. Many scientific hypotheses, on the other hand, refer to phenomena that can't be directly observed—that is, to populations with elements that are so numerous that viewing them all is impossible or impractical; for example, "all rats run under condition X." It is impossible to observe the entire population of rats under condition X. Likewise, it is impossible to observe all parents rearing their children, all student leaders, all smokers, or all insecure children. If a scientific hypothesis cannot be evaluated directly by observing all members of a population, it may be possible to evaluate the hypothesis indirectly by statistical inference. Recourse to statistical inference enables an experimenter to make a reasonable decision concerning the probable truth or falsity of the scientific hypothesis.

Statistical Hypotheses

Scientific hypotheses are statements about phenomena of nature and man, and are normally stated in fairly general terms—at least in the initial stages of an inquiry. As we have seen, if a scientific hypothesis can't be evaluated directly by observing all members of a population, it may be possible to evaluate the hypothesis indirectly by statistical inference. To do this, an experimenter must express a scientific hypothesis in the form of a *statistical hypothesis*. *A **statistical hypothesis** is a statement about one or more parameters of a population distribution that requires verification.* For example, $\mu > 115$ is a statistical hypothesis; it states that the population mean is greater than 115. Another statistical hypothesis can be formulated that states that the mean is equal to or less than 115—that is, $\mu \leq 115$.

These hypotheses, $\mu \leq 115$ and $\mu > 115$, are mutually exclusive and exhaustive; if one is true, the other must be false. They are examples, respectively, of the *null hypothesis H_0* and the *alternative hypothesis H_1*. The null hypothesis is the one whose tenability is actually tested. If on the basis of this test the null hypothesis is rejected, only the alternative hypothesis remains tenable. According to convention, the alternative hypothesis is formulated so that it corresponds to the experimenter's hunch.[3] The process of choosing between H_0 and H_1 is called *hypothesis testing*.

An example may help to explicate the nature of H_0 and H_1. Consider the scientific hypothesis that student leaders are more intelligent than average college students. Let's assume that the population mean of college students' IQs, denoted by μ_0, is known to equal 115. (This information could come from college entrance examinations.) The population mean of student leaders' IQs, denoted by μ, is unknown but can be estimated from a random sample. The null hypothesis, which is contrary to what we believe to be true about μ, is H_0: $\mu \leq 115$. This is read "the null hypothesis states that the population mean of student leaders' IQs is equal to or less than 115." The alternative hypothesis is H_1: $\mu > 115$. We have followed the convention of equating the alternative hypothesis with the situation we believe to be true—that student leaders have higher IQs than average college students. In this example the scientific hypothesis and its negation are expressed as two mutually exclusive and exhaustive statistical hypotheses concerning the value of μ, the population mean of student leaders.

Sometimes an experimenter's scientific hypothesis as originally formulated is identical in meaning to the alternative statistical hypothesis. This is most likely to occur in research areas that have been extensively investigated. In such cases it is necessary only to formulate an appropriate null hypothesis.

Hypothesis Testing and the Method of Indirect Proof

The reader may marvel at the circuitous procedure whereby an experimenter tests a null hypothesis that is believed to be untrue, in the hope of rejecting it and thereby accepting the alternative hypothesis that is believed to be true. On reflection, you may recall a similar technique taught in plane geometry and algebra—the method of indirect proof. This method consists of listing all possible answers or solutions to a problem and showing that all save one are contrary to known fact and lead to an absurdity. By a process of elimination, the one that doesn't lead to an absurdity must be true. The success of the method of indirect proof depends upon listing all possibilities and finding a contradiction for all but one. The comparable procedure in hypothesis testing consists of formulating H_0 and H_1 so that they exhaust all possibilities concerning a population parameter. A sample is obtained from the population and an appropriate statistic is computed from the sample. If it is highly improbable that the obtained value of the statistic would have occurred if the null hypothesis were true, then the null hypothesis

---

[3] The merits of this convention have been extensively debated by Binder (1963), Edwards (1965), Grant (1962), Wilson and Miller (1964), and Wilson, Miller, and Lower (1967). Articles setting forth the basic issues in the controversy are contained in Kirk (1972a, chap. 4).

must be considered a poor prediction of the parameter and should be rejected in favor of the alternative hypothesis.

There is one important difference between the methods of indirect proof and hypothesis testing. In indirect proof a possibility is rejected only if it is found to lead to a contradiction to known fact. In hypothesis testing the null hypothesis is rejected if the obtained value of a sample statistic would have been unlikely if the null hypothesis were indeed true. Therefore, hypothesis testing, unlike indirect proof, doesn't provide incontrovertible proof, since the null hypothesis is rejected because of the occurrence of an event that is improbable rather than impossible.

Rejection or Nonrejection of H_0—What Does It Mean?

If the null hypothesis isn't rejected, what conclusion can be drawn? Is the null hypothesis true? Not necessarily, since there are always two possibilities when the null hypothesis isn't rejected: (1) H_0 is true and shouldn't be rejected and (2) H_0 is false and should be rejected, but either the particular sample that was obtained belies this fact or the experimental methodology is not sufficiently sensitive to detect the true situation. An experimental methodology can lack sensitivity because the size of the sample is too small, the measuring procedure results in large random and/or systematic errors, and so on. Even if the sample size, measurement precision, and so forth are known to be adequate, the possibility always exists that one has obtained an unrepresentative sample. We know, for example, that a fair coin will, on occasion, produce ten or even 100 consecutive heads. If the null hypothesis isn't rejected, the experimenter has two options: to state that he or she failed to reject the null hypothesis, in which case it remains credible; or to suspend judgment about the null and scientific hypotheses pending completion of a new, improved experiment.

On the other hand, what does it mean if the null hypothesis is rejected? The experimenter can conclude that the alternative and scientific hypotheses are probably true. Here, too, the possibility always exists that one's sample isn't representative, but, as we will see later, the probability of erroneously rejecting a true null hypothesis is determined by the experimenter and can be made as small as desired.

The Role of Logic in Evaluating a Scientific Hypothesis

We have just dealt with the evaluation of statistical hypotheses. We turn now to the experimenter's ultimate objective—evaluation of the scientific hypothesis. This evaluation involves a chain of deductive and inductive logic that begins and ends with the scientific hypothesis. The chain is diagrammed in Figure 11.1-1. First, by means of deductive logic the scientific hypothesis and its negation are expressed as two mutually exclusive and exhaustive statistical hypotheses that make predictions concerning a population parameter. These predictions, denoted by H_0 and H_1, are made about the population mean, median, variance, and so on. If, as is usually the case, all the elements in the population can't be observed, a random

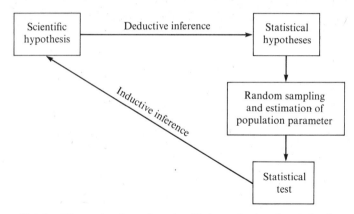

Figure 11.1-1.   The evaluation of a scientific hypothesis using deductive and inductive logic.

sample is obtained from the population. The sample provides an estimate of the unknown population parameter.

   *The process of deciding whether to reject the null hypothesis is called a **statistical test**. The decision is based on (1) a test statistic computed from a random sample from the population, (2) hypothesis testing conventions, and (3) a decision rule.* These are described in subsequent sections. The outcome of the statistical test is the basis for the final link in the chain shown in Figure 11.1-1, an inductive inference concerning the probable truth or falsity of the scientific hypothesis.

   Logic therefore plays a key role in hypothesis testing. It is the basis for arriving at both the statistical hypothesis that is tested and the final decision regarding the scientific hypothesis. If errors occur in the deductive or inductive links in the chain of logic, the statistical hypothesis subjected to test may have little or no bearing on the original scientific hypothesis, and/or the inference concerning the scientific hypothesis may be incorrect. Consider the scientific hypothesis that cigarette smoking is associated with high blood pressure. If this hypothesis is true, a measure of central tendency such as mean blood pressure should be higher for the population of smokers than for the population of nonsmokers. The statistical hypotheses are

$$H_0: \quad \mu_1 - \mu_2 \leq 0$$
$$H_1: \quad \mu_1 - \mu_2 > 0,$$

where μ_1 and μ_2 designate unknown population means for smokers and nonsmokers, respectively. The null hypothesis, H_0, states in effect that the mean blood pressure of smokers is equal to or less than that of nonsmokers; H_1 states that the mean blood pressure of smokers is greater than that of nonsmokers. These hypotheses follow logically from the original scientific hypothesis. Suppose that the experimenter formulated the statistical hypotheses in terms of population variances, for example,

$$H_0: \quad \sigma_1{}^2 - \sigma_2{}^2 \leq 0$$
$$H_1: \quad \sigma_1{}^2 - \sigma_2{}^2 > 0,$$

where $\sigma_1{}^2$ and $\sigma_2{}^2$ denote the population variances of smokers and nonsmokers, respectively. A statistical test of this null hypothesis, which states in effect that the variance of blood pressure for the population of smokers is equal to or less than the variance for nonsmokers, would have little bearing on the original scientific hypothesis. However, it would be relevant if the experimenter were interested in determining whether the two populations differed in dispersion.

The reader should not infer that for any scientific hypothesis there is only one pertinent null hypothesis. A null hypothesis stating that the correlation between number of cigarettes smoked and blood pressure is zero bears more directly on the scientific hypothesis than does the one involving population means. If cigarette smoking is associated with high blood pressure, there should be a positive correlation, $\rho > 0$, between cigarette consumption and blood pressure. The statistical hypotheses are

$$H_0: \quad \rho \le 0$$

$$H_1: \quad \rho > 0,$$

where ρ denotes the population correlation coefficient for cigarette consumption and blood pressure. So we see that both creativity and deductive skill are required to formulate pertinent statistical hypotheses.

Review Exercises for Section 11.1

*1. Which of the following are scientific hypotheses?
  a. Right-handed people tend to be taller than left-handed people.
  b. Behavior therapy is more effective than hypnosis in helping smokers kick the habit.
  c. Most clairvoyant people are able to communicate with beings from outer space.
  d. Rats are likely to fixate an incorrect response if it is followed by an intense noxious stimulus.

2. Why might it be necessary to use the techniques of statistical inference in evaluating a scientific hypothesis?

*3. Which of the following are examples of statistical hypotheses?
  a. $H_0: \quad \mu = 100$ b. $H_0: \quad S^2 \le 50$
  c. $H_1: \quad \rho > 0$ d. $H_0: \quad \bar{X} \ge 100$
  e. $H_1: \quad \sigma^2 > 0$ f. $H_1: \quad \bar{X} < 15$
  g. $H_0: \quad \mu \ge 60$ h. $H_0: \quad r = 0$
  i. $H_0: \quad \sigma^2 = 225$ j. $H_0: \quad \rho = 0$

4. a. According to convention, which statistical hypothesis corresponds to the experimenter's scientific hunch?
  b. Which is the hypothesis that actually is tested?

5. Why might an experimenter fail to reject H_0?

6. If H_0 is rejected, what does this imply about the experimental methodology?

*7. Assume that an experimenter has a hunch that insecure children engage in overt aggression more frequently than children who feel secure. Let μ and μ_0 denote the mean daily number of aggressive acts, respectively, of insecure and secure children, where it is known that $\mu_0 = 8$. State H_0 and H_1 for the experiment.

8. Terms to remember:
  a. Scientific hypothesis   b. Statistical inference
  c. Statistical hypothesis   d. Null hypothesis
  e. Alternative hypothesis   f. Statistical test

Section 11.2   Hypothesis Testing

We will now describe the procedures for testing statistical hypotheses. For the sake of clarity they are stated in five steps. This should not suggest that hypothesis testing is a formal or a rigid procedure—it isn't. However, as the experimenter plans research, each of the items in the following steps must be considered.

Step 1.   State the null and alternative hypotheses.

Step 2.   Specify the test statistic.

Step 3.   Specify the size n of the sample to be obtained and make assumptions that permit specification of the sampling distribution of the test statistic given that H_0 is true.

Step 4.   Specify an acceptable risk of rejecting H_0 when it is true.

Step 5.   Obtain a random sample of size n from the population, compute the test statistic, and make a decision about the null and alternative hypotheses.

Rejection of H_0 leads to the inductive inference that the scientific hypothesis is true. These five steps and the conventions they summarize require explication.

Step 1: Stating the Statistical Hypotheses

Let's assume that we are interested in testing the scientific hypothesis "student government leaders have higher IQs than average college students." The corresponding statistical hypothesis is $\mu > \mu_0$, where μ denotes the unknown population mean of student leaders and μ_0 denotes the population mean of college students. Assume also that μ_0 and the population standard deviation, σ, of college students are known to equal 115 and 15, respectively. We can now state the null and alternative hypotheses,

$$H_0: \quad \mu \leq 115$$

$$H_1: \quad \mu > 115,$$

where μ_0 has been replaced by 115, the known population mean of college students. As written, the null hypothesis is inexact because it states a whole range of possible values for the population mean. However, one exact value is specified, $\mu = 115$, and that is the value actually tested. If the null hypothesis $\mu = 115$ can be rejected, then the hypothesis $\mu < 115$ is rejected automatically. Obviously, if $\mu = 115$ is considered improbable because the sample statistic exceeds 115, then any $\mu < 115$ would be considered even less probable.

Step 2: Specifying the Test Statistic

A relatively small number of test statistics are used to evaluate hypotheses about population parameters. The principal ones are denoted by z, t, χ^2, and F. A test statistic is called a *z statistic* if its sampling distribution is the normal distribution; a test statistic is called a *t statistic* if its sampling distribution is a t distribution; and so forth. As we will see, the choice of a test statistic is determined by (1) the hypothesis to be tested, (2) the information that is known about the population, and

(3) the assumptions about the population that appear to be tenable. Which test statistic should be used to test the hypothesis H_0: $\mu \leq 115$? Since the hypothesis concerns the mean of a single population, since the population standard deviation is known, and since the population is assumed to be approximately normal, the appropriate test statistic is

$$z = \frac{\bar{X} - \mu_0}{\sigma_{\bar{X}}} = \frac{\bar{X} - \mu_0}{\sigma/\sqrt{n}}.$$

The factors that lead an experimenter to choose one test statistic over another are discussed in Chapters 12, 13, and 14 and are summarized in Tables 12.6-1, 13.6-1, and 14.3-1.

Step 3: Specifying n and the Sampling Distribution

In specifying the sample size, we must keep in mind that to use the standard normal distribution table, Table D.2, the sampling distribution of our z test statistic must be normal. To be more precise we should say *approximately normal*, since in practice random variables do not range from $-\infty$ to ∞ and, hence, they are never normally distributed. For simplicity we will often omit the qualifier "approximately." The sampling distribution of z will be normal if the sampling distribution of $\bar{X}$ is normal, since z is a linear transformation of $\bar{X}$; that is, $z = a + b\bar{X}$, where $a = -\mu_0/(\sigma/\sqrt{n})$ and $b = 1/(\sigma\sqrt{n})$. The sampling distribution of $\bar{X}$ will be normal if the population distribution of X is normal or if the sample is fairly large, say $n \geq 100$. This follows from the central limit theorem, discussed in Section 10.3. In the case of student leaders' IQs, the population distribution is probably nearly normal, since college students' IQs are approximately normally distributed, but to be on the safe side we will rely on a large sample and obtain an n of 100. Later, we will describe a more rational way to determine the value of n. For practical reasons (limitations on the availability of research funds and time, and the difficulty of securing subjects) we don't want to specify a larger n than is needed.

We can specify the sampling distributions of the random variables $\bar{X}$ and z, given that H_0 is true, on the basis of the information that (1) $\mu_0 = 115$, (2) $\sigma = 15$, and (3) $n = 100$ and the assumptions that (1)[4] $\mu = 115$ and (2) the population distribution of X is approximately normal. The sampling distribution of $\bar{X}$ will be normal, with mean equal to 115 and standard error equal to $\sigma/\sqrt{n} = 15/\sqrt{100} = 1.5$. This follows from the central limit theorem, discussed in Section 10.3. Because z is simply a linear transformation of $\bar{X}$, the sampling distribution of z also will be normal, with (as we saw in Chapter 10) mean equal to 0 and standard deviation equal to 1. We now have carried out the third step in testing the null hypothesis: We have specified the sample size, n, and the sampling distribution of the z test statistic.

Step 4: Specifying α

If we decided that $\mu > 115$ when in fact $\mu \leq 115$, we would make a decision error. The fourth step is to specify an acceptable risk of making this kind of error—that

---

[4] Recall that this "assumption" is the null hypothesis to be tested and, hopefully, rejected.

is, rejecting H_0 when it is true. We will touch on this subject here and return to it later. Considering the sample-to-sample variability of random variables, we wouldn't expect the mean of a single random sample, $\bar{X}$, to equal exactly the predicted value, μ_0, although $\mu = \mu_0$. We probably would be willing to attribute a small discrepancy between $\bar{X}$ and μ_0 to chance, but if the discrepancy is large enough, we would be inclined to believe that μ_0 is incorrect and that the null hypothesis should be rejected. According to hypothesis-testing conventions, a discrepancy between $\bar{X}$ and μ_0 that would be expected to occur five or fewer times in a hundred replications of the experiment is considered to be sufficiently large to warrant rejecting the hypothesis $\mu = \mu_0$. Stated another way, the null hypothesis $\mu = \mu_0$ should be rejected if the probability is equal to or less than .05 of observing a discrepancy between $\bar{X}$ and μ_0 as large as or larger than that observed.

A probability of .05 is by convention the largest risk an experimenter is willing to take of rejecting a true null hypothesis—declaring, for example, that $\mu > 115$ when in fact $\mu \le 115$. Such a probability, called a *level of significance*, is designated by the Greek letter α. For $\alpha = .05$ and H_1: $\mu > 115$, the region for rejecting H_0, called the *critical region*, is shown in Figure 11.2-1. The location and the size of the critical region are determined, respectively, by H_1 and α.

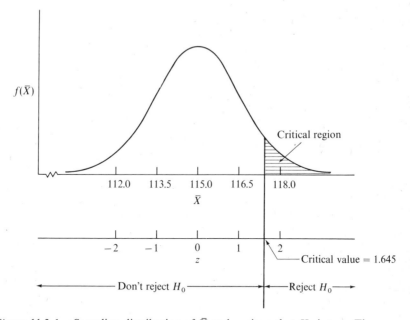

Figure 11.2-1. Sampling distribution of $\bar{X}$ and z, given that H_0 is true. The critical region, which corresponds in this example to the upper .05 portion of the sampling distribution, defines values of $\bar{X}$ and z that are improbable if the null hypothesis, H_0: $\mu \le 115$, is true. Hence, if the z test statistic falls in the critical region, the null hypothesis is rejected. The value of z, called the critical value, that cuts off the upper .05 portion can be found in the standard normal distribution table, Table D.2, and is 1.645. It can be shown that the corresponding $\bar{X}$ value is given by $\mu_0 + z\sigma_{\bar{X}} = 115 + 1.645(1.5) = 117.47$.

A decision to adopt the .05, .01, or any other level of significance generally is based on hypothesis-testing conventions which have evolved over the past 50 years. Unfortunately, these conventions don't always lead to decisions that are optimal for an experimenter's purposes. In Section 11.4 we will return to the problem of selecting a level of significance.

Step 5: Making a Decision

The fifth step in testing a statistical hypothesis is to obtain a sample from the population of interest, compute the test statistic, and make a decision. The decision rule is to reject H_0 if the test statistic falls in the critical region; otherwise, don't reject H_0.

The value of z that cuts off the critical region of the sampling distribution of z is called the *critical value*. In our example this value is denoted by $z_{.05}$, where .05 represents the area of the critical region. According to the decision rule, H_0 is rejected if $z \geq z_{.05}$; otherwise H_0 is not rejected. In other words, the null hypothesis is rejected only if the value of the z test statistic equals or exceeds the critical value.

Review Exercises for Section 11.2

9. For the past several years the mean arithmetic-achievement score for a population of ninth-grade students has been 45 with σ equal to 15. After participating in an experimental teaching program, a random sample of 100 students had a mean score of 50. (a) List the five steps you would follow in testing the hypothesis that the new program is superior to the old program, and supply the required information. (b) State the the decision rule.

10. For the data in Exercise 9 draw the sampling distribution associated with the null hypothesis, and indicate the regions that lead to rejection and nonrejection of the null hypothesis.

*11. Which of the following statistical hypotheses actually will be tested?

$$H_0: \quad \mu \leq 15$$

$$H_1: \quad \mu > 15$$

12. What determines the size of the critical region and its location?

13. Can you think of some reasons why an experimenter should always specify H_0, H_1, α, and n before collecting data?

14. Terms to remember:
   a. Level of significance
   b. Critical region
   c. Critical value

Section 11.3   One-Sample z Test for μ when σ^2 Is Known

We will now illustrate the use of

$$z = \frac{\bar{X} - \mu_0}{\sigma\sqrt{n}}$$

in testing a hypothesis about a population mean. Recall that $\bar{X}$ is the mean of a random sample from the population of interest, μ_0 is the mean specified in the null hypothesis, σ is the standard deviation of the population, and n is the number of elements in the sample used to compute $\bar{X}$. To use this formula we must know the population variance, σ^2. In Chapter 12 we will treat the case in which σ^2 is unknown but can be estimated from sample data.

Let's assume that a random sample of 100 student leaders has been obtained from the population of leaders at the Big Ten universities and that the mean IQ of this sample is 117. The number 117 is a point estimate of μ; it is the best guess we can make concerning the unknown value of μ. How improbable is a sample mean of 117 if the population mean is 115? Would it occur five or fewer times in 100 by chance? Stated another way, does the sample statistic $\bar{X} = 117$ fall in the critical region, which for our example is the upper 5% of the sampling distribution? To answer this question we first will transform the random variable $\bar{X}$ into a z random variable, with sampling distribution (if H_0 is true) identical to the standard normal distribution listed in Appendix Table D.2. If z falls in the upper 5% of the sampling distribution of z, we know that $\bar{X}$ also falls in the upper 5% of the sampling distribution of $\bar{X}$. Figure 11.2-1 illustrates the correspondence between the two sampling distributions. The linear transformation of $\bar{X}$ into z is a convenience that enables us to use the standard normal distribution table.

The steps to be followed in testing the null hypothesis and the decision rule are as follows.

Step 1.   State the statistical hypotheses: H_0: $\mu \leq 115$
$\qquad\qquad\qquad\qquad\qquad\qquad\qquad\quad$ H_1: $\mu > 115$.

Step 2.   Specify the test statistic: $$z = \frac{\bar{X} - \mu}{\sigma/\sqrt{n}}.$$

Step 3.   Specify the sample size: $n = 100$;
$\qquad\quad$ and the sampling distribution: standard normal distribution because σ^2 is known and the population distribution of X is approximately normal in form.

Step 4.   Specify the level of significance: $\alpha = .05$.

Step 5.   Obtain a random sample of size n, compute z, and make a decision.

Decision rule:

$\qquad\quad$ Reject H_0 if z falls in the upper 5% of the sampling distribution of z; otherwise don't reject H_0.

The z statistic for our example is

$$z = \frac{\bar{X} - \mu_0}{\sigma/\sqrt{n}} = \frac{117 - 115}{15/\sqrt{100}} = \frac{2}{1.5} = 1.33.$$

According to the standard normal distribution table, Table D.2, the value of z that cuts off the upper .05 region of the sampling distribution is 1.645. This value of z, called the critical value, is denoted by $z_{.05}$, where the subscript .05 refers to the proportion of the sampling distribution that falls above the z value. Since $z = 1.33$ is less than $z_{.05} = 1.645$, the observed z falls short of the upper .05 critical region. This is illustrated in Figure 11.3-1. According to our decision rule, we fail to reject H_0 and therefore conclude that our sample data don't indicate that student leaders at the Big Ten universities have higher IQs than average college students. Two points need to be emphasized. First, we haven't proven that the null hypothesis is true—only that the evidence doesn't warrant its rejection. Second, our conclusion has been restricted to the population from which we sampled, namely, the Big Ten universities.

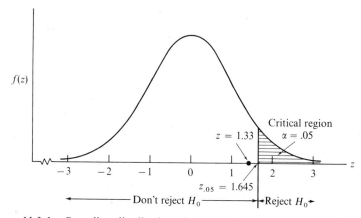

Figure 11.3-1.   Sampling distribution of z under the null hypothesis. Since $z = 1.33$ falls short of the critical region, H_0 is not rejected.

Review Exercises for Section 11.3

*15. Under what conditions is the sampling distribution of $z = (\bar{X} - \mu_0)/\sigma_{\bar{x}}$ the same as the standard normal distribution?

16. (a) Assume that the Pd (Psychopathic deviate) scale of the Minnesota Multiphasic Personality Inventory has been given to a random sample of 30 men classified as habitual criminals. The experimenter wants to test the hypothesis that habitual criminals have higher Pd scores than noncriminals. The latter population is known to be normally distributed, with mean and standard deviation equal to 50 and 10, respectively. List the five steps you would follow in testing the scientific hypothesis. Let $\alpha = .05$. (b) State the decision rule.

*17. Assume that the data in the table have been obtained for the habitual criminals in

Exercise 16. (a) Compute a z statistic for these data. (b) What conclusion can be drawn about the scientific hypothesis in Exercise 16?

Subject	Pd score	Subject	Pd score
1	50	16	55
2	51	17	56
3	54	18	48
4	55	19	45
5	25	20	41
6	61	21	82
7	64	22	65
8	55	23	67
9	55	24	75
10	52	25	40
11	71	26	61
12	57	27	35
13	59	28	56
14	54	29	56
15	55	30	55

18. If $\alpha = .001$ in Exercise 16, what conclusion would have been drawn about the scientific hypothesis?
19. One of the prison guards confessed that for a lark he had filled out the Pd scale and used a prisoner's name, subject number 22. (a) Recompute the z statistic for the data in Exercise 16, eliminating subject 22's score. (b) What conclusion can be drawn about the scientific hypothesis?

Section 11.4   More about Hypothesis Testing

In Section 11.2 we described the steps used in testing a hypothesis; we illustrated these steps in Section 11.3 by means of a one-sample z test. We now turn to several new concepts that round out our discussion of hypothesis testing.

One-Tailed and Two-Tailed Tests

*A statistical test for which the critical region is in either the upper or lower tail of the sampling distribution is called a **one-tailed test**. If the critical region is in both the upper and lower tails of the sampling distribution, the statistical test is called a **two-tailed test.***

A one-tailed test is used whenever the experimenter makes a *directional* prediction concerning the phenomenon of interest, for example, that student government leaders have higher IQs than average college students. We know from Section 11.3 that the statistical hypotheses corresponding to this scientific hypothesis are

$$H_0: \quad \mu \leq \mu_0$$

$$H_1: \quad \mu > \mu_0.$$

The region for rejecting the null hypothesis is shown in Figure 11.3-1. If the scientific hypothesis stated that student leaders have lower IQs than average college students, the following statistical hypotheses would be appropriate.

$$H_0: \quad \mu \geq \mu_0$$

$$H_1: \quad \mu < \mu_0$$

The region for rejecting this null hypothesis is shown in Figure 11.4-1(a). To be significant, an observed z test statistic would have to be less than the critical value $-z_{.05} = -1.645$.

Often, we don't have sufficient information to make a directional prediction about a population parameter; we simply believe that the parameter is not equal to the value specified by the null hypothesis. This situation calls for a two-tailed,

a.

b.

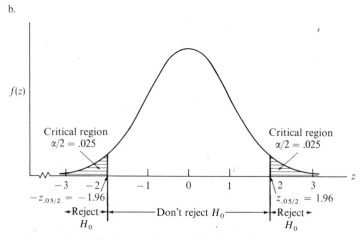

Figure 11.4-1. (a) Critical region for one-tailed test; $H_0: \mu \geq \mu_0$; $H_1: \mu < \mu_0$; $\alpha = .05$. (b) Critical regions for two-tailed test; $H_0: \mu = \mu_0$; $H_1: \mu \neq \mu_0$; $\alpha = .025 + .025 = .05$.

or *nondirectional*, test. The statistical hypotheses for a two-tailed test have the following form.

$$H_0: \quad \mu = \mu_0$$

$$H_1: \quad \mu \neq \mu_0$$

For a two-tailed test the region for rejecting the null hypothesis lies in both the upper and lower tails of the sampling distribution. The two-tailed critical region is shown in Figure 11.4-1(b).

   In summary, a one-tailed, or directional, hypothesis is called for when the experimenter's original hunch is expressed in such terms as "more than," "less than," "increased," or "decreased." Such a hunch indicates that the experimenter has quite a bit of knowledge about the research area. The knowledge could come from previous research, a pilot study, or perhaps theory. If the experimenter is interested in determining only whether something happened or whether there is a difference, without specifying the direction of the difference, a two-tailed, or nondirectional, hypothesis should be used. Generally, significance tests in the behavioral sciences are nondirectional, because most experimenters lack the information necessary to make directional predictions.

   How does the choice of a one- or two-tailed test affect the probability of rejecting a false null hypothesis? We will answer this question by means of an illustration. Assume that $\alpha = .05$ and the following hypotheses have been advanced.

$$H_0: \quad \mu = \mu_0$$

$$H_1: \quad \mu \neq \mu_0$$

If the z statistic falls in either the upper or lower .025 region of the sampling distribution, the result is said to be significant at the .05 level of significance because $.025 + .025 = .05$. The values of z that cut off the upper and lower $.05/2 = .025$ region are $-z_{.05/2} = -1.96$ and $z_{.05/2} = 1.96$. An observed z test statistic is significant at the 0.5 level if its value is greater than 1.96 or less than -1.96 or, more simply, if its absolute value $|z|$ is greater than 1.96. Now consider the hypotheses

$$H_0: \quad \mu \geq \mu_0$$

$$H_1: \quad \mu < \mu_0;$$

again $\alpha = .05$. If the z statistic falls in the appropriate tail of the sampling distribution, the result is said to be significant at the .05 level of significance. The critical regions and critical values for the two cases are shown in Figure 11.4-1. From an inspection of this figure it should be apparent that the size of the difference $\bar{X} - \mu_0$ necessary to reach the critical region for a two-tailed test is larger than that required for a one-tailed test. Consequently, an experimenter is less likely to reject a false null hypothesis with a two-tailed test than with a one-tailed test. The term *power* refers to the probability of rejecting a false null hypothesis. A one-tailed test is more powerful than a two-tailed test if the experimenter's hunch about the true difference $\mu - \mu_0$ is correct—that is, if the alternative hypothesis places the critical region in the correct tail of the sampling distribution. If the directional hunch is incorrect, the rejection region will be in the wrong tail and the experimenter will most certainly fail to reject the null hypothesis, even though it is false. An experimenter is rewarded for making a correct directional prediction and is penalized for making an incorrect directional prediction. In the absence of sufficient information for using a one-tailed test, the experimenter should play it safe and use a two-tailed test.

In one kind of research situation it is customary to use a one-tailed test. Consider a manufacturer who is evaluating a modified version of a product. If the modified version is better than the old product, the manufacturer will begin producing the modified version; otherwise, production of the old product will continue. In this example only one research outcome will lead to a product change— the case in which the modified version is superior. In this and similar research situations, where an experimenter is interested in a change in only one direction, a one-tailed test is appropriate.[5]

Exact versus Inexact Hypotheses

A statistical hypothesis can be either exact or inexact. The hypothesis $H_0: \mu = 100$ is an exact hypothesis. The hypothesis $H_0: \mu \leq 100$ is inexact because instead of specifying a single value for μ, it specifies a range of values—those equal to or less than 100.

If the null hypothesis is exact, the alternative hypothesis can be either exact or inexact; for example,

$$H_0: \quad \mu = 100 \quad \text{(exact } H_0\text{)} \qquad H_0: \quad \mu = 100 \quad \text{(exact } H_0\text{)}$$
$$\text{or}$$
$$H_1: \quad \mu = 115 \quad \text{(exact } H_1\text{)} \qquad H_1: \quad \mu \neq 100 \quad \text{(inexact } H_1\text{)}.$$

If the null hypothesis is inexact, the alternative hypothesis must also be inexact; for example,

$$H_0: \quad \mu \leq 100 \quad \text{(inexact } H_0\text{)} \qquad H_0: \quad \mu \geq 100 \quad \text{(inexact } H_0\text{)}$$
$$\text{or}$$
$$H_1: \quad \mu > 100 \quad \text{(inexact } H_1\text{)} \qquad H_1: \quad \mu < 100 \quad \text{(inexact } H_1\text{)}.$$

In the behavioral sciences inexact alternative hypotheses are the rule rather than the exception, because it is rarely possible to make a precise numerical prediction about the experimental outcome.

As we have seen from the previous examples, null hypotheses can be either exact or inexact, but the hypothesis-testing procedure is the same in both cases. If the null hypothesis is inexact, say $\mu \geq \mu_0$, it still embodies at least one exact statement, namely $\mu = \mu_0$, and this is the hypothesis that is actually tested. An examination of Figure 11.4-1(a) should convince you that if the null hypothesis $\mu = \mu_0$ is rejected at $\alpha = .05$, any null hypothesis for which $\mu > \mu_0$ would be automatically rejected.

---

[5] For an in-depth discussion of the issues involved in choosing between one- and two-tailed tests, see Kirk (1972a, chap. 8).

Type I and Type II Errors

When the null hypothesis is tested, an experimenter's decision will be correct or incorrect. *An incorrect decision can be made in two ways. The experimenter can reject the null hypothesis when it is true; this is called a **type I error**. Or the experimenter can fail to reject the null hypothesis when it is false; this is called a **type II error**. Likewise, a correct decision can be made in two ways. If the null hypothesis is true and the experimenter does not reject it, a **correct acceptance** has been made; if the null hypothesis is false and the experimenter rejects it, a **correct rejection** has been made.* The two kinds of correct decisions and the two kinds of errors are summarized in Table 11.4-1.

Table 11.4-1.   Decision outcomes categorized

		True situation	
		H_0 *true*	H_0 *false*
Experimenter's decision	*Fail to reject* H_0	Correct acceptance Probability $= 1 - \alpha$	Type II error Probability $= \beta$
	Reject H_0	Type I error Probability $= \alpha$	Correct rejection Probability $= 1 - \beta$

The probability of making a type I error is determined by the experimenter when the level of significance, α, is specified. If $\alpha = .05$, the probability of making a type I error is .05. The level of significance also determines the probability of a correct acceptance of a true null hypothesis, since this probability is equal to $1 - \alpha$.

The probability of making a type II error, symbolized by β, and the probability of making a correct rejection, which is equal to $1 - \beta$, are determined by a number of variables: (1) the level of significance, (2) the size of the sample, (3) the size of the population standard deviation, (4) the magnitude of the difference between μ and μ_0, and (5) whether a one- or two-tailed test is used. *The probability of making a correct rejection, $1 - \beta$, is called the **power** of a statistical test.* To compute β and power, it is necessary either to know μ or to specify a value of μ that would be of interest. The latter approach is usually necessary, since in any practical situation μ is unknown.

Perhaps an example will help clarify the meaning of α, $1 - \alpha$, β, and $1 - \beta$. In Section 11.3 we tested the hypothesis that student leaders have higher IQs than average college students, and we failed to reject the null hypothesis. Suppose for purposes of exposition that the sample mean equals 117.47 instead of 117. In this case, the z statistic is

$$z = \frac{\bar{X} - \mu_0}{\sigma/\sqrt{n}} = \frac{117.47 - 115}{15/\sqrt{100}} = 1.647.$$

Also assume that the population mean of student leaders' IQs is equal to 118. This information is illustrated in Figure 11.4-2. Two sampling distributions are shown in this figure, one associated with the null hypothesis and the other with

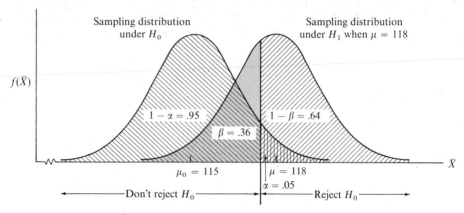

Figure 11.4-2. Regions corresponding to probabilities of making a type I error (α) and a type II error (β). The size and location of the region corresponding to a type I error are determined by α and H_1, respectively. If for a given n, σ, H_0, and true H_1, α is made smaller, the probability of making a type II error is increased.

the true IQ of student leaders. The region corresponding to a type II error (labeled β) is determined from

$$z = \frac{\bar{X} - \mu}{\sigma/\sqrt{n}} = \frac{117.47 - 118}{15/\sqrt{100}} = \frac{-0.53}{1.50} = -0.35,$$

where μ, the true population mean, is substituted for μ_0 in the z formula. According to the standard normal distribution table, Table D.2, the area below a z of -0.35 is .36. Thus, if the true parameter is 118, the probability of making a type II error (β) is equal to .36 and the probability of making a correct rejection (power) is equal to $1 - \beta = 1 - .36 = .64$. A power of .64 is considerably less than the minimum usually considered acceptable, which is .80.[6] Table 11.4-2 summarizes the probabilities associated with the possible decision outcomes when $\mu = 115$ and when $\mu = 118$.

Table 11.4-2.   Probabilities associated with the decision process

		True situation	
		$\mu = 115$	$\mu = 118$
Experimenter's decision	$\mu \leq 115$	Correct acceptance $1 - \alpha = .95$	Type II error $\beta = .36$
	$\mu > 115$	Type I error $\alpha = .05$	Correct rejection $1 - \beta = .64$

[6] The selection of .80 as the minimum acceptable power is a convenient rule of thumb and reflects the view that type I errors are more serious than type II errors. Consider the value of the ratio $p(\text{Type II error})/p(\text{Type I error})$ when the conventional .05 level of significance is adopted and power is equal to .80. Under these conditions the ratio is $.20/.05 = 4$. The probability of a type II error is four times larger than that for a type I error. An experimenter who adopts $\alpha = .05$ and $1 - \beta = .80$ is saying in effect that a type I error is considered to be four times more serious than a type II error.

For the example we have described, the probability of making a correct decision is larger when the null hypothesis is true (Probability = .95) than when the
null hypothesis is false (Probability = .64). It is also apparent that the probability
of making a type I error (.05) is much smaller than the probability of making a
type II error (.36). In most research situations the experimenter follows the convention of setting $\alpha = .05$ or $\alpha = .01$. This convention of choosing a small numerical value for α is based on the notion that a type I error is very bad and should
be avoided. Unfortunately, as the probability of a type I error is made smaller and
smaller, the probability of a type II error increases and vice versa. This can be seen
from an examination of Figure 11.4-2; if the vertical line cutting off the upper α
region is moved to the right or to the left, the region designated β is made, respectively, larger or smaller. In our example the decision rule, reject H_0 if z falls in
the upper .05 region of the sampling distribution, is weighted in favor of deciding
that the population mean is less than or equal to 115 rather than greater than 115,
because the associated probabilities are .95 and .64, respectively.

Determining the n Required to Achieve an Acceptable α,
$1 - \beta$, and $\mu - \mu_0$

As Figure 11.4-2 shows, power can be increased by adopting a large numerical
value for α, but this increases the probability of making a type I error and may
therefore be undesirable. Often the simplest way to increase the power of a statistical test is to increase the sample size. Until now we haven't said very much about
specifying n, except that it should be large enough—but not too large. There is a
more rational approach to specifying sample size. The factors we have been dis-
cussing (α, $1 - \beta$, σ, n, and $\mu - \mu_0$) are interrelated; if we specify any four of them,
the fifth is determined. The appropriate size of n can be estimated, therefore, if we
know or specify acceptable values for α, $1 - \beta$, σ, and $\mu - \mu_0$. The student leader
experiment, in which $\alpha = .05$, $\sigma = 15$, and $\mu_0 = 115$, will be used to illustrate the
procedure. Suppose that we want the power of the experiment to be .80. That
leaves one factor, $\mu - \mu_0$, unspecified. There is no way of knowing $\mu - \mu_0$ without measuring all the student leaders in the population. However, we can specify
the minimum IQ difference between μ and μ_0 that we would be interested in
finding—if in fact $\mu \neq \mu_0$. Suppose that this difference is 3 IQ points; then
$\mu - \mu_0 = 118 - 115 = 3$. By specifying that $\mu - \mu_0 = 3$ we are saying that any
difference less than 3 points is too small to be of practical interest. The formula
for determining the sample size[7] is

$$n = \frac{(z_\alpha - z_\beta)^2}{(\mu - \mu_0)^2/\sigma^2},$$

where z_α is the value of z that cuts off the α region of the sampling distribution of
μ_0, and z_β is the value that cuts off the β region of the sampling distribution of μ.
These regions are shown in Figure 11.4-2. If the alternative hypothesis is nondirectional, z_α is replaced by $z_{\alpha/2}$. For our example, $z_{.05} = 1.645$ and $-z_{.20} = -0.84$.

---

[7] The derivation of the formula is given in Kirk (1978, pp. 235–236).

Substituting in the formula gives

$$n = \frac{[1.645 - (-0.84)]^2}{(118 - 115)^2/(15)^2} = 154.4,$$

which when rounded to the next larger integer value is 155. Thus, the minimum sample size required to detect a 3-point IQ difference with $\alpha = .05$ and power equal to .80 is 155. This sample size gives the experimenter a fighting chance of detecting a difference considered worth finding. We can be more concrete. Suppose that we repeated the experiment 100 times using random samples of 155 student leaders. If the null hypothesis is true, we would expect to reject it five times (a type I error) and to fail to reject it 95 times (a correct acceptance). If, however, the null hypothesis is false and the true difference between μ and μ_0 is 3 IQ points, we would expect to reject the null hypothesis 80 times (a correct rejection) and to fail to reject it 20 times (a type II error).

Suppose that we consider type I and type II errors to be equally serious. If we set both errors at .05, the required sample size is

$$n = \frac{[1.645 - (-1.645)]^2}{(118 - 115)^2/(15)^2} \simeq 271.$$

To increase the power from .80 to .95, other things being equal, we have to increase the sample size from 155 to 271—a 75% increase. Research always involves a series of tradeoffs, as this example illustrates for power and sample size. Another tradeoff, as we will see, involves n and $\mu - \mu_0$; the larger $\mu - \mu_0$ is, the smaller is the n required to reject the null hypothesis.

In many experiments the dependent variable is a new, untried measure or one with which the experimenter has had little experience. For example, one may have developed a new test of assertiveness for which there are no norms or a new apparatus for measuring complex reaction time. In such cases it is difficult if not impossible to specify the minimum difference $\mu - \mu_0$ that would be worth detecting. Fortunately, there is an alternative procedure for estimating n that doesn't require either an estimate of $\mu - \mu_0$ or a knowledge of σ. In Section 10.2 we saw that a standard score $(X - \mu)/\sigma$ is a pure number free of the original unit of measurement. If the difference we want to detect is divided by the population standard deviation, $(\mu - \mu_0)/\sigma$, the resulting score is similar to a standard score. Cohen (1969, pp. 18–25) calls this relative measure an *effect size* and denotes it by the symbol d. It expresses the magnitude of the difference $\mu - \mu_0$ in standard deviation units. The value of d for our student leader experiment is

$$d = \frac{\mu - \mu_0}{\sigma} = \frac{118 - 115}{15} = \frac{3}{15} = 0.2.$$

Hence, the difference we wanted to detect in that experiment was 0.2 of a standard deviation. Cohen refers to a d value of 0.2 as a *small-effect size*. A medium-effect size is one for which $d = 0.5$, and a large-effect size is one for which $d = 0.8$. Using Cohen's rule of thumb, a medium-effect size for the IQ data is one for which

$\mu - \mu_0 = 122.5 - 115 = 7.5$, since

$$d = \frac{7.5}{15} = 0.5.$$

Similarly, a large-effect size corresponds to $\mu - \mu_0 = 127 - 115 = 12$, since

$$d = \frac{12}{15} = 0.8.$$

The specification of the minimum difference between μ and μ_0 that one wants to detect in terms of effect size simplifies the formula for estimating n. If d is substituted for $(\mu - \mu_0)/\sigma$, the formula becomes

$$n = \frac{(z_\alpha - z_\beta)^2}{(\mu - \mu_0)^2/\sigma^2} = \frac{(z_\alpha - z_\beta)^2}{d^2}.$$

To compute n, all we have to specify is (1) d, (2) the probability of a type I error, and (3) power. Using Cohen's rule of thumb concerning the interpretation of $d = 0.2$, 0.5, and 0.8, the n's necessary to detect small-, medium-, and large-effect sizes for the student leader experiment are, respectively,

$$n = \frac{[1.645 - (-0.84)]^2}{(0.2)^2} \simeq 155,$$

$$n = \frac{[1.645 - (-0.84)]^2}{(0.5)^2} \simeq 25, \quad \text{and}$$

$$n = \frac{[1.645 - (-0.84)]^2}{(0.8)^2} \simeq 10.$$

As in the earlier computation of n, the probability of a type I error is .05 ($z_{.05} = 1.645$) and the power is equal to .80 ($-z_{.20} = -0.84$).

It is obvious that one's sample size can be too small, resulting in insufficient power. But n can also be too large, resulting in wasted time and resources. An experimenter can avoid these problems by using the formulas just described to make a rational choice of sample size. This procedure has two other less obvious benefits: it focuses attention on the interrelationships among n, α, $1 - \beta$, σ, and $\mu - \mu_0$; and it forces the experimenter to think about the size of the difference $\mu - \mu_0$ that would be worth detecting. Critics of hypothesis-testing procedures have observed that the population mean is rarely equal to the exact value specified by the null hypothesis, and hence by obtaining a large enough sample virtually any H_0 can be rejected.[8] For this reason it is important to distinguish between *statistical significance*, which leads to the decision that $\mu \neq \mu_0$, and *practical significance*, which means that the difference $\mu - \mu_0$ is large enough to be useful in the real world. By estimating the n required to reject a difference of interest, an experimenter increases the chances of detecting practical differences. One of the challenges of research is to design an experiment that (1) has adequate power for

---

[8] For a discussion of this issue see Bakan (1966), Nunnally (1960), and Rozeboom (1960).

detecting a meaningful or practical difference between μ and μ_0, (2) uses minimum resources, n, (3) provides adequate protection against making a type I error, and (4) minimizes the effects of extraneous variables.

More about Type I and Type II Errors

In many research situations the cost of committing a type I error can be large relative to that of a type II error. It is a serious matter to commit a type I error by falsely deciding that a new medication is more effective than conventional therapies in halting the production of cancer cells and therefore can be used in place of other medical procedures. On the other hand, falsely deciding that the new medication is not more effective, a type II error, would result in withholding the medication from the public and would lead to further research. The further research eventually would demonstrate the effectiveness of the new medication. In this example a type I error is more costly and should be avoided more than a type II error. The use of the .01, or even the .001, level of significance seems warranted. However, in other research situations that don't involve life and death, a type I error may be less costly than a type II error. For example, an experimenter who makes a type II error may discontinue a promising line of research, whereas a type I error would lead to further exploration into a blind alley. Faced with these two alternatives, many experimenters would set the level of significance equal to .05 or even .20, preferring to make a type I error rather than a type II error.

It is apparent that the loss functions associated with type I and type II errors must be known before a rational choice of α can be made. Unfortunately, experimenters in the behavioral sciences generally are unable to specify the losses associated with the two kinds of errors, and therein lies the problem. The problem is resolved by using the conventional but arbitrary .05 or .01 level of significance.

Hopefully, this discussion has dispelled the magical aura that surrounds the .05 and .01 levels of significance—their use in hypothesis testing is simply a convention. A test of significance yields the probability of committing an error in rejecting the null hypothesis. It embodies no information concerning the importance or usefulness of the result obtained. It is just one bit of information used in making a decision concerning a scientific hypothesis.

Review Exercises for Section 11.4

*20. For each of the following statistical hypotheses, sketch the sampling distribution, designate the critical region(s), and indicate their size.

*a. H_0: $\mu = 60$ b. H_0: $\mu \leq 100$ c. H_0: $\mu \geq 25$
   H_1: $\mu \neq 60$ H_1: $\mu > 100$ H_1: $\mu < 25$
   $\alpha = .01$ $\alpha = .05$ $\alpha = .005$

21. Under what condition is a one-tailed test less powerful than a two-tailed test?
*22. Which of the hypotheses in Exercise 20 are exact and which are inexact?
23. Indicate the type of error or correct decision for each of the following.
   a. A true null hypothesis was rejected.
   b. The experimenter failed to reject a false null hypothesis.
   c. The null hypothesis is false and the experimenter rejected it.
   d. The experimenter did not reject a true null hypothesis.

e. A false null hypothesis was rejected.

f. The experimenter rejected the null hypothesis when he or she should have failed to reject it.

*24. The calculation of power was illustrated in this section by means of the student leaders' IQ data. We saw that if $\bar{X}$ is equal to 117.47, if σ is equal to 15, and if the student leaders' mean IQ is really equal to 118.0, the probability of correctly rejecting the null hypothesis is only .64.

*a. If a sample of $n = 150$ instead of $n = 100$ had been obtained, what would the power have been?

*b. How many subjects would be required to achieve a power of .90 in part a? Assume that μ_0 is equal to 115.

c. If a sample of $n = 100$ had been obtained but $\mu = 118.5$ instead of 118.0, what would the power have been?

d. How many subjects would be required to achieve a power of .80 in part c? Assume that μ_0 is equal to 115.

25. Set up a table that summarizes the probabilities associated with the four possible decision outcomes in Exercise 24a.

26. List the ways in which an experimenter can increase the power of an experimental methodology. What are their relative merits?

27. Terms to remember:

a. One-tailed test
c. Power
e. Inexact statistical hypothesis
g. Type II error
i. Correct rejection
k. Statistical significance

b. Two-tailed test
d. Exact statistical hypothesis
f. Type I error
h. Correct acceptance
j. Effect size
l. Practical significance

Section 11.5   Summary

Hypothesis-testing procedures, one form of statistical inference, use sample data in making a decision about a scientific hypothesis when it is impossible or impractical to observe all the elements in the population. The main features of hypothesis testing are as follows. An experimenter formulates from a scientific hypothesis two mutually exclusive and exhaustive statistical hypotheses—the null hypothesis, H_0, and the alternative hypothesis, H_1. The null hypothesis is a testable prediction about one or more parameters of a population distribution. A test of H_0 consists of determining whether the obtained value of a sample statistic would be improbable if H_0 is true. If the value would be improbable, then H_0 is a poor prediction and should be rejected in favor of H_1. The criterion for what constitutes improbable values of the test statistic is specified prior to the test. This criterion is expressed as a probability and is referred to as the level of significance. By convention, improbable values of the sample statistic are those whose probability of occurring if the null hypothesis is true is .05 or less. Such a range of improbable values defines the critical region(s). If a sample statistic falls in the critical region, H_0 is rejected.

How does one determine whether the sample statistic falls in the critical region? The answer to this question is a bit involved. An experimenter must specify H_0, H_1, a test statistic, n, and α. The population standard deviation, σ, must be known; a random sample from the population must be obtained; and, with good cause, the population distribution must be assumed to be normal. If the latter assumption is not tenable, the sample n must be fairly large. The above is sufficient to enable the experimenter to (1) completely specify the sampling distribution of the sample statistic, given that H_0 is true, (2) specify the size and the location(s) of the critical region for rejecting H_0, and (3) transform the sample statistic $\bar{X}$ into a z test statistic whose sampling distribution is normal, with $\mu = 0$ and $\sigma = 1$. Now, to answer the question posed earlier about the critical region. The z test statistic (and the sample statistic) falls in the critical region if $|z| \geq z_{\alpha/2}$ in the case of a two-tailed test or if $z \geq z_\alpha$ for $H_0: \mu < \mu_0$ (or $-z \leq -z_{\alpha/2}$ for $H_0: \mu > \mu_0$) in the case of a one-tailed test. The critical value, $z_{\alpha/2}$, z_α, or $-z_\alpha$, can be determined from the standard normal distribution table, Table D.2. If z falls in the critical region, the test statistic is said to be statistically significant, in which case the experimenter rejects the null hypothesis.

There is a tendency among experimenters to impart surplus meaning to the term statistical significance. All it really means is that a result has been obtained that is improbable if the null hypothesis is true. Statistical significance doesn't connote importance or usefulness, and it shouldn't be confused with practical significance. In the simplest terms, a significant result is one for which chance is an unlikely explanation.

12

Statistical Inference: Other One-Sample Test Statistics

Section 12.1 Introduction to Other One-Sample Test Statistics

This chapter could also be entitled "Theme with Variations"—it applies the five-step hypothesis testing format introduced in the last chapter to six new test statistics. In Chapter 11 procedures for testing hypotheses about μ when σ^2 is known were described. These same procedures, with slight modifications, are used in testing hypotheses about μ when σ^2 is unknown. And they can be used to make decisions about other population parameters such as variance (σ^2), proportion (p), and correlation (ρ). The statistics that accomplish this and the associated null hypotheses are (1) t for H_0: $\mu = \mu_0$, (2) χ^2 for H_0: $\sigma^2 = \sigma_0^2$, (3) z for H_0: $p = p_0$, (4) t for H_0: $\rho = 0$, and (5) z for H_0: $\rho = \rho_0$.

Section 12.2 One-Sample t Test for μ when σ^2 Is Unknown

When an experimenter tests a hypothesis about μ, the value of σ^2 rarely is known, and, as we have seen, σ^2 is required to compute the z test statistic. *The t test statistic that we will describe now does not require a knowledge of σ^2. Instead it requires only an estimate of σ^2 computed from a random sample and the assumption that the population distribution of X is approximately normal.* The formula for the one-sample t statistic is

$$ t = \frac{\bar{X} - \mu_0}{\hat{\sigma}_{\bar{X}}} = \frac{\bar{X} - \mu_0}{\hat{\sigma}/\sqrt{n}}, $$

where $\bar{X}$ is the mean of a random sample, μ_0 is the mean specified in the null hypothesis, $\hat{\sigma}_{\bar{X}}$ is an estimator of the population standard error of the mean, $\hat{\sigma}$ is an estimator of the population standard deviation, and n is the number of elements in the sample used to compute $\bar{X}$. The estimator $\hat{\sigma}$ is given by

$$ \hat{\sigma} = \sqrt{\frac{\sum (X_i - \bar{X})^2}{n - 1}} \quad \text{or} \quad \sqrt{\frac{\sum X_i^2 - (\sum X_i)^2/n}{n - 1}}. $$

A caret (or hat) over σ signifies that the statistic is a sample estimator of the population parameter.

The sampling distribution of t was derived by William Sealey Gossett, who published under the pseudonym *Student*; hence, the distribution is often referred to as *Student's t distribution*. The shape of the *t* distribution is similar to that of the *z* distribution in that it is symmetrical about a mean of zero. However, its variance depends on sample size, or, more specifically, degrees of freedom. To discuss this we must understand the concept of degrees of freedom, abbreviated df or v (Greek nu). The term comes from the physical sciences, where it refers to the number of planes or directions in which an object is free to move. In statistics the term *degrees of freedom* refers to the number of scores whose values are free to vary, as we will now see. Consider a sample of size 3, with mean equal to 5—that is, $(X_1 + X_2 + X_3)/3 = 5$. If we arbitrarily specify that $X_1 = 4$ and $X_2 = 5$, then X_3 must equal 6, since $(4 + 5 + 6)/3 = 15/3 = 5$. Given the statement $\bar{X} = 5$, we are free to assign any values to $n - 1 = 2$ of the scores, but having done so, the value of the *n*th score is determined. Thus, the number of degrees of freedom associated with $\bar{X}$ is $n - 1$. Similarly, the number of degrees of freedom associated with $\hat{\sigma} = \sqrt{\sum(X_i - \bar{X})^2/(n - 1)}$ is $n - 1$. This follows because once $n - 1$ of the n deviations $(X_i - \bar{X})$ have been arbitrarily specified, the *n*th deviation is not free to vary, since $\sum(X_i - \bar{X})$ must equal 0 (as shown in Technical Note 3.9-2). The number of degrees of freedom for the one-sample *t* test is equal to $n - 1$.

It can be shown that for $v > 2$ the variance of the *t* sampling distribution is

$$\text{Var}(t) = \frac{v}{v - 2},$$

where $v = n - 1$. According to the formula, if random samples of size $n = 5$ are obtained from a population, the variance of the resulting *t* sampling distribution is

$$\text{Var}(t) = \frac{v}{v - 2} = \frac{4}{2} = 2.$$

When $v = 4$, the variance is 2, which is twice the variance of the *z* sampling distribution. As the number of degrees of freedom increases, the variance of the sampling distribution of *t* approaches more and more closely that of the standard normal distribution. For example, when $n = 30$,

$$\text{Var}(t) = \frac{29}{29 - 2} = 1.0741,$$

which differs only slightly from the variance of the standard normal distribution. When $v = \infty$, the two sampling distributions are identical. Because the two sampling distributions are so similar for samples equal to or larger than 30, an n of 30 is often taken as a dividing point between large and small samples.

The *t* distribution is actually a family of distributions whose shapes depend on the associated number of degrees of freedom. A comparison of three members of the *t* family and the standard normal distribution is shown in Figure 12.2-1. As this figure illustrates, the *t* and *z* distributions are alike in that both have a mean of 0, are symmetrical, and are unimodal. The distributions differ when $v < \infty$ in that the distribution of *t* is more leptokurtic and has a larger variance.

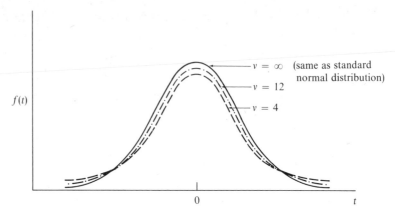

Figure 12.2-1. Graph of the distribution of t for 4, 12, and ∞ degrees of freedom. The t distribution for $v = \infty$ is identical to the z distribution.

At first glance the t and z test statistics look alike, but a difference can be seen on close inspection.

$$z = \frac{\bar{X} - \mu_0}{\sigma/\sqrt{n}} = \frac{\text{(Random variable)} - \text{(Constant)}}{\text{(Constant)}}$$

$$t = \frac{\bar{X} - \mu_0}{\hat{\sigma}/\sqrt{n}} = \frac{\text{(Random variable)} - \text{(Constant)}}{\text{(Random variable)}}$$

The z statistic is the ratio of a random variable to a constant; t is the ratio of two random variables. This follows, since when $v < \infty$, both $\bar{X}$ and $\hat{\sigma}$ in the t statistic vary from sample to sample and hence are random variables. The difference in the nature of the z and t denominators has an important ramification, which we will examine. Earlier we saw that the sampling distribution of the z test statistic is the standard normal distribution when H_0 is true and the population distribution of X is normal in form or n is sufficiently large. Note that the normality assumption is not required if n is sufficiently large. The normality assumption, however, is required for the t test statistic. It serves two purposes. First, it permits us to specify the sampling distribution of the numerator of the t statistic without regard to sample size. It is the normal distribution. Second, it is a necessary condition for the numerator and denominator (both random variables) of the t statistic to be statistically "independent," which means that the information contained in $\bar{X}$ does not affect the value of $\hat{\sigma}$ and vice versa. Independence was a simplifying assumption that Gossett made when he originally derived the sampling distribution of t. If the numerator and denominator of the t statistic are not independent, specifying the sampling distribution of t exactly is extremely difficult. This problem does not occur with the z test statistic, since its denominator is a constant rather than a random variable. The important point is that the normality assumption plays a more important role in the derivation and use of t than in z.

The t test statistic can be used to test hypotheses of the form

$$H_0: \ \mu \geq \mu_0 \qquad H_0: \ \mu \leq \mu_0 \qquad H_0: \ \mu = \mu_0$$

$$H_1: \ \mu < \mu_0 \qquad H_1: \ \mu > \mu_0 \qquad H_1: \ \mu \neq \mu_0,$$

where μ denotes the unknown population mean and μ_0 denotes the hypothesized value of the population mean. The assumptions associated with using t to test these hypotheses are that (1) the population distribution of X is normal in form, (2) the X's are a random sample from the population of interest, (3) the population variance isn't known, and (4) the null hypothesis is true, which is the hypothesis the experimenter hopes to reject. The null hypothesis is rejected if an observed t falls in the critical region of its sampling distribution. The critical value of t that cuts off the upper α region of the sampling distribution for v degrees of freedom is given in Appendix Table D.3 and is denoted by $t_{\alpha,v}$; the value that cuts off the lower α region is denoted by $-t_{\alpha,v}$. If α is to be divided equally between the two tails, as in performing a two-tailed test, the critical values are denoted by $t_{\alpha/2,v}$ and $-t_{\alpha/2,v}$. To illustrate, suppose that an experimenter wants to test the one-tailed hypothesis $H_0: \mu \leq 50$ versus $H_1: \mu > 50$, and suppose that α is set at .05 and $v = 30$. According to Table D.3 the value of t that cuts off the upper .05 region is 1.697. The null hypothesis would be rejected if $t \geq 1.697$. For $H_0: \mu \geq 50$ and $H_1: \mu < 50$, the null hypothesis would be rejected if $t \leq -1.697$. The critical value for a two-tailed hypothesis is 2.042, in which case the null hypothesis would be rejected if $|t| \geq 2.042$.

Computational Example for One-Sample t Test

Consider the scientific hypothesis that a new class registration procedure at Idle-On-In College will reduce the time required for a student to register. Over the past several years the mean time required to register has been 3.1 hr. A trial run to test the new procedure was conducted with a random sample of 21 undergraduates. The decision rule and the steps to be followed in testing the null hypothesis are as follows.

Step 1. State the statistical hypotheses: $\quad H_0: \ \mu \geq 3.1$
$$H_1: \ \mu < 3.1.$$

Step 2. Specify the test statistic: $\qquad t = \dfrac{\bar{X} - \mu_0}{\hat{\sigma}/\sqrt{n}}.$

Step 3. Specify the sample size: $\qquad n = 21;$
and the sampling distribution: $\quad t$ distribution with $v = n - 1$, because σ^2 is unknown and must be estimated, and we assume the population distribution of X is approximately normal in form.

Step 4. Specify the significance level: $\quad \alpha = .05.$
Step 5. Obtain a random sample of size n, compute t, and make a decision.

Decision rule:

> Reject H_0 if t falls in the lower 5% of the sampling distribution of t; otherwise, don't reject H_0.

The data for the trial run are shown in Table 12.2-1. According to Table D.3, a t of -1.725 with $n-1=20$ degrees of freedom cuts off the lower .05 region of the sampling distribution—that is, $-t_{.05,20} = -1.725$. Since $-5.797 < -1.725$, the null hypothesis is rejected and the experimenter concludes that registration can be completed in less time with the new procedure than with the old procedure.

Table 12.2-1.   Registration time data

(i) *Data:*

Student	Registration time, X_i (hours)	X_i^2	Student	Registration time, X_i (hours)	X_i^2
1	2.9	8.41	12	3.0	9.00
2	2.7	7.29	13	2.8	7.84
3	2.8	7.84	14	3.1	9.61
4	3.0	9.00	15	2.9	8.41
5	2.6	6.76	16	2.7	7.29
6	2.9	8.41	17	3.2	10.24
7	3.1	9.61	18	2.8	7.84
8	2.9	8.41	19	2.8	7.84
9	3.0	9.00	20	3.1	9.61
10	2.7	7.29	21	3.0	9.00
11	2.9	8.41		$\sum X_i = 60.9$	$\sum X_i^2 = 177.11$

(ii) *Computation:*

$$\bar{X} = \frac{\sum X_i}{n} = \frac{60.9}{21} = 2.9$$

$$\hat{\sigma} = \sqrt{\frac{\sum X_i^2 - (\sum X_i)^2/n}{n-1}} = \sqrt{\frac{177.11 - (60.9)^2/21}{21-1}} = \sqrt{\frac{0.50}{20}} = 0.1581$$

$$t = \frac{\bar{X} - \mu_0}{\hat{\sigma}/\sqrt{n}} = \frac{2.9 - 3.1}{0.1581/\sqrt{21}} = \frac{-0.2}{0.0345} = -5.797$$

$$v = n - 1 = 21 - 1 = 20$$

$$-t_{.05,20} = -1.725$$

Some Experimental Design Considerations

Note that there are alternative explanations for why the sample of 21 students completed registration in less time with the new procedure. The school administrators would like to believe that the new procedure is extremely efficient and, if adopted for all students, would cut down registration time. But what are some alternative explanations? Since they were selected for the trial run, the 21 students may have felt they should make a special effort to complete registration quickly—an effort they wouldn't make once the new procedure was adopted and they were no longer under scrutiny. It is also possible that the personnel assisting in registration were more alert and tried to expedite the registration because they too were under scrutiny and because the procedure was a break from the usual routine. Other explanations could be advanced, and unless these explanations can be ruled out, the administrators may be disappointed if they adopt the new procedure. Once the novelty of the new procedure wears off, the procedure may be no better, or even poorer, than the old registration procedure.

Designing an experiment whose outcome can be unambiguously interpreted requires careful planning. *It is customary in behavioral science research to use one or more* **control groups;** *these groups contain subjects that are not subjected to the treatment in order to provide data on the effects of extraneous variables that affect the interpretation of the experiment.*[1] For example, the design of our registration experiment could be improved by drawing a sample of 42 students, half to be assigned to the new procedure and half, to the old. This would provide data on the effects of being specially selected to participate in the trial run. The appropriate test statistic if this modification is adopted is the two-sample *t* statistic for independent samples discussed in Section 13.4.

Statistical Significance and Practical Significance

Let us assume for purposes of discussion that results obtained in the trial run would also be obtained if the new registration procedure were adopted for all students. The difference in registration time between the procedures was $2.9 - 3.1 = -0.2$ hr, or -12 min. Before adopting the new procedure, the school administrators would have to decide whether the statistically significant difference in registration time is significant from a practical point of view. Is a saving of 12 min worth the cost of changing the registration procedures? If the change involves hiring new personnel, scrapping costly registration forms, or extensively modifying physical facilities, the administrators might decide the change isn't worthwhile.

Statistical significance is concerned with whether a result could have occurred by chance; it says nothing about the practical significance of the result, which must be decided on the basis of other considerations.

Determining the n Required to Achieve an Acceptable α, $1 - β$, and $(μ - μ_0)/σ$

Procedures for making a rational choice of *n* were discussed in Section 11.4. Similar procedures can be used with *t*, but the computations are difficult to perform. If the difference one is interested in detecting is expressed as an effect size, *d*, the required *n* can be looked up in Appendix Table D.9.[2] Cohen (1969, pp. 18–25) refers to a *d* value of 0.2 as a small-effect size; *d* values of 0.5 and 0.8 are called, respectively, medium- and large-effect sizes. To use Table D.9, it is necessary to specify *d*, α, $1 - β$, and whether the test is directional or nondirectional. In the class registration experiment, suppose that we wanted to detect a large-effect size $(d = 0.8)$ and that we wanted α to equal .05 and $1 - β$ to equal 0.80. The required

---

[1] For an excellent discussion of these procedures see Campbell (1957). This article is reprinted in Kirk (1972a).

[2] More extensive tables are provided by Cohen (1969), Dixon and Massey (1957), and Minium (1970).

sample size according to Table D.9 is 12, which is approximately half the number used in the example.

More about the Normality Assumption

One of the assumptions of the t test is that the population distribution of X is normal in form, in which case the sampling distribution of $\bar{X}$ is normal for any size n and the numerator and denominator of t are statistically independent. What are the consequences when X is not normal? Research has suggested that when n is fairly large, and provided that the population distribution of X is unimodal and symmetrical, the t distribution still gives an adequate approximation to the exact probability of a type I error (Boneau, 1960). We can find comfort in the fact that when n is large, the t test is not very sensitive to departures from normality; that is, it is *robust* with respect to nonnormality. If, however, n is small and the population distribution of X deviates appreciably from normality, the approximation yielded by t can be seriously in error. For less marked deviations from normality the actual probability of a type I error generally will be fairly close to the nominal or specified probability of a type I error, even when n is small.

Research has also indicated that power and the probability of a type II error are not significantly affected by lack of normality of the population distribution, provided n is large. However, when n is small, an experimenter may fail to reject a false null hypothesis due to lack of power resulting from nonnormality of the population distribution.

Review Exercises for Section 12.2

1. List the similarities and differences between the t and z tests.
2. Suppose that several first-grade teachers have complained that their classes this year are unusually slow in learning to read. The school principal has asked you to determine if the children are below average in intelligence—have a mean IQ below 100. Since there are 362 first-grade children, giving each of them an individual intelligence test is not feasible. Instead, you administer the Wechsler Intelligence Scale for Children to a random sample of 16 children. List the steps you would follow in testing the scientific hypothesis. Let $\alpha = .05$.
*3. Assume that the data in the table have been obtained for the children in Exercise 2. (a) Compute a t statistic for these data and make a decision about the scientific hypothesis. (b) Use Table D.9 to estimate the sample size needed to detect a large-effect size for $\alpha = .05$ and $1 - \beta = .80$.

Child	IQ	Child	IQ
1	89	9	86
2	96	10	88
3	86	11	92
4	92	12	101
5	78	13	87
6	110	14	93
7	82	15	97
8	69	16	74

4. Make a frequency distribution for the data in Exercise 3. Use ten class intervals, with a class interval size of five. From a visual inspection of the frequency distribution, is it reasonable to assume that the population distribution is normal in form?

5. A random sample of 65 freshman college students was selected to participate in a new look–say teaching program designed to increase reading speed in French. The final exam consisted of a French passage that the students translated. The time required for each student to complete the translation was recorded. The sample statistics were $\bar{X} = 302$ sec and $\hat{\sigma} = 56$ sec. According to departmental records, the mean for students in conventional classes was 320 sec. Let $\alpha = .05$.

   a. List the steps you would use in testing the scientific hypothesis that the look–say program resulted in a decrease in time required to translate the French passage.

   b. Compute a t statistic and make a decision about the scientific hypothesis.

   c. How could the design of the experiment be improved?

   d. Is the sample adequate to detect a medium-effect size if a power of .95 is desired?

6. Terms to remember:

   a. Degrees of freedom

   b. Robust

Section 12.3   One-Sample Chi-Square Test for a Population Variance

Most scientific hypotheses are concerned with central tendency, but hypotheses concerning population variance, skewness, kurtosis, and correlation also can be of interest. In this section we describe a chi-square (χ^2) statistic for evaluating a hypothesis about a single population variance. *The chi-square test statistic is given by*

$$\chi^2 = \frac{(n-1)\hat{\sigma}^2}{\sigma_0^2},$$

where n is the number of elements in a random sample from a normally distributed population, $\hat{\sigma}^2$ is an unbiased estimator of the population variance, and σ_0^2 is the value of the population variance specified in the null hypothesis.

The chi-square statistic can be used to test hypotheses such as the following.

$$H_0: \ \sigma^2 = \sigma_0^2 \qquad H_0: \ \sigma^2 \leq \sigma_0^2 \qquad H_0: \ \sigma^2 \geq \sigma_0^2$$
$$H_1: \ \sigma^2 \neq \sigma_0^2 \qquad H_1: \ \sigma^2 > \sigma_0^2 \qquad H_1: \ \sigma^2 < \sigma_0^2$$

The assumptions associated with using the chi-square statistic to test these hypotheses are that (1) the population distribution of X is normal in form, (2) the X's are a random sample from the population of interest, and (3) the null hypothesis is true, which is the hypothesis that the experimenter hopes to reject.

The sampling distribution of chi square was derived in 1876 by F. R. Helmert; in 1900 it was used by Karl Pearson to test hypotheses. The chi-square distribution, like the t distribution, is actually a family of distributions whose shapes depend on the associated degrees of freedom. The number of degrees of freedom, v, for χ^2 is equal to $n - 1$. Unlike the z and t distributions, the chi-square distribution is positively skewed for small degrees of freedom, but as v increases, the distribution approaches the normal form with mean and variance, respectively,

$$E(\chi_v^2) = v \qquad \text{and} \qquad \text{Var}(\chi_v^2) = 2v.$$

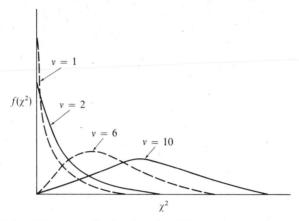

Figure 12.3-1.   Chi-square distributions for different degrees of freedom.

Since χ^2 is a squared quantity, it can range over only nonnegative numbers, zero to positive infinity, whereas t and z can range over all real numbers. The chi-square distributions for several different degrees of freedom are shown in Figure 12.3-1. The critical value of χ^2 that cuts off the upper α region of the sampling distribution for v degrees of freedom is given in Appendix Table D.4 and is denoted by $\chi^2_{\alpha,v}$; the value that cuts off the lower α region is denoted by $\chi^2_{1-\alpha,v}$. If α is to be divided equally between the two tails, as in performing a two-tailed test, the critical values are denoted by $\chi^2_{\alpha/2,v}$ and $\chi^2_{1-\alpha/2,v}$.

Computational Example for One-Sample Chi-Square Test

It has been shown that students who take a driver education course in high school receive on the average higher scores on the state licensing test than do students who haven't taken the course. One might also expect less variability among the scores of students who have taken driver education than among students who haven't.

We can make an inference regarding this scientific hypothesis by means of the one-sample chi-square test. Let us assume that the variance on the licensing test for high school seniors who haven't taken the driver education course is known to equal 324. The decision rule and the steps to be followed in testing the null hypothesis are as follows.

Step 1.   State the statistical hypotheses: H_0: $\sigma^2 \geq 324$
$\qquad\qquad\qquad\qquad\qquad\qquad\qquad\quad$ H_1: $\sigma^2 < 324$.

Step 2.   Specify the test statistic: $\qquad\quad$ $\chi^2 = \dfrac{(n-1)\hat{\sigma}^2}{\sigma_0^2}$.

Step 3.   Specify the sample size: $\qquad\quad$ $n = 16$;
$\qquad\quad$ and the sampling distribution: chi-square distribution with $v = n - 1$, because the population distribution of X is assumed to be normal in form.

Step 4.   Specify the significance level: $\alpha = .05$.

Step 5.   Obtain a random sample of
size n, compute χ^2, and make a decision.

Decision rule:

Reject H_0 if χ^2 falls in the lower 5% of the sampling distribution of χ^2; otherwise, don't reject H_0.

Assume that driver license scores for a random sample of 16 high school seniors who have taken the driving course have been obtained from the Bureaucracy of Motor Vehicles. The data are given in Table 12.3-1. For this sample the variance is equal to 150.7. Is this variance small enough to warrant the conclusion that the population parameter is not greater than or equal to 324?

According to the chi-square table, Table D.4, a χ^2 of 7.261 cuts off the lower .05 region of the sampling distribution—that is, $\chi^2_{.95,15} = 7.261$. Since $6.977 < 7.261$, the null hypothesis is rejected and the experimenter concludes that the population variance of driver license scores is smaller for students who have had a driving course in high school than for students who haven't had a course.

More about the Normality Assumption for χ^2

How important is the assumption that the population distribution of X is normal in form? In Section 12.2 we learned that the t test is robust with respect to departures from normality. Unfortunately, this isn't true for this chi-square test.

Table 12.3-1.   Driver license test scores for a random sample of high school seniors who have completed a driving course

(i) *Data:*

Student number	Score, X_i	X_i^2	Student number	Score, X_i	X_i^2
1	70	4900	9	57	3249
2	68	4624	10	61	3721
3	81	6561	11	72	5184
4	95	9025	12	78	6084
5	63	3969	13	71	5041
6	69	4761	14	82	6724
7	47	2209	15	87	7569
8	79	6241	16	59	3481
				$\sum X_i = 1139$	$\sum X_i^2 = 83{,}343$

(ii) *Computation:*

$$\hat{\sigma}^2 = \frac{\sum X_i^2 - (\sum X_i)^2/n}{n-1} = \frac{83{,}343 - (1139)^2/16}{16 - 1} = 150.6958$$

$$\chi^2 = \frac{(n-1)\hat{\sigma}^2}{\sigma_0^2} = \frac{(16-1)150.6958}{(18)^2} = \frac{2260.437}{324.000} = 6.977$$

$$v = n - 1 = 16 - 1 = 15$$

$$\chi^2_{.95,15} = 7.261$$

If the sampling distribution of χ^2 is to have the same form as the chi-square distribution, the population distribution of X must be normal in form. Deviations from the normal form can result in large discrepancies between the actual and nominal type I and II errors.

Review Exercises for Section 12.3

7. List the similarities and differences between the χ^2 and t sampling distributions.
*8. For the data in Exercise 3 (page 266), test the hypothesis $H_0: \sigma^2 = (15)^2$. Let $\alpha = .10$.
9. The mean number of movies attended per month by 14-year-old boys in the United States is 4.8, with a standard deviation of 1.3. According to juvenile court records in Houston, Texas, the corresponding statistics for a random sample of 31 boys who appeared in court are $\bar{X} = 4.9$ and $\hat{\sigma} = 1.6$. Evaluate the hypothesis that the dispersion of the movie attendance distribution for boys appearing in the Houston juvenile court is different from that for the nation at large. Let $\alpha = .02$.
10. List the assumptions involved in testing the null hypothesis in Exercise 9.

Section 12.4   One-Sample z Test for a Population Proportion

An experimenter is often interested in testing a hypothesis about a population proportion. For example, an opinion pollster may want to know if a majority of the voters favor a certain candidate, an automobile manufacturer may want to know if at least 0.70 of new car buyers are willing to pay $150 for a safety device, or the United States Marine Corps may want to know if at least 0.35 of its volunteers plan to reenlist. Each of these examples has a large number of occasions, or independent trials, in which one of two outcomes can occur, and the probabilities associated with the two outcomes remain constant from trial to trial. For convenience the outcomes are designated "success" and "failure," with probabilities p and $1 - p = q$, respectively.

We saw in Section 9.3 that a Bernoulli trial results in success with probability p or failure with probability q and that the probability of exactly r successes in n independent trials for the binomial random variable X is given by

$$p(X = r) = {}_nC_r p^r q^{n-r}.$$

If n is very small, this binomial function rule can be used to determine the probabilities associated with values of the random variable X—that is, the probabilities that various numbers of successes will occur. If, however, n is large, this procedure is not computationally feasible. In this case, the normal distribution approximation to the binomial distribution can be used. This approximation is excellent if n is large and $p = 0.5$; as n becomes smaller and/or p approaches either 0 or 1, the approximation becomes poorer. As a rule of thumb, if np_0 (the sample size multiplied by the population proportion specified in the null hypothesis) and $n(1 - p_0)$ are both greater than 5, the normal approximation is satisfactory.

The z statistic for testing the null hypothesis about a population proportion is given by

$$z = \frac{\hat{p} - p_0}{\sqrt{p_0 q_0 / n}} \qquad \text{or} \qquad z = \frac{X - np_0}{\sqrt{np_0 q_0}}.$$

where $\hat{p}$ is the sample estimator of the population proportion, p_0 is the hypothesized value of the population proportion specified in the null hypothesis, q_0 is equal to $1 - p_0$, n is the size of the sample used to compute $\hat{p}$, and X is equal to $n\hat{p}$ (the number of elements in the success category). The z statistic can be used to test hypotheses such as the following.

$$H_0:\ \ p = p_0 \qquad H_0:\ \ p \le p_0 \qquad H_0:\ \ p \ge p_0$$
$$H_1:\ \ p \ne p_0 \qquad H_1:\ \ p > p_0 \qquad H_1:\ \ p < p_0$$

Here, p is the population proportion.

Computational Example for One-Sample z Test

Suppose the League for Better Housing has made a survey to determine whether the proportion of substandard dwelling units in a large city has changed since the last census five years ago. At that time 0.30 of the dwelling units were classified as substandard. A random sample of 900 dwelling units is surveyed, and the proportion of substandard units is found to be 0.34. The decision rule and the steps to be followed in testing the null hypothesis are as follows.

Step 1. State the statistical hypotheses: $H_0:\ \ p = 0.30$
$$H_1:\ \ p \ne 0.30$$

Step 2. Specify the test statistic: $\quad z = \dfrac{\hat{p} - p_0}{\sqrt{p_0 q_0 / n}}.$

Step 3. Specify the sample size:[3] $n = 900$;
and the sampling distribution: standard normal distribution, because z is a satisfactory approximation to the binomial distribution, since both $np_0 = 270$ and $n(1 - p_0) = 630$ are greater than 5.

Step 4. Specify the significance level: $\alpha = .05$.

Step 5. Obtain a random sample of size n, compute z, and make a decision.

Decision rule:
Reject H_0 if z falls in either the lower 2.5% or the upper 2.5% of the sampling distribution of z; otherwise, don't reject H_0.

For a $\hat{p} = 0.34$, the proportion of substandard dwelling units in the last survey, the z statistic is

$$z = \frac{0.34 - 0.30}{\sqrt{(0.30)(0.70)/900}} = \frac{0.04}{\sqrt{0.0002333}} = 2.62.$$

According to Appendix Table D.2, z's of $+1.96$ and -1.96, respectively, cut off the upper and lower 0.025 regions of the sampling distribution. Since $2.62 > 1.96$,

[3] Cohen (1969, chap. 6) provides tables for estimating the sample size required for various values of α, $1 - \beta$, and effect size.

the null hypothesis is rejected and it is concluded that the housing situation hasn't improved, but rather has deteriorated.

Correction for Continuity

Many variables, such as the number of substandard dwelling units, have a discrete probability distribution. *When the continuous normal distribution is used to estimate probabilities for a discrete distribution and n is small, it is desirable to apply a correction for continuity. In making a correction for continuity we treat a number, say $X = 6$, as representing the interval 5.5–6.5; that is, a number is considered to have an actual lower limit equal to $X - (Unit\ of\ measurement)/2$ and an actual upper limit equal to $X + (Unit\ of\ measurement)/2$. The addition or subtraction of $\frac{1}{2}$ the unit of measurement is the correction for continuity.* The rule for applying the correction is simple. Add $\frac{1}{2}$ the unit of measurement to X in the formula $z = (X - np_0)/\sqrt{np_0q_0}$ if $X < np_0$; if $X > np_0$, subtract $\frac{1}{2}$ the unit of measurement from X.

If n is large, the use of the continuity correction has little effect on the z test and can be dispensed with. The correction has little effect on the z statistic for the housing survey data, as the following computation shows.

$$z = \frac{(X - \frac{1}{2}) - np_0}{\sqrt{np_0q_0}} = \frac{(306 - 0.5) - 900(0.30)}{\sqrt{900(0.30)(0.70)}} = \frac{35.5}{\sqrt{189}} = 2.58.$$

Computational Procedure when the Sample Size Is Equal to or Greater than 10% of the Population

If the population is finite and if sampling is carried out without replacement, the random variable X is distributed as the hypergeometric distribution (see Section 9.3) rather than as the binomial distribution. For this situation the z statistic should be modified if the ratio $n/n_{Pop} \geq 0.10$, where n is the number of elements in the sample and n_{Pop} is the number of elements in the population. The modified z statistic is

$$z = \frac{\hat{p} - p_0}{\sqrt{(p_0q_0/n)[(n_{Pop} - n)/(n_{Pop} - 1)]}}.$$

The denominator adjustment $(n_{Pop} - n)/(n_{Pop} - 1)$ can be ignored if $n/n_{Pop} < 0.10$.

Review Exercises for Section 12.4

*11. If $p_0 = 0.20$, how large should n be in order to use the normal approximation to the binomial distribution?

*12. Suppose that you are interested in testing babies' color preferences. On each of $n = 20$ trials you offer a baby a choice between two balls—one red and one green. The baby chooses a red ball on 12 of the 20 trials. Can you conclude that the baby has a preference for red over green? Use the correction for continuity. Let $\alpha = .05$.

13. The probability of recovery for schizophrenic patients after receiving 6 months of conventional therapy at Happyfarm Hospital was 0.60. A token economy program was introduced for a random sample of 30 schizophrenic patients. At the end of the 6-month trial period, 21 patients had recovered. Test the hypothesis that the recovery probability for the token economy program would be higher than that for the conventional therapy. Let $\alpha = .05$.

14. Sketch the sampling distribution for z in Exercise 13 and label the critical region.
15. Assume that in Exercise 13 there were only 90 schizophrenic patients in the hospital. Test the null hypothesis using the z statistic for the finite population case.
16. Term to remember:
    Correction for continuity

Section 12.5   One-Sample t and z Tests for a Population Correlation

t Test for H_0: $\rho = 0$

In many correlation problems the hypotheses of interest are

$$H_0: \quad \rho = 0$$

$$H_1: \quad \rho \neq 0.$$

A sample correlation coefficient, r, can differ from 0 due to chance sampling variability. Thus, just because $r \neq 0$ it can't be concluded that the population correlation parameter, ρ, differs from 0. In this section we describe a t test that helps an experimenter decide on the basis of sample data whether ρ is or isn't equal to 0.

Computational Example for t Test

Consider an experimenter who is interested in determining whether a correlation exists between college grades and income 10 years after graduation. Assume that grade point averages and income data for a random sample of 62 male graduates of Boston Central College have been obtained. The product-moment correlation between grade point average and income for this sample is 0.16. How likely is it that a sample correlation coefficient of this size would have been obtained if the correlation between income and grades really is equal to 0?
   A test of the hypothesis H_0: $\rho = 0$ is given by the t statistic

$$t = \frac{r\sqrt{n-2}}{\sqrt{1-r^2}},$$

with $n - 2$ degrees of freedom. The sampling distribution of r can be regarded as approximately normal when ρ is close to 0; for other values of ρ, the sampling distribution of r tends to be very skewed and the t statistic shouldn't be used.
   The t test of the hypothesis $\rho = 0$ assumes that the population of paired X and Y scores is *bivariate normal* in form. *A **bivariate normal population** is one in which the distributions of X and Y are normal, the relationship between X and Y is linear, and the distribution of Y for any value of X is normal with variance that does not depend on the X value selected and vice versa.*
   The decision rule and the steps to be followed in testing the hypothesis are as follows.

Step 1.   State the statistical hypotheses: H_0: $\rho = 0$
$\qquad\qquad\qquad\qquad\qquad\qquad\qquad\quad H_1$: $\rho \neq 0.$

Step 2.   Specify the test statistic: $\qquad t = \dfrac{r\sqrt{n-2}}{\sqrt{1-r^2}}.$

Step 3.   Specify the sample size:[4] $n = 62$;
          and the sampling distribution: t distribution with $v = n - 2$, because the population is assumed to be bivariate normal in form.

Step 4.   Specify the significance level: $\alpha = .05$

Step 5.   Obtain a random sample of size n, compute t, and make a decision.

Decision rule:

Reject H_0 if t falls in either the upper or lower 2.5% of the sampling distribution of t; otherwise, don't reject H_0.

Since $r = 0.16$ for the random sample of $n = 62$ male graduates, the t statistic is

$$t = \frac{0.16\sqrt{62 - 2}}{\sqrt{1 - (0.16)^2}} = \frac{1.239}{0.987} = 1.26,$$

with $v = 62 - 2 = 60$. According to Appendix Table D.3, $t_{.0.5/2.60} = 2.00$. Since $1.26 < 2.00$, the null hypothesis isn't rejected. The data do not warrant the conclusion that the population correlation coefficient is not equal to 0.

There is a much simpler way to test the null hypothesis when it specifies that $\rho = 0$. The critical values of r and r_s (Spearman's rank correlation coefficient) necessary to reject the null hypothesis are tabulated in Tables D.6 and D.7, respectively. If $|r|$ is equal to or greater than the critical value in the table, the hypothesis that $\rho = 0$ can be rejected.

z Test for Correlations Other than Zero

The t test just described can be used to test the null hypothesis when the predicted value equals 0. But what if the experimenter wants to test a hypothesis in which the predicted value is equal to some number other than 0, for example, $\rho = 0.50$? A procedure due to R. A. Fisher can be used in such cases. It uses a particular function of r rather than r; the function is called the *Fisher r-to-Z' transformation*. By means of Table D.8, r can be transformed easily into Fisher's Z'. This table gives for each value of r the corresponding Fisher Z' statistic. For example, if $r = 0.50$, $Z' = 0.549$. Fisher showed that the sampling distribution of Z' is approximately normal in form if (1) ρ is not too close to 1 or -1, (2) the population is bivariate normal in form, and (3) the sample size is at least moderately large, say $n > 50$.

The one-sample z statistic used to test hypotheses about ρ is given by

$$z = \frac{Z' - Z_0'}{\sqrt{1/(n - 3)}},$$

where Z' is the transformed sample r, Z_0' is the transformed value of ρ_0 specified by the null hypothesis, and n is the size of the sample used to estimate the population correlation parameter.

---

[4] Cohen (1969, chap. 4) provides tables for estimating the sample size required for various values of α, $1 - \beta$, and effect size.

Computational Example for z Test

Assume that an experimenter believes that the population correlation between two variables is greater than 0.25. A random sample of 83 subjects from the population of interest has been obtained, and r is found to equal 0.47. The decision rule and the steps to be followed in testing the hypothesis are as follows.

Step 1.   State the statistical hypotheses: $H_0: \rho \leq 0.25$
$H_1: \rho > 0.25$.

Step 2.   Specify the test statistic: $z = \dfrac{Z' - Z_0'}{\sqrt{1/(n - 3)}}$.

Step 3.   Specify the sample size: $n = 83$;
and the sampling distribution: z distribution, because the population is assumed to be bivariate normal in form and the parameter specified in H_0 is not 0.

Step 4.   Specify the significance level: $\alpha = .05$.

Step 5.   Obtain a random sample of size n, compute z, and make a decision.

Decision rule:
Reject H_0 if z falls in the upper 5% of the sampling distribution of z; otherwise, don't reject H_0.

The test statistic is given by

$$z = \frac{Z' - Z_0'}{\sqrt{1/(n - 3)}} = \frac{0.510 - 0.255}{\sqrt{1/(83 - 3)}} = \frac{0.255}{\sqrt{0.0125}} = 2.28.$$

According to Table D.2, a z of 1.645 cuts off the upper .05 region of the sampling distribution of z—that is, $z_{.0.5} = 1.645$. Since $2.28 > 1.645$, we reject the null hypothesis and conclude that the population correlation coefficient is greater than 0.25.

Review Exercises for Section 12.5

*17. Assume that $r = 0.24$ has been computed for a random sample of 26 subjects. Test the hypothesis $H_0: \rho = 0$. Let $\alpha = .05$.

*18. Convert r into Z'.
   *a. $r = 0.46$ *b. $r = -0.23$
    c. $r = -0.96$ d. $r = 0.15$

*19. Convert Z' into r.
   *a. $Z' = 0.549$ *b. $Z' = -0.203$
    c. $Z' = 0.121$ d. $Z' = -1.256$

20. Assume that $r = 0.63$ has been computed for a random sample of 96 subjects. Test the hypothesis $H_0: \rho \leq 0.50$. Let $\alpha = .05$.

*21. a. If r equaled 0.48 in Exercise 20, would it be necessary to compute a test of significance?
    b. Why?

22. The sampling distribution of r is not likely to be normal when ρ deviates appreciably from 0. From what you know about r, why is this true?

23. Terms to remember:
    a. Bivariate normal
    b. Fisher r-to-Z' transformation

Section 12.6   Summary

We have covered much ground in Chapters 11 and 12—the basic concepts of hypothesis testing and a variety of test statistics for the one-sample case. Despite the fact that the test statistics have different formulas and are used to test hypotheses about different parameters, they all use the same five steps in arriving at a decision about a hypothesis. We will see that the logic underlying hypothesis-testing procedures described in these chapters generalizes to the two-sample case and to more complex decision-making situations.

The test statistics for the one-sample case are presented in Table 12.6-1, along with a summary of the parameters they are used to test, their formulas, and their assumptions.

Table 12.6-1.   Summary of one-sample test statistics

Chapter section	Statistic	Statistical hypotheses	Formula	Assumptions
11.3	z	H_0: $\mu = \mu_0$ H_1: $\mu \neq \mu_0$	$z = \dfrac{\bar{X} - \mu_0}{\sigma/\sqrt{n}}$	1. Random sampling 2. Normality or large sample 3. Variance is known
12.2	t	H_0: $\mu = \mu_0$ H_1: $\mu \neq \mu_0$	$t = \dfrac{\bar{X} - \mu_0}{\hat{\sigma}/\sqrt{n}}$	1. Random sampling 2. Normality 3. Variance is unknown
12.3	χ^2	H_0: $\sigma^2 = \sigma_0{}^2$ H_1: $\sigma^2 \neq \sigma_0{}^2$	$\chi^2 = \dfrac{(n-1)\hat{\sigma}^2}{\sigma_0{}^2}$	1. Random sampling 2. Normality
12.4	z	H_0: $p = p_0$ H_1: $p \neq p_0$	$z = \dfrac{\hat{p} - p_0}{\sqrt{p_0 q_0/n}}$ $z = \dfrac{X - np_0}{\sqrt{np_0 q_0}}$	1. Random sampling 2. Binomial distribution 3. $np_0 > 5$, $n(1 - p_0) > 5$
12.4	z	H_0: $p = p_0$ H_1: $p \neq p_0$	$z = \dfrac{\hat{p} - p_0}{\sqrt{(p_0 q_0/n)[(n_{Pop} - n)/(n_{Pop} - 1)]}}$	1. Random sampling 2. Hypergeometric distribution 3. $np_0 > 5$, $n(1 - p_0) > 5$ 4. $n/n_{Pop} \geq 0.10$
12.5	t	H_0: $\rho = 0$ H_1: $\rho \neq 0$	$t = \dfrac{r\sqrt{n-2}}{\sqrt{1-r^2}}$	1. Random sampling 2. Bivariate normality
12.5	z	H_0: $\rho = \rho_0$ H_1: $\rho \neq \rho_0$	$z = \dfrac{Z' - Z_0'}{\sqrt{1/(n-3)}}$	1. Random sampling 2. Bivariate normality 3. Moderately large sample

†Section 12.7   Computer Supplement

An interactive microcomputer package developed by Brad Wilson (1984) was described in Section 3.10. The one-sample *t* test program of this package was used to analyze the registration-time data in Table 12.2-1. The printout is shown in Table 12.7-1.

Table 12.7-1.   Printout from the Wilson Stat Package: One-sample *t* test program (Data from Table 12.2-1)

```
*****************************************
*                                       *
*              Brad Wilson's            *
*              Stat Package             *
*                                       *
******************************************
*                                       *
*      Behavioral Statistics Program    *
*            Baylor University          *
*****************************************

Press RETURN to continue>>>>>

When you respond to a prompt just
type in your response and press
the RETURN key.

For all yes-no (Y/N) questions
it is not necessary to press the
RETURN key.

Press RETURN to continue>>>>>

Would you like to

   1. Perform statistical analyses

   2. See a demonstration of the
      Central Limit Theorem

   3. End the session?

Select a number:  1

Please wait a few moments.

Here are the available
statistical programs
   1. Frequency distribution and
      histogram
   2. Mean, median, and mode
   3. Standard deviation
   4. Index of dispersion
   5. Correlation
   6. Regression
   7. One-sample t test
```

(continued)

---

† This section can be omitted without loss of continuity.

Table 12.7-1.   (Continued)

```
 8.  One-sample chi-square test
 9.  Two-sample independent t test
10.  Two-sample dependent t test
11.  Analysis of variance for a
     completely randomized design
12.  Pearson's chi-square for
     testing goodness of fit
13.  Pearson's chi-square for
     testing independence
14.  Pearson's chi-square for
     testing equality of proportions
```

```
Indicate the program you want
by selecting a number:  7

Please wait a few moments.

Do you want to enter data from

 1.  The terminal

 2.  A file, or

 3.  Use the data from the
     previous session, or do
     you want to

 4.   Return to the menu?

Select a number:  1

What is the label for
sample 1? TIME

How many values in TIME? 21
```

sample	# of cases
TIME	21

```
Do you want to
 1.  Go on to the next step
 2.  Examine your data specifications
     again
 3.  Change your sample name(s)
 4.  Change your # of cases?
Select one:  1

Now you will be prompted for
the individual values

TIME   1:  2.9
TIME   2:  2.7
```

(continued)

Table 12.7-1.   (Continued)

```
TIME   3:  2.8
TIME   4:  3.0
TIME   5:  2.6
TIME   6:  2.9
TIME   7:  3.1
TIME   8:  2.9
TIME   9:  3.0
TIME  10:  2.7
TIME  11:  2.9
TIME  12:  3.0
TIME  13:  2.8
TIME  14:  3.1
TIME  15:  2.9
TIME  16:  2.7
TIME  17:  3.2
TIME  18:  2.8
TIME  19:  2.8
TIME  20:  3.1
TIME  21:  3.0
```

Press RETURN to continue>>>>>

Would you like to check the data for errors? (Y/N) N

If you have a second disk drive, you can store your data on a separate diskette. Would you like to store the data? (Y/N) N

Please wait a few moments.

Do you want the results sent to the

1. Screen

2. Printer?

Select a number: 1

NOTE: Use CTRL and S simultaneously to stop and start the scrolling of the screen.

Press RETURN to continue>>>>>

Select your alternative hypothesis:

1. That the sample mean is greater than the population mean
2. That the sample mean is less than the population mean

(continued)

Table 12.7-1.   (Continued)

---

3. That the sample mean
    is different from the
    population mean
    (a two-tailed test)

Select one: 2

```
                                  STANDARD
                  N     MEAN      DEVIATION
****************************************
  TIME           21    2.900        0.158
```

One-sample test for a population mean

```
  t Value        df    Standard error
****************************************
  −5.797         20       0.035

*********
  p<0.001
*********
```
Press RETURN to continue⟩⟩⟩

Would you like to

1. Perform statistical analyses

2. See a demonstration of the
    Central Limit Theorem

3. End the session?

Select a number: 3

Just turn off the computer, remove your
diskettes, and have a nice day!!

13

<hr>

Statistical Inference: Two Samples

Section 13.1   Introduction to Hypothesis Testing for Two Samples

Are men able to withstand weightlessness better than women? Do disadvantaged children learn more quickly in a contingency management classroom than in a traditional classroom? Do people who jog have fewer heart attacks than those who don't? Is one antilitter slogan more effective than another? Each of these questions involves a comparison of two population distributions. Population distributions can differ in central tendency, dispersion, skewness, and kurtosis. Most questions in the behavioral sciences and education are concerned with central tendency and, more specifically, with whether the means of two populations differ.

We learned in Chapter 11 that scientific hypotheses often involve predictions (1) about populations whose elements are so numerous that viewing them all is impossible (all men and women in a weightless environment, all disadvantaged school children, all joggers and nonjoggers) or (2) about phenomena that can't be directly observed (the effectiveness of two antilitter slogans). In such cases we can obtain random samples from the populations, and by using hypothesis testing procedures, we can make inferences as to whether the means, variances, and so on, of the populations differ. These procedures are a straightforward extension of those for the one-sample case described in Chapters 11 and 12.

Section 13.2   Two-Sample z Test for $\mu_1 - \mu_2$ Using Independent Samples

We now will describe a z test statistic for determining whether the means of two populations differ. The statistic can be used to test any of the following null hypotheses.

$$H_0: \ \mu_1 - \mu_2 = \delta_0 \qquad H_0: \ \mu_1 - \mu_2 \leq \delta_0 \qquad H_0: \ \mu_1 - \mu_2 \geq \delta_0$$
$$H_1: \ \mu_1 - \mu_2 \neq \delta_0 \qquad H_1: \ \mu_1 - \mu_2 > \delta_0 \qquad H_1: \ \mu_1 - \mu_2 < \delta_0$$

Here, δ_0 is the predicted difference between the population means. Usually, an experimenter is interested in testing the hypothesis that two population means are equal, in which case $\delta_0 = 0$.

Consider populations with unknown means μ_1 and μ_2. Assume that the variances of the populations, σ_1^2 and σ_2^2, are known but are not necessarily equal and that random samples of size n_1 and n_2 have been obtained from the respective populations. If the population distributions of X_1 and X_2 are normal in form or if the samples are sufficiently large, the sampling distribution of the difference between sample means, $\bar{X}_1 - \bar{X}_2$, is normal. It can be shown that the expectation of the difference between sample means, $E(\bar{X}_1 - \bar{X}_2)$, is equal to $\mu_1 - \mu_2$ and that the variance of the difference between sample means, $\sigma^2_{\bar{X}_1 - \bar{X}_2}$, for the case in which the selection of elements in one sample is not affected by (is independent of) the selection of elements in the other sample, is given by

$$\sigma^2_{\bar{X}_1 - \bar{X}_2} = \frac{\sigma_1{}^2}{n_1} + \frac{\sigma_2{}^2}{n_2}.$$

The dependent samples case is discussed in Section 13.5. The square root of $\sigma^2_{\bar{X}_1 - \bar{X}_2}$ is called the *standard error of the difference between two means. The z statistic for testing the null hypothesis when, as in this example, the scores in the two samples are independent is given by*

$$z = \frac{(\bar{X}_1 - \bar{X}_2) - \delta_0}{\sigma_{\bar{X}_1 - \bar{X}_2}} = \frac{(\bar{X}_1 - \bar{X}_2) - \delta_0}{\sqrt{\sigma_1{}^2/n_1 + \sigma_2{}^2/n_2}}.$$

If the null hypothesis specifies that $\delta_0 = 0$, the formula for z simplifies to

$$z = \frac{\bar{X}_1 - \bar{X}_2}{\sqrt{\sigma_1{}^2/n_1 + \sigma_2{}^2/n_2}}.$$

When the null hypothesis is true, the sampling distribution of z is approximately normal in form, with mean equal to 0, $E(z) = 0$, and variance equal to 1, $\sigma_z{}^2 = 1$. Hence, the sampling distribution of z has approximately the same shape, mean, and standard deviation as the standard normal distribution whose values are listed in Table D.2. These similarities enable us to use the standard normal distribution in determining whether an obtained z, our transformed difference between sample means, lies in the critical region of the sampling distribution of z.

In the preceding discussion we assumed that $\sigma_1{}^2$ and $\sigma_2{}^2$ were somehow known so that we could compute the standard error of the difference between means. In the real world such omniscience is rare; the fact of the matter is that population variances are seldom known. We can, however, compute unbiased estimates of the population variances from the samples used to estimate the population means. When the population variances are estimated, the appropriate test statistic is t. But before describing this statistic we will turn briefly to an issue concerning the design of experiments: selection of the appropriate randomization procedure.

Review Exercises for Section 13.2

1. The null hypothesis is sometimes written $H_0\colon \mu_1 = \mu_2$. What does this indicate about δ_0?
2. Under what conditions does the sampling distribution of $z = [(\bar{X}_1 - \bar{X}_2) - \delta_0]/\sigma_{\bar{X}_1 - \bar{X}_2}$ approximate the standard normal distribution?
3. An experimenter is interested in testing the hypothesis that college freshmen who are on probation have lower academic aptitude scores than those not on probation. Random samples of $n_1 = 50$ and $n_2 = 50$ are obtained from the populations of probationers and

nonprobationers, respectively. The populations are known to be normally distributed, with $\sigma_1 = 15$ and $\sigma_2 = 15$. List the five steps you would follow in testing the null hypothesis and state the decision rule. Let $\alpha = .05$.

*4. Suppose that for Exercise 3, $\bar{X}_1 = 112$, $\bar{X}_2 = 116$, and α has been set at .05. Compute the test statistic and make a decision.

5. Discuss the statement "The absolute magnitude of the z test statistic is indicative of the importance or practical significance of the difference between two sample means."

Section 13.3   Two Randomization Strategies: Random Sampling and Random Assignment

Two randomization strategies can be used in investigating scientific hypotheses. An experimenter can obtain random samples from two existing populations of interest or can randomly assign elements of a population to two experimental conditions. In rare cases the two methods can be combined—that is, the experimenter can sample randomly from a population and assign elements of the sample randomly to the experimental conditions. The choice of a strategy affects an experimenter's conclusions, as we will now see.

The Strategy of Random Sampling

Consider the scientific hypothesis that males who jog have fewer heart attacks than those who don't. The statistical hypotheses are

$$H_0: \quad \mu_1 - \mu_2 \geq 0$$

$$H_1: \quad \mu_1 - \mu_2 < 0,$$

where μ_1 and μ_2 designate the mean number of heart attacks of the populations of joggers and nonjoggers, respectively. The alternative hypothesis, which corresponds to the experimenter's hunch, states that the mean number of heart attacks is smaller for joggers than for nonjoggers. To test the null hypothesis, an experimenter could obtain a random sample of 70-year-old men who have jogged regularly since they were 40 and a second sample of men the same age who have never jogged. Suppose that the mean number of heart attacks is $\bar{X}_1 = 0.2$ for the joggers and $\bar{X}_2 = 1.1$ for the nonjoggers and that the difference between the means, $0.2 - 1.1 = -0.9$, is significant at the .01 level. It can be concluded that the population of joggers has fewer attacks than the nonjoggers, and hence the scientific hypothesis is supported.

Can the experimenter conclude that the difference between the population means is due to jogging per se? Unfortunately, the answer is no, because in all likelihood the two populations of men differ in other ways besides jogging. More than likely, males who jog are concerned about health and about staying in good physical shape. Joggers are probably less often obese, have better muscle tone, and have more healthful diets than nonjoggers. If our experimenter had sampled from populations of obese and nonobese males, or from males with good and poor muscle tone, or from males who are and aren't diet conscious, a significant difference in mean number of heart attacks would probably also have been found.

The Strategy of Random Assignment

Suppose that a population of 40-year-old prisoners at Oops Penitentiary is available and that it is possible to exercise some control over their lives for a period of 30 years. The prisoners are randomly assigned to one of two groups, which we will call the experimental and control groups. Those assigned to the experimental group participate in a jogging program for 30 years; those in the control group don't participate in the jogging program. Suppose that at the end of 30 years the mean numbers of heart attacks for those in the experimental and control groups are, respectively, $\bar{X}_1 = 0.3$ and $\bar{X}_2 = 1.4$, and that the difference, $0.3 - 1.4 = -1.1$, is significant at the .01 level. As in the previous experiment, the scientific hypothesis is supported.

Can the experimenter conclude that the difference between the experimental and control groups is due to jogging per se? Again the answer is no. What have we accomplished by using random assignment? This procedure ensures that idiosyncratic characteristics of the subjects, such as their interest in staying in shape, their tendency toward obesity, the adequacy of their diets, and so forth, will be randomly distributed over the two groups and therefore will not selectively bias the dependent measure for either group. Stated another way, random assignment ensures the comparability of the experimental and control groups, since prior to the experiment they should differ no more than would be expected by chance. If at the conclusion of the experiment a significant difference exists between them in the incidence of heart attacks, the experimenter can be confident that the difference is due to events that occurred after the experiment began rather than to idiosyncratic characteristics of the subjects that existed prior to the experiment. And if during the experiment all conditions except the independent variable of jogging are held constant, differences between the groups in number of heart attacks must be due to jogging per se. Unfortunately, it is often difficult and sometimes impossible to verify that all conditions except the independent variable have been held constant. This difficulty certainly would exist in a 30-year experiment.

Advantages and Disadvantages of the Two Research Strategies

Most experiments in the behavioral sciences and education are designed to establish **concomitant relationships** *rather than* **causal relationships.** *To establish that an independent variable X causes an effect Y, it is necessary to demonstrate that X is both necessary and sufficient for the occurrence of Y. To establish a concomitant relationship, it is only necessary to demonstrate that the occurrence or nonoccurrence of one event is accompanied by the occurrence or nonoccurrence of the other event.* The distinction between causal and concomitant relationships is important because the type of relationship affects the way in which an experimenter interprets his or her results. Neither the random-sampling nor the random-assignment experiments just described have established that jogging per se results in fewer heart attacks—a causal relationship—but they have established that males who jog have, on the average, fewer attacks than nonjoggers—a concomitant relationship.

The strategy of drawing random samples from two existing populations that are known to differ in X can't be used to establish causality, because the two populations may differ also on other variables. One or more of the other variables could be responsible for the effect *Y*. An experimenter samples randomly from two existing populations so that conclusions can be generalized to the populations. In many research situations, most notably opinion polling, the discovery of a concomitant relationship is sufficient for the experimenter's purposes.

In the behavioral sciences and education most experimenters have neither the time nor the resources to obtain random samples. In the rare cases in which random samples are obtained, the populations are often so narrowly defined that they are of little interest. For example, human subjects frequently are randomly sampled from a population of sophomore students enrolled at a college, or from volunteers, and so forth. And experimenters who work with animal subjects generally do not attempt to obtain random samples.

The second strategy of randomly assigning subjects to experimental conditions can be used to establish the existence of a causal relationship if all conditions except the independent variable can be held constant. This is a big if, since the requirement is difficult to satisfy in the behavioral sciences and education.

Random assignment of a large number of subjects to the experimental and control conditions equalizes the effects of extraneous subject differences across experimental conditions and allows the experimenter to rule out prior differences as an explanation for results. This is obviously an important advantage of this research strategy. A disadvantage is that the conclusions apply only to the population of subjects used in the experiment—in our example, prisoners at Oops Penitentiary. To the extent that these prisoners resemble males in the United States, the findings logically can be generalized to the larger population. It should be emphasized that any generalization to populations not actually sampled is an exercise in logic rather than statistical inference.[1] In this example it seems unlikely that the resemblance is great enough to permit such a generalization.

If an experimenter wants to generalize findings to some population and also to ensure the comparability of subjects in the experimental and control groups, the two research strategies can be combined. A random sample of subjects from the population of interest can be obtained and then these subjects, randomly assigned to the two conditions. This combined strategy obviously cannot be used when sampling from two populations that differ with respect to the independent variable, for example, populations of joggers and nonjoggers. Such populations are referred to as *intact populations*.

A final point: If an experimenter wants to use statistical inference, the experimental design must include some form of randomization. Which randomization procedure is used will depend upon the objectives of the experiment.

Review Exercises for Section 13.3

6. An experimenter in Conception, Iowa, wished to determine whether there is a relationship between children's IQs and their mothers' ages when they were born. Using school records, a list was compiled of 10-year-olds whose mothers were over 35 at parturition and a second list was compiled of 10-year-olds whose mothers were 20 or under at

---

[1] This point is thoroughly discussed by Edgington (1966). This article is reprinted in Kirk (1972a).

parturition. The experimenter randomly sampled 50 children from each list and administered the Stanford–Binet intelligence test to them. The IQs were found to be
considerably higher for the children of older mothers, and the difference was significant
beyond the .001 level. The experimenter concluded that a woman should postpone
childbearing until later life to ensure a high IQ for her offspring. (a) Comment on the
appropriateness of the experimenter's conclusion. (b) List some alternative explanations
for the observed difference in IQs.

7. a. In Exercise 6, which sampling strategy was used?
   b. Would this strategy enable the experimenter to establish a causal relationship between the IQs of children and the ages of their mothers at parturition?

*8. In Exercise 6, what does the fact that the test statistic was significant at the .001 level
tell you about the magnitude of the difference between the population means?

9. What condition in the random assignment strategy must be satisfied to establish a
causal relationship between the independent and dependent variables?

10. What are the advantages and disadvantages of random sampling and random
assignment?

*11. For each of the following research topics, indicate the research strategy that seems
most appropriate. Justify your choice.

   *a. Relative resistance to extinction of a bar-pressing response acquired under 100%
reinforcement and 50% reinforcement.
   b. Difference between adult males and females in the incidence of alcohol use.
   c. Relationship between grades in college and number of hours studied per week.
   d. Difference in reaction time to the onset of a light versus the onset of a tone.

12. Terms to remember:
   a. Concomitant relationship
   b. Causal relationship
   c. Intact populations

Section 13.4   Two-Sample t Test for $\mu_1 - \mu_2$ Using Independent Samples

If an experimenter wants to test the null hypothesis that the difference between
two population means is equal to some value, say 0, in all likelihood the population variances, σ_1^2 and σ_2^2, will not be known. Hence, the z test statistic described in Section 13.2 can't be computed. The solution to this problem is to
estimate the population variances from sample data and to use a t test instead of
the z test. For the t test, one assumes that the population variances are unknown
but equal and that random samples of size n_1 and n_2 have been obtained from
the respective populations. If the population distributions of X_1 and X_2 are normal in form or if the samples are sufficiently large, the sampling distribution of
the difference between sample means, $\bar{X}_1 - \bar{X}_2$, is normal in form. It can be shown
that the expectation of the difference between sample means, $E(\bar{X}_1 - \bar{X}_2)$, is equal
to $\mu_1 - \mu_2$ and that an unbiased estimator of the variance of the the difference
between sample means, $\sigma_{\bar{X}_1 - \bar{X}_2}^2$, for the case in which the selection of elements in
one sample is independent of the selection of elements in the other sample, is
given by

$$\hat{\sigma}_{\bar{X}_1 - \bar{X}_2}^2 = \frac{\hat{\sigma}_{Pooled}^2}{n_1} + \frac{\hat{\sigma}_{Pooled}^2}{n_2},$$

where

$$\hat{\sigma}^2_{Pooled} = \frac{(n_1 - 1)\hat{\sigma}_1{}^2 + (n_2 - 1)\sigma_2{}^2}{(n_1 - 1) + (n_2 - 1)}.$$

The t test statistic is given by

$$t = \frac{(\bar{X}_1 - \bar{X}_2) - \delta_0}{\hat{\sigma}_{\bar{X}_1 - \bar{X}_2}} = \frac{(\bar{X}_1 - \bar{X}_2) - \delta_0}{\sqrt{(\hat{\sigma}^2_{Pooled}/n_1) + (\hat{\sigma}^2_{Pooled}/n_2)}} = \frac{(\bar{X}_1 - \bar{X}_2) - \delta_0}{\sqrt{\hat{\sigma}^2_{Pooled}(1/n_1 + 1/n_2)}},$$

where δ_0 is the predicted difference between the population means, which is usually equal to 0, and $\hat{\sigma}_{\bar{X}_1 - \bar{X}_2}$ is an estimator of the standard error of the difference between two means. The number of degrees of freedom, v, for the t statistic is equal to $n_1 + n_2 - 2$. Sample 1 contributes $n_1 - 1$ degrees of freedom, 1 less than the number of independent observations, and, likewise, sample 2 contributes $n_2 - 1$ degrees of freedom.

The use of $\hat{\sigma}^2_{Pooled}$ in the formula for t requires a word of explanation. We assumed that the variances of populations 1 and 2 are equal; hence, $\hat{\sigma}_1{}^2$ and $\hat{\sigma}_2{}^2$ are both estimators for the same parameter, σ^2. Whenever two independent estimators of σ^2 are available, a pooled estimator is likely to be better than either of the sample estimators taken alone; $\hat{\sigma}^2_{Pooled}$ is simply a weighted mean of $\hat{\sigma}_1{}^2$ and $\hat{\sigma}_2{}^2$ where the weights are the respective v's. This can be seen from the equation

$$\hat{\sigma}^2_{Pooled} = \frac{(n_1 - 1)\hat{\sigma}_1{}^2 + (n_2 - 1)\hat{\sigma}_2{}^2}{(n_1 - 1) + (n_2 - 1)}.$$

But what if the variances of populations 1 and 2 are unequal? It has been shown that the two-sample t test for independent samples is robust with respect to violation of the assumption of equal population variances, provided that the sample n's are equal. This is an excellent reason for always using equal sample sizes. If the population variances are unequal and the sample n's are markedly different, the sample variances should not be pooled in computing a t statistic. Several approximate solutions have been proposed for this situation; the interested reader is referred to Hays (1981, pp. 286–287) or Kirk (1982, pp. 100–101). Some computer packages perform two t tests: one based on the use of pooled variances and the other, an approximate t test, based on separate (unpooled) variances. Experimenters then can choose the test that appears to be most appropriate for their data.

As in the case of the two-sample z test statistic, t can be used to test any of the following null hypotheses.

$$H_0: \; \mu_1 - \mu_2 = \delta_0 \qquad H_0: \; \mu_1 - \mu_2 \leq \delta_0 \qquad H_0: \; \mu_1 - \mu_2 \geq \delta_0$$

$$H_1: \; \mu_1 - \mu_2 \neq \delta_0 \qquad H_1: \; \mu_1 - \mu_2 > \delta_0 \qquad H_1: \; \mu_1 - \mu_2 < \delta_0$$

Computational Example for Two-Sample t Test
(Independent Samples)

Let's suppose that a student in an experimental psychology course is investigating the hypothesis that distributed practice is superior to massed practice in developing skill on a mirror-tracing task. The task requires subjects to trace a star pattern on a sheet of paper with their nonpreferred hand; they can see themselves tracing the pattern only by looking in a mirror. Forty students, an entire introductory psychology class, are randomly assigned to the practice conditions, with the restriction that 20 students are assigned to each condition. Although the subjects are assigned to the conditions without replacement (once selected, a subject is no longer available for another assignment), the samples can be regarded for all practical purposes as independent. Subjects in the distributed condition have a 3-min rest period at the end of each practice trial. Subjects in the massed condition have only a 5-sec pause at the end of each trial—just long enough to permit the experimenter to place a new sheet of paper in the tracing apparatus. Both groups receive 15 practice trials. Since the groups may differ in the amount of fatigue at the conclusion of practice, the dependent variable is measured the following day. The subjects are given two warm-up trials; the dependent variable is the time required to trace the star pattern on the next three trials. The decision rule and the steps to be followed in testing the null hypothesis are as follows.

Step 1. State the statistical hypotheses: H_0: $\mu_1 - \mu_2 \geq 0$
H_1: $\mu_1 - \mu_2 < 0$,
where μ_1 and μ_2 denote the population means, respectively, for the distributed and massed conditions.

Step 2. Specify the test statistic: $$t = \frac{\bar{X}_1 - \bar{X}_2}{\hat{\sigma}_{\bar{X}_1 - \bar{X}_2}}.$$

Step 3. Specify the sample sizes:[2] $n_1 = 20$ and $n_2 = 20$;
and the sampling distribution: t distribution, because the population variances are estimated from sample data and the population distributions of X_1 and X_2 are approximately normal in form.

Step 4. Specify the significance level: $\alpha = .05$.

Step 5. Obtain random samples of size n_1 and n_2, compute t, and make a decision.

Decision Rule:
Reject H_0 if t falls in the lower .05 portion of the sampling distribution of t; otherwise, don't reject H_0.

The data for the experiment are shown in Table 13.4-1. According to Table D.3, a t of -1.69 cuts off the lower .05 region of the sampling distribution—that is, $-t_{.05,38} = -1.69$. Since $-1.46 > -1.69$, the null hypothesis is not rejected.

---

[2] Table D.9 can be used to estimate the sample sizes required for various values of α, $1 - \beta$, and effect size, $(\mu_1 - \mu_2)/\sigma$. The concept of effect size is discussed in Section 11.4; use of the table is discussed in Sections 12.2 and 13.5.

Table 13.4-1.   Mirror-tracing data

(i) *Data:*

| | Distributed practice | | Massed practice | |
Student	Time, X_1 (seconds)	Student	Time, X_2 (seconds)
1	17	21	19
2	18	22	20
3	16	23	22
4	18	24	24
5	12	25	10
6	20	26	25
7	18	27	20
8	20	28	22
9	20	29	21
10	22	30	23
11	20	31	20
12	10	32	10
13	8	33	12
14	12	34	14
15	16	35	12
16	16	36	20
17	18	37	22
18	20	38	24
19	18	39	23
20	21	40	17
$n_1 = 20$	$\sum X_1 = 340$	$n_2 = 20$	$\sum X_2 = 380$
	$\sum X_1{}^2 = 6054$		$\sum X_2{}^2 = 7662$

(ii) *Computation of preliminary statistics:*

$$\bar{X}_1 = \frac{\sum X_1}{n_1} = \frac{340}{20} = 17 \qquad \bar{X}_2 = \frac{\sum X_2}{n_2} = \frac{380}{20} = 19$$

$$\hat{\sigma}_1{}^2 = \frac{\sum X_1{}^2 - (\sum X_1)^2/n_1}{n_1 - 1} \qquad \hat{\sigma}_2{}^2 = \frac{\sum X_2{}^2 - (\sum X_2)^2/n_2}{n_2 - 1}$$

$$= \frac{6054 - (340)^2/20}{20 - 1} \qquad = \frac{7662 - (380)^2/20}{20 - 1}$$

$$= 14.421 \qquad\qquad = 23.263$$

$$\hat{\sigma}^2_{Pooled} = \frac{(n_1 - 1)\hat{\sigma}_1{}^2 + (n_2 - 1)\hat{\sigma}_2{}^2}{(n_1 - 1) + (n_2 - 1)} = \frac{(20 - 1)(14.421) + (20 - 1)(23.263)}{(20 - 1) + (20 - 1)}$$

$$= 18.842$$

(iii) *Computation of t:*

$$t = \frac{\bar{X}_1 - \bar{X}_2}{\sqrt{\hat{\sigma}^2_{Pooled}(1/n_1 + 1/n_2)}} = \frac{17 - 19}{\sqrt{18.842(1/20 + 1/20)}}$$

$$= \frac{-2}{1.373} = -1.46$$

$$-t_{.05,38} = -1.69$$

The test doesn't warrant the inference that distributed practice leads to better performance on the tracing task than massed practice.

Review Exercises for Section 13.4

13. Discuss the meaning of the following statement: $t = (\bar{X}_1 - \bar{X}_2)/\sqrt{\hat{\sigma}^2_{Pooled}(1/n_1 + 1/n_2)}$ is the ratio of two random variables, but $z = (\bar{X}_1 - \bar{X}_2)/\sqrt{\sigma_1^2/n_1 + \sigma_2^2/n_2}$ is the ratio of a random variable to a constant.

*14. Under what condition is it appropriate to pool $\hat{\sigma}_1^2$ and $\hat{\sigma}_2^2$ in estimating $\sigma_{\bar{X}_1 - \bar{X}_2}$?

*15. It has been reported that employment interviewers spend more time talking to applicants who are hired than to applicants who are rejected. To determine whether this is true for college students seeking summer employment through the University Placement Center, an experimenter posing as an applicant accompanied a random sample of referees to their job interviews. A record of the duration and outcome of $n = 49$ interviews was kept. Compute a t test statistic and make a decision about the experimenter's hypothesis. Let $\alpha = .05$.

Duration of interview
(minutes)

Hired		Rejected	
30	23	19	17
21	24	18	18
24	26	22	19
25	27	13	22
29	24	15	15
24	22	18	19
23	25	17	17
24	26	20	20
28	23	18	18
25	24	19	17
24	27	23	
19	26	12	
25	25	18	

16. A college dean believed that car ownership among students leads to lower grades. To test this hypothesis, she obtained a random sample of student car owners and non-owners and looked up their grades. Compute a t test statistic for the data in the table and make a decision about the dean's hypothesis. Let $\alpha = .05$.

Grade point averages

Students owning cars			Students not owning cars		
2.6	2.5	2.4	2.7	2.9	3.0
2.4	2.6	2.5	2.9	2.5	2.9
2.9	2.8	2.8	2.6	3.1	2.7
2.6	2.7	2.6	2.8	2.8	3.2
2.7	3.0	2.5	3.0	2.9	2.9
2.2	2.3	2.6	2.8	3.0	3.0

17. In exercise 16, the dean decided to prohibit freshmen from bringing cars to campus. (a) Do you think this action was justified by the data? (b) What other kinds of data about car owners and nonowners would be useful in helping the dean arrive at a rational car policy?

*18. The effect of early language experience on the discrimination of speech sounds was investigated. Twenty-eight 6- to 8-month-old infants raised in English- or Spanish-speaking homes were trained to turn their heads when they detected a change in a sound stimulus. Following the discrimination training, Spanish contrasts involving a tapped and a trilled "r" were presented. The dependent measure was the number of head turns to stimuli involving a change minus the number of head turns on control trials divided by the number of experimental trials. The following data were obtained. (Suggested by Eilers, Rebecca E., Gavin, William J., and Oller, D. Kimbrough. Cross-linguistic perception in infancy: early effects of linguistic experience. *Journal of Child Language*, 1981, 9, 289–302.)

English-speaking home	Spanish-speaking home
.0421	.1081
.0941	.0986
.1064	.1566
.0242	.1961
.1331	.1125
.0773	.1942
.0243	.1079
.0815	.1021
.1186	.1583
.0356	.1673
.0728	.1675
.0999	.1856
.0614	.1688
.0479	.1512

a. Use a t statistic to test the hypothesis that $\mu_1 - \mu_2 = 0$, where μ_1 and μ_2 denote, respectively, the population means for infants raised in English- and Spanish-speaking homes. Let $\alpha = .001$.

b. What conclusion would you draw concerning the effect of early language experience?

Section 13.5   The z and t Tests for $\mu_1 - \mu_2$ Using Dependent Samples

Introduction to Dependent Samples

The tests we have described so far have required the use of independent samples in which the selection of elements in one sample is not affected by the selection of elements in the other. Samples are independent if, for example, an experimenter samples randomly from two populations or uses a random procedure to assign elements to two samples. If it is feasible to do so, an experimenter may prefer to use dependent samples because this results in a more powerful test of a false null hypothesis. Dependent samples can be obtained by any of the following research procedures.

1. Observing subjects under both the experimental condition and the control condition—that is, obtaining *repeated measures* on the subjects.
2. Matching each subject in the experimental condition with a subject in the control condition on some variable that is correlated with the dependent variable. This is called *subject matching*.
3. Obtaining sets of identical twins or litter mates and assigning one member of the pair randomly to the experimental condition and the other member to the control condition.
4. Obtaining pairs of subjects who are matched by mutual selection, for example, husband-and-wife pairs or business partners.

Let us consider these procedures in more detail. The first procedure, observing a set of subjects under two conditions, can be used only with independent variables that have relatively short-duration effects. The nature of the independent variable should be such that the effects of one condition dissipate before the subject is observed under the other condition. Otherwise, the second dependent measure will reflect the cumulative effects of two conditions rather than only the effects of the second condition. There is no such restriction, of course, when carryover effects such as learning or fatigue are the principal interest of the experimenter. If repeated measures are obtained, the order of presentation of the two conditions should be randomized independently for each subject if possible. It is customary to restrict randomization so that half the subjects receive one condition first while the other half receive the other condition first.

The remaining three procedures for obtaining dependent samples involve forming pairs of subjects who are matched on some basis. In subject matching, a matching variable is used to pair up otherwise unrelated subjects; this matching variable must correlate highly with the dependent variable. For example, IQ and ability to learn verbal material are highly correlated; hence, subjects can be assigned to pairs so that members of each pair have similar IQs and therefore similar verbal learning abilities. The higher the correlation between the matching variable and the dependent variable, the more effective the matching. If identical twins or litter mates are used, it can be assumed that subjects within a pair are matched with respect to genetic characteristics. The aptitudes and abilities of identical twins, fraternal twins to some extent, and even siblings are more similar than those of unrelated subjects. When subjects are matched by mutual selection, the experimenter always must ascertain that the subjects within pairs are in fact more similar with respect to the dependent variable than are unmatched subjects. Knowing a husband's attitudes on certain issues, for example, may provide considerable information about his wife's attitudes on the issues, and vice versa. However, knowing the husband's mechanical aptitude or his recreational interests may provide no information about his wife's aptitude or interests.

z Test for Dependent Samples

You probably wonder what difference it makes whether samples are dependent or independent. If the same subjects are observed twice or if subjects in one sample are paired with subjects in another sample, the outcomes X_1 and X_2 for each pair are not statistically independent. This doesn't affect the expectation of the difference between sample means; the expectation $E(\bar{X}_1 - \bar{X}_2)$ is equal to $\mu_1 - \mu_2$. However, dependence within pairs affects the variance of the difference between means. Let us consider first the case in which σ_1^2, σ_2^2, and ρ_{12} are known. The variance of $\bar{X}_1 - \bar{X}_2$ when samples are dependent is

$$\sigma^2_{\bar{X}_1 - \bar{X}_2} = \sigma^2_{\bar{X}_1} + \sigma^2_{\bar{X}_2} - 2\rho_{12}\sigma_{\bar{X}_1}\sigma_{\bar{X}_2},$$

where $\sigma^2_{\bar{X}_1} = \sigma_1^2/n_1$, $\sigma^2_{\bar{X}_2} = \sigma_2^2/n_2$, and ρ_{12} is the product-moment correlation between X_1 and X_2. The corresponding formula for independent samples given in Section 13.2 is

$$\sigma^2_{\bar{X}_1 - \bar{X}_2} = \sigma^2_{\bar{X}_1} + \sigma^2_{\bar{X}_2}.$$

A comparison of these formulas indicates that the larger the positive correlation between X_1 and X_2, the smaller is the variance error for dependent samples relative to that for independent samples. Hence, if $\rho > 0$, an experimenter will underestimate z if

$$z = \frac{(\bar{X}_1 - \bar{X}_2) - \delta_0}{\sqrt{\sigma^2_{\bar{X}_1} + \sigma^2_{\bar{X}_2}}}$$

is used instead of

$$z = \frac{(\bar{X}_1 - \bar{X}_2) - \delta_0}{\sqrt{\sigma^2_{\bar{X}_1} + \sigma^2_{\bar{X}_2} - 2\rho_{12}\sigma_{\bar{X}_1}\sigma_{\bar{X}_2}}}.$$

The computational procedures for the dependent-samples z test will not be illustrated because in practice we rarely know the values of σ_1^2, σ_2^2, and ρ_{12}. When these parameters are unknown, it is customary to use the t statistic, discussed next, instead of z.

t Test for Dependent Samples

We will now consider the more usual case in which σ_1^2, σ_2^2, and ρ_{12} are unknown. If the population distributions are normal in form or if the samples are sufficiently large, a t statistic for dependent samples can be used to test any of the following null hypotheses.

$$H_0: \ \mu_1 - \mu_2 = \delta_0 \qquad H_0: \ \mu_1 - \mu_2 \leq \delta_0 \qquad H_0: \ \mu_1 - \mu_2 \geq \delta_0$$
$$H_1: \ \mu_1 - \mu_2 \neq \delta_0 \qquad H_1: \ \mu_1 - \mu_2 > \delta_0 \qquad H_1: \ \mu_1 - \mu_2 < \delta_0.$$

One form of the t statistic is

$$t = \frac{(\bar{X}_1 - \bar{X}_2) - \delta_0}{\sqrt{\hat{\sigma}^2_{\bar{X}_1} + \hat{\sigma}^2_{\bar{X}_2} - 2r_{12}\hat{\sigma}_{\bar{X}_1}\hat{\sigma}_{\bar{X}_2}}}.$$

This formula is rarely used because a simpler computational formula—one that doesn't require the computation of a correlation coefficient, r_{12}—is available. The formula to be described is simpler because it replaces each pair of scores, X_1 and

X_2, with one difference score, D, where $D = X_1 - X_2$. In effect this converts the two-sample dependent t test for X_1 and X_2 into a one-sample t test for D. The null hypothesis for this analysis is

$$H_0: \quad \mu_D = \delta_0,$$

where μ_D is the mean of differences between paired scores. This null hypothesis is equivalent to $H_0: \mu_1 - \mu_2 = \delta_0$, since $\mu_D = \mu_1 - \mu_2$.

The t test statistic for dependent samples using the difference score approach is

$$t = \frac{\bar{X}_D}{\hat{\sigma}_{\bar{X}_D}} = \frac{\dfrac{\sum D_i}{n}}{\dfrac{\sqrt{\dfrac{\sum D_i^2 - (\sum D_i)^2/n}{n}}}{\sqrt{n-1}}},$$

where $\bar{X}_D$ is the mean of difference scores, $\hat{\sigma}_{\bar{X}_D}$ is used to estimate the standard error of the mean of difference scores, $D_i = X_{i1} - X_{i2}$ for the ith pair of scores, and n is the number of pairs of scores. The number of degrees of freedom, v, for this test statistic is equal to $n - 1$. On reflection, it seems reasonable that $v = n - 1$. Recall that $\sum(X_i - \bar{X}) = 0$. Also, it can be shown that $\sum(D_i - \bar{X}_D) = 0$. Thus, when any of the $n - 1$ deviations $D_i - \bar{X}_D$ are known, the remaining deviation is also known, since the $n - 1$ deviations plus the remaining deviation must sum to zero. The following example should help to clarify the meaning of the terms in this t test statistic.

Computational Example for Two-Sample t Test (Dependent Samples)

The scientific hypothesis for the mirror-tracing task described in Section 13.4 could have been investigated using matched subjects. Suppose that subjects are tested on the mirror-tracing task using their preferred hand. The time required to trace the star pattern on the last three of five trials is used to form pairs of subjects having comparable tracing times and hence similar motor skills. One subject of each pair is randomly assigned to the distributed practice condition and the other subject, to the massed condition. Following the matching procedure, the mirror-tracing experiment is carried out as described previously. Data for the experiment are shown in Table 13.5-1. According to Table D.3, a t of -1.729 with $v = 20 - 1 = 19$ cuts off the lower .05 region of the sampling distribution. Since $-3.04 < -1.729$, the null hypothesis is rejected, and it is concluded that distributed practice leads to better performance on the task than massed practice. Of course, this inference applies only to the population represented by the subjects in the experiment and to the particular practice conditions and task that were used.

Has the experimenter gained anything by using matched subjects? To answer this question we must compare the results of using independent samples with those

Table 13.5-1.   Mirror-tracing data (dependent samples)

(i) *Data:*

Student pair, i	Distributed practice time, X_{i1} (seconds)	Massed practice time, X_{i2} (seconds)	Difference, $D_i = X_{i1} - X_{i2}$
1	17	19	-2
2	18	20	-2
3	16	22	-6
4	18	24	-6
5	12	10	2
6	20	25	-5
7	18	20	-2
8	20	22	-2
9	20	21	-1
10	22	23	-1
11	20	20	0
12	10	10	0
13	8	12	-4
14	12	14	-2
15	16	12	4
16	16	20	-4
17	18	22	-4
18	20	24	-4
19	18	23	-5
20	21	17	4
$n = 20$	$\bar{X}_1 = 17$	$\bar{X}_2 = 19$	$\sum D_i = -40$
			$\sum D_i^2 = 244$

(ii) *Computation of preliminary statistics:*

$$\bar{X}_D = \frac{\sum D_i}{n} = \frac{-40}{20} = -2$$

Computational check:

$$\bar{X}_D = \bar{X}_1 - \bar{X}_2 = 17 - 19 = -2$$

$$\hat{\sigma}_{\bar{X}_D} = \frac{\sqrt{\dfrac{\sum\limits_{i=1}^{n} D_i^2 - \left(\sum\limits_{i=1}^{n} D_i\right)^2 / n}{n}}}{\sqrt{n-1}} = \frac{\sqrt{\dfrac{244 - (-40)^2/20}{20}}}{\sqrt{20-1}} = \frac{2.8636}{4.3589} = 0.657$$

(iii) *Computation of t:*

$$t = \frac{\bar{X}_D}{\hat{\sigma}_{\bar{X}_D}} = \frac{-2}{0.657} = -3.04$$

$$-t_{.05,19} = -1.729$$

for dependent samples. This comparison can be made from Tables 13.4-1 and 13.5-1, which illustrate the t computational procedures for the same set of data. The null hypothesis is rejected for the dependent samples analysis but not for the independent samples analysis. The respective t test statistics are $-2/0.657 = -3.04$ and $-2/1.373 = -1.46$. An examination of the data for the two practice conditions suggests that they are positively correlated; the product-moment correlation coefficient r is actually 0.793. This example illustrates an important principle: whenever the correlation between samples is positive, the t for dependent samples is larger than the t for independent samples. Hence, the use of dependent samples results in a more powerful test of a false null hypothesis. This statement must be qualified. The number of degrees of freedom for the independent t test, $n_1 + n_2 - 2 = 38$, is larger than that for the dependent t test, $n - 1 = 19$. The values of t that cut off the critical region are $-t_{.05,38} = -1.69$ and $-t_{.05,19} = -1.73$. Now for the qualification: In order for a t test with dependent samples to be more powerful than a t test with independent samples, the correlation between dependent samples must be large enough to more than compensate for the smaller degrees of freedom and for the larger absolute value of t required for significance.

Several assumptions are associated with the t statistic for dependent samples.

1. The population distribution of differences, $D_i = X_{i1} - X_{i2}$, is approximately normal in form or the number of pairs of differences is large enough that the sampling distribution of $\bar{D} = \sum D_i/n$ is normal in form. D_i will be normally distributed if X_1 and X_2 are normally distributed.

2. The population variance of the mean of the difference scores, $\sigma^2_{\bar{X}_D}$, is unknown.

3. The pairs of matched subjects are a random sample from a population of matched subjects and/or subjects in a pair are randomly assigned to experimental and control conditions.

Determining the n Required to Achieve an Acceptable α,
$1 - \beta$, *and* $(\mu_1 - \mu_2)/\sigma$

In the previous chapter (Section 12.2) we saw how to use Appendix Table D.9 to make a rational choice of sample size for the one-sample t test. Table D.9 also can be used to choose a sample size for the two-sample t test with either independent or dependent samples. As before, it is necessary to specify α, $1 - \beta$, and d. If the samples are dependent, it is also necessary to specify ρ, the correlation between two populations. Since ρ is rarely known, its estimation must be based on previous research or informed judgment. Consider the mirror-tracing task described in Section 13.4. Suppose that we wanted to detect a medium-effect size ($d = 0.5$) and that we wanted $\alpha = .05$ and $1 - \beta = .80$. If we estimate that the population correlation between the distributed and massed practice conditions is approximately 0.70, the required n according to Table D.9 is 16.

Group Matching: A Research Strategy to Be Avoided

A procedure called **group matching** *is sometimes seen in the literature. It involves matching samples on one or more relevant characteristics so that the means and the variances of the samples are approximately equal. No attempt is made to match individuals in one sample with those in another sample.*

Group matching instead of individual matching is often used in *ex post facto* experiments. In an *ex post facto* experiment the independent variable has occurred prior to the experiment. Thus, the independent variable is not under an experimenter's control; rather, records or other information are used to construct two samples that differ with respect to the independent variable. The experimenter can then determine whether the two constructed samples differ on the dependent variable. For example, an experimenter might be interested in determining whether a difference exists in amount of community service (the dependent variable) between females who did or did not participate in girl scouting (the independent variable). Scout records can be used to identify those who were scouts. In all likelihood, two constructed samples—those women who did and those who did not participate in scouting—differ on a variety of variables besides the independent variable. Group matching is used in an attempt to equate the constructed samples on extraneous variables such as IQ, school advancement, socioeconomic background, and so on. High school records, for example, can be used to adjust the membership of the constructed samples so that the sample means and variances are identical on the extraneous variables.

Unfortunately, several problems are inherent in the group-matching procedure.[3] Although it results in dependent samples, the t statistic for dependent samples can't be used because $D_i = X_{i1} - X_{i2}$ can't be computed. This follows since group means and variances are matched, but individual subjects aren't. The only "solution" to this dilemma is to use the t statistic for independent samples. This is a poor solution because (1) group matching restricts the ordinary variation between treatment means that is expected on the basis of random sampling and (2) the denominator of the t statistic for independent samples, $\sqrt{\hat{\sigma}^2_{Pooled}(1/n_1 + 1/n_2)}$, overestimates the standard error of the difference between means when the samples are dependent. Hence, the t statistic for independent samples gives a value that is too small, resulting in a less powerful test of a false null hypothesis. An important experimental design principle emerges from this discussion—the sampling, randomization, and control procedures used in an experiment must be reflected in the statistical analysis and interpretation of data. If this is not possible, presumed refinements such as group matching should not be used in the design of an experiment.

Review Exercises for Section 13.5

19. The use of repeated measures isn't appropriate with some kinds of independent variables. List three variables with which you think it would be appropriate and three with which it wouldn't.

*20. If repeated measures are obtained, what restriction customarily is placed on the order of presentation of the conditions in the experiment?

21. (a) List three matching variables that you believe could be used to form pairs of subjects in a learning experiment using nonsense syllables. (b) Which matching variable do you

---

[3] For an in-depth discussion of these problems see Boneau and Pennypacker (1961). This article is reproduced in Kirk (1972a).

think would have the highest correlation with number of trials required to learn non-sense syllables?

*22. (a) How is the size of the correlation between dependent samples related to the size of the standard error of the difference between means? (b) How is the size of the correlation between dependent samples related to the probability of rejecting a false null hypothesis?

*23. Before and after seeing a film about marijuana, 16 subjects completed a questionnaire designed to assess their attitudes toward legalization of the drug. (a) For the data in the table, compute a t test statistic and decide whether viewing the film can be expected to result in more favorable attitudes toward legalization of marijuana. Let $\alpha = .05$. (b) Use Appendix Table D.9 to estimate the total number of subjects required to reject a false null hypothesis for $\alpha = .05$, $1 - \beta = .95$, $d = 0.80$, and $\rho = .70$.

Favorableness of attitude

Subject	Before	After	Subject	Before	After
1	13	16	9	19	20
2	16	18	10	16	18
3	10	12	11	15	18
4	14	18	12	14	15
5	15	18	13	12	12
6	12	15	14	13	17
7	11	12	15	14	16
8	18	20	16	15	17

24. It is well known that increasing room illumination up to some level increases reading speed. A random sample of 14 sixth-grade students read standardized passages under two levels of ambient room illumination, 5 foot-candles and 15 foot-candles. The order in which the conditions were presented was randomized independently for each subject, with the restriction that the conditions were presented first or second equally often. The reading sessions were separated by an interval of 2 hr. (a) For the data in the table, compute a t test statistic and decide whether reading speed is faster under the 15 foot-candle condition than under the 5 foot-candle condition. Let $\alpha = .05$. (b) Use Table D.9 to estimate the total number of subjects required to reject a false null hypothesis for $\alpha = .05$, $1 - \beta = .90$, $d = 0.80$, and $\rho = .60$.

Reading speed (words/minute)

Subject	5 Foot-candles	15 Foot-candles	Subject	5 Foot-candles	15 Foot-candles
1	88	92	8	90	92
2	92	91	9	84	88
3	86	88	10	82	88
4	84	89	11	86	84
5	90	95	12	84	87
6	86	86	13	86	89
7	88	95	14	86	87

*25. Expanding technology and the growth of knowledge in medicine require that nurses continuously upgrade their skills. One way to accomplish this upgrading is through continuing-education workshops. The present study investigated the impact of a 60-hr workshop on a measure of the participants' cognitive knowledge. Twenty-two staff nurses took a paper-and-pencil pretest to evaluate their basic knowledge of cancer and cancer nursing prior to the ten-day workshop. The same instrument was used to measure their knowledge immediately after the workshop. The following data were obtained. (Suggested by Donovan, Marilee, Wolpert, Patricia, and Yasko, Joyce. Gaps and contracts. *Nursing Outlook*, 1981, 467–471.)

Knowledge score

Subject	Pretest score	Posttest score
1	29	35
2	20	41
3	24	33
4	32	41
5	33	39
6	19	20
7	17	29
8	32	42
9	16	36
10	28	37
11	35	36
12	19	27
13	31	50
14	28	33
15	23	23
16	18	35
17	24	34
18	25	30
19	28	39
20	32	45
21	25	36
22	27	29

a. Use a t statistic to test the null hypothesis $\mu_1 - \mu_2 = 0$, where μ_1 and μ_2 denote, respectively, the population means for the pretest and the posttest. Let $\alpha = .01$.

b. Use Table D.9 to estimate the total number of subjects required to reject a false null hypothesis for $\alpha = .01$, $1 - \beta = .80$, $d = 0.8$, and $\rho = .5$.

c. For purposes of comparison, compute a t statistic for independent samples. Compare the result with the t statistic for dependent samples. Was the use of repeated measures an effective experimental design strategy?

d. In this experiment the order of presentation of the pretest and the posttest obviously could not be randomized. Describe how a control group could be used in the experiment. How could the use of a control group help to clarify the interpretation of the results of the experiment?

26. The effect of a curriculum designed to develop in children a critical viewing attitude toward television programs was investigated. Eighteen second-grade children participated in the curriculum that dealt with such topics as the portrayal of violence on TV, commercials, stereotypes about sex and race, and the comprehension of magical effects on TV. The curriculum was presented in six 30-to-45-minute lessons and used brief videotape excerpts, class play activities, and homework assignments. A specially developed TV Comprehension Test was administered prior to the introduction of the curriculum and at its conclusion. The following data on the "impossible" characters

subtest were obtained. (Suggested by Rapaczynski, Wanda, and Singer, Dorothy G. Teaching television: A curriculum for young children. *Journal of Communication*, 1982, 32(2), 46–55.)

Score on "impossible" characters subtest

Subject	Pretest score	Posttest score
1	1	1
2	3	3
3	0	3
4	2	4
5	1	2
6	2	4
7	3	3
8	3	2
9	2	4
10	2	3
11	1	4
12	3	3
13	1	2
14	2	2
15	3	4
16	3	4
17	1	2
18	2	4

   a. Use a t statistic to test the null hypothesis $\mu_1 - \mu_2 \geq 0$, where μ_1 and μ_2 denote, respectively, the population means for the pretest and the posttest. Let $\alpha = .01$.

   b. Use Table D.9 to estimate the number of subjects required to reject a false null hypothesis for $\alpha = .01$, $1 - \beta = .80$, $d = 0.8$, and $\rho = .40$.

   c. For purposes of comparison, compute a t statistic for independent samples. Compare the result with the t statistic for dependent samples. Was the use of repeated measures an effective experimental design strategy?

 *27. Assume that a t statistic will be used to test the following null hypotheses. For (a), (b), and (c), estimate the total number of subjects required; for (d), (e), and (f), estimate the number of pairs of dependent subjects required.

*a. H_0: $\mu_1 - \mu_2 \geq 0$	*b. H_0: $\mu_1 - \mu_2 = 0$	c. H_0: $\mu_1 - \mu_2 = 0$
$\alpha = .05$	$\alpha = .01$	$\alpha = .05$
$1 - \beta = .80$	$1 - \beta = .90$	$1 - \beta = .95$
$d = 0.5$	$d = 0.2$	$d = 0.8$

*d. H_0: $\mu_1 - \mu_2 \geq 0$	*e. H_0: $\mu_1 - \mu_2 = 0$	f. H_0: $\mu_1 - \mu_2 = 0$
$\alpha = .05$	$\alpha = .01$	$\alpha = .05$
$1 - \beta = .80$	$1 - \beta = .90$	$1 - \beta = .95$
$d = 0.5$	$d = 0.2$	$d = 0.8$
$\rho = .6$	$\rho = .7$	$\rho = .5$

28. If the correlation between matched samples equals 0, the t test for dependent samples will be less powerful than the t test for independent samples. Explain why this assertion is true.

29. Why should group matching be avoided?

30. Terms to remember:
   a. Independent samples   b. Dependent samples
   c. Group matching   d. *Ex post facto* experiment

Table 13.6-1. Summary of two-sample test statistics

Chapter section	Statistic	Statistical hypotheses	Formula	Assumptions
13.2	z	$H_0: \mu_1 - \mu_2 = \delta_0$ $H_1: \mu_1 - \mu_2 \neq \delta_0$	$z = \dfrac{(\bar{X}_1 - \bar{X}_2) - \delta_0}{\sqrt{\sigma_1^2/n_1 + \sigma_2^2/n_2}}$	1. Random sampling or random assignment 2. Normality or large samples 3. Variances known but not necessarily equal 4. Independence between samples
13.4	t	$H_0: \mu_1 - \mu_2 = \delta_0$ $H_1: \mu_1 - \mu_2 \neq \delta_0$	$t = \dfrac{(\bar{X}_1 - \bar{X}_2) - \delta_0}{\sqrt{\hat{\sigma}_{Pooled}^2(1/n_1 + 1/n_2)}}$ $\nu = n_1 + n_2 - 2$	1. Random sampling or random assignment 2. Normality 3. Variances unknown but assumed equal 4. Independence between samples
13.5	z	$H_0: \mu_1 - \mu_2 = \delta_0$ $H_1: \mu_1 - \mu_2 \neq \delta_0$	$z = \dfrac{(\bar{X}_1 - \bar{X}_2) - \delta_0}{\sqrt{\sigma_{\bar{X}_1}^2 + \sigma_{\bar{X}_2}^2 - 2\rho_{12}\sigma_{\bar{X}_1}\sigma_{\bar{X}_2}}}$	1. Random sampling or random assignment 2. Normality or large samples 3. Variances and population correlation known 4. Dependence between samples
13.5	t	$H_0: \mu_1 - \mu_2 = \delta_0$ $H_1: \mu_1 - \mu_2 \neq \delta_0$	$t = \dfrac{\sum\limits_{i=1}^{n} D_i/n}{\sqrt{\left\{\left[\sum\limits_{i=1}^{n} D_i^2 - \left(\sum\limits_{i=1}^{n} D_i\right)^2 / n\right] / n\right\} / \sqrt{n-1}}}$ $\nu = n - 1$	1. Random sampling or random assignment 2. Normality 3. Variances and population correlation unknown 4. Dependence between samples

Section 13.6 Summary

Two-sample tests for means are described in this chapter. The associated z and t test statistics are presented within the now familiar five-step hypothesis testing format.

Two important topics related to the design of experiments are also discussed. The first concerns the relative merits of two randomization strategies—random sampling of elements from two populations versus random assignment of elements to experimental and control conditions. An experimenter's research objectives determine whether one or the other procedure is sufficient or if both procedures are required. Remember that an experiment must contain some randomization procedure to justify using statistical inference.

The other topic related to the design of experiments concerns the use of independent samples versus dependent samples. It is to an experimenter's advantage to use dependent samples whenever the nature of the independent variable permits it. Matching subjects on some variable that correlates positively with the dependent variable or observing the same subjects under both the experimental and control conditions results in a more powerful test of a false null hypothesis than obtaining independent samples. However, the use of group matching instead of individual matching is not recommended because the presumed refinement cannot be taken into account in the statistical analysis. This suggests an important general principle—the sampling, randomization, and control procedures used in an experiment must be reflected in the statistical analysis and interpretation.

The test statistics we have described in this chapter are presented in Table 13.6-1, along with a summary of the parameters they are used to test, their formulas, and their assumptions.

Table 13.7-1. Printout from the SPSS subprogram T-Test (Data from Table 13.4-1)

MIRROR TRACING DATA

FILE   NONAME   (CREATION DATE = 02/01/83)

GROUP 1 — PRACTICE EQ   1.
GROUP 2 — PRACTICE EQ   2.

						- - - - - - T - T E S T - - - - - -	
VARIABLE	NUMBER OF CASES	MEAN	STANDARD DEVIATION	STANDARD ERROR	*	* F VALUE	* 2-TAIL PROB.
TIME					*	*	*
GROUP 1	20	17.0000	3.798	0.849	*	* 1.61	* 0.306
GROUP 2	20	19.0000	4.823	1.078	*	*	*

POOLED VARIANCE ESTIMATE			*	SEPARATE VARIANCE ESTIMATE		
T VALUE	DEGREES OF FREEDOM	2-TAIL PROB.	*	T VALUE	DEGREES OF FREEDOM	2-TAIL PROB.
-1.46	38	0.153	*	-1.46	36.02	0.154

†*Section 13.7   Computer Supplement*

The Statistical Package for the Social Sciences (SPSS), version M, Release 9.1, was used to perform an independent samples t test for the data in Table 13.4-1. These data are tracing times for subjects who practiced under either a distributed or massed condition. The T-TEST subprogram was used to obtain the printout in Table 13.7-1.

The T-TEST subprogram provides (1) number of cases in each group, (2) means, (3) standard deviations, (4) standard errors, (5) F test of the hypothesis that the two population variances are equal, $H_0: \sigma^2_{\text{Group 1}} = \sigma^2_{\text{Group 2}}$, (6) t test using pooled variances, and (7) approximate t test using separate (unpooled) variances. The latter test is appropriate when the sample n's are markedly different and the experimenter has reason to believe that the population variances are unequal. In the present example the sample n's are equal and the F test is not even significant at $\alpha = .25$; hence the t test using pooled variances is appropriate.[4]

The alternative hypothesis for the mirror-tracing data is

$$H_0: \quad \mu_1 - \mu_2 < 0,$$

with $\alpha = .05$. The sample means, Group 1 = 17 and Group 2 = 19, are consistent with the alternative hypothesis. The T-TEST subprogram provides a two-tailed p-value; a one-tailed p-value can be obtained from

$$(2\text{-}Tail\ Prob.)/2 = 0.153/2 = 0.076.$$

This one-tailed p-value, 0.076, exceeds $\alpha = .05$. Thus the null hypothesis cannot be rejected.

[†] This section can be omitted without loss of continuity.
[4] The F statistic for testing the hypothesis $H_0: \sigma_1{}^2 = \sigma_2{}^2$ is discussed in Section 14.1.

14

Statistical Inference: Other Two-Sample Test Statistics

Section 14.1 Two-Sample Tests for Equality of Variances

F Test for Equality of Variances (Independent Samples)

Frequently, an experimenter's research interest focuses on whether two populations differ in dispersion. An experimenter might want to know if placing disadvantaged children in a contingency management classroom results in less variability in their English-achievement scores than does placing them in a traditional classroom. Or the experimenter might want to test one of the assumptions of the t test for independent samples—that two unknown population variances are equal.[1]

An F statistic for testing the null hypothesis

$$H_0: \quad \sigma_1{}^2 = \sigma_2{}^2$$

$$H_1: \quad \sigma_1{}^2 \neq \sigma_2{}^2$$

is

$$F = \frac{\hat{\sigma}_1{}^2}{\hat{\sigma}_2{}^2},$$

where $\hat{\sigma}_1{}^2$ and $\hat{\sigma}_2{}^2$ are unbiased estimators of the population variances.[2] The degrees of freedom for the numerator and the denominator are, respectively, $v_1 = n_1 - 1$ and $v_2 = n_2 - 1$. The sampling distribution of the F statistic was derived by R. A. Fisher in 1924 and was given the name F in his honor by G. W. Snedecor. The F distribution, like the t and chi-square distributions, is actually a family of distributions whose shape is dependent on its degrees of freedom. The F distribution is positively skewed, as shown in Figure 14.1-1, but it approaches the normal distribution shape for very large values of v_1 and v_2. Since F is a ratio of nonnegative numbers, it can take values only from zero to positive infinity; F values around

---

[1] Some books recommend always testing the assumption of equality of variances before performing a t test for $\mu_1 - \mu_2 = \delta_0$. If this advice is followed, it should be borne in mind that the t test is robust with respect to violation of the assumption of normalcy. However, the F test for $\sigma_1{}^2 = \sigma_2{}^2$ described in this section is almost as sensitive to nonnormality as it is to nonequality of variances. Hence, an experimenter may be dissuaded from using a t test when it is actually appropriate.

[2] This statistic is based on the definition of F as the ratio of two independent chi-square random variables, each divided by its degrees of freedom: $F = \chi^2_{v_1}/v_1 / \chi^2_{v_2}/v_2$. According to Section 12.3, $\chi_v{}^2 = v\hat{\sigma}^2/\sigma_0{}^2$ and $\chi_v{}^2/v = \hat{\sigma}^2/\sigma_0{}^2$; hence, $F = \hat{\sigma}_1{}^2/\sigma_1{}^2 \hat{\sigma}_2{}^2/\sigma_2{}^2 = \sigma_2{}^2 \hat{\sigma}_1{}^2 / \sigma_1{}^2 \hat{\sigma}_2{}^2$. If $H_0: \sigma_1{}^2 = \sigma_2{}^2$ is true, then $F = \hat{\sigma}_1{}^2/\hat{\sigma}_2{}^2$.

1 are expected if the null hypothesis that $\sigma_1^2 = \sigma_2^2$ is true. The assumptions associated with using the F statistic to test this hypothesis are (1) that the population distributions of X_1 and X_2 are independent and normal, (2) that the two collections of X's are random samples from the populations of interest, and (3) that the population variances are equal, which is the hypothesis that the experimenter hopes to reject. This F test, unlike the t test, is not robust with respect to violation of the normality assumption. Hence unless the assumption is fulfilled, the probability of making a type I error will not be equal to the preselected value of α.

The critical value of F that cuts off the upper α region of the sampling distribution for v_1 and v_2 degrees of freedom is given in Appendix Table D.5 and is denoted by $F_{\alpha;v_1,v_2}$. To use the table, we locate the column corresponding to the numerator degrees of freedom (v_1) along the top of the table and the row corresponding to the denominator degrees of freedom (v_2) along the side. The column–row intersection gives the critical values of F. The critical value that cuts off the lower α region is denoted by $F_{1-\alpha;v_1,v_2}$. These values are not given in the table, but they can be determined from $F_{\alpha;v_2,v_1}$, since

$$F_{1-\alpha;v_1,v_2} = \frac{1}{F_{\alpha;v_2,v_1}}.$$

That is, the value of F in the lower portion of the F distribution can be found by computing the reciprocal of the corresponding value in the upper portion, with degrees of freedom for numerator and denominator reversed.

In testing the one-tailed null hypothesis

$$H_0: \quad \sigma_1^2 \leq \sigma_2^2$$
$$H_1: \quad \sigma_1^2 > \sigma_2^2,$$

the hypothesis is rejected if $F \geq F_{\alpha;v_1,v_2}$. For

$$H_0: \quad \sigma_1^2 \geq \sigma_2^2$$
$$H_1: \quad \sigma_1^2 < \sigma_2^2,$$

the null hypothesis is rejected if $F \leq F_{1-\alpha;v_1,v_2}$. Consider next the two-tailed null hypothesis

$$H_0: \quad \sigma_1^2 = \sigma_2^2$$
$$H_1: \quad \sigma_1^2 \neq \sigma_2^2,$$

in which the α is divided equally between the two tails of the F distribution; the critical values that cut off the upper and lower $\alpha/2$ regions are denoted by $F_{\alpha/2;v_1,v_2}$ and $F_{1-\alpha/2;v_1,v_2}$, respectively. For this test, the null hypothesis is rejected if $F \leq F_{1-\alpha/2;v_1,v_2}$ or $F \geq F_{\alpha/2;v_1,v_2}$. An experimenter can avoid having to compute the critical values in the lower tail, $F_{1-\alpha;v_1,v_2}$ and $F_{1-\alpha/2;v_1,v_2}$, by always placing the larger variance in the numerator of $F = \sigma_1^2/\sigma_2^2$. This is illustrated at the end of the following computational example.

Computational Example for F Test with Independent Samples

Suppose that 50 disadvantaged children are assigned randomly to contingency management and traditional classrooms, with the restriction that 25 children are placed in each classroom. At the end of the school year an English-achievement test is administered to the two samples. The steps to be followed in testing the hypothesis that the population dispersion is smaller for children in the contingency management classroom and the decision rule are as follows.

Step 1. State the statistical hypotheses: $H_0: \sigma_1{}^2 \geq \sigma_2{}^2$
$H_1: \sigma_1{}^2 < \sigma_2{}^2$,
where $\sigma_1{}^2$ and $\sigma_2{}^2$ denote the population variances, respectively, for the contingency management and traditional classrooms.

Step 2. Specify the test statistic: $F = \dfrac{\hat{\sigma}_1{}^2}{\hat{\sigma}_2{}^2}.$

Step 3. Specify the sample sizes: $n_1 = 25$ and $n_2 = 25$;
and the sampling distribution: F distribution, because the population distributions of X_1 and X_2 are independent and approximately normal in form.

Step 4. Specify the significance level: $\alpha = .05$.

Step 5. Obtain random samples of size n_1 and n_2, compute F, and make a decision.

Decision rule:
Reject H_0 if F falls in the lower .05 portion of the sampling distribution of F; otherwise, don't reject H_0.

Assume that unbiased estimates of the population variances are $\hat{\sigma}_1{}^2 = 64$ and $\hat{\sigma}_2{}^2 = 196$, where $\hat{\sigma}_1{}^2$ and $\hat{\sigma}_2{}^2$ are computed from

$$\hat{\sigma}^2 = \frac{\sum X_i{}^2 - (\sum X_i)^2/n}{(n-1)}.$$

The F test statistic is

$$F = \frac{64}{196} = 0.33.$$

The value of F that cuts off the upper α proportion of the sampling distribution of F is given in Table D.5. The value that cuts off the lower α proportion can be determined from the F table by means of the relation $F_{1-\alpha;v_1,v_2} = 1/F_{\alpha;v_2,v_1}$. Thus, $F_{1-.05;24,24} = 1/1.98 = 0.51$. The critical region for rejecting H_0 is shown in Figure 14.1-1. Since the observed F falls in the critical region, $0.33 < 0.51$, the null

Figure 14.1-1. Sampling distribution of F given that H_0 is true. Since the observed $F = 0.33$ falls in the critical region, H_0 is rejected.

hypothesis is rejected. It is concluded that placing disadvantaged children in a contingency management classroom results in a smaller variance in English-achievement scores than does placing them in a traditional classroom.

We could have avoided having to compute $F_{1-.05;24,24}$, the lower critical value, by placing the larger variance estimate in the numerator of the F statistic and determining if $F > F_{\alpha;v_2,v_1}$, where v_2 and v_1 now denote the F numerator and denominator degrees of freedom, respectively. For this case

$$F = \frac{\hat{\sigma}_2{}^2}{\hat{\sigma}_1{}^2} = \frac{196}{64} = 3.06.$$

The critical value of F is $F_{.05;24,24} = 1.98$. The one-tailed null hypothesis can be rejected, since $3.06 > 1.98$ and the relative size of $\sigma_1{}^2$ and $\sigma_2{}^2$ is consistent with the alternative hypothesis; that is, $64 < 196$ is consistent with $H_1: \sigma_1{}^2 < \sigma_2{}^2$. For a two-tailed test, the latter requirement isn't applicable.

t Test for Equality of Variances (Dependent Samples)

When the variances to be compared arise from dependent samples, for example, matched or repeated-measures samples, the F test statistic is not appropriate. Instead, a t test statistic should be used. The formula for t is

$$t = \frac{\hat{\sigma}_1{}^2 - \hat{\sigma}_2{}^2}{\sqrt{[4\hat{\sigma}_1{}^2\hat{\sigma}_2{}^2/(n-2)](1 - r_{12}^2)}}$$

with degrees of freedom equal to $n - 2$, where n is the number of pairs of scores and r_{12} is the product-moment correlation coefficient.

To illustrate, assume that an arithmetic-achievement test is given to a random sample of $n = 26$ sixth-grade students at the beginning and end of the school year. The experimenter hypothesizes that sixth-grade students become more homogeneous in arithmetic achievement during the year. Suppose that achievement test variances at the beginning and end of the year are, respectively, $\hat{\sigma}_1^2 = 256$ and $\hat{\sigma}_2^2 = 100$, with $n = 26$ and $r = 0.60$. A test of the null hypothesis

$$H_0: \quad \sigma_1^2 \le \sigma_2^2$$
$$H_1: \quad \sigma_1^2 > \sigma_2^2$$

is given by

$$t = \frac{256 - 100}{\sqrt{[4(256)(100)/(26-2)][1-(0.60)^2]}} = \frac{156}{52.256} = 2.99,$$

with $v = 26 - 2 = 24$. According to Appendix Table D.3, a t of 1.711 cuts off the upper .05 region of the sampling distribution of t. Since $t > t_{.05,24}$, the null hypothesis is rejected and the experimenter concludes that the population of sixth-grade students is more homogeneous in arithmetic achievement at the end of the school year than it was at the beginning.

Review Exercises for Section 14.1

*1. Can $F = \hat{\sigma}_1^2/\hat{\sigma}_2^2$ be used to test hypotheses of the form $H_0: \sigma_1^2 - \sigma_2^2 = \delta_0$, where $\delta_0 \ne 0$?

2. What are the main factors an experimenter should keep in mind when using an F test to determine the tenability of the assumption $\sigma_1^2 = \sigma_2^2$ prior to testing $H_0: \mu_1 - \mu_2 = \delta_0$?

3. In testing the tenability of the assumption $\sigma_1^2 = \sigma_2^2$ prior to testing $H_0: \mu_1 - \mu_2 = \delta_0$, it is common practice to set $\alpha = .20$ or .25. What justification for this practice can you offer?

*4. For the data in Chapter 13, Exercise 15 (page 291), test the hypothesis that the population variances are equal versus the alternative that they are not equal. Let $\alpha = .20$.

5. For the data in Table 13.4-1 (page 290), test the tenability of the t test assumption that $\sigma_1^2 = \sigma_2^2$. Let $\alpha = .20$.

6. It is reasonable to expect 13-year-old boys to exceed 12-year-old boys in strength of grip. In all likelihood the dispersion of strength of grip is greater for 13-year-olds than for 12-year-olds. To test this hypothesis, strength of grip was measured by means of a hand dynamometer for a random sample of 42 boys who had just turned 12. One year later the same boys were remeasured. The variances for the first and second sets of measurements are 196 and 289, respectively. The correlation between the two sets of measurements is 0.83. Compute a t test statistic and make a decision about the scientific hypothesis. Let $\alpha = .05$.

Section 14.2   Two-Sample Tests for $p_1 - p_2$

z Test for $p_1 = p_2$ (Independent Samples)

Many variables in the behavioral sciences and education can be partitioned into nonoverlapping and exhaustive classes and are qualitative in character, for example, sex, year in school, and pass or fail. In such cases p, the proportion in each class, is a useful descriptive measure. *A statistic for testing any of the following null hypotheses*

$$H_0: \quad p_1 = p_2 \qquad H_0: \quad p_1 \leq p_2 \qquad H_0: \quad p_1 \geq p_2$$
$$H_1: \quad p_1 \neq p_2 \qquad H_1: \quad p_1 > p_2 \qquad H_1: \quad p_1 < p_2,$$

where p_1 and p_2 are parameters of two independent binomial distributions, is

$$z = \frac{\hat{p}_1 - \hat{p}_2}{\sqrt{\hat{p}_{Pd}\hat{q}_{Pd}/n_1 + \hat{p}_{Pd}\hat{q}_{Pd}/n_2}}.$$

Here, $\hat{p}_1$ and $\hat{p}_2$ are the sample estimators of the population proportions, n_1 and n_2 are the sizes of the samples used to estimate the population proportions, $\hat{p}_{Pooled}$ $(\hat{p}_{Pd}) = (n_1\hat{p}_1 + n_2\hat{p}_2)/(n_1 + n_2)$, and $\hat{q}_{Pooled}$ $(\hat{q}_{Pd}) = 1 - \hat{p}_{Pd}$.

It can be shown that the sampling distribution of z approaches a standard normal distribution if all the products $n_1\hat{p}_{Pd}$, $n_1\hat{q}_{Pd}$, $n_2\hat{p}_{Pd}$, and $n_2\hat{q}_{Pd}$ are greater than 5. If any one of these products is between 5 and 10, it is desirable to apply a correction for continuity. As discussed in Section 12.4, the correction should be applied when the continuous normal distribution is used to estimate probabilities for a discrete distribution and n is small. The correction consists of subtracting the quantity $(1/2)(1/n_1 + 1/n_2)$ from the absolute value of the z numerator—that is, $|\hat{p}_1 - \hat{p}_2| - (1/2)(1/n_1 + 1/n_2)$.

z Test for $p_1 - p_2 = \delta_0$, where $\delta_0 \neq 0$ (Independent Samples)

In the preceding discussion the hypothesis that is tested in each case is $p_1 = p_2$. If the hypothesis is true, $\hat{p}_1$ and $\hat{p}_2$ are estimators of the same population parameter p, and the best estimator of p is obtained by computing a weighted mean, $\hat{p}_{Pd}$, of $\hat{p}_1$ and $\hat{p}_2$.[3] *If an experimenter wishes to test H_0: $p_1 - p_2 = \delta_0$, where $\delta_0 \neq 0$, a weighted mean should not be computed, since $\hat{p}_1$ and $\hat{p}_2$ are estimators for different population parameters. For this hypothesis the appropriate test statistic is*

$$z = \frac{(\hat{p}_1 - \hat{p}_2) - \delta_0}{\sqrt{\hat{p}_1\hat{q}_1/n_1 + \hat{p}_2\hat{q}_2/n_2}}.$$

z Test for $p_1 = p_2$ (Dependent Samples)

If two samples are dependent, a statistic developed by McNemar (1947) should be used to test any of the following null hypotheses:

$$H_0: \quad p_1 = p_2 \qquad H_0: \quad p_1 \leq p_2 \qquad H_0: \quad p_1 \geq p_2$$
$$H_1: \quad p_1 \neq p_2 \qquad H_1: \quad p_1 > p_2 \qquad H_1: \quad p_1 < p_2.$$

---

[3] Section 13.4 describes a similar procedure for a t test statistic in which the best estimator of the unknown population variance, σ^2, is a weighted mean of the two sample estimators of σ^2.

For these tests the data are cast into a 2×2 table, as shown.

Sample 2

		Category 0	Category 1	
Sample 1	Category 1	a	b	$a + b$
	Category 0	c	d	$c + d$
		$a + c$	$b + d$	n

The cell entry a denotes the number of elements classified in category 1 for sample 1 and in category 0 for sample 2; b denotes the number of elements that is classified in category 1 for both samples 1 and 2; and so on. The number of elements in each sample is n.

An estimator of the population proportion of individuals in category 1 for sample 1 is $\hat{p}_1 = (a + b)/n$. Similarly, the proportion in category 1 for sample 2 is given by $\hat{p}_2 = (b + d)/n$. The difference between the two populations can be expressed either as a proportion, $p_1 - p_2$, or as a frequency, $a - d$. A little algebra is all that is required to show that $n(\hat{p}_1 - \hat{p}_2) = a - d$:

$$\hat{p}_1 - \hat{p}_2 = \frac{a + b}{n} - \frac{b + d}{n} \qquad \text{By definition}$$

$$= \frac{1}{n}(a + b - b - d)$$

$$n(\hat{p}_1 - \hat{p}_2) = a - d.$$

If the null hypothesis is true, the test statistic

$$z = \frac{a - d}{\sqrt{a + d}}$$

is approximately distributed as the standard normal distribution, provided that $(a + d) \geq 10$. If $10 \leq (a + d) < 20$, a correction for continuity should be applied to the z statistic as follows.

$$z = \frac{|a - d| - 1}{\sqrt{a + d}}$$

Computational Example for z Test with Dependent Samples

Suppose that a random sample of 100 students at Thanatos University are polled about whether they approve or disapprove of capital punishment. Following the survey the students are shown a film depicting the effects of crime and acts of violence on the victims and their families. The 100 students are again polled about capital punishment. It is hypothesized that the proportion of students who approve of capital punishment will be higher after seeing the film than before. The statistical hypotheses are H_0: $p_1 \geq p_2$ versus H_1: $p_1 < p_2$. The data, number of students who approve or disapprove of capital punishment before and after seeing the film, are listed in Table 14.2-1. Since $z > z_{.05}$ (2.02 > 1.645) and $\hat{p}_1 < \hat{p}_2$, the null hypothesis is rejected. It is concluded that if the population of students were polled, a higher proportion of them would approve of capital punishment after seeing the film than before seeing the film.

Table 14.2-1.   Capital punishment data

(i) *Data:*

Sample 2

		Disapprove	Approve	
Sample 1	Approve	$a = 2$	$b = 12$	$a + b = 14$
	Disapprove	$c = 76$	$d = 10$	$c + d = 86$
		$a + c = 78$	$b + d = 22$	$n = 100$

(ii) *Computation:*

$$z = \frac{|a - d| - 1}{\sqrt{a + d}} = \frac{|2 - 10| - 1}{\sqrt{2 + 10}} = \frac{7}{3.46} = 2.02$$

$$z_{.05} = 1.645$$

$$\hat{p}_1 = \frac{a + b}{n} = \frac{14}{100} = 0.14$$

$$\hat{p}_2 = \frac{b + d}{n} = \frac{22}{100} = 0.22$$

Review Exercises for Section 14.2

*7. In a 1972 Shuffle Poll of $n_2 = 500$ Americans over 18 years old, 11% said they had smoked pot. In 1969 the figure for a sample of $n_1 = 600$ was 4%. Test the hypothesis that $p_1 = p_2$. Let $\alpha = .05$.

8. In the 1972 survey cited in Exercise 7, 81% of the interviewees opposed legalization of marijuana. The figure in 1969 was 84%. Test the hypothesis that $p_1 = p_2$. Let $\alpha = .05$.

9. According to the 1969 survey cited in Exercise 7, 8% more men than women had tried marijuana. In the 1972 survey, 11% of 300 men and 8% of 200 women reported that they had smoked pot. For these data, test the hypothesis that the difference between the proportion of pot smokers among men and women is 8%.

*10. A national survey of 3000 college and university students conducted by the American Council of Day-Care Centers found that 78% of west coast freshmen return to college for their second year. The comparable figure for freshmen at southern schools is 85%. The percentages are based on $n_1 = 1800$ and $n_2 = 1200$ students, respectively. Test the hypothesis that $p_1 = p_2$. Let $\alpha = .001$.

*11. Attitudes of a sample of college students toward taking a required course in music appreciation were measured prior to taking the course and after completing the course.

For the data in the table, test the hypothesis that the proportions in favor of the requirement before and after the course are equal. Let $\alpha = .05$.

		Postcourse attitude		
		Unfavorable	*Favorable*	
	Favorable	16	24	40
Precourse attitude				
	Unfavorable	19	27	46
		35	51	86

12. Learning one task often enhances the learning of a similar task; this phenomenon is called *learning to learn.* To investigate this phenomenon, an experimenter asked students to learn 20 lists of nonsense syllables. For the data in the table, test the hypothesis that p_1, the population proportion of students who learned lists 2 to 6 in less than 25 trials, and p_2, the population proportion who learned lists 16 to 20 in less than 25 trials, are equal. Let $\alpha = .05$.

		Number of students who learned lists 16 to 20		
		In 25 trials or more	*In less than 25 trials*	
Number of students who learned lists 2 to 6	*In less than 25 trials*	4	6	10
	In 25 trials or more	13	13	26
		17	19	36

Section 14.3 Summary

Procedures for testing hypotheses about two variances and two proportions are described in this chapter. Also, a new F statistic and sampling distribution are introduced.

The z, t, χ^2, and F test statistics presented in Chapters 11 through 14 have a number of common characteristics that might be overlooked because of differences in their formulas. Each statistic (1) assumes random sampling or random assignment of X's, (2) is used to test hypotheses about a parameter(s) of the sampled population(s), and (3) assumes that the sampled population is normal with the exception of the z statistics that are used to test hypotheses about proportions.

The test statistics described in this chapter are presented in Table 14.3-1, along with a summary of the parameters they are used to test, their formulas, and their assumptions.

Table 14.3-1.   Summary of two-sample test statistics

Chapter section	Statistic	Statistical hypotheses	Formula	Assumptions
14.1	F	H_0: $\sigma_1^2 = \sigma_2^2$ H_1: $\sigma_1^2 \neq \sigma_2^2$	$F = \dfrac{\hat{\sigma}_1^2}{\hat{\sigma}_2^2}$ $v_1 = n_1 - 1, \quad v_2 = n_2 - 1$	1. Random sampling or random assignment 2. Normality 3. Independence between samples
14.1	t	H_0: $\sigma_1^2 = \sigma_2^2$ H_1: $\sigma_1^2 \neq \sigma_2^2$	$t = \dfrac{\hat{\sigma}_1^2 - \hat{\sigma}_2^2}{\sqrt{[4\hat{\sigma}_1^2 \hat{\sigma}_2^2/(n-2)](1 - r_{12}^2)}}$ $v = n - 2$	1. Random sampling or random assignment 2. Normality 3. Dependence between samples
14.2	z	H_0: $p_1 = p_2$ H_1: $p_1 \neq p_2$	$z = \dfrac{\hat{p}_1 - \hat{p}_2}{\sqrt{\hat{p}_{Pd}\hat{q}_{Pd}/n_1 + \hat{p}_{Pd}\hat{q}_{Pd}/n_2}}$	1. Random sampling or random assignment 2. Binomial distributions 3. Independence between samples 4. $n_1\hat{p}_{Pd} > 5$, $n_1\hat{q}_{Pd} > 5$, $n_2\hat{p}_{Pd} > 5$, $n_2\hat{q}_{Pd} > 5$
14.2	z	H_0: $p_1 - p_2 = \delta_0$ H_1: $p_1 - p_2 \neq \delta_0$ where $\delta_0 \neq 0$	$z = \dfrac{(\hat{p}_1 - \hat{p}_2) - \delta_0}{\sqrt{\hat{p}_1\hat{q}_1/n_1 + \hat{p}_2\hat{q}_2/n_2}}$	1. Random sampling or random assignment 2. Binomial distributions 3. Independence between samples 4. $n_1\hat{p}_1 > 5$, $n_1\hat{q}_1 > 5$, $n_2\hat{p}_2 > 5$, $n_2\hat{q}_2 > 5$
14.2	z	H_0: $p_1 = p_2$ H_1: $p_1 \neq p_2$	$z = \dfrac{a - d}{\sqrt{a + d}}$	1. Random sampling or random assignment 2. Binomial distributions 3. Dependence between samples 4. $a + d > 10$

15

Interval Estimation

Section 15.1 Introduction to Interval Estimation

The election is only days away and the latest Giddyup poll gives Mr. Jerry Mander 54% of the vote. In the language of the statistician, 54% is a *point estimate*, a single point on the number line that is the best estimate of the population parameter that can be made from sample data. Between periods of euphoria Mr. Jerry Mander ponders the question, should he or should he not cancel the paid political announcement planned for election eve? If he is winning, the expenditure of campaign funds is unnecessary, but what if he is really losing? Is it possible that less than 51% of the voters favor him even though the highly respected poll gives him a majority? Now, Mr. Mander is no statistician, but he does know that polls are subject to sampling error. With anxiety mounting, he decides to forego a vacation to Hawaii and to use the campaign funds for their intended purpose. The use of *confidence interval* procedures, the subject of this chapter, might have saved Mr. Mander's vacation.

We noted in Section 11.1 that two complementary topics are categorized under classical statistical inference—hypothesis testing and interval estimation. In many research situations an experimenter isn't interested in testing a hypothesis but rather in obtaining an estimate of some population parameter. *Although we can never know the value of the parameter except by measuring all the elements in the population, a random sample can be used to specify a segment or an interval on the number line such that the parameter has a high probability of lying on the segment. The segment is called a* **confidence interval**.

Interval-estimation procedures are used much less frequently in the behavioral sciences than hypothesis-testing procedures. This is true despite the fact that an interval estimate provides more information regarding the outcome of an experiment than simply reporting that the null hypothesis was or wasn't rejected.[1] Behavioral scientists have been slow to adopt the recommendation of statisticians that research reports should include confidence interval information.

---

[1] For an in-depth discussion of the relative merits of the two procedures, see Natrella (1960). This article is reproduced in Kirk (1972a).

Section 15.2   Confidence Intervals for μ and $\mu_1 - \mu_2$ when σ^2 Is Unknown

Confidence Interval for μ

Here we will construct a confidence interval for μ in such a way that the interval has a probability equal to $1 - \alpha$ of including μ. The underlying logic is presented in some detail, since it is applicable to confidence intervals for any parameter, such as σ^2, p, or ρ. Suppose that we plan to draw a random sample of size n from a normal population and that we want $1 - \alpha$ to equal .95. Here .95, called the *confidence coefficient*, reflects the degree of our confidence that μ does indeed lie in the specified interval. We will assume that σ^2 is unknown but can be estimated from the sample so that the appropriate statistic is a one-sample t. The values of t that cut off the lower .05/2 and upper .05/2 regions of the t sampling distribution (population) are denoted by $-t_{.05/2,v}$ and $t_{.05/2,v}$, respectively. The region of the t population from $-t_{.05/2,v}$ to $t_{.05/2,v}$ includes 95% of the t statistics; this is illustrated in Figure 15.2-1. If we randomly draw one t statistic from this population of t's, the probability is .95 that the obtained t will lie in the interval between $-t_{.05/2,v}$ and $t_{.05/2,v}$. We can state this more formally:

$$\text{Prob}(-t_{.05/2,v} \leq t \leq t_{.05/2,v}) = .95.$$

The first step in deriving the confidence interval for μ is to replace t by its formula, $t = (\bar{X} - \mu)/(\hat{\sigma}/\sqrt{n})$, which gives

$$\text{Prob}\left(-t_{.05/2,v} \leq \frac{\bar{X} - \mu}{\hat{\sigma}/\sqrt{n}} \leq t_{.05/2,v}\right) = .95.$$

Figure 15.2-1.   Sampling distribution of $t = (\bar{X} - \mu)/(\hat{\sigma}/\sqrt{n})$. If one t statistic is randomly sampled from this population of t's, the probability is .95 that the obtained t will come from the interval $-t_{.05/2,v}$ to $t_{.05/2,v}$.

Multiplying each term in the inequalities by $\hat{\sigma}/\sqrt{n}$, we have[2]

$$\text{Prob}\left(\frac{-t_{.05/2,v}\hat{\sigma}}{\sqrt{n}} \leq \bar{X} - \mu \leq \frac{t_{.05/2,v}\hat{\sigma}}{\sqrt{n}}\right) = .95.$$

Subtracting $\bar{X}$ from each term, we obtain

$$\text{Prob}\left(-\bar{X} - \frac{t_{.05/2,v}\hat{\sigma}}{\sqrt{n}} \leq -\mu \leq -\bar{X} + \frac{t_{.05/2,v}\hat{\sigma}}{\sqrt{n}}\right) = .95,$$

and multiplying by -1, which reverses the direction of the inequalities and the signs of the terms, gives

$$\text{Prob}\left(\bar{X} + \frac{t_{.05/2,v}\hat{\sigma}}{\sqrt{n}} \geq \mu \geq \bar{X} - \frac{t_{.05/2,v}\hat{\sigma}}{\sqrt{n}}\right) = .95.$$

For convenience, this is rearranged to form the confidence statement

$$\text{Prob}\left(\bar{X} - \frac{t_{.05/2,v}\hat{\sigma}}{\sqrt{n}} \leq \mu \leq \bar{X} + \frac{t_{.05/2,v}\hat{\sigma}}{\sqrt{n}}\right) = .95.$$

In words, this says the probability is .95 that a random interval from

$$\bar{X} - t_{.05/2,v}\hat{\sigma}/\sqrt{n} \text{ to } \bar{X} + t_{.05/2,v}\hat{\sigma}/\sqrt{n}$$

contains the parameter μ. This random interval is the confidence interval for μ, with *confidence coefficient* equal to $1 - .05 = .95$. The values of $\bar{X} - t_{.05/2,v}\hat{\sigma}/\sqrt{n}$ and $\bar{X} + t_{.05/2,v}\hat{\sigma}/\sqrt{n}$ are the *lower* and *upper boundaries*, respectively, of the confidence interval; the boundaries also are called *confidence limits* and are designated L_1 and L_2. *The general form of a $100(1 - \alpha)\%$ confidence interval for μ is given by*

$$\bar{X} - \frac{t_{\alpha/2,v}\hat{\sigma}}{\sqrt{n}} \leq \mu \leq \bar{X} + \frac{t_{\alpha/2,v}\hat{\sigma}}{\sqrt{n}},$$

where $t_{\alpha/2,v}$ is the value that cuts off the upper $\alpha/2$ proportion of Student's t distribution for $n - 1$ degrees of freedom.

To summarize, it is impossible to know the value of a parameter without measuring all the elements in the population. However, for a given parameter it is possible to find two functions L_1 and L_2 of sample values such that before the sample is drawn the probability that the interval from L_1 to L_2 will contain the parameter is equal to $1 - \alpha$. We can be $100(1 - \alpha)\%$ confident that this interval contains the parameter.

---

[2] A review of inequalities is given in Appendix Section A.6.

Computational Example of Confidence Interval for μ when σ^2 Is Unknown

In Section 12.2 we illustrated a test of the hypothesis that a new class registration procedure at Idle-On-In College would reduce the time required for students to register. A $100(1 - .05)\%$ confidence interval for these data in which $\bar{X} = 2.9$, $\hat{\sigma} = 0.1581$, $n = 21$, and $t_{.05/2,20} = 2.086$ is given by

$$2.9 - 2.086 \frac{0.1581}{\sqrt{21}} \leq \mu \leq 2.9 + 2.086 \frac{0.1581}{\sqrt{21}}$$

$$2.9 - 0.072 \leq \mu \leq 2.9 + 0.072$$

$$2.828 \leq \mu \leq 2.972.$$

We can feel quite confident that the value of μ is between $L_1 = 2.828$ and $L_2 = 2.972$. The measure of our confidence is .95 because before the sample was drawn, .95 was the probability that the interval we were going to construct would cover the true mean. If we want to feel even more confident that we have specified L_1 and L_2 so that they capture μ, we can compute a 99% confidence interval. This is accomplished by substituting $t_{.01/2,20} = 2.845$ for $t_{.05/2,20} = 2.086$. The 99% confidence interval is given by

$$2.9 - 2.845 \frac{0.1581}{\sqrt{21}} \leq \mu \leq 2.9 + 2.845 \frac{0.1581}{\sqrt{21}}$$

$$2.9 - 0.098 \leq \mu \leq 2.9 + 0.098$$

$$2.802 \leq \mu \leq 2.998.$$

Note that as our confidence that we have captured μ increases, so does the size of the interval $L_1 - L_2$.

Interpretation of a Confidence Interval

Earlier we stated that the probability is .95 that a random interval from $\bar{X} - t_{.05/2,v}\hat{\sigma}/\sqrt{n}$ to $\bar{X} + t_{.05/2,v}\hat{\sigma}/\sqrt{n}$ contains the parameter μ; in symbols this can be written

$$\text{Prob}\left(\bar{X} - \frac{t_{.05/2,v}\hat{\sigma}}{\sqrt{n}} \leq \mu \leq \bar{X} + \frac{t_{.05/2,v}\hat{\sigma}}{\sqrt{n}}\right) = .95.$$

This statement refers to the set of all confidence intervals that could be constructed. To clarify, consider obtaining an infinite number of random samples of size n from a normally distributed population and computing $\bar{X}$ for each sample. One can conceive of constructing a confidence interval based on each $\bar{X}$. Ninety-five percent of these intervals will contain μ between L_1 and L_2 and 5% will not. The probability that a randomly selected interval from this infinite set will contain μ is .95. However, once one of the $\bar{X}'s$ has been computed and found to equal, say, 2.9, it is incorrect to state that the probability is .95 that the interval from

$2.9 - t_{.05/2,v}\hat{\sigma}/\sqrt{n}$ to $2.9 + t_{.05/2,v}\hat{\sigma}/\sqrt{n}$ contains μ; in symbols, the statement

$$\text{Prob}\left(2.9 - \frac{t_{.05/2,v}\hat{\sigma}}{\sqrt{n}} \leq \mu \leq 2.9 + \frac{t_{.05/2,v}\hat{\sigma}}{\sqrt{n}}\right) = .95$$

is incorrect. Such a probability statement is incorrect because 2.9 is not a random variable. Once a value of $\bar{X}$ has been inserted in the confidence statement, the interval either does or doesn't contain μ. Hence one should not say, for example, that the probability is .95 that μ lies in the interval 2.828–2.972. It is correct to say that a 95% confidence interval for μ is 2.828–2.972, or that the degree of one's confidence that μ lies in the interval 2.828–2.972 is .95, or, more simply, I am 95% confident that μ falls between 2.828 and 2.972.

As noted earlier, many statisticians prefer confidence interval procedures to hypothesis-testing procedures because the former provide more information than that a particular H_0 was or wasn't rejected at α level of significance. A confidence interval for μ not only specifies the unbiased estimate of the population mean, but it also includes a range of values, the error variation, qualifying that estimate. One can think of a confidence interval as containing all values of μ_0 for which the null hypothesis would not be rejected at α level of significance. Only if μ_0 lies outside the $100(1 - \alpha)\%$ confidence interval can the hypothesis $H_0: \mu = \mu_0$ be rejected. Also any null hypothesis can be tested simply by looking at the confidence interval for μ. For example, a 95% confidence interval for the mean time required for students to register using a new procedure at Idle-On-In College was computed previously and found to be 2.828–2.972. A test of the hypothesis $H_0: \mu = \mu_0$ will be rejected only if μ_0 lies outside the interval 2.828–2.972. Thus, $H_0: \mu = 2.85, 2.90, 2.95$, et cetera, would not be rejected, but $H_0: \mu = 2.75, 3.00$, or 3.15 would be rejected.

It is hoped that reporting confidence intervals in research reports eventually will become standard practice. Certainly, a confidence interval communicates far more information than a test of a null hypothesis.

One-Sided Confidence Interval

The confidence intervals computed earlier would be appropriate if we were interested in the possibility that the new registration time is either quicker or slower than the old procedure. In this sense it is analogous to a two-tailed test of the null hypothesis. The statistical hypotheses for the registration example in Section 12.2 are

$$H_0: \quad \mu \geq 3.1$$
$$H_1: \quad \mu < 3.1,$$

where 3.1 is the mean registration time associated with the old procedure. Here the hypotheses are one-tailed. An analogous $100(1 - .05)\%$ confidence limit is

$$L_2 = \bar{X} + \frac{t_{\alpha,v}\hat{\sigma}}{\sqrt{n}} = 2.9 + 1.725\frac{0.1581}{\sqrt{21}} = 2.9 + 0.0595 = 2.96,$$

where $t_{\alpha,v}$ is the value of t that cuts off the upper $\alpha = .05$ portion of the sampling distribution of t. The experimenter can be fairly confident that μ for the new registration procedure is not greater than 2.96 hr. Since the mean registration time for the old procedure is 3.1 hr, it can be concluded that the new procedure requires less time. In Section 12.2 we reached the same conclusion using hypothesis-testing procedures. However, confidence-interval procedures allow us to test all hypotheses of interest, not just $H_0\colon \mu = 3.1$. For example, since $L_2 = 2.96$, we know that a one-tailed test of significance at the .05 level would reject $H_0\colon \mu \geq 3.06$ and $H_0\colon \mu \geq 2.98$ but not $H_0\colon \mu \geq 2.94$ and $H_0\colon \mu \geq 2.91$.

Confidence Interval for $\mu_1 - \mu_2$ (Independent Samples)

Confidence-interval procedures for a single population mean can be generalized to the difference between the means of two independent populations, $\delta = \mu_1 - \mu_2$. This is accomplished by substituting

$$t = \frac{[(\bar{X}_1 - \bar{X}_2) - (\mu_1 - \mu_2)]}{\sqrt{\hat{\sigma}^2_{Pooled}\left(\dfrac{1}{n_1} + \dfrac{1}{n_2}\right)}}$$

for $t = (\bar{X} - \mu_0)/(\hat{\sigma}/\sqrt{n})$ in the probability statement $\mathrm{Prob}(-t_{\alpha/2,v} \leq t \leq t_{\alpha/2,v}) = 1 - \alpha$. A $100(1 - \alpha)\%$ confidence interval for the difference between means of independent samples is given by

$$(\bar{X}_1 - \bar{X}_2) - t_{\alpha/2,v}\sqrt{\hat{\sigma}^2_{Pooled}\left(\frac{1}{n_1} + \frac{1}{n_2}\right)} \leq \mu_1 - \mu_2$$

$$\leq (\bar{X}_1 - \bar{X}_2) + t_{\alpha/2,v}\sqrt{\hat{\sigma}^2_{Pooled}\left(\frac{1}{n_1} + \frac{1}{n_2}\right)},$$

where $t_{\alpha/2,v}$ is the value of t that cuts off the upper $\alpha/2$ portion of the sampling distribution of t for $v = n_1 + n_2 - 2$ and

$$\hat{\sigma}^2_{Pooled} = \frac{[(n_1 - 1)\hat{\sigma}_1{}^2 + (n_2 - 1)\hat{\sigma}_2{}^2]}{[(n_1 - 1) + (n_2 - 1)]}.$$

We will use the data in Table 13.4-1 (page 290) to illustrate the use of this two-sided confidence interval. The data were obtained in an experiment designed to compare the effectiveness of distributed practice and massed practice in developing skill on a mirror-tracing task. The information from Table 13.4-1 necessary to construct a confidence interval is $\bar{X}_1 = 17$, $\hat{\sigma}_1{}^2 = 14.421$, $n_1 = 20$, $\bar{X}_2 = 19$, $\hat{\sigma}_2{}^2 = 23.263$, $n_2 = 20$, and $\hat{\sigma}^2_{Pooled} = 18.842$. The value of t that cuts off the upper $\alpha/2$ portion of the sampling distribution of t for $n_1 + n_2 - 2 = 38$ degrees of freedom is $t_{.05/2,38} = 2.025$. A $100(1 - .05)\%$ confidence interval for $\mu_1 - \mu_2$ is

$$(17\text{--}19) - 2.025 \sqrt{18.842\left(\frac{1}{20} + \frac{1}{20}\right)} \le \mu_1 - \mu_2$$

$$\le (17\text{--}19) + 2.025 \sqrt{18.842\left(\frac{1}{20} + \frac{1}{20}\right)}$$

$$-2 - 2.780 \le \mu_1 - \mu_2 \le -2 + 2.780$$

$$-4.780 \le \mu_1 - \mu_2 \le 0.780.$$

Thus, the mean for the distributed practice condition μ_1 could be larger than that for the massed practice condition μ_2, or it could be equal to it or smaller. Because the confidence interval for $\mu_1 - \mu_2$ includes zero, it can't be concluded that performance is better under either condition. Any difference between -4.780 and 0.780 is likely to be a reflection of sampling variation and/or other chance factors.

Confidence Interval for $\mu_1 - \mu_2$ (Dependent Samples)

A $100(1 - \alpha)\%$ confidence interval for the difference between means of dependent samples is given by

$$\bar{X}_D - t_{\alpha/2,v}\hat{\sigma}_{\bar{X}_D} \le \mu_D \le \bar{X}_D + t_{\alpha/2,v}\hat{\sigma}_{\bar{X}_D},$$

where $\bar{X}_D = \sum D_i/n = \bar{X}_1 - \bar{X}_2$, $D_i = X_{i1} - X_{i2}$, $t_{\alpha/2,v}$ is the value of t that cuts off the upper $\alpha/2$ portion of the sampling distribution of t for $v = n - 1$,

$$\hat{\sigma}_{\bar{X}_D} = \frac{\sqrt{[\sum D_i^2 - (\sum D_i)^2/n]/n}}{\sqrt{n - 1}},$$

and

$$\mu_D = \mu_1 - \mu_2.$$

We will illustrate the use of this two-sided confidence interval with the mirror-tracing data in Table 13.5-1 (page 296). These data are for matched subjects; one subject of each pair was assigned to the distributed practice condition and the other, to the massed practice condition. The information from Table 13.5-1 necessary to construct a confidence interval is $\bar{X}_D = -2$, $\hat{\sigma}_{\bar{X}_D} = 0.657$, and $n = 20$. The value of $t_{.05/2,19}$ is 2.093. A $100(1 - .95)\%$ confidence interval for $\mu_1 - \mu_2$ is

$$-2 - 2.093(0.657) \le \mu_D \le -2 + 2.093(0.657)$$

$$-2 - 1.375 \le \mu_D \le -2 + 1.375$$

$$-3.375 \le \mu_D \le -0.625.$$

Since the interval -3.375 to -0.625 doesn't include zero, the experimenter can be confident that performance, the time required to trace the star pattern, is superior under the distributed practice condition.

Review Exercises for Section 15.2

1. What assumptions are associated with the statement

$$\text{Prob}\left(\bar{X} - \frac{t_{\alpha/2,v}\hat{\sigma}}{\sqrt{n}} \leq \mu \leq \bar{X} + \frac{t_{\alpha/2,v}\hat{\sigma}}{\sqrt{n}}\right) = 1 - \alpha?$$

*2. Derive a confidence interval for μ from

$$\text{Prob}(-z_{\alpha/2} \leq z \leq z_{\alpha/2}) = 1 - \alpha,$$

where $z = (\bar{X} - \mu)/(\sigma/\sqrt{n})$.

3. What are the advantages of confidence-interval procedures over hypothesis-testing procedures?

*4. Ordinarily, $L_1 < L_2$. Can you think of two conditions under which L_1 and L_2 would be equal?

*5. A soft-drink machine is designed to dispense a measured amount of a popular drink. Construct a 99% confidence interval for μ if a random sample of 49 drinks has $\bar{X} = 7.2$ oz. Assume that the distribution is approximately normal, with $\hat{\sigma} = 0.42$ oz.

*6. If 23–36 is a 95% confidence interval for μ, indicate which of the following statements are correct (C) and which are incorrect (I).
   a. The probability is .95 that the interval from 23 to 36 contains the population mean.
   b. The probability that the interval $\bar{X} \pm 1.96\hat{\sigma}/\sqrt{n}$ contains μ is .95.
   c. $\text{Prob}(23 \leq \mu \leq 36) = .95$.
   d. An experimenter can be 95% confident that the interval 23–36 contains μ.
   e. A 95% confidence interval for μ is 23–36.
   f. $\text{Prob}(\bar{X} - 1.96\hat{\sigma}/\sqrt{n} \leq \mu \leq \bar{X} + 1.96\hat{\sigma}/\sqrt{n}) = .95$, where $\bar{X} = 29.5$.

*7. Students desiring to enter graduate school at Kandykane Technical Institute (KTI) are required to submit Graduate Record Examination (GRE) scores with their applications. The verbal scores for the first 20 applications received this year are given in the table. Construct a 95% confidence interval for these data.

GRE scores for verbal section of test

402	390	429	391
381	407	410	403
430	413	406	398
376	424	382	410
395	360	410	404

8. (a) If the first 20 applicants in Exercise 7 can be considered a random sample of applicants who will apply, what is the best estimate of the mean for this year's applicants? (b) Last year the mean GRE verbal score of all KTI applicants was 428. Is the mean verbal aptitude for this year's applicants different from that for last year's?

*9. The nicotine content of random samples of two brands of cigarettes was measured. The $\bar{X}$'s for brands 1 and 2 were, respectively, 18.6 and 16.1 milligrams, with $\hat{\sigma}$'s 2.4 and 2.6 milligrams. Both sample n's were 36. Construct a 99% confidence interval for $\mu_1 - \mu_2$.

10. An experiment was performed to compare simple and disjunctive reaction times. In the former condition a subject responded to a single light by pressing a button below the light; the disjunctive condition required a subject to press the right button if the right light was illuminated and the left button if the left light was illuminated. Twenty-four subjects were randomly assigned to the two conditions with the restriction that

an equal number participate under each condition. Construct a 95% confidence interval for $\mu_1 - \mu_2$ for the data in the table.

Reaction time (hundredths of a second)

Simple			Disjunctive		
24	24	24	27	34	33
27	26	22	31	32	31
23	21	25	29	30	32
25	23	24	35	28	31

*11. Learning one task enhances the learning of a different but similar task; this phenomenon is called *learning to learn*. To investigate this phenomenon, 18 students learned 15 lists of nonsense syllables with equal association value. The total time required to learn the first three lists and the last three lists is given in the table. Construct a 99% confidence interval for $\mu_1 - \mu_2$.

Time to learn lists of nonsense syllables (minutes)

Subject	First three lists	Last three lists	Subject	First three lists	Last three lists
1	31	26	10	32	30
2	34	32	11	35	28
3	27	26	12	22	22
4	38	34	13	26	21
5	30	25	14	30	27
6	26	25	15	34	31
7	33	26	16	29	26
8	30	28	17	30	24
9	23	20	18	32	27

*12. How is the size of a confidence interval related to the following?
   a. Size of population standard deviation
   b. Sample size
   c. Confidence coefficient

13. Terms to remember:
   a. Point estimate
   b. Confidence interval
   c. Confidence coefficient
   d. Confidence limits

---

†Section 15.3   Confidence Intervals for σ^2 and σ_1^2/σ_2^2

Confidence Interval for σ^2

A confidence interval for the unknown variance of a population, σ^2, can be derived from the statement

$$\text{Prob}(\chi^2_{1-\alpha/2,v} \leq \chi^2 \leq \chi^2_{\alpha/2,v}) = 1 - \alpha,$$

† This section can be omitted without loss of continuity.

where $\chi^2 = [(n-1)\hat{\sigma}^2]/\sigma^2$ and $\chi^2_{1-\alpha/2,\nu}$ and $\chi^2_{\alpha/2,\nu}$ are the values of the chi-square distribution for $\nu = n-1$ degrees of freedom that cut off the lower and upper $\alpha/2$ regions, respectively. Following the derivation described in Section 15.2, we substitute $[(n-1)\hat{\sigma}^2]/\sigma^2$ for χ^2 in the preceding probability statement:

$$\text{Prob}\left[\chi^2_{1-\alpha/2,\nu} \leq \frac{(n-1)\hat{\sigma}^2}{\sigma^2} \leq \chi^2_{\alpha/2,\nu}\right] = 1 - \alpha.$$

Dividing each term in the inequality by $(n-1)\hat{\sigma}^2$ and then inverting each term, thereby reversing the direction of the inequalities, we obtain

$$\text{Prob}\left[\frac{(n-1)\hat{\sigma}^2}{\chi^2_{\alpha/2,\nu}} \leq \sigma^2 \leq \frac{(n-1)\hat{\sigma}^2}{\chi^2_{1-\alpha/2,\nu}}\right] = 1 - \alpha.$$

Thus, a $100(1-\alpha)\%$ *confidence interval for* σ^2 *is given by*

$$\frac{(n-1)\hat{\sigma}^2}{\chi^2_{\alpha/2,\nu}} \leq \sigma^2 \leq \frac{(n-1)\hat{\sigma}^2}{\chi^2_{1-\alpha/2,\nu}}.$$

The assumptions associated with this interval are that (1) the population distribution of X is normal in form and (2) the collection of X's is a random sample from the population of interest.

Computational Example of Confidence Interval for σ^2

To illustrate the construction of a confidence interval for σ^2 we will use the driver license test score data presented in Table 12.3-1 (page 269), in which $\hat{\sigma}^2 = 150.6958$ and $n = 16$. A $100(1-.10)\%$ confidence interval for these data is

$$\frac{(16-1)150.6958}{24.996} \leq \sigma^2 \leq \frac{(16-1)150.6958}{7.261}$$

$$90.432 \leq \sigma^2 \leq 311.312.$$

Thus, an experimenter can be confident that σ^2 is in the interval from 90.432 to 311.312.

Confidence Interval for σ_1^2/σ_2^2 (Independent Samples)

A point estimator for the ratio of two population variances, σ_1^2/σ_2^2, is the ratio $\hat{\sigma}_1^2/\hat{\sigma}_2^2$ of sample variances. If σ_1^2 and σ_2^2 are the variances of independent normal populations, we can construct a confidence interval for σ_1^2/σ_2^2 by using the statistic[3] $F = \sigma_2^2\hat{\sigma}_1^2/\sigma_1^2\hat{\sigma}_2^2$. Following the pattern of previous derivations, we begin with the statement

$$\text{Prob}(F_{1-\alpha/2;\nu_1,\nu_2} \leq F \leq F_{\alpha/2;\nu_1,\nu_2}) = 1 - \alpha$$

---

[3] This statistic is described in footnote 2 in Section 14.1 (page 307).

and substitute $\sigma_2^2\hat{\sigma}_1^2/\sigma_1^2\hat{\sigma}_2^2$ for F to obtain

$$\text{Prob}\left(F_{1-\alpha/2;v_1,v_2} \leq \frac{\sigma_2^2\hat{\sigma}_1^2}{\sigma_1^2\hat{\sigma}_2^2} \leq F_{\alpha/2;v_1,v_2}\right) = 1 - \alpha,$$

where $F_{1-\alpha/2;v_1,v_2}$ and $F_{\alpha/2;v_1,v_2}$ are the values of F that cut off the lower and upper $\alpha/2$ portions of the sampling distribution of F, respectively, with $v_1 = n_1 - 1$ and $v_2 = n_2 - 1$. Multiplying each term in the inequality by $\hat{\sigma}_2^2/\hat{\sigma}_1^2$ and then inverting each term, which changes the direction of the inequality, we obtain

$$\text{Prob}\left(\frac{\hat{\sigma}_1^2}{\hat{\sigma}_2^2}\frac{1}{F_{\alpha/2;v_1,v_2}} \leq \frac{\sigma_1^2}{\sigma_2^2} \leq \frac{\hat{\sigma}_1^2}{\hat{\sigma}_2^2}\frac{1}{F_{1-\alpha/2;v_1,v_2}}\right) = 1 - \alpha.$$

As discussed in Section 14.1, $1/F_{1-\alpha/2;v_1,v_2} = F_{\alpha/2;v_2,v_1}$. *Making this substitution, we obtain the following* $100(1-\alpha)\%$ *confidence interval for* σ_1^2/σ_2^2:

$$\frac{\hat{\sigma}_1^2}{\hat{\sigma}_2^2}\frac{1}{F_{\alpha/2;v_1,v_2}} \leq \frac{\sigma_1^2}{\sigma_2^2} \leq \frac{\hat{\sigma}_1^2}{\hat{\sigma}_2^2}F_{\alpha/2;v_2,v_1}.$$

Computational Example of Confidence Interval for σ_1^2/σ_2^2

In Section 14.1 English-achievement data were presented for 25 disadvantaged children who had been placed in a contingency management classroom and 25 who had been placed in a traditional classroom. At the end of the school year unbiased estimates of the population variances for the two samples were $\hat{\sigma}_1^2 = 64$ and $\hat{\sigma}_2^2 = 196$. A $100(1 - .02)\%$ confidence interval for the ratio σ_1^2/σ_2^2 is

$$\frac{64}{196}\frac{1}{2.66} \leq \frac{\sigma_1^2}{\sigma_2^2} \leq \frac{64}{196}2.66$$

$$(0.3265)(0.3759) \leq \frac{\sigma_1^2}{\sigma_2^2} \leq (0.3265)(2.66)$$

$$0.1227 \leq \frac{\sigma_1^2}{\sigma_2^2} \leq 0.8685.$$

Since the interval doesn't include 1, we can be confident that the two variances are not equal. The best guess that we can make regarding the ratio σ_1^2/σ_2^2 is $\hat{\sigma}_1^2/\hat{\sigma}_2^2 = 0.3265$; we can be 98% confident that the ratio is not less than 0.1227 or greater than 0.8685.

Review Exercises for Section 15.3

*14. For the data in Exercise 7 (page 325), construct a 98% confidence interval for σ^2.

15. Data for this year's KTI graduate school applicants are presented in Exercise 7 (page 325). Suppose that the variance for last year's 20 applicants, sample 1, is 306.1. (a) Construct a 98% confidence interval for the ratio of the variances. (b) Is the variance for this year's applicants different from that for last year's?

*16. A 99% confidence interval for σ_1^2/σ_2^2 is -0.6 to 2.7. Could the population variances be equal? Why?

*17. (a) For the data in Exercise 9 (page 325), construct a 98% confidence interval for σ_1^2/σ_2^2. (b) Could the population variances be equal? (c) If, prior to performing a t test, you wanted to test the assumption that the population σ^2's are equal, what level of significance would you adopt?

18. (a) For the data in Exercise 10 (page 326), construct a 98% confidence interval for σ_1^2/σ_2^2. (b) Could the population variances be equal?

---

†Section 15.4   Confidence Intervals for p and $p_1 - p_2$

Confidence Interval for p

We learned in Section 10.1 that the binomial distribution is adequately approximated by the normal distribution if n is large and if both np and $n(1 - p)$ are greater than 5. *A 100(1 − α)% confidence interval for p, the parameter of a binomial distribution, is given by*

$$\hat{p} - z_{\alpha/2}\sqrt{\frac{\hat{p}\hat{q}}{n}} \le p \le \hat{p} + z_{\alpha/2}\sqrt{\frac{\hat{p}\hat{q}}{n}}.$$

If n is small, a correction for continuity should be applied to the confidence interval as follows:

$$\left(\hat{p} - \frac{1}{2n}\right) - z_{\alpha/2}\sqrt{\frac{\hat{p}\hat{q}}{n}} \le p \le \left(\hat{p} + \frac{1}{2n}\right) + z_{\alpha/2}\sqrt{\frac{\hat{p}\hat{q}}{n}}.$$

Let us return to our politician, Mr. Jerry Mander, who in Section 15.1 had decided not to cancel a political television announcement scheduled for election eve. Was Mr. Mander's decision a wise one? According to the Giddyup poll, a majority (54%) of the voters said they planned to vote for Mr. Mander. Suppose that this percentage is based on a random sample of $n = 1500$ registered voters. A 100(1 − .01)% confidence interval for p, the proportion of voters favoring Mr. Mander, is given by

$$0.54 - 2.576\sqrt{\frac{(0.54)(0.46)}{1500}} \le p \le 0.54 + 2.576\sqrt{\frac{(0.54)(0.46)}{1500}}$$

$$0.54 - 0.033 \le p \le 0.54 + 0.033$$

$$0.507 \le p \le 0.573.$$

We can be 99% confident that the true proportion is covered by the interval 0.507–0.573. Had Mr. Jerry Mander known this, he undoubtedly would have decided to listen to election returns while vacationing in Hawaii.

Confidence Interval for $p_1 - p_2$ (Independent Samples)

The most efficient point estimator for $p_1 - p_2$, the difference between the parameters of two binomial distributions, is the statistic $\hat{p}_1 - \hat{p}_2$. If $n_1\hat{p}_1$, $n_1\hat{q}_1$, $n_2\hat{p}_2$,

---

† This section can be omitted without loss of continuity.

and $n_2\hat{q}_2$ are all greater than 5 and if the samples are statistically independent, a $100(1 - \alpha)\%$ confidence interval for $p_1 - p_2$ is given by

$$(\hat{p}_1 - \hat{p}_2) - z_{\alpha/2}\sqrt{\frac{\hat{p}_1\hat{q}_1}{n_1} + \frac{\hat{p}_2\hat{q}_2}{n_2}} \le p_1 - p_2 \le (\hat{p}_1 - \hat{p}_2) + z_{\alpha/2}\sqrt{\frac{\hat{p}_1\hat{q}_1}{n_1} + \frac{\hat{p}_2\hat{q}_2}{n_2}}.$$

This confidence interval, like that for p, is approximate, since the standard error of p depends on the true p—that is, $\sigma_p = \sqrt{pq/n}$. Since p is unknown, a sample estimate of p must be used in the confidence interval. This is satisfactory for all practical purposes if n is large, say at least 100.

Review Exercises for Section 15.4

*19. A national survey of unmarried women between the ages of 15 and 19 found that 46% of the 19-year-olds had experienced sexual intercourse. If $n = 200$, construct a 95% confidence interval for p.

20. In the survey cited in Exercise 19, 26.6% of the 17-year-olds reported they had experienced sexual intercourse. If the sample contained $n = 150$ 17-year-olds, construct a 95% confidence interval for p.

*21. One hundred men who had suffered one heart attack participated in a supervised physical fitness program. Only seven of the men had a second attack during the 12 months after beginning the program. Construct a 99% confidence interval for p.

22. According to national statistics, the chances of a man's having a second heart attack are 1 in 20 each year after the first seizure. Was the physical fitness program cited in Exercise 21 effective? Why?

23. In a random sample of 50 homes in Junction City, it was found that 42 have color television sets. Construct a 99% confidence interval for p; use a correction for continuity.

*24. For the data reported in Exercises 19 and 20, construct a 95% confidence interval for the difference between the proportion of 19-year-olds, p_1, and 17-year-olds, p_2, who had experienced sexual intercourse.

25. Twenty-four of 100 men and 13 of 100 women surveyed at a southwest college reported they had a car on campus. Construct a 99% confidence interval for $p_1 - p_2$.

26. A test comparing the detectability of two hues of stoplights under simulated fog conditions found that the relative frequencies of detection for red and yellow lights were 0.56 and 0.62, respectively. The subjects were randomly assigned to view one or the other condition; 321 viewed the red light and 315 viewed the yellow light. Construct a 95% confidence interval for $p_1 - p_2$.

---

†Section 15.5   Confidence Intervals for ρ and $\rho_1 - \rho_2$

Confidence Interval for ρ

The Fisher r-to-Z' transformation described in Section 12.5 is used to construct a confidence interval for a population correlation coefficient ρ. Recall that the sampling distribution of Z' is approximately normal in form if (1) ρ is not too

---

† This section can be omitted without loss of continuity.

close to 1 or -1, (2) the population is bivariate normal in form, and (3) the sample size is at least moderately large, say $n > 50$. The r-to-Z' transformation is accomplished by means of Appendix Table D.8.

A $100(1 - \alpha)\%$ confidence interval for Z'_{Pop} is given by

$$Z' - z_{\alpha/2} \sqrt{\frac{1}{n - 3}} \le Z'_{Pop} \le Z' + z_{\alpha/2} \sqrt{\frac{1}{n - 3}},$$

where Z' is the transformed sample r, $z_{\alpha/2}$ is the value of z that cuts off the upper $\alpha/2$ portion of the sampling distribution of z, n is the size of the sample used to estimate ρ, and Z'_{Pop} is the transformed population correlation coefficient. Once the lower and upper limits of Z'_{Pop} have been determined, they can be transformed into limits for ρ by means of Table D.8.

Computational Example of Confidence Interval for ρ

Suppose that an experimenter has correlated college grades and income 10 years after graduation for a random sample of 62 graduates of Florida State U. The sample estimate of the population correlation coefficient is 0.161. A $100(1 - .05)\%$ confidence interval is given by

$$0.161 - 1.96 \sqrt{\frac{1}{62 - 3}} \le Z'_{Pop} \le 0.161 + 1.96 \sqrt{\frac{1}{62 - 3}}$$

$$0.161 - 0.255 \le Z'_{Pop} \le 0.161 + 0.255$$

$$-0.094 \le Z'_{Pop} \le 0.416.$$

Transforming the lower and upper limits of Z' into correlations yields the 95% confidence interval for ρ, which is $-0.09 \le \rho \le 0.39$. Since the confidence interval includes 0, it is apparent that a test of the null hypothesis $H_0: \rho = 0$ or any other hypothesis in which ρ_0 is in the interval -0.09 to 0.39 could not be rejected.

Confidence Interval for $\rho_1 - \rho_2$

The Fisher r-to-Z' transformation can be used to construct a confidence interval for $\rho_1 - \rho_2$, the difference between population correlation coefficients. A $100(1 - \alpha)\%$ confidence interval for $Z'_{Pop_1} - Z'_{Pop_2}$ is given by

$$(Z'_1 - Z'_2) - z_{\alpha/2} \sqrt{\frac{1}{n_1 - 3} + \frac{1}{n_2 - 3}} \le Z'_{Pop_1} - Z'_{Pop_2}$$

$$\le (Z'_1 - Z'_2) + z_{\alpha/2} \sqrt{\frac{1}{n_1 - 3} + \frac{1}{n_2 - 3}},$$

where the terms are as defined on page 330.

Review Exercises for Section 15.5

27. Derive a confidence interval for Z' from $\text{Prob}(-z_{\alpha/2} \leq z \leq z_{\alpha/2}) = 1 - \alpha$, where $z = (Z' - Z'_{Pop})/(1/\sqrt{n-3})$.

*28. The correlation between scores on the TAC (a college entrance test) and grade point averages for a random sample of $n = 84$ freshmen was 0.54. Construct a 95% confidence interval for ρ.

29. Psychological Associates, a consulting firm, has developed a test to select managers for a large chain of hamburger restaurants. The test was given to a random sample of 52 managers. The correlation between their test scores and a measure of their stores' net incomes was 0.20. (a) Construct a 95% confidence interval for ρ. (b) Should the test be used in selecting future managers for the chain? Why?

30. The correlation between the recreational interests of a random sample of $n = 67$ pairs of husbands and wives who had contacted a large travel agency was 0.52. Construct a 99% confidence interval for ρ.

*31. The correlation between scores on the BEEC (a college entrance test) and grade point averages for a random sample of $n = 52$ freshmen was 0.46. Assume that this sample was obtained from the college cited in Exercise 28. Construct a 95% confidence interval for the difference between the TAC and BEEC correlations.

32. The correlations between IQ and school achievement for random samples of male and female high school seniors were 0.37 and 0.46, respectively. Each sample contained 103 students. Construct a 95% confidence interval for $\rho_1 - \rho_2$.

Section 15.6   Summary

Estimation procedures are used when an experimenter is interested in estimating an unknown population parameter rather than in testing a particular hypothesis about the parameter. Point estimation provides us with a single numerical value that is the best guess we can make concerning the unknown parameter. Interval estimation, which is the subject of this chapter, provides us with a range of values that have a specified probability of capturing the parameter. They indicate the reliability of our point estimate.

Given a parameter, say θ, we can find two functions L_1 and L_2 of the sample values such that before the sample is drawn the probability that the interval from L_1 to L_2 will contain θ is equal to $1 - \alpha$. The interval from L_1 to L_2 is called a confidence interval for θ, with a confidence coefficient equal to $1 - \alpha$. We can be $100(1 - \alpha)\%$ confident that θ is contained within the confidence limits L_1 and L_2.

A confidence interval indicates the values of the parameter that are consistent with the observed sample statistic. Thus, a $100(1 - \alpha)\%$ confidence interval contains a range of values of μ_0 for which the null hypothesis would not be rejected at α level of significance.

The size of a confidence interval is determined by (1) the level of the confidence coefficient, $(1 - \alpha)$, (2) the size of the sample, and (3) the size of the population variance.

The confidence intervals we have described in this chapter are presented in Table 15.6-1, along with a summary of the parameters with which they are used, their formulas, and their assumptions.

Table 15.6-1. Summary of one- and two-sample confidence intervals

Chapter section	Parameter	Confidence interval	Assumptions
15.2	μ	$\bar{X} - t_{\alpha/2,\nu}\hat{\sigma}_{\bar{X}} \leq \mu \leq \bar{X} + t_{\alpha/2,\nu}\hat{\sigma}_{\bar{X}}$ where $\hat{\sigma}_{\bar{X}} = \hat{\sigma}/\sqrt{n}$	1. Random sampling 2. Normality 3. Variance unknown
15.2	$\mu_1 - \mu_2$	$(\bar{X}_1 - \bar{X}_2) - t_{\alpha/2,\nu}\hat{\sigma}_{\bar{X}_1-\bar{X}_2} \leq \mu_1 - \mu_2 \leq (\bar{X}_1 - \bar{X}_2) + t_{\alpha/2,\nu}\hat{\sigma}_{\bar{X}_1-\bar{X}_2}$ where $\hat{\sigma}_{\bar{X}_1-\bar{X}_2} = \sqrt{\hat{\sigma}^2_{Pooled}\left(\dfrac{1}{n_1} + \dfrac{1}{n_2}\right)}$	1. Random sampling or random assignment 2. Normality 3. Variances unknown but assumed equal 4. Independence between samples
15.2	$\mu_1 - \mu_2$	$(\bar{X}_1 - \bar{X}_2) - t_{\alpha/2,\nu}\hat{\sigma}_{\bar{X}_D} \leq \mu_1 - \mu_2 \leq (\bar{X}_1 - \bar{X}_2) + t_{\alpha/2,\nu}\hat{\sigma}_{\bar{X}_D}$ where $\hat{\sigma}_{\bar{X}_D} = \dfrac{\sqrt{\dfrac{\sum D_i^2 - (\sum D_i)^2/n}{n}}}{\sqrt{n-1}}$	1. Random sampling or random assignment 2. Normality 3. Dependence between samples
15.3	σ^2	$\dfrac{(n-1)\hat{\sigma}^2}{\chi^2_{\alpha/2,\nu}} \leq \sigma^2 \leq \dfrac{(n-1)\hat{\sigma}^2}{\chi^2_{1-\alpha/2,\nu}}$	1. Random sampling 2. Normality
15.3	$\dfrac{\sigma_1^2}{\sigma_2^2}$	$\dfrac{\hat{\sigma}_1^2}{\hat{\sigma}_2^2}\dfrac{1}{F_{\alpha/2;\nu_1,\nu_2}} \leq \dfrac{\sigma_1^2}{\sigma_2^2} \leq \dfrac{\hat{\sigma}_1^2}{\hat{\sigma}_2^2} F_{\alpha/2;\nu_2,\nu_1}$	1. Random sampling or random assignment 2. Normality 3. Independence between samples

continued

Table 15.6-1. (Continued)

Chapter section	Parameter	Confidence interval	Assumptions
15.4	p	$\hat{p} - z_{\alpha/2}\hat{\sigma}_p \le p \le \hat{p} + z_{\alpha/2}\hat{\sigma}_p$ where $\hat{\sigma}_p = \sqrt{\dfrac{\hat{p}\hat{q}}{n}}$	1. Random sampling 2. Binomial distribution 3. $n\hat{p} > 5,\ n(1 - \hat{p}) > 5$
15.4	$p_1 - p_2$	$(\hat{p}_1 - \hat{p}_2) - z_{\alpha/2}\hat{\sigma}_{p_1 - p_2} \le p_1 - p_2 \le (p_1 - p_2) + z_{\alpha/2}\hat{\sigma}_{p_1 - p_2}$ where $\hat{\sigma}_{p_1 - p_2} = \sqrt{\dfrac{\hat{p}_1\hat{q}_1}{n_1} + \dfrac{\hat{p}_2\hat{q}_2}{n_2}}$	1. Random sampling or random assignment 2. Binomial distributions 3. Independence between samples 4. $n_1\hat{p}_1 > 5,\ n_1\hat{q}_1 > 5,$ $n_2\hat{p}_2 > 5,\ n_2\hat{q}_2 > 5$
15.5	ρ	$Z' - z_{\alpha/2}\sigma_{Z'} \le Z'_{Pop} \le Z' + z_{\alpha/2}\sigma_{Z'}$ where $\sigma_{Z'} = \sqrt{\dfrac{1}{n - 3}}$	1. Random sampling 2. Bivariate normality 3. ρ not too close to ± 1 4. n at least moderately large
15.5	$\rho_1 - \rho_2$	$(Z'_1 - Z'_2) - z_{\alpha/2}\sigma_{Z'_1 - Z'_2} \le Z_{Pop_1} - Z_{Pop_2} \le (Z'_1 - Z'_2) + z_{\alpha/2}\sigma_{Z'_1 - Z'_2}$ where $\sigma_{Z'_1 - Z'_2} = \sqrt{\dfrac{1}{n_1 - 3} + \dfrac{1}{n_2 - 3}}$	1. Random sampling or random assignment 2. Bivariate normality 3. ρ_1 and ρ_2 not too close to ± 1 4. n_1 and n_2 at least moderately large

[†]Section 15.7   Computer Supplement

The Minitab Statistical Computing System, Release 81.1 (Ryan, Joiner, and Ryan, 1976), was used to construct a confidence interval and to perform a one-sample *t* test for the data in Table 12.2-1. These data are registration times for a new class registration procedure at Idle-On-In College.

Minitab can be used in either batch or interactive mode. It is a user-friendly system in the sense that the commands resemble English sentences. Only seven commands, the lines beginning with "--," were required to obtain the printout in Table 15.7-1.

Table 15.7-1. Printout from the Minitab Statistical Computing System (Data from Table 12.2-1)

```
MINITAB RELEASE 81.1 *** COPYRIGHT — PENN STATE UNIV. 1981
JULY  4, 1776 *** HANKAMER SCHOOL OF BUSINESS — LOCAL VERSION 1.0
STORAGE AVAILABLE    4800

-- READ IN C1
COLUMN        C1
COUNT        21
     2.90000     2.70000     2.80000     3.00000  .  .  .

-- AVERAGE C1
   AVERAGE =       2.9000

-- STANDARD DEVIATION C1
   ST.DEV. =      0.15811

-- COUNT C1
   COUNT   =      21.000

-- TINTERVAL 95 PERCENT CONFIDENCE FOR DATA IN C1
   C1          N =  21     MEAN =      2.9000     ST.DEV. =      0.158

   A 95.00  PERCENT C.I. FOR MU IS (      2.8280,      2.9720)

-- TTEST OF MU = 3.1 ON DATA IN C1
   C1          N =  21     MEAN =      2.9000     ST.DEV. =      0.158

   TEST OF MU =      3.1000 VS. MU N.E. =    3.1000
   T = -5.797
   THE TEST IS SIGNIFICANT AT   0.0000

-- STOP
```

The key features of the printout are as follows. The analysis session began with the command

--*READ* THE FOLLOWING DATA IN COLUMN *C1*

Only the first four characters of the command and the column number (the italic characters) are actually used by the computer. Thus for simplicity, the command

---

[†] This section can be omitted without loss of continuity.

was shortened to

READ IN C1.

Twenty-one scores then were typed into the computer and stored in a file called a worksheet. The program acknowledges that the data are in column 1 of the worksheet and that it has counted 21 scores; the first four scores in the worksheet are then displayed. The second command,

--AVERAGE THE VALUES IN COLUMN *C1*,

was shortened to

--AVERAGE C1.

In response to this command, the program displayed the mean of the 21 scores, AVERAGE = 2.9000. Similar commands,

--STANDARD DEVIATION OF COLUMN *C1*

and

--COUNT THE NUMBER OF VALUES IN COLUMN *C1*,

led to the displays ST. DEV. = 0.15811 and COUNT = 21.000.

A 95% confidence interval for μ was obtained by the following command.

--TINTERVAL WITH *95* PERCENT CONFIDENCE FOR DATA IN *C1*

This command produced the following information: (1) column for which the analysis was performed, C1, (2) sample size, 21, (3) sample mean, 2.9000, (4) standard deviation, 0.158, and (5) the lower and upper bounds of the 95% confidence interval for μ, 2.8280 and 2.97201.

A one-sample t test for the null hypothesis $\mu = 3.1$ was obtained by the following command:

--TTEST OF POPULATION MU = *3.1*, DATA IN *C1*.

This command produced the same printout as that for the 95% confidence interval except that the confidence interval was replaced by (1) a statement of the hypothesis tested, $\mu = 3.1$ versus $\mu \neq 3.1$, (2) the value of the t test statistic, -5.797, and (3) the p value of the statistic, $p < .0000$. The analysis session was terminated with the command *--STOP*.

16

Introduction to the Analysis of Variance

Section 16.1 Purposes of Analysis of Variance

Testing the Null Hypothesis that $\mu_1 = \mu_2 = \cdots = \mu_p$

One of the most frequently used statistical procedures in the behavioral sciences is analysis of variance (see Edgington, 1974), often referred to as ANOVA. The procedure was developed by R. A. Fisher in the early 1920s to test null hypotheses of the form

$$H_0: \quad \mu_1 = \mu_2 = \cdots = \mu_p,$$

where $\mu_1, \mu_2, \ldots, \mu_p$ denote the means of $p \geq 2$ populations. If H_0 is rejected, the alternative hypothesis that at least two of the population means are not equal is tenable—that is, $\mu_j \neq \mu_{j'}$ for some pair of means μ_j and $\mu_{j'}$.

In Chapter 13 we described z and t statistics for testing hypotheses of the form $H_0: \mu_j - \mu_{j'} = 0$. Since these statistics can be used to test hypotheses about the equality of means, you may wonder why we need a procedure like ANOVA that utilizes an F statistic. To see that such a procedure is needed, let's consider an experiment with $p = 3$ experimental conditions. How many hypotheses can be tested using a t statistic? There are $p(p - 1)/2 = 3$ hypotheses of the form $\mu_j - \mu_{j'} = \delta_0$, where $\delta_0 = 0$; they are $H_0: \mu_1 - \mu_2 = 0$, $H_0: \mu_1 - \mu_3 = 0$, and $H_0: \mu_2 - \mu_3 = 0$. In addition to these three hypotheses, we could also use a t statistic to test hypotheses involving three means,[1] such as

$$H_0: \quad \mu_1 - \frac{\mu_2 + \mu_3}{2} = 0 \qquad H_0: \quad \mu_2 - \frac{\mu_1 + \mu_3}{2} = 0 \qquad H_0: \quad \mu_3 - \frac{\mu_1 + \mu_2}{2} = 0.$$

These hypotheses state that one population mean is equal to the average of two other population means. Unlike the set of two-mean hypotheses, there is an infinite number of three-mean hypotheses that can be formulated by selecting weights[2] other than 1, $-\frac{1}{2}$, and $-\frac{1}{2}$, as in

$$H_0: \quad \mu_1 - \frac{0.5\mu_2 + 1.5\mu_3}{2} = 0 \qquad \text{or} \qquad H_0: \quad \mu_1 - \frac{0.2\mu_2 + 1.8\mu_3}{2}.$$

---

[1] The t formula for testing these hypotheses is equal to the square root of Scheffé's statistic (see Section 16.5).

[2] The only restriction on the weights is that they must sum to 0, for example,
$$1 + (-\tfrac{1}{2}) + (-\tfrac{1}{2}) = 0.$$

Here the weights are 1, $-.5/2$, $-1.5/2$ and 1, $-.2/2$, $-1.8/2$. If an experimenter using a t statistic tests a large number of such hypotheses, each at α level of significance, the probability that one or more null hypotheses will be falsely rejected is larger than α—and the probability increases as the number of t tests increases. The advantage of the ANOVA approach is that by performing one F test an experimenter tests simultaneously all possible null hypotheses for p means, and the probability of making a type I error is equal to α for the collection of tests. If the overall null hypothesis $H_0: \mu_1 = \mu_2 = \cdots = \mu_p$ is rejected, the experimenter can conclude that at least two of the population means are not equal. Then the next step is to use the multiple comparison procedure described in Section 16.5 to determine which means are not equal.

The principal difference between the overall ANOVA F approach and the multiple t approach can be summarized as follows. For ANOVA, the probability of making a type I error is equal to α for the overall null hypothesis $\mu_1 = \mu_2 = \cdots = \mu_p$. For t, the probability of a type I error is equal to α for each null hypothesis; this allows the probability of erroneously rejecting one or more true null hypotheses to increase as a function of the number of hypotheses tested. Thus, a persistent experimenter who performs enough t tests, each at α level of significance, will certainly reject too many null hypotheses.

For the special case in which the experiment contains only two experimental conditions and the null hypothesis is $\mu_j - \mu_{j'} = 0$, the F and t tests have the same probability of making a type I error, since in both cases only one hypothesis is tested.

Other Applications

Although ANOVA is used most frequently to test hypotheses about means, it also can be used, for example, to determine (1) the strength of the association between independent and dependent variables, (2) whether two independent variables interact, and (3) whether the regression of the dependent variable on the independent variable is linear or nonlinear.[3]

Review Exercises for Section 16.1

*1. Suppose that five methods of teaching foreign language vocabulary are compared in an experiment. The dependent variable is performance on a 25-item vocabulary test. (a) State the null hypothesis. (b) How many t tests would be required to test just hypotheses of the form $\mu_j - \mu_{j'} = 0$?

*2. If the overall null hypothesis in Exercise 1a is rejected by means of an ANOVA F test, what does this tell the experimenter?

3. Four colors of warning lights on an automobile instrument panel were compared. The dependent measure was reaction time to the onset of a light. (a) State the null hypothesis. (b) If the overall null hypothesis is rejected, what does this tell the experimenter?

4. Under what conditions do the F and t approaches lead to the same probability of making a type I error?

[3] For a discussion of these topics see Edwards (1972), Kirk (1982), Myers (1979), and Winer (1971).

5. For experiments in which the number of experimental conditions is greater than two, what advantage does the ANOVA F approach have over the multiple t approach?

*6. (a) Give two examples of independent variables for which the ANOVA F and multiple t approaches would lead to identical conclusions. *(b) What characteristic do the examples have in common?

Section 16.2   Basic Concepts of ANOVA

The material in this section provides a glimpse of some of the basic concepts associated with a completely randomized ANOVA, the simplest of all the ANOVA designs. The rationale underlying ANOVA is somewhat involved, although the computations are straightforward. An acquaintance with, rather than a mastery of, this section is sufficient for understanding Section 16.3, which deals with the computational procedures for ANOVA. It should be helpful to review Section 16.2 after working through an ANOVA problem.

The Composite Nature of a Score

The value of a score in an experiment is determined by a variety of variables. A score can be thought of as a composite, reflecting, for example, the effects of the (1) independent variable, (2) individual characteristics of the subject or experimental unit, (3) chance fluctuations in the subject's performance, and (4) environmental conditions. Similarly, the variance of scores is also a composite. ANOVA is a procedure for determining how much of the total variability among scores to attribute to various sources of variation and for testing hypotheses concerning some of them.

The composite nature of a score will be illustrated by an example. Consider an experiment to determine the effectiveness of three diets for obese teenage girls. Thirty girls who want to lose weight are randomly assigned to the diets, with ten girls using each diet. The independent variable is type of diet; the dependent variable is weight loss in pounds. *For notational convenience the diets are called treatment[4] A. The levels of treatment A, corresponding to the specific diets, are designated by the lowercase letter a and numeric subscripts—a_1, a_2, and a_3. A particular but unspecified score is denoted by X_{ij}, where the first subscript designates one of the i = 1, ..., n subjects in a treatment level and the second subscript designates one of the j = 1, ..., p levels of treatment A.* Thus, if Bella Ablipid's score is denoted by X_{72}, we know that she is subject seven and that she used diet a_2. What factors have affected the value of her score? If she stuck to her diet, one major factor is the efficacy of diet a_2. Other factors are degree of obesity, day-to-day fluctuations in eating and exercise habits, time of day that weight loss was measured, and so on. In summary, Bella's weight loss, X_{72}, reflects (1) the effect of treatment level a_2, (2) effects unique to her, (3) effects attributable to chance fluctuations in her behavior, and (4) effects attributable to environmental conditions.

---

[4] Some writers use the term *factor* instead of treatment.

The Linear Model Equation Associated with a Score

Our conjectures about X_{72} or any other score can be expressed more formally by a *linear model equation*. We assume that each of the 30 scores is the sum of three components in the equation

$$X_{ij} = \mu + \alpha_j + \varepsilon_{ij},$$

where

X_{ij} is the score for subject i in treatment level j;

μ is the grand mean of μ_1, μ_2, and μ_3, the means of the three treatment populations (μ is a constant for the 30 scores and reflects their general elevation);

α_j is the treatment effect of population j and is equal to $\mu_j - \mu$, the deviation of
the jth population mean from the grand mean (α_j is a constant for the ten
scores in population j and reflects the elevation or depression of these scores
relative to μ resulting from the use of the jth diet); and

ε_{ij} is the "error" effect associated with X_{ij} and is equal to $X_{ij} - \mu - \alpha_j$, the portion
of a score that remains after the grand mean and jth treatment effect are
subtracted from it (ε_{ij} reflects effects unique to subject i, effects attributable
to chance fluctuations in subject i, and effects attributable to environmental
conditions—in other words, all effects not attributable to treatment level j).

The values of the parameters μ, α_j, and ε_{ij} are unknown, but we will see that
they can be estimated from sample data and that the variability among the
X_{ij}'s can be partitioned by ANOVA so as to test the null hypothesis that $\mu_1 = \mu_2 = \cdots = \mu_p$. If the null hypothesis is true, each treatment effect, α_j, must equal
0. This follows, since in that case all the population means equal the grand mean,
$\mu_1 = \mu_2 = \cdots = \mu_p = \mu$, and $\alpha_1 = \mu_1 - \mu = 0$, $\alpha_2 = \mu_2 - \mu = 0$, and so on. Thus,
a test of $\mu_1 = \mu_2 = \cdots = \mu_p$ is equivalent to a test of the hypothesis $\alpha_1 = \alpha_2 = \cdots = \alpha_p = 0$—that is, that all population treatment effects are equal to 0.

Estimating the Parameters of the Linear Model Equation

Suppose that the data in Table 16.2-1 have been obtained in our diet experiment.
The treatment means, $\bar{X}_{.1}$, $\bar{X}_{.2}$, and $\bar{X}_{.3}$, and the grand mean, $\bar{X}_{..}$, are shown in
the table. The use of a dot in the notation indicates that the mean was obtained
by averaging over the subscript replaced by the dot. For example, treatment means
are obtained by

$$\bar{X}_{.1} = \sum_{i=1}^{n} \frac{X_{i1}}{n} = \frac{X_{11} + X_{21} + X_{31} + \cdots + X_{10,1}}{n} = \frac{80}{10} = 8$$

$$\bar{X}_{.2} = \sum_{i=1}^{n} \frac{X_{i2}}{n} = \frac{X_{12} + X_{22} + X_{32} + \cdots + X_{10,2}}{n} = \frac{90}{10} = 9$$

$$\bar{X}_{.3} = \sum_{i=1}^{n} \frac{X_{i3}}{n} = \frac{X_{13} + X_{23} + X_{33} + \cdots + X_{10,3}}{n} = \frac{120}{10} = 12.$$

Table 16.2-1.   One-month weight losses measured to the nearest pound

(i) *Data and notation (X_{ij} denotes a score for subject i in treatment level j; $i = 1, \ldots, n$ subjects; $j = 1, \ldots, p$ levels of treatment A):*

	Treatment levels (diets)	
a_1	a_2	a_3
7	10	12
9	13	11
8	9	15
12	11	7
8	5	14
7	9	10
4	8	12
10	10	12
9	8	13
6	7	14

Sum of $i = 1, \ldots, n$ scores in each treatment level $\rightarrow \sum_{i=1}^{n} X_{i1} = 80 \quad \sum_{i=1}^{n} X_{i2} = 90 \quad \sum_{i=1}^{n} X_{i3} = 120 \quad$ Sum of all scores $\rightarrow \sum_{j=1}^{p} \sum_{i=1}^{n} X_{ij} = 290$

Mean of each treatment level $\rightarrow \bar{X}_{.1} = 8 \qquad \bar{X}_{.2} = 9 \qquad \bar{X}_{.3} = 12 \qquad$ Grand mean $\rightarrow \bar{X}_{..} = 9.67$

The grand mean is obtained by summing all the np scores and dividing by np:

$$\bar{X}_{..} = \sum_{j=1}^{p} \sum_{i=1}^{n} \frac{X_{ij}}{np} = [(X_{11} + X_{21} + \cdots + X_{10,1}) + \cdots$$
$$+ (X_{13} + X_{23} + \cdots + X_{10,3})]/np$$
$$= \frac{80 + 90 + 120}{(10)(3)} = 9.67.$$

Table 16.2-1 contains scores and means, but the values of the parameters μ, α_j, and ε_{ij} in the linear model equation are unknown. However, unbiased estimators of the parameters can be obtained from the sample data, as shown in Table 16.2-2. We can rewrite the linear model equation $X_{ij} = \mu + \alpha_j + \varepsilon_{ij}$ using these estimators, as follows:

$$X_{ij} = \mu + \alpha_j + \varepsilon_{ij}$$
$$X_{ij} = \bar{X}_{..} + (\bar{X}_{.j} - \bar{X}_{..}) + (X_{ij} - \bar{X}_{.j}).$$

Table 16.2-2.   Estimators for parameters of the linear model

Statistic	Parameter estimated	Interpretation of parameter
$\bar{X}_{..}$	μ	General elevation of scores
$\bar{X}_{.j} - \bar{X}_{..}$	α_j	Elevation or depression of scores attributable to the jth treatment level
$X_{ij} - \bar{X}_{.j}$	ε_{ij}	Error effect unique to subject i in treatment level j

Perhaps an example using Bella's score, X_{72}, will help to clarify the nature of these estimators. According to the equation, this score is a composite, reflecting all the factors in the experiment; it can be expressed as

$$X_{72} = \bar{X}.. + (\bar{X}._2 - \bar{X}..) + (X_{72} - \bar{X}._2)$$
$$8 = 9.67 + (9 - 9.67) \quad + (8 - 9)$$
$$8 = 9.67 + (-0.67) \quad \quad + (-1)$$
$$8 = 9.67 - 1.67.$$

This model equation gives us a bit more insight into why Bella's weight loss, $X_{72} = 8$ lb, was 1.67 lb less than the average, $\bar{X}.. = 9.67$ lb. She used a less effective diet, $\bar{X}._2 - \bar{X}.. = -0.67$ lb, and in addition, the diet was not as effective for her as it was on the average for the ten girls who used it, $X_{72} - \bar{X}._2 = -1$ lb. This simple experiment allows us to estimate three factors that account for Bella's weight loss: the population average weight loss, estimated by $\bar{X}..$; the effect of diet a_2, estimated by $\bar{X}._2 - \bar{X}..$; and the error effect unique to Bella, estimated by $X_{72} - \bar{X}._2$. This error effect estimates that part of X_{72} not due to μ and α_2; in other words $\varepsilon_{72} = X_{72} - \mu - \alpha_2$. In fact, the estimator for ε_{72}, denoted by $\hat{\varepsilon}_{72}$, is obtained by replacing the parameters in $X_{72} - \mu - \alpha_2$ with statistics, as follows.

$$\varepsilon_{72} = X_{72} - \mu - \alpha_2$$
$$\hat{\varepsilon}_{72} = X_{72} - \bar{X}.. - (\bar{X}._2 - \bar{X}..)$$
$$= X_{72} - \bar{X}._2$$

The estimate $\hat{\varepsilon}_{72} = -1$ reflects all of the characteristics peculiar to this subject, such as her degree of obesity and her day-to-day fluctuations in eating and exercise habits, as well as those characteristics peculiar to the testing conditions, such as the time of day that her weight loss was measured. The name *error effect* is an apt one because it represents all of the effects not otherwise estimated in the experiment. More complex ANOVA designs[5], which are discussed in advanced books, enable an experimenter to estimate more of the factors that affect a score, which generally results in fewer factors being lumped together in the error effect.

We have illustrated the composite nature of a score and procedures for estimating the parameters of the model equation. Our ultimate purpose is to develop procedures for testing the null hypothesis that all of the treatment effects, α_j's, are equal to 0. To accomplish this purpose, several more concepts must be introduced.

Partition of the Total Sum of Squares

We now will show that a measure of the total variability among scores in an experiment, called the *total sum of squares* (SS_{Total}), can be partitioned into two parts—variability between treatment levels and variability within treatment levels.

[5] A discussion of complex ANOVA designs is beyond the scope of this book. For an elementary description, see Kirk (1972b); for an advanced discussion, including computational procedures, see Edwards (1972), Keppel (1973), Kirk (1982), Lee (1975), Lindman (1974), Myers (1979), and Winer (1971).

These two parts are called the *between-groups sum of squares* (SS_{BG}) and the *within-groups sum of squares* (SS_{WG}). We begin by subtracting $\bar{X}..$ from both sides of the equation $X_{ij} = \bar{X}.. + (\bar{X}._j - \bar{X}..) + (X_{ij} - \bar{X}._j)$ given earlier, and obtain

$$X_{ij} - \bar{X}.. = \bar{X}.. - \bar{X}.. + (\bar{X}._j - \bar{X}..) + (X_{ij} - \bar{X}._j)$$
$$= (\bar{X}._j - \bar{X}..) + (X_{ij} - \bar{X}._j).$$

This equation expresses the deviation of the grand mean from a score in terms of a treatment effect plus an error effect. Next, we square both sides of the equation:

$$(X_{ij} - \bar{X}..)^2 = [(\bar{X}._j - \bar{X}..) + (X_{ij} - \bar{X}._j)]^2.$$

This equation is for a single score, but we can treat each of the np scores in the same way and sum the resulting equations to obtain

$$\sum_{j=1}^{p} \sum_{i=1}^{n} (X_{ij} - \bar{X}..)^2 = \sum_{j=1}^{p} \sum_{i=1}^{n} [(\bar{X}._j - \bar{X}..) + (X_{ij} - \bar{X}._j)]^2.$$

The quantity on the left is the total sum of squares. Using elementary algebra and the summation rules in Technical Note 3.9-1, the quantity on the right can be shown to equal the sum of two sums of squares, denoted by SS_{BG} and SS_{WG}.

$$\sum_{j=1}^{p} \sum_{i=1}^{n} (X_{ij} - \bar{X}..)^2 = \sum_{j=1}^{p} \sum_{i=1}^{n} [(\bar{X}._j - \bar{X}..)^2 + 2(\bar{X}._j - \bar{X}..)(X_{ij} - \bar{X}._j) + (X_{ij} - \bar{X}._j)^2]$$

$$= n \sum_{j=1}^{p} (\bar{X}._j - \bar{X}..)^2 + 2 \sum_{j=1}^{p} (\bar{X}._j - \bar{X}..) \sum_{i=1}^{n} (X_{ij} - \bar{X}._j)$$

$$+ \sum_{j=1}^{p} \sum_{i=1}^{n} (X_{ij} - \bar{X}._j)^2.$$

Since $\sum_{i=1}^{n} (X_{ij} - \bar{X}._j) = 0$ (see Technical Note 3.9-2), the middle term on the right equals 0, and we are left with

$$\sum_{j=1}^{p} \sum_{i=1}^{n} (X_{ij} - \bar{X}..)^2 = n \sum_{j=1}^{p} (\bar{X}._j - \bar{X}..)^2 + \sum_{j=1}^{p} \sum_{i=1}^{n} (X_{ij} - \bar{X}._j)^2$$

$$SS_{Total} = \qquad SS_{BG} \qquad + \qquad SS_{WG}.$$

Now that we have partitioned SS_{Total} into SS_{BG} and SS_{WG}, we need to show what this has to do with testing the hypothesis that $\mu_1 = \mu_2 = \cdots = \mu_p$ or the equivalent hypothesis that $\alpha_1 = 0, \alpha_2 = 0, \ldots, \alpha_p = 0$. But before we can do this we must discuss the concepts of degrees of freedom, mean squares, and the F ratio.

Degrees of Freedom

The term *degrees of freedom* refers to the number of observations whose values can be assigned arbitrarily, as we saw in Section 12.2. We now will determine the degrees of freedom associated with SS_{BG}, SS_{WG}, and SS_{Total}.

Consider $SS_{BG} = n \sum_{j=1}^{p} (\bar{X}._j - \bar{X}..)^2$. If we take, say, $p = 4$ sample means, they are related to the grand mean by the equation

$$\frac{\bar{X}._1 + \bar{X}._2 + \bar{X}._3 + \bar{X}._4}{4} = \bar{X}...$$

If $\bar{X}_{..} = 6$ and we arbitrarily specify that $\bar{X}_{.1} = 6$, $\bar{X}_{.2} = 8$, and $\bar{X}_{.3} = 7$, then $\bar{X}_{.4}$ must equal 3, since $(6 + 8 + 7 + 3)/4 = 6$. Alternatively, if we specify that $\bar{X}_{.1} = 5$, $\bar{X}_{.2} = 7$, and $\bar{X}_{.3} = 4$, then $\bar{X}_{.4}$ must equal 8, since $(5 + 7 + 4 + 8)/4 = 6$. Given the value of the grand mean, we are free to assign any values to three of the four treatment means, but having done so, the fourth mean is determined. *Hence, the number of degrees of freedom for SS_{BG} is $p - 1$, one less than the number of treatment means.*

The degrees of freedom associated with SS_{WG} are $p(n - 1)$. To see why this is true, consider $SS_{WG} = \sum_{j=1}^{p} \sum_{i=1}^{n} (X_{ij} - \bar{X}_{.j})^2$ and let $p = 4$ and $n = 8$. For each of the $j = 1, \ldots, 4$ treatment levels, we can compute $n = 8$ deviations.

$$\overbrace{(X_{1j} - \bar{X}_{.j})^2}^{\text{1st}} + \overbrace{(X_{2j} - \bar{X}_{.j})^2}^{\text{2nd}} + \cdots + \overbrace{(X_{8j} - \bar{X}_{.j})^2}^{\text{8th}}$$

Each score in the jth treatment level is related to the jth mean by

$$\frac{X_{1j} + X_{2j} + \cdots + X_{8j}}{8} = \bar{X}_{.j}.$$

Seven of the scores can take any value, but the eighth is determined, since the sum of the scores divided by eight must equal $\bar{X}_{.j}$. Hence, there are $n - 1 = 8 - 1 = 7$ degrees of freedom associated with the jth treatment level. *If $n_1 = n_2 = \cdots = n_p$, each of the p treatment levels has $n - 1$ degrees of freedom; thus, $p(n - 1)$ degrees of freedom are associated with SS_{WG}. If the n_j's are not equal, the degrees of freedom for SS_{WG} are $(n_1 - 1) + (n_2 - 1) + \cdots + (n_p - 1) = N - p$, where N is the total number of scores.*

The same line of reasoning can be used to show that when $n_1 = n_2 = \cdots = n_p$, the total sum of squares has $pn - 1$ degrees of freedom. This follows, since each of the $pn = 4(8) = 32$ scores is related to the grand mean by

$$\frac{X_{11} + X_{21} + \cdots + X_{84}}{32} = \bar{X}_{...}$$

Hence, $np - 1 = 31$ of the scores can take any value, but the 32nd must be assigned so that the mean of the scores equals $\bar{X}_{...}$ If the n_j's are not equal, the number of degrees of freedom is $n_1 + n_2 + \cdots + n_p - 1 = N - 1$.

Mean Squares and the F Ratio

*A **mean square (MS)** is obtained by dividing a sum of squares (SS) by its degrees of freedom (df).* Thus, $MS_{Total} = SS_{Total}/(np - 1)$, $MS_{BG} = SS_{BG}/(p - 1)$, and $MS_{WG} = SS_{WG}/[p(n - 1)]$. The term MS is new, but the concept isn't; MS is simply another name for variance. *The hypothesis $\mu_1 = \mu_2 = \cdots = \mu_p$ is tested by means of an F test statistic, which is the ratio of the between-groups variance to the within-groups variance: $F = MS_{BG}/MS_{WG}$.* The degrees of freedom for the numerator and denominator of the statistic are, respectively, $v_1 = p - 1$ and $v_2 = p(n - 1)$. The F statistic is referred to the sampling distribution of F, which is tabled in Appendix Table D.5. If the F statistic is equal to or greater than the tabled value, $F_{\alpha;v_1,v_2}$, the null hypothesis is rejected.

The Nature of MS_{BG} and MS_{WG}

It may seem paradoxical to test a hypothesis about means by testing the ratio of two variances, so perhaps we should consider the nature of the population variances estimated by MS_{WG} and MS_{BG}. As we will see, MS_{WG} estimates the variation among subjects who have been treated alike; MS_{BG} estimates the variation among subjects who have been treated differently, that is, assigned to different treatment levels. We will examine MS_{WG} first and show that it is a pooled estimator of the population *error variance* (σ_ε^2), the variance of scores for subjects who have been treated alike. Suppose that random samples of size n are drawn[6] from p normally distributed populations having equal means $(\mu_1 = \mu_2 = \cdots = \mu_p = \mu)$ and equal variances $(\sigma_1^2 = \sigma_2^2 = \cdots = \sigma_p^2 = \sigma_\varepsilon^2)$. Unbiased estimators of the p population variances are given by

$$\hat{\sigma}_1^2 = \frac{\sum_{i=1}^{n}(X_{i1} - \bar{X}_{.1})^2}{n-1}, \quad \hat{\sigma}_2^2 = \frac{\sum_{i=1}^{n}(X_{i2} - \bar{X}_{.2})^2}{n-1}, \ldots, \quad \hat{\sigma}_p^2 = \frac{\sum_{i=1}^{n}(X_{ip} - \bar{X}_{.p})^2}{n-1}.$$

The best estimator of the population error variance, σ_ε^2, is obtained by combining the p variance estimators according to the formula

$$\frac{\sum_{i=1}^{n}(X_{i1} - \bar{X}_{.1})^2/(n-1) + \sum_{i=1}^{n}(X_{i2} - \bar{X}_{.2})^2/(n-1) + \cdots + \sum_{i=1}^{n}(X_{ip} - \bar{X}_{.p})^2/(n-1)}{p}$$

$$= \frac{\sum_{j=1}^{p}\sum_{i=1}^{n}(X_{ij} - \bar{X}_{.j})^2}{p(n-1)} = \hat{\sigma}_\varepsilon^2.$$

The term on the right of the equals sign is MS_{WG}, as previously defined. Thus, MS_{WG} is an estimator of σ_ε^2 and is simply the average of p variance estimators— that is, $MS_{WG} = [(\hat{\sigma}_1^2 + \cdots + \hat{\sigma}_p^2)/p] = \hat{\sigma}_\varepsilon^2$. If the n's are not equal, the variance estimators are weighted proportional to their respective degrees of freedom; the formula is

$$MS_{WG} = \frac{(n_1 - 1)\hat{\sigma}_1^2 + (n_2 - 1)\hat{\sigma}_2^2 + \cdots + (n_p - 1)\hat{\sigma}_p^2}{(n_1 - 1) + (n_2 - 1) + \cdots + (n_p - 1)} = \hat{\sigma}_\varepsilon^2.$$

Recall from Section 13.4 that pooling of estimators—in this example $\hat{\sigma}_1^2, \hat{\sigma}_2^2, \ldots, \hat{\sigma}_p^2$—is appropriate when it is reasonable to assume that they all estimate the same parameter, σ_ε^2.

We now will show that MS_{BG} also estimates σ_ε^2 when p random samples of size n are drawn from p normally distributed populations with equal means and equal variances. These are the same conditions under which we said that MS_{WG} estimates σ_ε^2. From Section 10.3 we know that the means of p random samples can be regarded as random variables from a theoretical sampling distribution of means, with mean and standard deviation μ and $\sigma_{\bar{X}}$, respectively. We also know

---

[6] The use of equal n_j's is a convenience, not a requirement.

from Section 10.3 that the standard error of a mean can be computed from $\sigma_{\bar{X}} = \sigma_\varepsilon/\sqrt{n}$; hence, $\sigma_{\bar{X}}^2 = \sigma_\varepsilon^2/n$. If we could estimate the variance of population means $\sigma_{\bar{X}}^2$, we then could estimate σ_ε^2 by $\hat{\sigma}_\varepsilon^2 = n\hat{\sigma}_{\bar{X}}^2$. Fortunately an unbiased estimator of the variance of population means can be computed from the sample data in ANOVA; the estimator is given by

$$\hat{\sigma}_{\bar{X}}^2 = \frac{\sum_{j=1}^{p} (\bar{X}_{\cdot j} - \bar{X}_{\cdot \cdot})^2}{p-1}.$$

Hence,

$$n\hat{\sigma}_{\bar{X}}^2 = \frac{n \sum_{j=1}^{p} (\bar{X}_{\cdot j} - \bar{X}_{\cdot \cdot})^2}{p-1} = \hat{\sigma}_\varepsilon^2.$$

The term in the middle is MS_{BG}, as previously defined. Thus, if p random samples are drawn from normally distributed populations with equal means and equal variances, MS_{BG} is also an estimator of the population error variance, σ_ε^2. It can be shown that under these conditions, MS_{BG} and MS_{WG} are independent estimators of σ_ε^2, and that the ratio $F = MS_{BG}/MS_{WG}$ is distributed as the F distribution with $p - 1$ and $p(n - 1)$ degrees of freedom. Small values of the F ratio (around 1) are likely if the null hypothesis is true—that is, if $\mu_1 = \mu_2 = \cdots = \mu_p$.

But what if the null hypothesis is false? Suppose that random samples are drawn from p normally distributed populations with two or more unequal means (the treatment effects are not all equal) but equal variances. We would expect more variation among the sample means than can be accounted for by chance sampling variability. However, variation among the sample means is independent of the variation within each of the groups, $\hat{\sigma}_1^2, \hat{\sigma}_2^2, \ldots, \hat{\sigma}_p^2$; this can be seen from Figure 16.2-1. As before, $MS_{WG} = [(\hat{\sigma}_1^2 + \cdots + \hat{\sigma}_p^2)/p] = \hat{\sigma}_\varepsilon^2$. However, the presence of unequal means will increase the variance of the sample means. Therefore, when the null hypothesis is false, $MS_{BG} = n\hat{\sigma}_{\bar{X}}^2$ tends to be larger than $\hat{\sigma}_\varepsilon^2$ and the ratio $F = MS_{BG}/MS_{WG}$ tends to be larger than 1.

How large must F be in order to conclude that the discrepancy between MS_{BG} and MS_{WG} should be attributed to unequal population means rather than to chance? The usual practice is to reject the null hypothesis of equal means if F falls in the upper .05 portion of the sampling distribution of F. Values of F that cut off various portions of the upper tail of this sampling distribution are given in Appendix Table D.5.

Expectations of MS_{BG} and MS_{WG}

We can express the central ideas in the previous section in terms of the expectations (expected values) of MS_{BG} and MS_{WG}. An expectation (E) is, according to the empirical relative-frequency view of probability, the long-run average of a random variable. We learned in Section 10.3 that $E(\bar{X}) = \mu$. This means that if we randomly sample from a population an indefinitely large number of times and for each sample compute a mean, $\bar{X}$, the long-run average of the sample means is μ.

Figure 16.2-1. A within-groups estimator of the population error variance is given by $\hat{\sigma}_\varepsilon^2 = (\hat{\sigma}_1^2 + \hat{\sigma}_2^2 + \cdots + \hat{\sigma}_p^2)/p$. This estimator is not affected by differences among the sample means. An estimator of the variance of the p population means is given by $\hat{\sigma}_{\bar{X}}^2 = \sum_{j=1}^{p} (\bar{X}_{\cdot j} - \bar{X}_{\cdot\cdot})^2/(p - 1)$. It is affected by differences among the sample means.

It can be shown (Kirk, 1982, pp. 64–70) that the expectation of MS_{WG} is the population error variance,

$$E(MS_{WG}) = \sigma_\varepsilon^2.$$

If the null hypothesis is true, the expectation of MS_{BG} is also the population error variance,

$$E(MS_{BG}) = \sigma_\varepsilon^2.$$

However, if at least two of the population means are not equal, the expectation of MS_{BG} is

$$E(MS_{BG}) = \sigma_\varepsilon^2 + \frac{n \sum_{j=1}^{p} (\mu_j - \mu)^2}{p - 1}$$

$$= \sigma_\varepsilon^2 + \frac{n \sum_{j=1}^{p} \alpha_j^2}{p - 1}.$$

The ratio $F = MS_{BG}/MS_{WG}$ is distributed as the F distribution, with $v_1 = p - 1$ and $v_2 = p(n - 1)$ if the null hypothesis is true. A ratio that is sufficiently large suggests that the null hypothesis is false. The probability of obtaining an F statistic as large as or larger than that observed if H_0 is true can be determined from the F table. If the probability is equal to or less than some preselected value, say $\alpha = .05$, the null hypothesis that all the means are equal, $\mu_1 = \mu_2 = \cdots = \mu_p$, or that all the treatment effects equal 0, $\alpha_j = 0$ for all j, is rejected.

Review Exercises for Section 16.2

*7. Identify the following.
  a. a_2 b. X_{24}
  c. $X_{16,1}$ d. $\bar{X}_{.4}$
  e. $\bar{X}_{..}$ f. $X_{73} = \mu + \alpha_3 + \varepsilon_{73}$
  g. $\mu_2 - \mu$ h. $X_{13} - \mu - \alpha_3$
 8. In the derivation of

$$\sum_{j=1}^{p}\sum_{i=1}^{n}(X_{ij} - \bar{X}_{..})^2 = n\sum_{j=1}^{p}(\bar{X}_{.j} - \bar{X}_{..})^2 + \sum_{j=1}^{p}\sum_{i=1}^{n}(X_{ij} - \bar{X}_{.j})^2,$$

 we assume that $n_1 = n_2 = \cdots = n_p = n$. Assume instead that the n_j's are not all equal, and derive $\sum_{j=1}^{p}\sum_{i=1}^{n_j}(X_{ij} - \bar{X}_{..})^2 = \sum_{j=1}^{p}n_j(\bar{X}_{.j} - \bar{X}_{..})^2 + \sum_{j=1}^{p}\sum_{i=1}^{n_j}(X_{ij} - \bar{X}_{.j})^2$.
*9. Which of the following alternative hypotheses is correctly stated? (a) $\mu_j - \mu_{j'} \neq 0$ for some j and j', (b) $\mu_1 \neq \mu_2 \neq \mu_3$, (c) $\sum_{j=1}^{p}(\mu_j - \mu) \neq 0$, (d) $\alpha_1 \neq 0, \alpha_2 \neq 0, \alpha_3 \neq 0$, (e) $\alpha_j \neq 0$ for some j.
*10. Express the following scores in terms of estimates of the parameters of the linear model equation: *(a) X_{83}, (b) X_{52}, (c) X_{24}.
 11. Express the hypothesis $\mu_1 = \mu_2 = \mu_3 = \mu_4$ in terms of treatment effects (α_j's).
*12. Calculate the degrees of freedom for MS_{Total}, MS_{BG}, and MS_{WG} with the following conditions. *(a) $p = 4$, $n = 21$, *(b) $p = 5$, $n = 11$, (c) $p = 4$, $n = 8$, *(d) $p = 3$, $n_1 = 6$, $n_2 = 5$, $n_3 = 6$, (e) $p = 4$, $n_1 = 10$, $n_2 = 10$, $n_3 = 9$, $n_4 = 8$.
*13. Under what conditions do both MS_{BG} and MS_{WG} estimate only the population error variance, σ_ε^2?
*14. Under what conditions is $F = MS_{BG}/MS_{WG}$ distributed as the F distribution with $p - 1$ and $p(n - 1)$ degrees of freedom?
*15. Under what conditions does $F = MS_{BG}/MS_{WG}$ tend to be larger than 1?
 16. Terms to remember:
  a. Treatment   b. Treatment level
  c. Linear model equation d. Treatment effect
  e. Error effect   f. Sum of squares
  g. Degrees of freedom   h. Mean square

Section 16.3   Completely Randomized Design

This section presents the computational procedures associated with the simplest of all ANOVA designs—the *completely randomized design*. Here you will see how nicely some of the complex ideas presented previously fit together to produce a decision about the null hypothesis. In fact, after pondering over the next two tables, you will be amazed at how simple the computational procedures for ANOVA really are.

   The completely randomized design is appropriate for experiments with one treatment (independent variable) with $p \geq 2$ treatment levels. The p treatment levels are randomly assigned to $n_1 + n_2 + \cdots + n_p = N$ subjects. If the treatment levels are of equal interest to the experimenter, he or she should assign each level to the same number of subjects, although this is not necessary. *The completely randomized design is so named because the assignment of treatment levels to subjects is completely random, with each subject participating under only one level. For convenience it is referred to as a type **CR-p design**,*[7] *where p designates the number of levels of treatment A.*

Computational Formulas

In Section 16.2 we showed that

$$SS_{Total} = SS_{BG} + SS_{WG}$$

is equal to

$$\sum_{j=1}^{p} \sum_{i=1}^{n} (X_{ij} - \bar{X}_{..})^2 = n \sum_{j=1}^{p} (\bar{X}_{.j} - \bar{X}_{..})^2 + \sum_{j=1}^{p} \sum_{i=1}^{n} (X_{ij} - \bar{X}_{.j})^2.$$

These sums of squares formulas aren't the most convenient for computational purposes. It can be shown[8] that they are equivalent to

$$SS_{Total} = \sum_{j=1}^{p} \sum_{i=1}^{n} X_{ij}^2 - \frac{\left(\sum_{j=1}^{p} \sum_{i=1}^{n} X_{ij} \right)^2}{np}$$

$$SS_{BG} = \sum_{j=1}^{p} \frac{\left(\sum_{i=1}^{n} X_{ij} \right)^2}{n} - \frac{\left(\sum_{j=1}^{p} \sum_{i=1}^{n} X_{ij} \right)^2}{np}$$

$$SS_{WG} = \sum_{j=1}^{p} \sum_{i=1}^{n} X_{ij}^2 - \sum_{j=1}^{p} \frac{\left(\sum_{i=1}^{n} X_{ij} \right)^2}{n}.$$

The use of these *raw-score formulas* is illustrated in Table 16.3-1.

Computational Example for Type CR-3 Design

The computational procedures associated with a completely randomized design will be illustrated for the data from the diet experiment. You will recall that three diets were randomly assigned to 30 girls, with ten girls using each diet. The steps to be followed in testing the null hypothesis and the decision rule are as follows.

Step 1. State the statistical hypotheses: H_0: $\mu_1 = \mu_2 = \mu_3$
H_1: $\mu_j \neq \mu_{j'}$ for some j and j'.

Step 2. Specify the test statistic: $F = \dfrac{MS_{BG}}{MS_{WG}}$.

---

[7] For a comprehensive design classification system and nomenclature, see Kirk (1982).
[8] See Exercise 18 (page 353).

Step 3.   Specify the sample size:[9] $np = 30$;
            and the sampling distribution: F distribution with $v_1 = p - 1$ and $v_2 = p(n - 1)$, because X is assumed to be approximately normal in form for the a_1, a_2, and a_3 population distributions.

Step 4.   Specify the significance level: $\alpha = .05$.

Step 5.   Obtain a random sample of size np subjects and/or randomly assign p treatment levels to np subjects, compute F, and make a decision.

Decision rule:
            Reject H_0 if F falls in the upper 5% of the sampling distribution of F; otherwise, don't reject H_0.

Table 16.3-1 presents the data for the diet experiment as well as details of the computational procedures. The results of the analysis are summarized in an ANOVA table; see Table 16.3-2. The sums of squares values in Table 16.3-2 were obtained from Table 16.3-1, and the MS's were obtained by dividing the SS's by their respective degrees of freedom. The F was obtained by dividing the MS_{BG} in row 1 by the MS_{WG} in row 2; this operation is indicated in the table by the symbol $[\frac{1}{2}]$.

According to Appendix Table D.5, an F of 3.36 with 2 and 27 degrees of freedom cuts off the upper .05 region of the sampling distribution—that is, $F_{.05;2,27} = 3.36$.[10] Since $8.60 > 3.36$, the null hypothesis is rejected, and we conclude that at least two of the diets aren't equally effective. The results of the F test can be presented either by means of a table as in Table 16.3-2 or in the text. Using the latter form, one might say, "We can infer from the analysis of variance that the weight-loss population means for the three diets are not all equal, $F(2, 27) = 8.60$, $p < .01$." When the results are presented in the text, it is customary to identify (1) the test statistic, (2) degrees of freedom, (3) value of the statistic, and (4) *p-value* associated with the statistic. *The **p-** or **probability** value of a test statistic is the probability of observing a sample result as deviant as or more deviant than the result actually obtained if H_0 is true.* According to Table D.5, $F = 8.60$ exceeds the critical value required for significance at the .01 level; hence its *p*-value is .01. A decision to reject the null hypothesis should always be based on the experimenter's preselected level of significance, .05 in our example. The inclusion of the *p*-value in the text or in a footnote to the ANOVA table permits a reader to, in effect, set his or her own level of significance.

In presenting the results of an experiment, a descriptive summary of the data—means, standard deviations, and perhaps a graph—always should precede the report of tests of significance. For the diet experiment, the descriptive statistics

---

[9] A discussion of procedures for making a rational specification of sample size for a type CR-p design is beyond the scope of this book. The interested reader is referred to Cohen (1969, chap. 8) and Kirk (1982, pp. 144–145).

[10] F values for 2,26 and 2,28 degrees of freedom are given in the table. The value for $F_{.05;2,27}$ was obtained by interpolation.

Table 16.3-1.   Computational procedures for a type CR-3 design

(i) *Data and notation* [X_{ij} denotes a score for subject i in treatment level j; i = 1, ..., n subjects (s_i); j = 1, ..., p treatment levels (a_j)]:

AS Summary table[a]

	a_1	a_2	a_3
	7	10	12
	9	13	11
	8	9	15
	12	11	7
	8	5	14
	7	9	10
	4	8	12
	10	10	12
	9	8	13
	6	7	14
$\sum_{i=1}^{n} X_{ij} =$	80	90	120
$\bar{X}_{\cdot j} =$	8	9	12

(ii) *Computational symbols:*[b]

$$\sum_{j=1}^{p} \sum_{i=1}^{n} X_{ij} = 7 + 9 + 8 + \cdots + 14 = 290.00$$

$$\sum_{j=1}^{p} \sum_{i=1}^{n} X_{ij}^2 = [AS] = (7)^2 + (9)^2 + (8)^2 + \cdots + (14)^2 = 3026.00$$

$$\frac{\left(\sum_{j=1}^{p} \sum_{i=1}^{n} X_{ij}\right)^2}{np} = [X] = \frac{(290)^2}{(3)(10)} = 2803.33$$

$$\sum_{j=1}^{p} \frac{\left(\sum_{i=1}^{n} X_{ij}\right)^2}{n} = [A] = \frac{(80)^2}{10} + \cdots + \frac{(120)^2}{10} = 2890.00$$

(iii) *Computational formulas:*

$$SS_{Total} = [AS] - [X] = 3026.00 - 2803.33 = 222.67$$
$$SS_{BG} = [A] - [X] = 2890.00 - 2803.33 = 86.67$$
$$SS_{WG} \doteq [AS] - [A] = 3026.00 - 2890.00 = 136.00$$

[a] A denotes treatment A and S denotes subjects; the table is so named because it reflects variation attributable to treatment levels (A) and subjects (S).
[b] The simplified symbols [AS], [X], and [A] are used in the computational formulas in place of the complex summation notation. The letters in the brackets denote the various sources of variation. For example, [AS] reflects variation due to treatment A and subjects; [A] reflects variation due to treatment A; and [X] is a correction term. The sum of squares attributable to treatment A and subjects (SS_{Total}) is computed from [AS] − [X]; that attributable to the treatment (SS_{BG}), from [A] − [X]; and that attributable to subjects (SS_{WG}), from [AS] − [A].

in Table 16.3-3 should provide the reader with an adequate summary of the data. According to this summary, some variation exists among the sample means—the greatest weight loss occurred for the girls who used diet a_3; the smallest, for those using a_1. The standard deviations appear to be fairly homogeneous.

Table 16.3-2.   ANOVA table for type CR-3 design

Source	SS	df	MS	F
1. Between groups (BG) (three diets)	86.67	$p - 1 = 3 - 1 = 2$	43.335	$[\frac{1}{2}]$ 8.60*
2. Within groups (WG)	136.00	$p(n - 1) = 3(10 - 1) = 27$	5.037	
3. Total	222.67	$np - 1 = (3)(10) - 1 = 29$		

* $p < .01$; $[\frac{1}{2}]$ indicates that F was obtained by dividing the value of MS in row 1 by the value of MS in row 2.

Table 16.3-3.   Descriptive statistics for weight-loss data

	Diet		
	a_1	a_2	a_3
$\bar{X}_{.j}$	8.00	9.00	12.00
$\hat{\sigma}_j$	2.21	2.21	2.31

If the overall null hypothesis $\mu_1 = \mu_2 = \cdots = \mu_p$ is rejected in ANOVA, we know that at least one difference among population means is not equal to 0. The next question is "Which difference(s) isn't equal to 0?" A procedure for answering this question is described in Section 16.5. But before turning to that topic we will examine the assumptions underlying the F test for a type CR-p design.

Review Exercises for Section 16.3

17. Under what conditions is a completely randomized design appropriate?

*18. Show that

$$= n \sum_{j=1}^{p} (\bar{X}_{.j} - \bar{X}_{..})^2 = \sum_{j=1}^{p} \frac{\left(\sum_{i=1}^{n} X_{ij}\right)^2}{n} - \frac{\left(\sum_{j=1}^{p} \sum_{i=1}^{n} X_{ij}\right)^2}{np}$$

and

$$\sum_{j=1}^{p} \sum_{i=1}^{n} (X_{ij} - \bar{X}_{.j})^2 = \sum_{j=1}^{p} \sum_{i=1}^{n} X_{ij}^2 - \sum_{j=1}^{p} \frac{\left(\sum_{i=1}^{n} X_{ij}\right)^2}{n}.$$

19. Fill in the blanks in the ANOVA table.

Source	SS	df	MS	F
Between groups	36	3	12	( )
Within groups	( )	( )	( )	
Total	164	35		

*20. An experiment was performed to investigate the effects of meaningfulness, or associa-
tion value, of nonsense syllables on learning. Four treatment levels were randomly
assigned to thirty-two subjects, with eight in each level. The nonsense syllables were
selected from the list compiled by C. E. Noble. The association values of the lists were
25% for a_1, 50% for a_2, 75% for a_3, and 100% for a_4. The dependent variable was time
(in minutes) to learn the list to the criterion of two correct recitations. The following
data were obtained.

a_1	a_2	a_3	a_4
22	22	18	18
21	20	20	17
20	18	17	16
21	21	16	18
22	20	18	19
24	19	19	15
22	21	18	16
23	19	17	17

a. Compute descriptive statistics for the data, $\bar{X}$'s and $\hat{\sigma}$'s, and construct a table similar
to Table 16.3-3.
b. Are the sample data consistent with the experimenter's hypothesis that $\mu_j \neq \mu_{j'}$ for
some j and j'?
c. Test the hypothesis $\mu_1 = \mu_2 = \mu_3 = \mu_4$. Let $\alpha = .05$. Construct an ANOVA summary table; include the p-value.
d. Summarize the results of the ANOVA in a sentence.

21. List the steps used in testing the null hypothesis in Exercise 20, and state the decision
rule.

22. The learning of one task enhances the learning of different but similar tasks. To investigate this phenomenon (called *learning to learn*), three conditions were randomly assigned to 30 subjects: subjects in condition a_1 learned two lists of nonsense syllables,
those in a_2 learned eight lists, and those in a_3 learned 14 lists. The next day all the subjects learned another list. The dependent variable was the number of trials required to
learn this list. The following data were obtained.

a_1	a_2	a_3
7	6	3
9	5	2
5	7	3
7	3	6
8	4	3
7	5	4
6	6	5
8	5	5
7	4	4
6	5	4

a. Compute descriptive statistics for the data, $\bar{X}$'s and $\hat{\sigma}$'s, and construct a table similar
to Table 16.3-3.

b. Are the sample data consistent with the experimenter's hypothesis that $\mu_j \neq \mu_{j'}$ for some j and j'?

c. Test the hypothesis $\mu_1 = \mu_2 = \mu_3$. Let $\alpha = .05$. Construct an ANOVA summary table; include the p-value.

d. Summarize the results of the ANOVA in a sentence.

*23. Reaction time to red, green, and yellow instrument-panel warning lights was investigated. The three colors of warning lights were randomly assigned to 31 subjects, who pressed a microswitch as soon as they noticed the warning light. The dependent variable was reaction time in hundredths of a second. The following data were obtained; decimal points have been deleted.

a_1 (Yellow)	a_2 (Red)	a_3 (Green)
20	23	21
20	20	21
21	21	20
22	21	23
21	23	22
20	22	20
19	22	21
21	21	22
19	22	22
20	22	20
		19

a. Compute descriptive statistics for the data, $\bar{X}$'s, and $\hat{\sigma}$'s, and construct a table similar to Table 16.3-3.

b. Are the sample data consistent with the experimenter's hypothesis that $\mu_j \neq \mu_{j'}$ for for some j and j'?

c. Test the hypothesis $\mu_1 = \mu_2 = \mu_3$. Let $\alpha = .05$. Construct an ANOVA summary table; include the p-value.

d. Summarize the results of the ANOVA in a sentence.

24. Presidents of companies employing between 5000 and 8000 employees were randomly sampled from five geographic areas: a_1 = southeast, a_2 = east, a_3 = midwest, a_4 = southwest, and a_5 = west. Use ANOVA to determine whether mean income for the presidents is different in different areas of the country. The following data representing thousands of dollars were obtained.

a_1	a_2	a_3	a_4	a_5
40	42	37	36	46
31	40	46	40	40
32	46	45	34	45
35	45	42	34	48
37	37	42	33	46
38	43	43	39	47
35	43	40	38	
33	44	39	37	
35	42		34	
37	39			

    a. Compute descriptive statistics for the data, $\bar{X}$'s and $\hat{\sigma}$'s, and construct a table similar to Table 16.3-3.

    b. Are the sample data consistent with the experimenters's hypothesis that $\mu_j \neq \mu_{j'}$ for some j and j'?

    c. Test the hypothesis $\mu_1 = \mu_2 = \mu_3 = \mu_4 = \mu_5$. Let $\alpha = .05$. Construct an ANOVA summary table; include the p-value.

    d. Summarize the results of the ANOVA in a sentence.

25. Term to remember:

    a. p-value

Section 16.4   Assumptions Associated with a Type CR-p Design

As in all statistical tests, the use of ANOVA to test a null hypothesis involves assumptions. The assumptions fall into two categories: those associated with the F random variable and those associated with the ANOVA linear model equation. The two sets of assumptions overlap to some degree, as can be seen from the following lists.

F Assumptions

A1. Observations are drawn from normally distributed populations.

A2. Observations are random samples from the populations.

A3. The numerator and denominator of the F ratio are independent.

A4. The numerator and denominator of the F ratio are estimates of the same population variance, σ_ε^2.

Model Assumptions

B1. The model equation $X_{ij} = \mu + \alpha_j + \varepsilon_{ij}$ reflects all the sources of variation that affect X_{ij}.

B2. The experiment contains all the treatment levels, a_j's, of interest.

B3. The error effect, ε_{ij}, (a) is independent of all other ε's, and (b) is normally distributed within each treatment population, with (c) mean equal to zero and (d) variance equal to σ_ε^2.

    The assumptions associated with the F random variable apply in testing hypotheses for any ANOVA design; those associated with the model equation vary from one design to the next. We now will briefly discuss those assumptions and the consequences of violating them. At the outset it should be observed that for real data some of the assumptions will always be violated. For example, the underlying populations from which samples are drawn are never exactly normally distributed with equal variances. The important question then is not whether the assumptions are violated, but rather whether minor violations have serious effects on the significance level and power of the F test. Fortunately, as we will see, the F test in ANOVA is robust with respect to violation of a number of assumptions. That is, the test is not very sensitive to departures from some of its assumptions.

Assumption that $X_{ij} = \mu + \alpha_j + \varepsilon_{ij}$ (B1)

Assumption B1 states that a score, X_{ij}, is the sum of three components: the overall elevation of scores, μ; the elevation or depression of scores attributable to the jth treatment level, α_j; and all effects not attributable to the jth treatment level, ε_{ij}, which includes effects often referred to as *individual differences* and *measurement error.*

A type CR-p design is appropriate for an experiment with one treatment in which the treatment levels are randomly assigned to the subjects. If, for example, an experiment contains two or more treatments or if an experimenter wants to observe the subjects under more than one treatment level, a different ANOVA design must be used. The choice of an incorrect design can seriously affect the probability of a type I error and the power of the F test.

Assumption that the Experiment Contains All Treatment Levels of Interest (B2)

Assumption B2 states that the experiment contains all the treatment levels about which an experimenter wants to draw conclusions. *The model $X_{ij} = \mu + \alpha_j + \varepsilon_{ij}$ for such an experiment is called a* **fixed-effects model** *or* **model I,** *and α_j is called a* **fixed-treatment effect.** Sometimes an experimenter wants to draw conclusions about more treatment levels than can be included in the experiment. This can be done by performing the experiment using a random sample of p treatment levels from the population of P levels. The results of the experiment apply to the population. The model for this case is called a *random-effects model* or *model II,* and α_j is called a *random-treatment effect.*

The expectation of MS_{BG} is different for models I and II, as is the nature of the null hypothesis tested. The expectations for MS_{BG} and MS_{WG} are

Fixed-Effects Model (I)	*Random-Effects Model (II)*

$$\frac{E(MS_{BG})}{E(MS_{WG})} = \frac{\sigma_\varepsilon^2 + n \sum_{j=1}^{p} \alpha_j^2/(p-1)}{\sigma_\varepsilon^2} \qquad \frac{E(MS_{BG})}{E(MS_{WG})} = \frac{\sigma_\varepsilon^2 + n\sigma_\alpha^2}{\sigma_\varepsilon^2}.$$

The null hypothesis for model I concerns only the p population treatment effects (α_j's) that are represented in the experiment. For model II the null hypothesis concerns all the P population treatment effects (α_j's), not just the p represented in the experiment. If the null hypothesis for model I is rejected, we can conclude that at least two of the p treatment effects are not equal; if the null hypothesis for model II is rejected, at least two of the treatment effects in the population of P effects are not equal. The computational procedures for both models are identical, as is the form of the F ratio. The difference is in the $E(MS_{BG})$'s and in the nature of the null hypothesis tested.[11]

---

[11] For a discussion of the random-effects model, the reader is referred to Edwards (1972), Kirk (1982), Myers (1979), and Winer (1971).

Assumptions of Independence of Experimental Errors (B3a) and Random Sampling (A2)

The validity of an experiment depends on random sampling and/or random assignment of treatment levels to subjects. Random sampling and/or assignment ensures that idiosyncratic characteristics of subjects will be randomly distributed over the treatment levels and thus will not selectively bias the outcome of the experiment.[12]

Random sampling (assumption A2) and assignment help to ensure that the experimental errors (ε_{ij}) are independent (assumption B3a). Applying the definition of statistical independence from Section 7.3, the errors for subjects ij and $i'j$ are independent if $p(\varepsilon_{ij}|\varepsilon_{i'j}) = p(\varepsilon_{ij})$. That is, errors are independent if knowing subject i's error tells us nothing about subject i''s error. Nonindependence of errors seriously affects both the probability of a type I error and the power of the F test.

Assumptions of Normally Distributed Errors (B3b) and Populations (A1), and Independence of MS_{BG} and MS_{WG} (A3)

The assumption (B3b) that the ε's within each of the populations are normally distributed is equivalent to the assumption that the X_{ij}'s in each of the populations are normally distributed (A1). This follows, since according to the linear model equation $X_{ij} = \mu + \alpha_j + \varepsilon_{ij}$, the ε's are the only source of variation within a particular treatment population. A rough check on the normality assumption can be made by constructing a frequency distribution for the scores in each treatment level and inspecting the distributions for evidence of skewness and kurtosis. Marked departures from normality raise questions concerning normality of the populations. A statistical test of the assumption ordinarily is not performed because the F test in ANOVA, like the t test, is robust with respect to departures from normality. Studies indicate that if the treatment populations have the same shape—for example, all positively skewed or all leptokurtic—the actual probability of making a type I error will be fairly close to the nominal or specified probability.[13]

It can be shown that the numerator and denominator of the F ratio are independent (assumption A3) if the populations are normally distributed or approximately so.[14]

Assumption that $E(\varepsilon_{ij}) = 0$ for Each Treatment Population (B3c)

Assumption B3c states that the expectation of the error effects within each treatment population is equal to zero. This follows from the fact that ε_{ij} is the deviation

---

[12] Sometimes factors beyond the investigator's control preclude the random assignment of subjects to treatment levels and the control of important extraneous variables. Campbell and Stanley (1963) refer to such experiments as "quasi-experimental designs." Campbell (1957) examines potential sources of bias inherent in these designs. His paper is reprinted in Kirk (1972a).

[13] The reader may want to take the robustness of the F test on faith, since the classic studies by Box and Anderson (1955), Norton as cited by Lindquist (1953), and Pearson (1931) were written for a technical audience.

[14] See Lindquist (1953, pp. 31–35).

of a population mean from a random variable $(\varepsilon_{ij} = X_{ij} - \mu_j)$ and the fact that the expectation or long-run average of this deviation is zero, which can be shown as follows. The expectation $E(X_{ij} - \mu_j)$ can be written as $E(X_{ij}) - E(\mu_j)$. By definition, the expectation of the random variable X_{ij} is its population mean, and the expectation[15] of the constant μ_j is μ_j. Hence,

$$E(X_{ij} - \mu_j) = E(X_{ij}) - E(\mu_j) = \mu_j - \mu_j = 0.$$

Assumption of Homogeneity of Variance (B3d)

Assumption B3d states that the $j = 1, \ldots, p$ population variances are equal to σ_ε^2—that is, that $\sigma_1^2 = \sigma_2^2 = \cdots = \sigma_p^2 = \sigma_\varepsilon^2$. This is referred to as the **homogeneity of variance assumption.** Fortunately, the ANOVA F test is robust with respect to heterogeneity of population variances provided there are an equal number of observations in the treatment levels.[16]

Assumption that $E(MS_{BG}) = E(MS_{WG}) = \sigma_\varepsilon^2$ (A4)

The assumption that the numerator and denominator of the F ratio are estimates of the same population variance, σ_ε^2, corresponds to the null hypothesis. It is advanced in the hope that it can be rejected.

Review Exercises for Section 16.4

*26. Qualify the statement "The F test in ANOVA is robust with respect to departures from normality."

*27. A rough but adequate check on the tenability of the normality assumption consists of making a frequency distribution of the scores in each treatment level and inspecting them for evidence of skewness and kurtosis. Decide on the tenability of this assumption for the data in exercises *(a) 20 (page 354), (b) 22 (page 354), and (c) 23 (page 355).

28. What is meant by the assumption of homogeneity of variance?

29. Qualify the statement "The F test in ANOVA is robust with respect to heterogeneity of variance."

30. Distinguish between fixed-effects and random-effects models in terms of (a) expectations of the MS's and (b) the null hypothesis tested.

31. Terms to remember:
   a. Homogeneity of variance   b. Model I
   c. Fixed-treatment effect   d. Model II
   e. Random-treatment effect

Section 16.5   Introduction to Multiple Comparisons

The ANOVA F test is used to determine the tenability of the hypothesis $\mu_1 = \mu_2 = \cdots = \mu_p$. If this hypothesis is rejected, interest centers on determining which of the means are not equal. A number of test statistics have been developed for

---

[15] For a discussion of the algebra of expected values, see Hays (1981, pp. 625–630) and Kirk (1982, pp. 768–772).

[16] The robustness of the test to heterogeneity of population variances has been shown by Cochran (1947), Godard and Lindquist (1940), Horsnell (1953), and Norton as cited by Lindquist (1953).

ferreting out significant differences among means, or, as it is often called, *data snooping*. The one described here—Scheffé's S test[17]—is referred to as an a posteriori or post hoc procedure because it tests comparisons suggested by an inspection of one's data. If the overall null hypothesis in ANOVA is rejected, at least one significant difference among means will be detectable using Scheffé's test.

We observed in Section 16.1 that if an experiment contains more than two treatment levels and if Student's t statistic is used to test the significance of comparisons, the probability of erroneously rejecting one or more true null hypotheses increases as a function of the number of comparisons tested. The persistent experimenter who performs enough t tests, each at α level of significance, will reject too many true H_0's. Scheffé's multiple comparison procedure, described next, was developed to solve this problem.

Scheffé's Multiple Comparison Procedure

Scheffé's S test was designed to test all 2, 3, ..., p mean null hypotheses if the overall null hypothesis has been rejected. Regardless of the number of tests performed, it sets the probability of falsely rejecting one or more null hypotheses equal to or less than α for the collection of tests. The test assumes that the $j = 1$, ..., p populations are normally distributed and that their variances are equal. However, the S test, which is based on the F distribution, is robust with respect to violation of these assumptions. Before describing Scheffé's statistic we need to introduce some concepts associated with contrasts or comparisons among means.

Contrasts Among Means

A **contrast** or **comparison** among means is a difference among the means, with appropriate algebraic signs. We will use the symbols ψ_i and $\hat{\psi}_i$ to denote, respectively, the ith contrast among population means and a sample estimate of the ith contrast. For example, $\psi_i = \mu_j - \mu_{j'}$ is the ith contrast between population means for treatment levels j and j'. If experiment contains p equal to three means, contrasts involving two and three population means may be of interest:

$$\psi_1 = \mu_1 - \mu_2 \qquad \psi_4 = (\mu_1 + \mu_2)/2 - \mu_3$$

$$\psi_2 = \mu_1 - \mu_3 \qquad \psi_5 = (\mu_1 + \mu_3)/2 - \mu_2$$

$$\psi_3 = \mu_2 - \mu_3 \qquad \psi_6 = (\mu_2 + \mu_3)/2 - \mu_1.$$

The contrasts on the right involve the average of two means versus a third mean.

More formally, a contrast or comparison among population means is a linear combination of means with known coefficients, c_j's, such that (1) at least one coefficient is not equal to 0 and (2) the coefficients sum to 0. That is,

$$\psi_i = c_1\mu_1 + c_2\mu_2 + \cdots + c_p\mu_p,$$

---

[17] See Kirk (1982, chap. 3) for an in-depth discussion of this and other multiple comparison procedures, including Tukey's test, Dunnett's test, the Newman–Keuls test, and the controversial test proposed by Duncan. An excellent but more technical discussion is given by Games (1971).

where $c_j \neq 0$ for some j and $\sum_{j=1}^{p} c_j = 0$. The six contrasts listed previously can be expressed as linear combinations of the μ_j's by the appropriate choice of coefficients.

$\psi_i =$	$c_1\mu_1 +$	$c_2\mu_2 +$	$c_3\mu_3$
$\psi_1 =$	$(1)\mu_1 + (-1)\mu_2 +$	$(0)\mu_3 = \mu_1 - \mu_2$	
$\psi_2 =$	$(1)\mu_1 +$	$(0)\mu_2 + (-1)\mu_3 = \mu_1 - \mu_3$	
$\psi_3 =$	$(0)\mu_1 +$	$(1)\mu_2 + (-1)\mu_3 = \mu_2 - \mu_3$	
$\psi_4 =$	$(\frac{1}{2})\mu_1 +$	$(\frac{1}{2})\mu_2 + (-1)\mu_3 = (\mu_1 + \mu_2)/2 - \mu_3$	
$\psi_5 =$	$(\frac{1}{2})\mu_1 + (-1)\mu_2 +$	$(\frac{1}{2})\mu_3 = (\mu_1 + \mu_3)/2 - \mu_2$	
$\psi_6 = (-1)\mu_1 +$	$(\frac{1}{2})\mu_2 +$	$(\frac{1}{2})\mu_3 = (\mu_2 + \mu_3)/2 - \mu_1$	

Note that for each contrast, $c_j = 0$ for some j and $\sum_{j=1}^{p} c_j = 0$. For convenience, coefficients of contrasts often are chosen so that the sum of their absolute values is equal to 2. That is,

$$\sum_{j=1}^{p} |c_j| = 2,$$

where $|c_j|$ indicates that the sign of c_j is always taken to be plus. All six preceding contrasts satisfy this property. For example, the sum of the absolute value of the coefficients for ψ_1 and ψ_4 are, respectively,

$$|1| + |-1| + |0| = 1 + 1 + 0 = 2$$
$$|\tfrac{1}{2}| + |\tfrac{1}{2}| + |-1| = \tfrac{1}{2} + \tfrac{1}{2} + 1 = 2.$$

Scheffé's Test Statistic

Scheffé's statistic for testing hypotheses of the form

$$H_0: \quad c_1\mu_1 + c_2\mu_2 + \cdots + c_p\mu_p = 0$$

is given by

$$F = \frac{(c_1\bar{X}_{.1} + c_2\bar{X}_{.2} + \cdots + c_p\bar{X}_{.p})^2}{MS_{WG}\left(\dfrac{c_1^2}{n_1} + \dfrac{c_2^2}{n_2} + \cdots + \dfrac{c_p^2}{n_p}\right)}.$$

The null hypothesis is rejected if $F \geq (p-1)F_{\alpha;v_1,v_2}$, where

p is the number of treatment levels in the experiment,
$F_{\alpha;v_1,v_2}$ is obtained from Appendix Table D.5,
α is the level of significance adopted for $F = MS_{BG}/MS_{WG}$ in the ANOVA,
v_1 is the degrees of freedom associated with MS_{BG},
v_2 is the degrees of freedom associated with MS_{WG}.

Computational Example for Scheffé's Test

Scheffé's test will be illustrated using the diet data in Table 16.3-1. For these data the overall ANOVA F statistic was significant at the .05 level. After examining the data, an experimenter might be interested in the following null hypotheses: $\mu_1 - \mu_2 = 0$, $\mu_1 - \mu_3 = 0$, $\mu_2 - \mu_3 = 0$, and $\mu_3 - (\frac{1}{2})\mu_1 - (\frac{1}{2})\mu_2 = 0$. The test statistics are

$$F = \frac{[(1)\bar{X}_{.1} + (-1)\bar{X}_{.2} + (0)\bar{X}_{.3}]^2}{MS_{WG}\left[\dfrac{(1)^2}{n_1} + \dfrac{(-1)^2}{n_2} + \dfrac{(0)^2}{n_3}\right]} = \frac{(8 - 9 + 0)^2}{5.037\left[\dfrac{1}{10} + \dfrac{1}{10} + \dfrac{0}{10}\right]} = 0.99$$

$$F = \frac{[(1)\bar{X}_{.1} + (0)\bar{X}_{.2} + (-1)\bar{X}_{.3}]^2}{MS_{WG}\left[\dfrac{(1)^2}{n_1} + \dfrac{(0)^2}{n_2} + \dfrac{(-1)^2}{n_3}\right]} = \frac{(8 + 0 - 12)^2}{5.037\left[\dfrac{1}{10} + \dfrac{0}{10} + \dfrac{1}{10}\right]} = 15.88$$

$$F = \frac{[(0)\bar{X}_{.1} + (1)\bar{X}_{.2} + (-1)\bar{X}_{.3}]^2}{MS_{WG}\left[\dfrac{(0)^2}{n_1} + \dfrac{(1)^2}{n_2} + \dfrac{(-1)^2}{n_3}\right]} = \frac{(0 + 9 - 12)^2}{5.037\left[\dfrac{0}{10} + \dfrac{1}{10} + \dfrac{1}{10}\right]} = 8.94$$

$$F = \frac{[(1)\bar{X}_{.1} + (-\frac{1}{2})\bar{X}_{.2} + (-\frac{1}{2})\bar{X}_{.3}]^2}{MS_{WG}\left[\dfrac{(1)^2}{n_1} + \dfrac{(-\frac{1}{2})^2}{n_2} + \dfrac{(-\frac{1}{2})^2}{n_3}\right]} = \frac{[12 - (8 + 9)/2]^2}{5.037\left[\dfrac{1}{10} + \dfrac{\frac{1}{4}}{10} + \dfrac{\frac{1}{4}}{10}\right]} = 16.21.$$

The critical value that F must equal or exceed to be significant at the .05 level is $(3 - 1)3.36 = 6.72$. The only F that doesn't exceed 6.72 is that for $\mu_1 - \mu_2 = 0$; all the other null hypotheses are rejected.

   Regardless of the number of tests performed among p means using Scheffé's statistic, the probability of falsely rejecting one or more null hypotheses is always equal to or less than α for the collection of tests. Scheffé's test is probably the most versatile of the data-snooping procedures. It can be used with unequal n's, and to construct confidence intervals, and it is robust with respect to violation of its normality and equal variance assumptions. But this versatility is accompanied by a disadvantage. The power of Scheffé's test is lower than that of more specialized multiple comparison procedures. For a discussion of this point, see Games (1971) and Kirk (1982, chap. 3).

Review Exercises for Section 16.5

32. List the requirements for using Scheffé's test.
*33. List the coefficients, c_j, for the following population contrasts: *(a) μ_1 versus μ_2, *(b) μ_1 versus mean of μ_2 and μ_3, (c) μ_1 versus mean of μ_2, μ_3, and μ_4, (d) mean of μ_1 and μ_2 versus mean of μ_3 and μ_4, (e) mean of μ_1 and μ_2 versus mean of μ_3, μ_4, and μ_5, *(f) μ_1 versus the weighted mean of μ_2 and μ_3, where μ_2 is weighted twice as much as μ_3, (g) the weighted mean of μ_1 and μ_2 versus the weighted mean of μ_3 and μ_4, where μ_1 and μ_3 are weighted twice as much as μ_2 and μ_4.
*34. Which of the following meet the three requirements for a contrast? *(a) $\mu_1 - \mu_2$, *(b) $2\mu_1 - \mu_2 - \mu_3$, (c) $\mu_1 - (\frac{1}{3})\mu_2 - (\frac{1}{3})\mu_3$, (d) $(1\frac{1}{2})\mu_1 - \mu_2 - (\frac{1}{2})\mu_3$, (e) $(\frac{1}{2})\mu_1 + (\frac{1}{2})\mu_2 - (\frac{1}{3})\mu_3 - (\frac{1}{3})\mu_4 - (\frac{1}{3})\mu_5$.
*35. Which contrasts in Exercise 34 satisfy $|c_1| + |c_2| + \cdots + |c_p| = 2$?
*36. Determine the value of $(p - 1)F_{\alpha; v_1, v_2}$ for *(a) $p = 4$, $n = 11$, $\alpha = .01$; (b) $p = 5$, $n = 13$,

$\alpha = .05$; *(c) $p = 3$, $n_1 = 6$, $n_2 = 7$, $n_3 = 8$, $\alpha = .05$; (d) $p = 4$, $n_1 = 5$, $n_2 = 5$, $n_3 = 6$, $n_4 = 8$, $\alpha = .01$.

*37. The effectiveness of three approaches to drug education in junior high school was investigated. The approaches were scare tactics, treatment level a_1; providing objective scientific information about physiological and psychological effects, a_2; and examining the psychology of drug use, a_3. Forty-one students who didn't use drugs were randomly assigned to each treatment level. At the conclusion of the education program the students evaluated its effectiveness; a high score signified effectiveness. The sample means were $\bar{X}_{.1} = 23.1$, $\bar{X}_{.2} = 23.8$, $\bar{X}_{.3} = 26.7$; $MS_{WG} = 16.4$; and $v_2 = 3(41 - 1) = 120$. The hypothesis $\mu_1 = \mu_2 = \mu_3$ was rejected at the .01 level of significance using a type CR-3 design. Use Scheffé's statistic to test $\mu_1 - \mu_2 = 0$, $\mu_1 - \mu_3 = 0$, $\mu_2 - \mu_3 = 0$, $(\mu_1 + \mu_2)/2 - \mu_3 = 0$.

*38. The effects of three dosages of ethylene glycol on the reaction time of chimpanzees was investigated. The dosage levels were randomly assigned to the animals so that five received 2 cc of the drug, treatment level a_1; five received 4 cc, a_2; and five received 6 cc, a_3. The sample means were $\bar{X}_{.1} = 0.29$ sec, $\bar{X}_{.2} = 0.31$ sec, $\bar{X}_{.3} = 0.39$ sec; $MS_{WG} = .002$ and $v_2 = 3(5 - 1) = 12$. The hypothesis that $\mu_1 = \mu_2 = \mu_3$ was rejected at the .05 level of significance using a type CR-3 design. Perform all two-mean contrasts using Scheffé's statistic.

39. The religious dogmatism of four church denominations in a large midwestern city was investigated. A random sample of 30 members from each denomination took a paper-and-pencil test of dogmatism. The sample means were $\bar{X}_{.1} = 64$, $\bar{X}_{.2} = 73$, $\bar{X}_{.3} = 61$, $\bar{X}_{.4} = 49$; $MS_{WG} = 120$ and $v_2 = 4(30 - 1) = 116$. The hypothesis that $\mu_1 = \mu_2 = \mu_3 = \mu_4$ was rejected at the .01 level of significance using a type CR-4 design. Use Scheffé's statistic to test $\mu_2 - \mu_3 = 0$, $\mu_2 - (\frac{1}{2})\mu_1 - (\frac{1}{2})\mu_3 = 0$, $\mu_4 - (\frac{1}{2})\mu_1 - (\frac{1}{2})\mu_3 = 0$.

40. A hospital questionnaire designed to measure concern about the recovery period was administered to a random sample of 19 coronary patients, a_1; 23 accident victims, a_2; and 21 pulmonary patients, a_3. The sample means were $\bar{X}_{.1} = 82$, $\bar{X}_{.2} = 83$, $\bar{X}_{.3} = 63$; $MS_{WG} = 5.1$ and $v_2 = 63 - 3 = 60$. A low score indicated insecurity about physical well-being and lowered self-esteem. The hypothesis that $\mu_1 = \mu_2 = \mu_3$ was rejected at the .05 level of significance using a type CR-3 design. Use Scheffé's statistic to test $\mu_1 - \mu_2 = 0$, $\mu_1 - \mu_3 = 0$, $\mu_2 - \mu_3 = 0$, and $\mu_3 - (\frac{1}{2})\mu_1 - (\frac{1}{2})\mu_2 = 0$.

41. For the data in Exercise 22 (page 354) use Scheffé's statistic to test the hypothesis that $\mu_1 - \mu_2 = -1$ and $\mu_1 - \mu_3 = -3$. (*Hint:* Examine the numerator of the t formula on page 288.)

42. Terms to remember:
   a. Contrast or comparison
   b. Data snooping
   c. A posteriori or post hoc comparisons

Section 16.6 Introduction to Factorial Designs

The term *analysis of variance* is apropos. As we have seen, the total variation among scores for a completely randomized design is partitioned into two sources of variation. Complex ANOVAs partition the total variation into more sources; designs with as many as 16 sources are common in behavioral and educational research. A discussion of the computational procedures associated with such designs is beyond the scope of this book. We will limit our discussion to the main features of one of the more widely used complex ANOVA designs—the completely randomized factorial design.[18]

A factorial design is distinguished from other ANOVA designs in that it has two or more treatments and the levels of each treatment are investigated in combination with those of other treatments. For example, a subject's performance on a learning task might be observed for the combined conditions of large monetary reward for good performance (a level of one treatment) and absence of distractions in the learning environment (a level of a second treatment). A subject in a factorial design is always simultaneously exposed to one level each of two or more treatments. The combination of conditions to which the subject is simultaneously exposed is called a *treatment combination.* By way of contrast, the CR-p design described in Section 16.3 can be used to investigate only one treatment at a time, and a subject is exposed to a single treatment level.

In a factorial design the levels of the treatments can be *completely crossed,* meaning that each level of one treatment occurs once with each level of the other treatments and vice versa, or *partially crossed,* meaning that each level of one treatment occurs with some of the levels of the other treatments but not with all of them. A completely randomized factorial design (CRF), the simplest of the factorial designs, is constructed by combining (completely crossing) the treatment levels of two or more completely randomized designs in a single experiment and randomly assigning the treatment combinations to the subjects. A design with two treatments, say A and B, is designated by the letters CRF-pq, where p refers to the number of levels of treatment A, and q, to the number of levels of treatment B.[19]

Consider an experiment to investigate the effects of treatments A and B on reading speed. Suppose that treatment A consists of two levels of room illumination: $a_1 = 5$ foot-candles and $a_2 = 30$ foot-candles. Treatment B consists of three levels of type size: $b_1 = 6$-point type, $b_2 = 12$-point type, and $b_3 = 18$-point type. Each level of A is combined with all levels of B to form $2 \times 3 = 6$ treatment combinations: $ab_{11}, ab_{12}, ab_{13}, \ldots, ab_{23}$. The layout for such a CRF-23 design with five subjects receiving each treatment combination is shown in Figure 16.6-1. This design partitions the total variation among the scores into four sources: treatment A, treatment B, AB interaction, and within-cell (error) variation. It enables the experimenter to test the hypotheses that the p population means for treatment A are equal, the q means for treatment B are equal, and the two treatments do not interact. The third hypothesis concerning interaction is unique to factorial designs. *Two treatments are said to* **interact** *if differences in performance under the levels of one treatment are different at two or more levels of the other treatment.* An interaction between treatments A and B for the data in Figure 16.6-1 is illustrated in Figure 16.6-2. Figure 16.6-2(a) shows that the mean for treatment level a_2 is considerably larger than that for a_1 when $b_1 = 6$-point type, is slightly

---

[18] The computational procedures for this design are discussed by Edwards (1972), Keppel (1973), Kirk (1982), Lee (1975), Lindman (1974), Myers (1979), and Winer (1971).

[19] Some writers refer to this design as a *two-way ANOVA,* but this designation is imprecise, since the name could refer to any one of ten different kinds of factorial designs.

Treatment combinations

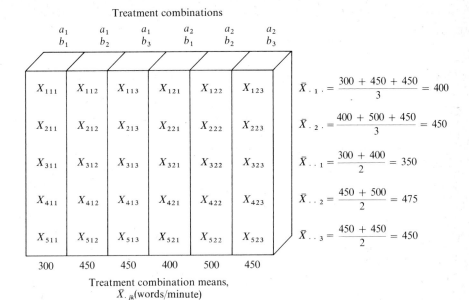

$$\bar{X}_{.1.} = \frac{300 + 450 + 450}{3} = 400$$

$$\bar{X}_{.2.} = \frac{400 + 500 + 450}{3} = 450$$

$$\bar{X}_{..1} = \frac{300 + 400}{2} = 350$$

$$\bar{X}_{..2} = \frac{450 + 500}{2} = 475$$

$$\bar{X}_{..3} = \frac{450 + 450}{2} = 450$$

Treatment combination means,
$\bar{X}_{.jk}$(words/minute)

Figure 16.6-1.   Layout for a type CRF-23 design with $2 \times 3 = 6$ treatment combinations randomly assigned to $N = 30$ subjects. Each subject receives one level of treatment A in combination with one level of treatment B. Scores are denoted by X_{ijk}, where the subscripts j and k specify the levels, respectively, of A and B, and i specifies one of the five subjects in cell jk. Treatment means for A are denoted by $\bar{X}_{.j.}$; means for B, by $\bar{X}_{..k}$; and treatment combination means, by $\bar{X}_{.jk}$.

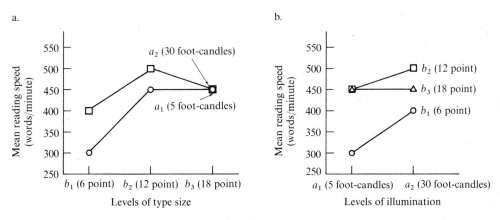

Figure 16.6-2.   Two graphs of the interaction between treatments A and B for the data in Figure 16.6-1. Nonparallelism of the lines signifies interaction.

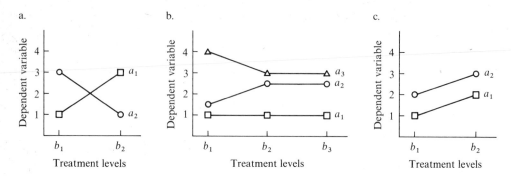

Figure 16.6-3. Interaction is illustrated in parts a and b; part c shows treatments that don't interact.

larger when b_2 = 12-point type, and is equal to that for a_1 when b_3 = 18-point type. A similar picture of interaction emerges in Figure 16.6-2(b) when the means for treatment B at the two levels of a_j are examined. An interaction always signals that the interpretation of treatment effects must be qualified. For example, although mean reading speed is faster under 30 foot-candles of illumination ($\bar{X}_{.2.}$ = 450) than under 5 foot-candles ($\bar{X}_{.1.}$ = 400), the former condition is superior only for the 6- and 12-point type conditions. If the experiment had been performed with only treatment A and 18-point type, the experimenter would have concluded that there was no difference in reading speed between the two levels of illumination; if either 6- or 12-point type had been used, the conclusion would have been that there was a difference.

If a test of significance indicates that two treatments interact, a graph like those in Figures 16.6-2 and 16.6-3 is helpful in interpreting the interaction.[20] When the interaction null hypothesis is rejected, we know that a graph will reveal at least two nonparallel lines between at least two levels of a treatment; it doesn't mean that all lines throughout their length are nonparallel. Such a case is shown in Figure 16.6-3(b), where the a_i lines are not parallel between b_1 and b_2 but are parallel between b_2 and b_3. We can think about this in another way. Consider the interaction shown in Figure 16.6-3(b). The value of some contrast among the A means, say $\hat{\psi} = (\bar{X}_{.1.} + \bar{X}_{.2.})/2 - \bar{X}_{.3.}$, at b_1 is different from its value at b_2, but the value of the contrast is the same at b_2 and b_3.

The ability to test interactions is an important feature of factorial designs. But the designs have other features that help account for their wide use. For example, they make efficient use of subjects. The CRF-23 design shown in Figure 16.6-1 uses all 30 subjects simultaneously in evaluating the effects of treatments A and B. If treatment A were evaluated by means of a type CR-2 design and treatment B, by means of a separate type CR-3 design, 60 subjects—30 in each experiment—would be required to achieve the precision of the CRF-23 design. In view of these advantages, it is easy to understand the popularity of factorial designs.

[20] Kirk (1982) discusses more advanced techniques for interpreting interactions.

The major advantages of factorial designs can be summarized as follows.

1. All subjects are used in simultaneously evaluating the effects of two or more treatments. Since each treatment is evaluated with the same precision as if the entire experiment had been devoted to that treatment alone, factorial experiments provide efficient use of resources.
2. They enable an experimenter to evaluate interaction effects.

The major disadvantages of factorial designs are the following.

1. If numerous treatments are included in an experiment, the number of subjects required may be prohibitive. For example, a four-treatment type CRF-2433 design has $2 \times 4 \times 3 \times 3 = 72$ treatment combinations. If each combination were assigned to only two subjects, the experiment would require 144 subjects.
2. The interpretation of a factorial design is complex if interaction effects are present. This is not a criticism of the design, but rather an acknowledgment of the fact that interactions always call for some qualification of tests of differences among means for the treatments.
3. The use of a factorial design commits an investigator to a relatively large experiment. Small one-treatment exploratory experiments may indicate much more promising lines of investigation than those originally envisioned. Relatively small experiments permit greater freedom in the pursuit of serendipity.

Section 16.7   Summary

Analysis of variance, ANOVA, is a statistical procedure for (1) determining how much of the total variability among scores to attribute to each source of variation in an experiment and (2) testing hypotheses about some of these sources. The principal application of ANOVA is testing the null hypothesis that two or more population means are equal. A completely randomized design (type CR-p), the simplest ANOVA design, is described in this chapter. It is appropriate for experiments that meet the following conditions.

1. One treatment or independent variable with two or more treatment levels. The levels of the treatment can differ either quantitatively or qualitatively.
2. Random assignment of treatment levels to subjects, with each subject designated to receive only one level. If the treatment levels are of equal interest, it may be advantageous to assign each level to the same number of subjects, although this is not necessary. Actually, flexibility in this respect is one of the advantages of the design.

Although ANOVA appears to be a complicated procedure, the basic notions are relatively simple. A score X_{ij} in a type CR-p design is assumed to be a composite that estimates the sum of three components, as is stated in the model equation

$$X_{ij} = \mu + \alpha_j + \varepsilon_{ij},$$

where μ is a constant that reflects the general elevation of scores; α_j is the effect of the jth treatment level; and ε_{ij} reflects any other effects on X_{ij}, which is apparent when the terms in the model are rearranged to $\varepsilon_{ij} = X_{ij} - \mu - \alpha_j$.

The total variation among the scores, designated by SS_{Total}, is also a composite and can be partitioned into two parts: the sum of squares between groups, SS_{BG}, and the sum of squares within groups, SS_{WG}. A variance or mean square is obtained by dividing a sum of squares by its degrees of freedom, for example, $SS_{BG}/df_{BG} = MS_{BG}$ and $SS_{WG}/df_{WG} = MS_{WG}$. The statistic for testing $\mu_1 = \mu_2 = \cdots = \mu_p$ is $F = MS_{BG}/MS_{WG}$. To use the ratio of two variances to test a hypothesis about means is not strange if you consider the expected values of MS_{BG} and MS_{WG} for the case in which the null hypothesis is true and the case in which it is false. If H_0 is true, all the squared treatment effects equal 0 $[\alpha_1{}^2 = (\mu_1 - \mu)^2 = 0, \alpha_2{}^2 = (\mu_2 - \mu)^2 = 0, \ldots, \alpha_p{}^2 = (\mu_p - \mu)^2 = 0]$, in which case

$$\frac{E(MS_{BG})}{E(MS_{WG})} = \frac{\sigma_\varepsilon{}^2}{\sigma_\varepsilon{}^2}.$$

If H_0 is false, at least two of the squared treatment effects are greater than 0 ($\alpha_j{}^2 \neq 0$ and $\alpha_{j'}{}^2 \neq 0$ for some j and j'), in which case

$$\frac{E(MS_{BG})}{E(MS_{WG})} = \frac{\sigma_\varepsilon{}^2 + n \sum_{j=1}^{p} \alpha_j{}^2/(p-1)}{\sigma_\varepsilon{}^2}.$$

It is apparent that the larger the ratio $F = MS_{BG}/MS_{WG}$, the less tenable the null hypothesis. How large should the ratio be in order to reject H_0? According to hypothesis-testing conventions, an F is considered large enough if it falls in the upper 5% of the sampling distribution of F. To determine whether F falls in this region it is referred to the F sampling distribution for degrees of freedom equal to $p - 1$ and $p(n - 1)$. This is why a statistic involving two variances is used to test a hypothesis about means.

If the overall null hypothesis is rejected, the experimenter must still decide which means aren't equal. This can be accomplished by using Scheffé's data-snooping test statistic. The principal advantage of this multiple comparison procedure over Student's t is that the probability of erroneously rejecting one or more null hypotheses doesn't increase as a function of the number of hypotheses tested. Regardless of the number of tests performed among p means, this probability remains equal to or less than α for the collection of tests.

†*Section 16.8   Computer Supplement*

A microcomputer package developed by Steinmetz, Romano, and Patterson (1981) was used to analyze the weight-loss data in Table 16.2-1. This interactive package was written in APPLESOFT BASIC for Apple II microcomputers with 48K of memory. It contains 12 programs and can be used to analyze data from the nine most frequently used analysis of variance designs.

The package begins by displaying the menu, a list of available programs. Once a program is selected, it prompts the user to enter information as required. The sequence of prompts and responses for obtaining a test of the null hypothesis $\mu_1 = \mu_2 = \mu_3$ for the weight-loss data in Table 16.2-1 is shown in Table 16.8-1. A double asterisk occurs after the F statistic in the ANOVA table, which indicates that the p-value for the F is .01. Hence the null hypothesis can be rejected at the experimenter's preselected .05 level of significance.

Table 16.8-1.   Printout from the Steinmetz, Romano, and Patterson One-Way ANOVA Program (Data from Table 16.2-1)

```
[ONE-WAY ANOVA]

  ***ONE-WAY ANALYSIS OF VARIANCE***

THIS PROGRAM EXECUTES A ONE-WAY ANOVA
(NON-REPEATED MEASURES).

AVAILABLE OPTIONS:

    (1) AN INPUT CORRECTION ROUTINE
    (2) INPUT DATA FROM DISK
    (3) SAVE DATA TO DISK
PRESS <RETURN> TO CONTINUE 1

ENTER TOTAL NUMBER OF OBSERVATIONS:  30

ENTER TOTAL NUMBER OF GROUPS:  3

ARE THE 3 GROUPS EQUAL IN SIZE (Y /N)
?Y

THIS ANALYSIS HAS 30 TOTAL OBSERVATIONS
EQUALLY DISTRIBUTED AMONG 3 GROUPS.
PRESS <RETURN> TO CONTINUE
WOULD YOU LIKE TO USE DATA THAT IS
STORED ON A DISK (Y / N) ? N
ENTER SCORES FOR GROUP 1

?7
?9
?8
?12
```

(continued)

---

† This section can be omitted without loss of continuity.

Table 16.8-1 (Continued)

?8
?7
?4
?10
?9
?6
ENTER SCORES FOR GROUP 2

?10
?13
?9
?11
?5
?9
?8
?10
?8
?7
ENTER SCORES FOR GROUP 3

?12
?11
?15
?7
?14
?10
?12
?12
?13
?14
WOULD YOU LIKE THE DATA
DISPLAYED (Y/N) ? N
ARE THERE ANY CHANGES (Y/N) ? N
PROCESSING. . . . . . .
THE CALCULATIONS HAVE BEEN COMPLETED.

HOW WOULD YOU LIKE YOUR RESULTS
REPORTED ?

   (1) VIDEO DISPLAY
   (2) PRINTER

(ENTER A 1 OR 2): 2

PRINTER SLOT ? 1

ENTER A LABEL FOR THE TREATMENT
VARIABLE: DIETS
SUMMARY TABLE FOR ONE-WAY ANOVA

(continued)

Table 16.8-1 (Continued)

===

TREATMENT VARIABLE = DIETS
===

BETWEEN: SS = 86.666667   DF = 2   MS = 43.3333335   F = 8.60294121   **

WITHIN: SS = 136   DF = 27   MS = 5.03703704
===

TOTAL: SS = 222.666667   DF = 29
===

** P < .01   * P < .05

*** SUMMARY OF GROUP MEANS ***

GROUP 1 MEAN = 8
GROUP 2 MEAN = 9
GROUP 3 MEAN = 12

17

Statistical Inference for Frequency Data

Section 17.1　Applications of Pearson's Chi-Square Statistic

So far our discussion has been limited to test statistics that are appropriate for measured characteristics such as IQ, time to learn a list of nonsense syllables, and age. Instead of measuring a variable we may choose to count the number of observations in two or more mutually exclusive categories of the variable, in which case the data are a set of frequencies. A statistic for testing hypotheses about frequency data was developed by Karl Pearson in 1900. It is called *Pearson's chi-square statistic* because it is approximately distributed as the chi-square distribution.[1] We will describe three applications of the statistic: testing goodness of fit, testing independence, and testing equality of c population proportions.[2] The three tests are often confused, since they all use Pearson's test statistic, but they can be distinguished as follows.

1. Testing goodness of fit. Pearson's chi-square statistic is used to determine whether the population distribution estimated by a single random sample containing n independent observations is identical to some hypothesized or expected population distribution. Depending on the experimenter's interests, expectations may be based on one of the theoretical distributions (such as the normal curve) or on the results of an earlier empirical investigation. Pearson's chi-square statistic is used to test the hypothesis that the observed frequencies $O_1, O_2, \ldots, O_k$ in k mutually exclusive categories of a population are equal to a set of expected frequencies $E_1, E_2, \ldots, E_k$.

2. Testing independence. Each of n independent observations for a single random sample is classified in terms of two variables, A and B. Variable A might represent a person's sex and variable B, political affiliation (Democrat, Republican, independent, or other). Pearson's chi-square statistic is used to determine whether the variables for some population are statistically independent. If the variables are independent, then $p(\text{Male}|\text{Democrat}) = p(\text{Male})$, $p(\text{Male}|\text{Republican}) = p(\text{Male})$, and so on, which means that a knowledge of political affiliation indicates nothing about the individual's sex and vice versa.

---

[1] Pearson's chi-square statistic is not related to $\chi^2 = (n-1)\hat{\sigma}^2/\sigma_0^2$, which is used to test a hypothesis about a population variance, although both are referred to the chi-square distribution.

[2] For an in-depth study of the use and misuse of Pearson's chi-square statistic see Edwards (1950), Lewis and Burke (1949, 1950), Pastore (1950), and Peters (1950).

3. Testing equality of $c \geq 2$ population proportions. An experimenter draws c random samples from c populations, and the proportion of elements possessing some characteristic is computed for each sample. Pearson's chi-square statistic is used to test the hypothesis that the population proportions are equal—that is,[3] $p_1 = p_2 = \cdots = p_c$.

Table 17.1-1. Comparison of three tests that use Pearson's chi-square statistic

Purpose	Null hypothesis	Number of samples
1. Testing goodness of fit	$O_{Pop_1} = E_{Pop_1}, O_{Pop_2} = E_{Pop_2}, \ldots, O_{Pop_k} = E_{Pop_k}$	One random sample; n observations assigned to k categories
2. Testing independence	$p(A \text{ and } B) = p(A)p(B)$	One random sample; each of n observations is classified in terms of two variables denoted by A and B
3. Testing equality of proportions	$p_1 = p_2 = \cdots = p_c$	c random samples, where $c \geq 2$

Section 17.2   Testing Goodness of Fit

The goodness-of-fit test was developed to test the hypothesis that a population distribution estimated by a random sample is identical to an hypothesized or expected distribution. If $O_1, O_2, \ldots, O_k$ represent observed frequencies and $E_1, E_2, \ldots, E_k$, expected frequencies, the hypothesis of equality is rejected if Pearson's statistic,

$$\chi^2 = \sum_{j=1}^{k} \frac{(O_j - E_j)^2}{E_j},$$

equals or exceeds the value of $\chi^2_{\alpha,v}$ (given in Table D.4) for $v = k - 1$ degrees of freedom and α level of significance. The test is approximate, since it uses the continuous chi-square distribution to estimate a probability for a discrete sampling distribution.[4] An alternative exact test can be used. When $k = 2$, the exact test is based on the binomial distribution; when $k > 2$, it is based on the multinomial distribution. Exact tests are rarely used because they involve a prohibitive amount of computation.[5]

Computational Example

Suppose we want to know whether the distribution of academic potential for this year's graduate school applicants, as measured by the Graduate Record Examination, is the same as that for past years. Data for previous applicants are given in Table 17.2-1. The statistical hypotheses for Pearson's test usually are stated in terms of proportions, although the test statistic uses frequencies. This is perfectly consistent because frequencies are readily converted into proportions and vice versa. An observed proportion in the jth category is denoted by $\hat{p}_j$ and the

---

[3] The letter p is used to denote a proportion as well as a probability. The meaning of the letter will be stated if it isn't clear from the context.

[4] For the case in which $k = 2$, a similar approximate test is based on the normal distribution. A comparison of this test and Pearson's test based on the chi-square distribution is presented in Technical Note 17.6-1.

[5] This point is discussed in Technical Note 17.6-2.

Table 17.2-1. Proportion of
former applicants with various
Graduate Record Examination
(GRE) Scores ($n = 200$)

GRE score	p_j'
1400–1499	0.04
1300–1399	0.14
1200–1299	0.20
1100–1199	0.32
1000–1099	0.22
900–999	0.08

expected proportion, by p_j'. Proportions can be computed from frequencies by $\hat{p}_j = O_j/n$ and $p_j' = E_j/n$, where n is the sample size. If proportions are known, the frequencies are given by $O_j = n\hat{p}_j$ and $E_j = np_j'$. The null and alternative hypotheses for the data in Table 17.2-1 are as follows.

H_0: $p_1 = 0.08$, $p_2 = 0.22$, $p_3 = 0.32$, $p_4 = 0.20$, $p_5 = 0.14$, $p_6 = 0.04$

H_1: $p_j \neq p_j'$ for one or more of the j categories

The hypotheses can also be expressed in terms of $O_{Pop\,j}$'s and $E_{Pop\,j}$'s, as follows.

H_0: $O_{Pop\,1} = 16$, $O_{Pop\,2} = 44$, $O_{Pop\,3} = 64$, $O_{Pop\,4} = 40$, $O_{Pop\,5} = 28$, $O_{Pop\,6} = 8$

H_1: $O_{Pop\,j} \neq E_{Pop\,j}$ for one or more of the j categories

We want to know whether the proportions in the six categories for the population represented by the sample differ from the proportions for the population of previous applicants. We are faced with a minor interpretation problem, because this year's applicants were not obtained by random sampling. We circumvent the problem by assuming that there is a population for which random sampling could have produced the sample we obtained. Conclusions about equality of the proportions apply only to this population.

The observed frequencies for 200 students who applied to graduate school this year, along with expected frequencies based on data for previous applicants, are given in Table 17.2-2. The computation of Pearson's statistic is illustrated in the table. Degrees of freedom equal one less than the number of categories, $k - 1 = 6 - 1 = 5$. If the .05 level of significance is adopted, the critical value of

Table 17.2-2. Computation of Pearson's chi-square statistic for 200 graduate school applicants

GRE scores	O_j	$np_j' = E_j$	$(O_j - E_j)$	$\dfrac{(O_j - E_j)^2}{E_j}$
1400–1499	13	$200(0.04) = 8$	5	3.125
1300–1399	35	$200(0.14) = 28$	7	1.750
1200–1299	49	$200(0.20) = 40$	9	2.025
1100–1199	57	$200(0.32) = 64$	−7	0.766
1000–1099	38	$200(0.22) = 44$	−6	0.818
900–999	8	$200(0.08) = 16$	−8	4.000

$$\chi^2 = 12.484$$

$$\chi^2_{.05,5} = 11.070$$

$\chi^2_{.05,5}$ is 11.070. Although the null hypothesis is nondirectional, the critical region always lies in the upper tail of the sampling distribution of χ^2. This seems reasonable when we recall that all $O - E$ discrepancies are squared; consequently, the test statistic is insensitive to the direction of the discrepancies. Large discrepancies can produce only large values of the test statistic. An examination of the chi-square table reveals that the upper tail contains the larger values. According to Table 17.2-2, the observed test statistic (12.484) exceeds the value for $\chi^2_{.05,5}$, so the null hypothesis is rejected and it is concluded that the frequencies in one or more of the population categories are not equal to the corresponding expected frequencies.[6]

Characteristics of Pearson's Statistic

An examination of the formula $\sum (O_j - E_j)^2/E_j$ for Pearson's statistic reveals the following.

1. The statistic is never negative, since all $O - E$ discrepancies are squared; and, since the statistic makes no distinction between positive and negative discrepancies, the hypothesis tested is nondirectional. This is true even though the critical region of the sampling distribution of χ^2 is always in the upper tail.
2. The only way the statistic can equal 0 is for each observed frequency to equal the corresponding expected frequency.
3. The larger the $O - E$ discrepancy, the larger χ^2. However, the contribution of a discrepancy to χ^2 is affected by the size of E_j, since $(O_j - E_j)^2$ is divided by E_j. This seems reasonable. If we tossed ten coins and observed nine heads where the expected number is five, the discrepancy of four would lead us to question the fairness of the coins. If the same discrepancy occurred when 100 coins were tossed where the expected frequency is 50, we wouldn't be surprised. A discrepancy of four is viewed one way when $E = 5$ and a different way when $E = 50$. The formula takes this into account by expressing the size of the discrepancy relative to the magnitude of the expected frequency.
4. The larger the number of categories, the larger the degrees of freedom and the observed χ^2. As the number of degrees of freedom increases, the chi-square value required for significance also increases, since degrees of freedom determine which member of the family of chi-square distributions is used in evaluating χ^2. Thus, the test procedure takes the number of categories into account.

Degrees of Freedom when E_j's Are Based on a Theoretical Distribution

A modification of the goodness-of-fit test is required if parameters of a theoretical distribution must be estimated in computing the expected frequencies. In the

---

[6] Procedures for determining which observed and expected frequencies are not equal are described by Marascuilo (1971, pp. 380–382) and Marascuilo and McSweeney (1977).

previous example, the E_j's were computed from data, and no distribution param-eters were estimated. For this case $v = k - 1$, where k is the number of categories. Suppose, however, that we compared a frequency distribution with that predicted by, say, the normal distribution. We would have to estimate two parameters, μ and σ, to compute the E_j's. For each parameter estimated, the degrees of freedom are reduced by one. The formula for degrees of freedom is $v = k - 1 - e$, where e is the number of theoretical distribution parameters estimated. Except for this modification, the test procedure is the same.

Assumptions of the Test

The goodness-of-fit test can be used for any hypothetical population, provided the population is discrete or can be grouped into a manageable number of categories. The assumptions of the test are minimal: (1) every sample observation must fall in one and only one category, (2) the observations must be independent, and (3) the sample n must be large. How large is large? This is difficult to specify because the satisfactoriness of Pearson's chi-square statistic in approximating exact multi-nominal probabilities depends on n, the true proportion in the k categories, and the number of degrees of freedom, among other things. Furthermore, Pearson's statistic is not exactly distributed as the chi-square distribution unless n is infinitely large. As a rule of thumb, the approximation to the exact multinomial probabili-ties is satisfactory if, when the degrees of freedom equal one, each expected fre-quency is at least ten. Each expected frequency should be at least five when the degrees of freedom are greater than one. One remedy if $k > 2$ and expected fre-quencies are below the minimum is to combine categories until all frequencies are large enough.[7]

When the test has one degree of freedom, *Yates' correction for continuity*[8] can be applied to make the discrete sampling distribution of the test statistic when the null hypothesis is true more consistent with the theoretical model—the con-tinuous chi-square distribution. The correction can be included in the test statistic as follows.

$$\chi^2 = \sum_{j=1}^{k} \frac{(|O_j - E_j| - 0.5)^2}{E_j}$$

It is good practice to apply the correction when the degrees of freedom equal one and any expected frequency is not appreciably greater than ten.

Review Exercises for Section 17.2

1. Why is the hypothesis tested by Pearson's statistic always nondirectional?
*2. Students in a random sample were asked if they favor a change from the semester system to the quarter system. Thirty-three said yes; 17 said no. (a) List the steps you would follow in testing the null hypothesis that opinion is equally divided on the issue. Let $\alpha = .05$. (b) Do the data suggest that opinion is not equally divided on this issue?

---

[7] An alternative approach is to use the Kolomogorov–Smirnov goodness-of-fit test, which has greater power than Pearson's test. For a discussion of this test see Korin (1975, pp. 297–298) and Taylor (1972, pp. 141–143). Cochran (1954) discusses other rules of thumb for using Pearson's statistic.
[8] Proposed by Frank Yates (1902), a British statistician.

3. College students ($n = 197$) voted at a student-center booth for a beauty queen from among six photographs equivalent in physical attractiveness as determined by rankings of 35 students at a nearby university. The pictures were randomly assigned names that the 35 students had previously judged to be desirable (Kathy, Jennifer, Christine) or undesirable (Ethel, Harriet, Gertrude). The total number of votes for the photographs having the desirable names was 158; the number for the photographs having the undesirable names was 39. (Experiment suggested by Garwood, S. C., Cox, L., Kaplan, V., Wasserman, N., and Sulzer, Jefferson L. Beauty is only "name" deep: The effect of first-name on ratings of physical attraction. *Journal of Applied Social Psychology*, 1980, *10*, 431–435.) (a) List the steps you would follow in testing the null hypothesis that for the population represented by the sample of students who voted, the number of votes is evenly divided between the photographs assigned the attractive and unattractive names. Let $\alpha = .01$. (b) Is it necessary to apply Yates' correction? Why? (c) Perform the test and make a decision.

*4. According to the most recent public opinion poll in Johnson County, 71 eligible voters were Democrats, 52 were Republicans, and 33 belonged to other parties. Traditionally, the ratio of Democrats to Republicans to others has been 4:3:2. Does the poll suggest a change in party affiliation? Let $\alpha = .05$.

5. The dean believes that students in M–W–F classes are more likely to be absent on Monday and Friday than on Wednesday. A random sample of 200 students revealed that 68 were absent on Monday, 48 on Wednesday, and 84 on Friday. (a) List the steps you would follow in testing the hypothesis that the ratio of absences is M:W:F = 3:2:3. Let $\alpha = .05$. (b) Perform the test and make a decision.

*6. A student in a statistics class tossed a die 300 times and obtained the results shown in the table. Is the die fair? Let $\alpha = .05$.

Outcome	1	2	3	4	5	6
Frequency	53	41	60	47	38	61

7. A random sample of 300 music majors took a test of creativity. (a) Test the hypothesis that their scores are normally distributed. Let $\alpha = .01$. The E column in the table presents expected frequencies, given that the null hypothesis is true; the O column presents the observed frequencies. (b) How many degrees of freedom does the test have?

Test score	O_j	E_j
140 and above	0	5.0
130–139	32	20.2
120–129	48	54.2
110–119	76	83.8
100–109	90	78.4
90–99	36	42.4
80–89	18	13.4
Less than 80	0	2.7

8. What is Yates' correction and why is it used?

9. Terms to remember:
   a. Approximate and exact tests
   b. Yates' correction for continuity

Section 17.3   Testing Independence

Pearson's statistic can be used to obtain an approximate test of the hypothesis that two variables are independent. The test is appropriate where n independent observations for a single random sample have been classified in terms of two variables.

Computational Example

Suppose that a random sample of 200 high school students has been classified in terms of sex, variable A, and use or nonuse of marijuana, variable B. *A table representing the classification of elements in terms of two or more variables is called a **contingency table**.* A contingency table has $i = 1, \ldots, r$ rows for variable A and $j = 1, \ldots, c$ columns for variable B. A partial summary of the sex and marijuana data is given in Table 17.3-1. If the two variables in the population are inde-

Table 17.3-1.   Partial summary of sex and marijuana data

	User, B_1	Nonuser, B_2	
Male, A_1			$n_{A_1} = 40;\ p(A_1) = \dfrac{40}{200} = 0.20$
Female, A_2			$n_{A_2} = 160;\ p(A_2) = \dfrac{160}{200} = 0.80$
	$n_{B_1} = 128$	$n_{B_2} = 72$	$n = 200$
	$p(B_1) = \dfrac{128}{200}$	$p(B_2) = \dfrac{72}{200}$	
	$= 0.64$	$= 0.36$	

pendent, what frequencies should be in the cells of the contingency table? We know from Section 7.3 that two events are independent if $p(A|B) = p(A)$, in which case $p(A \text{ and } B) = p(A)p(B)$. The population proportions are unknown, but we can use the marginal proportions, $p(A_1)$, $p(A_2)$, and so on, in Table 17.3-1 to estimate the expected cell frequencies $E_{A_i \text{ and } B_j}$. Using the relationship

$$E_{A_i \text{ and } B_j} = np(A_i \text{ and } B_j) = np(A_i)p(B_j),$$

we can estimate the expected cell values, as follows.

$$E_{A_1 \text{ and } B_1} = np(A_1 \text{ and } B_1) = np(A_1)p(B_1) = 200(0.20)(0.64) = \quad 25.6$$

$$E_{A_2 \text{ and } B_1} = np(A_2 \text{ and } B_1) = np(A_2)p(B_1) = 200(0.80)(0.64) = 102.4$$

$$E_{A_1 \text{ and } B_2} = np(A_1 \text{ and } B_2) = np(A_1)p(B_2) = 200(0.20)(0.36) = \quad 14.4$$

$$E_{A_2 \text{ and } B_2} = np(A_2 \text{ and } B_2) = np(A_2)p(B_2) = 200(0.80)(0.36) = \quad 57.6$$

Alternatively, the expected cell frequencies can be computed directly from the marginal frequencies by the formula

$$E_{A_i \text{ and } B_j} = \frac{n_{A_i} n_{B_j}}{n},$$

since $E_{A_i \text{ and } B_j} = np(A_i)p(B_j) = n(n_{A_i}/n)(n_{B_j}/n) = n_{A_i}n_{B_j}/n$. For example, $E_{A_1 \text{ and } B_1} = (40)(128)/200 = 25.6$. The expected cell frequencies, along with the observed cell frequencies, are given in Table 17.3-2. The computation of the chi-square statistic is illustrated in the table.[9] The advisability of using Yates' correction for a 2×2 contingency table is the subject of continuing debate among statisticians (Conover, 1974a, 1974b; Grizzle, 1967; Mantel, 1974; Miettinen, 1974; Plackett, 1964; Starmer, Grizzle, & Sen, 1974) and accordingly is not illustrated in Table 17.3-2. If the correction is desired, it can be incorporated in the chi-square formula as follows.

$$\chi^2 = \sum_{i=1}^{r} \sum_{j=1}^{c} \frac{(|O_{ij} - E_{ij}| - 0.5)^2}{E_{ij}}$$

Degrees of Freedom for a Contingency Table

How many degrees of freedom are associated with this χ^2? For the goodness-of-fit test we saw that the degrees of freedom are $k - 1 - e$, where e is the number of parameters estimated. A 2×2 table is a special case of an $r \times c$ contingency table, where r and/or c are greater than 2. We will develop the formula for degrees of freedom for the more general case. If we denote the number of rows by r and the number of columns by c, an $r \times c$ contingency table has $k = rc$ categories. Since there are r categories for variable A, we must estimate $r - 1$ expected row frequencies for this variable. Once we have estimated any $r - 1$ of the

Table 17.3-2.   Sex and marijuana data

(i) *Data (O and E denote observed and expected frequencies for the $i = 1, \ldots, r$ rows of variable A and $j = 1, \ldots, c$ columns of variable B):*

	User, B_1	Nonuser, B_2	
Male, A_1	$O = 32$ $E = 25.6$	$O = 8$ $E = 14.4$	$n_{A_1} = 40$
Female, A_2	$O = 96$ $E = 102.4$	$O = 64$ $E = 57.6$	$n_{A_2} = 160$
	$n_{B_1} = 128$	$n_{B_2} = 72$	$n = 200$

(ii) *Computation of χ^2:*

O_{ij}	E_{ij}	$O_{ij} - E_{ij}$	$\dfrac{(O_{ij} - E_{ij})^2}{E_{ij}}$
32	25.6	6.4	1.600
96	102.4	−6.4	0.400
8	14.4	−6.4	2.844
64	57.6	6.4	0.711
		$\chi^2 =$	5.555
		$\chi^2_{.05,1} =$	3.841

[9] A simpler computational formula for a 2×2 contingency table is given in Technical Note 17.6-3.

expected row frequencies, the remaining one can be obtained by subtraction, because the sum of the r expected row frequencies must equal n. Similarly, we must estimate $c - 1$ expected column frequencies for variable B. In all, we must make $e = (r - 1) + (c - 1)$ estimates. Thus, the number of degrees of freedom for an $r \times c$ contingency table is

$$\begin{aligned}
\text{df} &= k - 1 - e \\
&= rc - 1 - [(r - 1) + (c - 1)] \\
&= rc - 1 - r + 1 - c + 1 \\
&= rc - r - c + 1 \\
&= (r - 1)(c - 1).
\end{aligned}$$

For a 2×2 contingency table the degrees of freedom are $(2 - 1)(2 - 1) = 1$.

Statistical Hypotheses

The statistical hypotheses for the data in Table 17.3-2 are

$$H_0: \quad p(A \text{ and } B) = p(A)p(B)$$

$$H_1: \quad p(A \text{ and } B) \neq p(A)p(B).$$

The hypothesis of statistical independence is rejected if the observed χ^2 equals or exceeds the critical value at α level of significance for $(r - 1)(c - 1)$ degrees of freedom. For the data in Table 17.3-2, the critical value at the .05 level of significance ($\chi^2_{.05,1}$) is 3.841. Since the value of the test statistic (5.555) exceeds 3.841, the null hypothesis is rejected, and we conclude that sex and marijuana usage are not independent.

Knowing that two variables are related is useful, but it would be even more useful to know the strength of the association. Before describing an index of association, we will see how to apply a test of independence to contingency tables with more than two rows or columns.

Contingency Tables with Three or More Rows or Columns

The test for independence can be extended to the case in which the row variable, A, has $r > 2$ mutually exclusive and exhaustive categories, $A_1, A_2, \ldots, A_i, \ldots, A_r$, and the column variable, B, has $c > 2$ mutually exclusive and exhaustive categories, $B_1, B_2, \ldots, B_j, \ldots, B_c$.

Suppose we want to know if a college graduate's starting salary is independent of the size of the university from which he or she graduated. Data for a random sample of 200 graduates are given in Table 17.3-3. The expected frequencies are computed from $E_{A_i \text{ and } B_j} = n_{A_i} n_{B_j} / n$. The degrees of freedom are $(r - 1)(c - 1) = (2)(2) = 4$. For a test at the .05 level of significance, the value of $\chi^2_{.05,4}$ is 9.488. Since the value of the test statistic (17.978) exceeds 9.488, the null hypothesis is rejected, and it is concluded that starting salary and university size are not independent.

Finding a significant χ^2 is often just the first step in analyzing data. To better understand the data, it is helpful to look for large discrepancies between observed

Table 17.3-3.   University size and starting salaries of graduates

(i) Data (O and E denote observed and expected frequencies for the $i = 1,\ldots, r$ rows of variable A and $j = 1,\ldots,c$ columns of variable B):

	Less than $10,000, B_1$	$10,000–15,000, B_2$	Greater than $15,000, B_3$	
Small university, A_1	O = 13 E = 10.08	O = 28 E = 34.44	O = 15 E = 11.48	$n_{A_1} = 56$
Medium-sized university, A_2	O = 11 E = 13.14	O = 40 E = 44.90	O = 22 E = 14.96	$n_{A_2} = 73$
Large university, A_3	O = 12 E = 12.78	O = 55 E = 43.66	O = 4 E = 14.56	$n_{A_3} = 71$
	$n_{B_1} = 36$	$n_{B_2} = 123$	$n_{B_3} = 41$	$n = 200$

(ii) Computation of χ^2:

O_{ij}	E_{ij}	$O_{ij} - E_{ij}$	$\dfrac{(O_{ij} - E_{ij})^2}{E_{ij}}$
13	10.08	2.92	0.846
11	13.14	−2.14	0.349
12	12.78	−0.78	0.048
28	34.44	−6.44	1.204
40	44.90	−4.90	0.535
55	43.66	11.34	2.945
15	11.48	3.52	1.079
22	14.96	7.04	3.313
4	14.56	−10.56	7.659

$$\chi^2 = 17.978$$
$$\chi^2_{.05,4} = 9.488$$

and expected frequencies in the contingency table. From a visual inspection of Table 17.3-3, it appears that graduates of large universities are less likely than expected to have starting salaries over $15,000, while the converse is true for those from medium and small universities. Such hypotheses must be regarded as tentative, since they are not based on the outcome of significance tests.[10] The significant chi-square test statistic applies to the data taken as a whole and provides no clue as to which cells are responsible for significance.

In the following section we will describe a statistic for determining the strength of the association between the row and column variables.

Measuring Strength of Association

If the null hypothesis $p(A \text{ and } B) = p(A)p(B)$ is rejected, we know that the variables are correlated. Several correlation coefficients were discusssed in Chapter 5 (r, r_s, and η^2), but none of them is appropriate for unordered qualitative variables. One

---

[10] Procedures for determining the contribution to the overall chi-square statistic of a portion of the contingency table are described by Bresnahan and Shapiro (1966), Castellan (1965), Marascuilo (1966), and Marascuilo and McSweeney (1977).

coefficient that is appropriate is *Cramer's measure of association,*[11] denoted by V or ϕ'. The formula for estimating V is

$$\hat{V} = \frac{\sqrt{\hat{\phi}^2_{observed}}}{\sqrt{\hat{\phi}^2_{maximum}}} = \frac{\sqrt{\chi^2/n}}{\sqrt{s-1}} = \sqrt{\frac{\chi^2}{n(s-1)}},$$

where s is the smaller of the numbers r and c.[12] If $p(A \text{ and } B) = p(A)p(B)$, the parameter ϕ, which is estimated by $\hat{\phi}_{observed} = \sqrt{\chi^2/n}$, is equal to 0. When there is complete association, $\phi = s - 1$, which is its maximum value. Cramer's statistic is a relative measure, since it is the ratio of an observed statistic to its maximum possible value. The statistic can range from 0 (indicating complete independence) to 1 (indicating complete dependence, or perfect correlation). Cramer's statistic is computed only when the chi-square test is significant. A nonsignificant test suggests that any $\hat{V}$ greater than 0 is due to chance. Put another way, a test of the hypothesis H_0: $p(A \text{ and } B) = p(A)p(B)$ is equivalent to a test of H_0: $V = 0$. If the null hypothesis isn't rejected, it is meaningless to proceed further and compute $\hat{V}$. Unfortunately, Cramer's statistic does not have a simple intuitively useful interpretation as, say, the proportion of explained variance between two variables. It can only be thought of as reflecting magnitude of association on a scale of 0 to 1; the larger the number, the stronger the association. Cramer's statistic for the data in Tables 17.3-2 and 17.3-3 is, respectively,

$$\hat{V} = \sqrt{\frac{5.555}{200(2-1)}} = 0.17 \quad \text{and} \quad \hat{V} = \sqrt{\frac{17.978}{200(3-1)}} = 0.21.$$

The coefficients and the significance tests indicate a low but statistically significant association for both sets of variables. This is another illustration of the maxim that statistical significance doesn't mean practical significance. For example, the chi-square test statistic for the variables of university size and starting salaries of graduates is large enough to be significant beyond the .005 level, yet the strength of association is low. The point cannot be made too often that statistical significance only means that in all likelihood a relationship is not due to chance. Other procedures must be used to assess the usefulness or practical significance of a relationship.

Assumptions of the Test

The assumptions associated with both the test for independence and the estimation of strength of association are as follows: (1) Every observation must fall in one and only one cell of the contingency table. (2) The observations must be independent. One situation in which this assumption is likely to be violated occurs when an individual is represented more than once in a cell or in more than one cell. (3) The sample n should be large enough so that every expected frequency is

[11] For a discussion of other coefficients see Loether and McTavish (1974, chaps. 6 and 7).
[12] For a 2×2 contingency table, $\hat{V} = \sqrt{\chi^2/n}$, since $s - 1 = 1$. In this special case, $\hat{V}$ is identical to another measure of strength of association called the *phi coefficient,* $\hat{\phi}$, and both are related to the Pearson product-moment correlation coefficient. Suppose that the two categories of variables A and B are considered to be ordered and are assigned scores of 0 and 1, where 1 is assigned to the higher category. Under these conditions the formula for r is algebraically equivalent to that for $\hat{V}$ and $\hat{\phi}$ (Hays, 1981, pp. 556–557).

at least ten when there is one degree of freedom and at least five when there is more than one degree of freedom.[13]

Review Exercises for Section 17.3

10. State in your own words the meaning of the hypothesis H_0: $p(A \text{ and } B) = p(A)p(B)$. If H_0 is rejected, what do you know about the variables?

*11. List the similarities and differences between Pearson's product-moment correlation coefficient and Cramer's measure of association.

12. Why is Cramer's measure of association computed only when the hypothesis

$$p(A \text{ and } B) = p(A)p(B)$$

is rejected?

*13. Two hundred women between the ages of 19 and 25 were asked if they favored the use of birth-control pills. The women were classified according to attitude and religious preference. (a) For the data in the table, list the steps you would use in testing the null hypothesis. Let $\alpha = .001$. (b) What is your decision? (c) Compute Cramer's statistic.

	Protestants	Roman Catholics	Non-Christians	
Favor	58	30	28	116
Oppose	8	23	5	36
Undecided	10	15	23	48
	76	68	56	200

14. Three hundred divorced males were classified by age at time of first marriage and duration of first marriage. (a) List the steps you would use in testing the null hypothesis. Let $\alpha = .05$. (b) Perform the test and make a decision. (c) Compute Cramer's statistic.

		Duration of marriage (years)				
		<5	5–9	10–14	≥15	
Age at marriage	<19	41	31	15	15	102
	19–24	30	27	22	21	100
	25–34	11	7	16	16	50
	≥35	13	14	9	12	48
		95	79	62	64	300

15. A random sample of 200 college students was classifed according to class standing and political conservatism. (a) Are the variables independent? Let $\alpha = .05$. (b) Compute Cramer's statistic.

	Conservative	Neutral	Liberal	
Freshman	38	22	6	66
Sophomore	22	24	5	51
Junior	11	12	19	42
Senior	7	13	21	41
	78	71	51	200

16. Term to remember:
    Contingency table

---

[13] Fisher's exact test for a 2 × 2 contingency table can be used to test independence when n is small and expected cell frequencies are less than ten. The test is described in Hays (1981, pp. 552–554) and Taylor (1972, pp. 143–145).

Section 17.4   Testing Equality of $c \geq 2$ Proportions

The last application of Pearson's chi-square statistic that will be described is testing the equality of $c \geq 2$ population proportions. If a random sample is obtained from each of $j = 1, \ldots, c$ binomially distributed populations, the sample proportions $\hat{p}_1$, $\hat{p}_2, \ldots, \hat{p}_c$ can be used to test the null hypothesis

$$H_0: \quad p_1 = p_2 = \cdots = p_c$$

versus

$$H_1: \quad p_j \neq p_{j'} \text{ for some } j \text{ and } j'.$$

The null hypothesis is rejected if $\chi^2 = \sum_{i=1}^{r} \sum_{j=1}^{c} (O_{ij} - E_{ij})^2/E_{ij}$ equals or exceeds the critical value of χ^2 at α level of significance for $c - 1$ degrees of freedom. Rejection of H_0 means that some population proportions are not equal. It doesn't mean that they are all unequal; perhaps only one is discrepant. The test, like the two described earlier, is approximate, since the continuous chi-square distribution is used to estimate a probability for a discrete sampling distribution. The proportions $\hat{p}_1$, $\hat{p}_2, \ldots, \hat{p}_c$, which may be based on unequal sample sizes, are assumed to represent independent observations for c independent binomially distributed random variables. Like all binomial experiments, each trial must result in one of two outcomes. We will see later that the test can be extended to the multinomial case in which each trial can result in one of three or more outcomes.

Computational Example

Suppose we surveyed older people living in public housing in Borborygme, Texas, to determine whether satisfaction with living conditions is related to age heterogeneity of people in the neighborhood. Neighborhoods were categorized as high, medium, or low in heterogeneity of residents' ages. Random samples of 100 elderly females from each category were interviewed (recall that equal sample n's are not required) and asked, among other things, "Are you satisfied with your living conditions?" The answers were classified as "satisfied" or "not satisfied." It was anticipated that satisfaction would be different for the three neighborhood categories. Responses to the question are given in Table 17.4-1. The computation of Pearson's chi-square statistic is shown in the table; the procedure is identical to that for an $r \times c$ contingency table for testing independence (see Section 17.3). The degrees of freedom are equal to $c - 1 = 2$, the number of age heterogeneity categories minus one. Alternatively, the degrees of freedom can be computed from $(r - 1)(c - 1) = (1)(2) = 2$. For a test at the .05 level of significance, the critical value of $\chi^2_{.05, 2}$ is 5.991 (according to Table D.4). Since the value of the test statistic (9.708) exceeds 5.991, the null hypothesis is rejected, and it is concluded that at least two of the population proportions are not equal.[14]

Comparison of Independence Test and Equality of Proportions Test

The tests for equality of $c \geq 2$ proportions and independence use the same Pearson test statistic and the same formula for computing degrees of freedom, and

---

[14] Marascuilo and McSweeney (1977, pp. 141–147) describe a procedure for determining which population proportions are unequal. The procedure is based on Scheffé's method, which is described in Section 16.5.

Table 17.4-1.   Satisfaction and age heterogeneity

(i) Data (O and E denote observed and expected frequencies for the $i = 1, \ldots, r$ rows of variable A and $j = 1, \ldots, c$ columns of variable B):

Age heterogeneity

	Low, B_1	Medium, B_2	High, B_3	
Satisfied, A_1	$O = 56$ $E = 50.67$	$O = 58$ $E = 50.67$	$O = 38$ $E = 50.67$	$n_{A_1} = 152$
Not satisfied, A_2	$O = 44$ $E = 49.33$	$O = 42$ $E = 49.33$	$O = 62$ $E = 49.33$	$n_{A_2} = 148$
	$n_{B_1} = 100$	$n_{B_2} = 100$	$n_{B_3} = 100$	$n = 300$

(ii) Computation of χ^2:

O_{ij}	E_{ij}	$O_{ij} - E_{ij}$	$\dfrac{(O_{ij} - E_{ij})^2}{E_{ij}}$
56	50.67	5.33	0.561
44	49.33	−5.33	0.576
58	50.67	7.33	1.060
42	49.33	−7.33	1.089
38	50.67	−12.67	3.168
62	49.33	12.67	3.254

$$\chi^2 = 9.708$$
$$\chi^2_{.05,2} = 5.991$$

both are approximately distributed as the chi-square distribution. It is not surprising then that the tests are often confused. The key difference is in the sampling method. In testing independence, (1) a single random sample is obtained from a population and (2) the elements are classified in terms of two variables. On the other hand, in testing equality of proportions, (1) c independent random samples are obtained from c binomially distributed populations and (2) the sample elements are classified in terms of only one variable (such as satisfaction), since the experimenter has predetermined the elements' status with respect to the other variable (such as neighborhoods) by sampling from the c categories. Note that in a test of independence the status of the sample elements is not predetermined for either variable. In a test of equality an experimenter often attempts to sample elements so as to have an equal number from each of the c populations, although this isn't required. It is because of the different sampling procedures that the hypotheses tested are different; H_0: $p(A \text{ and } B) = p(A)p(B)$ for the test of independence and H_0: $p_1 = p_2 = \cdots = p_c$ for the test of equality.

Extension of the Test of Equality to More than Two Response Categories

The test illustrated in Table 17.4-1 can be extended to the case in which variable A has more than two categories, $A_1, A_2, \ldots, A_r$; for example, very satisfied, somewhat satisfied, somewhat dissatisfied, very dissatisfied. In this form the test is referred to as a *test of homogeneity*. The proportions $p_{A_1|B_1}, p_{A_2|B_1}, \ldots, p_{A_r|B_1}; p_{A_1|B_2}, p_{A_2|B_2}, \ldots, p_{A_r|B_2}; \ldots; p_{A_1|B_c}, p_{A_2|B_c}, \ldots, p_{A_r|B_c}$ are assumed to represent observations for c independent multinomially distributed random variables instead of c binomial variables. The null hypothesis states that the proportions in the r rows are equal across the c columns of B. This can be stated symbolically as

$$H_0: \begin{bmatrix} p_{A_1|B_1} \\ p_{A_2|B_1} \\ \vdots \\ p_{A_r|B_1} \end{bmatrix} = \begin{bmatrix} p_{A_1|B_2} \\ p_{A_2|B_2} \\ \vdots \\ p_{A_r|B_2} \end{bmatrix} = \cdots = \begin{bmatrix} p_{A_1|B_c} \\ p_{A_2|B_c} \\ \vdots \\ p_{A_r|B_c} \end{bmatrix}$$

versus

$$H_1: \begin{bmatrix} p_{A_1|B_j} \\ p_{A_2|B_j} \\ \vdots \\ p_{A_r|B_j} \end{bmatrix} \neq \begin{bmatrix} p_{A_1|B_{j'}} \\ p_{A_2|B_{j'}} \\ \vdots \\ p_{A_r|B_{j'}} \end{bmatrix} \quad \text{for some } j \text{ and } j'.$$

Ordinarily the values of $p_{A_i|B_j}$ in the null hypothesis are not specified, but if they were, the hypothesis for $i = 1, \ldots, 4$ and $j = 1, \ldots, 3$ might take the form

$$H_0: \begin{bmatrix} .20 \\ .40 \\ .30 \\ .10 \end{bmatrix} = \begin{bmatrix} .20 \\ .40 \\ .30 \\ .10 \end{bmatrix} = \begin{bmatrix} .20 \\ .40 \\ .30 \\ .10 \end{bmatrix}.$$

If A has only two categories (the binomial case), the null hypothesis can be written as

$$H_0: \quad p_{A_1|B_1} = p_{A_1|B_2} = \cdots = p_{A_1|B_c},$$

or simply,

$$H_0: \quad p_1 = p_2 = \cdots = p_c.$$

The computation of Pearson's statistic for a homogeneity test is like that for an $r \times c$ contingency table. The number of degrees of freedom is equal to $(r-1)(c-1)$. The general assumptions discussed in connection with an $r \times c$ contingency table apply as well to testing the equality of c population proportions.

Review Exercises for Section 17.4

17. The equality and independence chi-square tests are applicable to an $r \times c$ contingency table, and both use the same test statistic. How do they differ?

*18. Company executives were classified as smokers or nonsmokers. Random samples from the two populations were obtained and tested for presence of lung cancer. State the null hypothesis and test it at the .01 level of significance.

	Smoker	Nonsmoker	
Cancer present	16	5	21
Cancer absent	14	25	39
	30	30	60

19. Students were given a choice of writing or not writing a paper for extra credit. Half of them, selected randomly, were made to feel coerced; the other half received no pressure. The number who chose to write or not to write a paper is given in the table. (a) List the steps you would use to test the null hypothesis. Let $\alpha = .05$. (b) Did the two conditions affect paper writing?

	Coerced	Not Coerced	
Wrote paper	13	9	22
Didn't write paper	17	21	38
	30	30	60

*20. State in words the meaning of the hypothesis

$$H_0: \begin{bmatrix} 0.30 \\ 0.60 \\ 0.10 \end{bmatrix} = \begin{bmatrix} 0.30 \\ 0.60 \\ 0.10 \end{bmatrix} = \begin{bmatrix} 0.30 \\ 0.60 \\ 0.10 \end{bmatrix}.$$

*21. New employees on an assembly line were randomly assigned to one of four groups and were given different amounts of training. After 2 weeks on the job, their performance was rated by their supervisor. For the data in the table (a) state the statistical hypotheses and (b) test the null hypothesis at the .05 level of significance.

		\multicolumn{4}{c}{Amount of training (days)}	
		1	2	5	10	
	Excellent	4	4	6	6	20
Rating	Good	5	3	10	11	29
	Fair	9	9	5	8	31
	Poor	7	9	4	0	20
		25	25	25	25	100

22. Random samples of seventh-, eighth-, and ninth-grade students were interviewed following 1 year of busing. They were asked, "Did black and white students mix more than, the same as, or less than last year?" State the null hypothesis and test it at the .05 level of significance.

		\multicolumn{3}{c}{Grade}	
		7	8	9	
	More	26	12	4	42
Response	Same	18	25	35	78
	Less	6	13	11	30
		50	50	50	150

Section 17.5   Summary

Pearson's chi-square test statistic is one of the few that is appropriate for frequency data. It provides an approximate test when an exact test based on the binomial or multinomial distributions would require a prohibitive amount of computation. The statistic is versatile; the three applications we have described are testing goodness of fit, independence, and equality of proportions. These are summarized in Table 17.5-1. In each application a set of observed frequencies is compared with a set of expected frequencies. The apparent simplicity of the statistic $\sum (O - E)^2/E$ is deceptive. Its use involves important assumptions that are often overlooked or misunderstood by the novice investigator. The problem is compounded because all three applications use the same statistic for testing different hypotheses. The main points to consider in using Pearson's chi-square statistic are the following.

1. A single random sample is used in testing goodness of fit and independence; more than one random sample is used in testing equality of population proportions.
2. Each observation must be assigned to one and only one category or one cell of a contingency table.
3. Each subject or observational unit should be represented only once. Multiple observations on the same subject almost always result in violation of the assumption of independence of observations.
4. The Pearson chi-square approximation to exact binomial or multinomial probabilities is generally unsatisfactory for small samples. As a rule of thumb, if $v = 1$, all expected frequencies should exceed 10; if $v > 1$, all expected frequencies should exceed 5.
5. If the chi-square test statistic is not significant for an $r \times c$ contingency table, no further analyses such as Cramer's $\hat{V}$ should be performed.
6. The hypothesis tested by Pearson's statistic is nondirectional, even though the region for rejection always lies in the upper tail of the sampling distribution.

†*Section 17.6   Technical Notes*

17.6-1   *Two Ways to Test the Hypothesis that* $p = p'$ *(or* $p = p_0$*)*

A goodness-of-fit test with one degree of freedom is formally equivalent to the large-sample z test for a proportion described in Section 12.4. We will show by an example that

$$z = \frac{\hat{p} - p_0}{\sqrt{p_0 q_0/n}}$$

leads to the same decision about the population proportion, p, as Pearson's statistic and, in addition, that $|z| = \sqrt{\chi^2}$.

In Section 12.4 we tested the hypothesis that the results of a random sample of 900 dwelling units in which 0.34 were found to be substandard don't differ from the results of an earlier survey in which the proportion was 0.30. To test the hypothesis that the number of substandard units had not changed, $H_0: p = 0.30$, the z test statistic was computed:

$$z = \frac{\hat{p} - p_0}{\sqrt{p_0 q_0/n}} = \frac{0.34 - 0.30}{\sqrt{(0.30)(0.70)/900}} = \frac{0.04}{\sqrt{0.0002333}} = 2.62.$$

† These technical notes can be skipped without loss of continuity.

Table 17.5-1.   Three applications of Pearson's chi-square test statistic

Purpose	Null hypothesis	Degrees of freedom	Requirements
1. Testing goodness of fit	$p_1 = p_1', p_2 = p_2', \ldots, p_k = p_k'$	$k - 1 - e$	1. One random sample. 2. If $v = 1$, every expected frequency should exceed 10. If $v > 1$, every expected frequency should exceed 5. 3. Random variable is binomially distributed for $k = 2$ and multinomially distributed for $k > 2$.
2. Testing independence	$p(A \text{ and } B) = p(A)p(B)$	$(r - 1)(c - 1)$	1. Same as for test of goodness of fit. 2. Same as for test of goodness of fit. 3. Random variable is binomially distributed when A and B have two categories, and is multinomially distributed otherwise.
3a. Testing equality of proportions	$p_1 = p_2 = \cdots = p_c$	$c - 1$	1. $c \geq 2$ random samples. 2. If $v = 1$, every expected frequency should exceed 10. If $v > 1$, every expected frequency should exceed 5. 3. Random variable is binomially distributed.
3b. Testing homogeneity of proportions	$$\begin{bmatrix} p_{A_1|B_1} \\ p_{A_2|B_1} \\ \vdots \\ p_{A_r|B_1} \end{bmatrix} = \begin{bmatrix} p_{A_1|B_2} \\ p_{A_2|B_2} \\ \vdots \\ p_{A_r|B_2} \end{bmatrix} = \cdots = \begin{bmatrix} p_{A_1|B_c} \\ p_{A_2|B_c} \\ \vdots \\ p_{A_r|B_c} \end{bmatrix}$$	$(r - 1)(c - 1)$	1. $c \geq 2$ random samples. 2. Every expected frequency should exceed 5. 3. Random variable is multinomially distributed.

The statistic exceeded the critical value required to reject the null hypothesis at the .05 level; hence, we concluded that the proportion of substandard dwelling units had increased.

The same decision is reached if the data are analyzed by Pearson's chi-square statistic. The computation is shown in Table 17.6-1. The critical value of χ^2 for $v = 1$ degree of freedom at the .05 level of significance is $\chi^2_{.05,1} = 3.841$. Since the observed value (6.86) exceeds 3.841, the null hypothesis is rejected. We note also that both the value of the z test statistic and the square root of the chi-square test statistic equal 2.62—that is, $|z| = \sqrt{\chi^2} = 2.62$.

Table 17.6-1.   Computation of Pearson's chi-square statistic for $k = 2$ categories

(i) *Computation of observed* (O_j) *and expected* (E_j) *frequencies* $[\hat{p} = 0.34, 1 - \hat{p} = 0.66$ *(observed proportions) and* $p' = 0.30, 1 - p' = 0.70$ *(expected proportions according to the null hypothesis)]*:

$$O_1 = \hat{p}n = 0.34(900) = 306$$
$$O_2 = (1 - \hat{p})n = 0.66(900) = 594$$
$$E_1 = p'n = 0.30(900) = 270$$
$$E_2 = (1 - p')n = 0.70(900) = 630$$

(ii) *Computation of chi-square statistic:*

O_j	E_j	$O_j - E_j$	$\dfrac{(O_j - E_j)^2}{E_j}$
306	270	36	$(36)^2/270 = 4.80$
594	630	-36	$(-36)^2/630 = 2.06$
			$\chi^2 = 6.86$
			$\chi^2_{.05,1} = 3.841$

17.6-2   Outline of an Exact Test of $p_1 = p_1', p_2 = p_2', \ldots, p_k = p_k'$

The binomial distribution is the appropriate theoretical model when (1) samples are randomly selected from a population whose elements belong to one of two categories, (2) the probability of obtaining an element in a category remains constant, and (3) the trials are independent.[15] The probability of observing exactly r successes in n trials is given by the function rule

$$p(X = r) = \frac{n!}{r!(n - r)!} p'^r q'^{n-r}.$$

An exact test of the hypothesis $H_0: p = p'$, where p is the population proportion estimated by $\hat{p}$, p' is the expected proportion, and $q' = 1 - p'$, is obtained by computing the sum

$$p(X \geq r) = p(X = r) + p(X = r + 1) + p(X = r + 2) + \cdots + p(X = n).$$

[15] The binomial distribution is discussed in Section 9.3.

This tells us the probability of observing r or more successes if the population proportion is equal to p'. For large n's this involves a prohibitive amount of computation. Fortunately, the normal distribution or the chi-square distribution can be used to approximate the exact probabilities when n is large.

Consider next the more general theoretical model in which the population elements are classified as belonging to one of three or more categories. The population is referred to as a *multinomial* population.[16] The probability of observing exactly n_1 elements of kind 1, n_2 elements of kind 2, ..., n_k elements of kind k, where $n_1 + n_2 + \cdots + n_k = n$, is given by the multinomial function rule

$$p(X_1 = n_1 \text{ and } X_2 = n_2 \text{ and } \ldots \text{ and } X_k = n_k)$$

$$= \frac{n!}{n_1! n_2! \cdots n_k!} (p_1)^{n_1}(p_2)^{n_2} \cdots (p_k)^{n_k}.$$

To test the hypothesis

$$H_0: \quad p_1 = p_1', p_2 = p_2', \ldots, p_k = p_k'$$

versus

$$H_1: \quad p_j \neq p_{j'} \quad \text{for one or more of the } j \text{ categories}$$

we would need to compute the probability of this result plus those for all results more deviant from expectation.

When n is small, it isn't too difficult to compute an exact test based on the multinomial distribution. Suppose that we want to compare students' preferences for three soft drinks, brands A, B, and C. A random sample of seven students taste the drinks and indicate their preferences. Five prefer brand A; one prefers brand B; and one prefers brand C. If the expected proportions are $p_A' = p_B' = p_C' = \frac{1}{3}$, what is the probability of a result as extreme or more extreme than that observed? We can observe five or more preferences for brand A as follows. The associated probabilities are computed using the multinomial function rule.

$$p(5 \ A \text{ and } 1 \ B \text{ and } 1 \ C) = \frac{7!}{5!1!1!} \left(\frac{1}{3}\right)^5 \left(\frac{1}{3}\right)^1 \left(\frac{1}{3}\right)^1 = \frac{42}{2187} = 0.0192$$

$$p(5 \ A \text{ and } 2 \ B \text{ and } 0 \ C) = \frac{7!}{5!2!0!} \left(\frac{1}{3}\right)^5 \left(\frac{1}{3}\right)^2 \left(\frac{1}{3}\right)^0 = \frac{21}{2187} = 0.0096$$

$$p(5 \ A \text{ and } 0 \ B \text{ and } 2 \ C) = \frac{7!}{5!0!2!} \left(\frac{1}{3}\right)^5 \left(\frac{1}{3}\right)^0 \left(\frac{1}{3}\right)^2 = \frac{21}{2187} = 0.0096$$

$$p(6 \ A \text{ and } 1 \ B \text{ and } 0 \ C) = \frac{7!}{6!1!0!} \left(\frac{1}{3}\right)^6 \left(\frac{1}{3}\right)^1 \left(\frac{1}{3}\right)^0 = \frac{7}{2187} = 0.0032$$

$$p(6 \ A \text{ and } 0 \ B \text{ and } 1 \ C) = \frac{7!}{6!0!1!} \left(\frac{1}{3}\right)^6 \left(\frac{1}{3}\right)^0 \left(\frac{1}{3}\right)^1 = \frac{7}{2187} = 0.0032$$

$$p(7 \ A \text{ and } 0 \ B \text{ and } 0 \ C) = \frac{7!}{7!0!0!} \left(\frac{1}{3}\right)^7 \left(\frac{1}{3}\right)^0 \left(\frac{1}{3}\right)^0 = \frac{1}{2187} = 0.0005$$

$$\Sigma = \overline{0.0453}$$

[16] The multinomial distribution is discussed in Section 9.3.

The probability of observing five or more preferences for brand A is 0.0453. Similarly, the probability of observing five or more preferences for B is 0.0453, and for C is 0.0453. Thus, the exact probability of observing any outcome as extreme as or more extreme than five preferences for one brand is $3(0.0453) = 0.136$.

An approximation to the exact probability is given by Pearson's chi-square statistic, where $O_A = 5$, $O_B = 1$, $O_C = 1$; and $E_A = np_A' = 7(\frac{1}{3}) = 2.333$, $E_B = 2.333$, $E_C = 2.333$. We will illustrate the procedure even though all the expected frequencies are below the minimum recommended, which is five. Substituting the frequencies in the formula $\chi^2 = \sum (O_j - E_j)^2/E_j$, we obtain

$$\chi^2 = \frac{(5 - 2.333)^2}{2.333} + \frac{(1 - 2.333)^2}{2.333} + \frac{(1 - 2.333)^2}{2.333} = 4.572.$$

The probability associated with this statistic, which has two degrees of freedom, is approximately 0.10. For $n = 7$, the chi-square approximation to the exact multinomial probability (0.136) isn't very good. For larger n's typically encountered in research, the approximation is quite satisfactory.

17.6-3   Special Computational Procedure for a 2 × 2 Contingency Table

A simpler procedure for computing Pearson's chi-square statistic can be used for a 2 × 2 contingency table. Consider the table shown, in which observed

a	b	$a + b$
c	d	$c + d$
$a + c$	$b + d$	

frequencies are denoted by the letters a, b, c, and d. The formula for Pearson's statistic using only observed cell frequencies is

$$\chi^2 = \frac{n(ad - bc)^2}{(a + b)(c + d)(a + c)(b + d)}.$$

For the data in Table 17.3-2 (page 380) the formula yields

$$\chi^2 = \frac{200[(32)(64) - (8)(96)]^2}{(32 + 8)(96 + 64)(32 + 96)(8 + 64)} = 5.555,$$

which is identical to the answer obtained using the conventional formula. If desired, Yates' correction for continuity can be included in the formula:

$$\chi^2 = \frac{n(|ad - bc| - n/2)^2}{(a + b)(c + d)(a + c)(b + d)}.$$

†*Section 17.7   Computer Supplement*

The Statistical Analysis System (SAS), Release 79.6, was used to perform a test for independence for the data in Table 17.3-2. The data represent 200 high school students who have been classified in terms of sex (male = 0 and female = 1) and marijuana usage (user = 0 and nonuser = 1). The Procedure Freq of SAS, which is a batch mode package, produced the printout in Table 17.7-1.

The first number in each cell of the contingency table is the observed frequency, O_{ij}, where i denotes the ith row of variable A (sex) and j, the jth column of variable B (marijuana usage). The second, third, and fourth numbers in the cells are, respectively, $(O_{ij}/N) \times 100$, $(O_{ij}/n_{A_i}) \times 100$, and $(O_{ij}/n_{B_j}) \times 100$. The program provides a variety of statistics for the 2×2 contingency table. The two of particular interest are chi-square ($\chi^2 = 5.556$) and Cramer's V ($V = 0.167$). These statistics are discussed in Section 17.3. A discussion of the other statistics is beyond the scope of this book; the interested reader is referred to the *SAS User's Guide* (Helwig and Council, 1979).

The SAS package was developed for mainframe computers and minicomputers. For purposes of comparison, the interactive microcomputer package developed by Brad Wilson (1984) was used to analyze the sex and marijuana data in Table 17.3-2. The printout is shown in Table 17.7-2.

A comparison of Tables 17.7-1 and 17.7-2 reveals that the SAS program provides a greater variety of statistics for these data. Of course the microcomputer program could be revised to include all of the statistics provided by SAS. What is not apparent from this comparison is that SAS requires the user to have much more computer expertise.

---

† This section can be omitted without loss of continuity.

Table 17.7-1. Printout from the SAS Procedure Freq Program (Data from Table 17.3-2)

```
TABLE OF SEX BY USE                    STATISTICS FOR 2-WAY TABLES

SEX      USE                    CHI-SQUARE                  5.556  DF= 1  PROB=0.0184
                                PHI                         0.167
FREQUENCY                       CONTINGENCY COEFFICIENT     0.164
PERCENT                         CRAMER'S V                  0.167
ROW PCT                         LIKELIHOOD RATIO CHISQUARE  5.971  DF= 1  PROB 0.0145
COL PCT    0      1    TOTAL    CONTINUITY ADJ. CHI-SQUARE  4.721  DF= 1  PROB=0.0298
                                FISHER'S EXACT TEST (1-TAIL)              PROB=0.0129
0         32      8     40                          (2-TAIL)              PROB=0.0261
       16.00   4.00   20.00
       80.00  20.00            GAMMA                       0.455  ASE1= 0.169
       25.00  11.11            KENDALL'S TAU-B             0.167
                               STUART'S TAU-C              0.128  ASE1= 0.049
1         96     64    160
       48.00  32.00   80.00    SOMER'S D C|R               0.200  ASE1= 0.074
       60.00  40.00            SOMER'S D R|C               0.139  ASE1= 0.053
       75.00  88.89
                               PRODUCT MOMENT CORRELATION  0.167
TOTAL    128     72    200     SPEARMAN CORRELATION        0.167
       64.00  36.00  100.00
                               LAMBDA ASYMMETRIC C|R       0.000
                               LAMBDA ASYMMETRIC R|C       0.000
                               LAMBDA SYMMETRIC            0.000

                               UNCERTAINTY COEFFICIENT C|R  0.023
                               UNCERTAINTY COEFFICIENT R|C  0.030
                               UNCERTAINTY COEFFICIENT SYM  0.026

                               ASE1 IS THE ASYMPTOTIC STANDARD ERROR.
                               R|C MEANS ROW VAR DEPENDENT ON COLUMN VAR.
```

Table 17.7-2.   Printout from the Wilson Stat Package:
Pearson's chi-square for testing independence
(Data from Table 17.3-2)

```
**********************************************
*                                            *
*              Brad Wilson's                 *
*               Stat Package                 *
**********************************************
*                                            *
*      Behavioral Statistics Program         *
*             Baylor University              *
*                                            *
**********************************************
```

Press RETURN to continue>>>>>

When you respond to a prompt just
type in your response and press
the RETURN key.

For all yes-no (Y/N) questions
it is not necessary to press the
RETURN key.

Press RETURN to continue>>>>>

Would you like to

1. Perform statistical analyses

2. See a demonstration of the
   Central Limit Theorem

3. End the session?

Select a number: 1

Please wait a few moments.

Here are the available
statistical programs
    1.   Frequency distribution and
         histogram
    2.   Mean, median, and mode
    3.   Standard deviation
    4.   Index of dispersion
    5.   Correlation
    6.   Regression
    7.   One-sample t test
    8.   One-sample chi-square test
    9.   Two-sample independent t test
   10.   Two-sample dependent t test
   11.   Analysis of variance for a
         completely randomized design
   12.   Pearson's chi-square for
         testing goodness of fit

(continued)

Table 17.7-2.   (Continued)

```
13.  Pearson's chi-square for
     testing independence
14.  Pearson's chi-square for
     testing equality of proportions
```

```
Indicate the program you want
by selecting a number: 13
```

```
Please wait a few moments.
```

```
Do you want to enter data from

1. The terminal

2. A file, or

3. Use the data from the
   previous session, or do
   you want to

4. Return to the menu?
```

```
Select a number: 1
```

```
How many column categories
do you have? 2
```

```
What is the label for
column category 1? USER
```

```
What is the label for
column category 2? NONUSER
```

```
How many row categories are there? 2
```

```
The name of row category 1? MALE
The name of row category 2? female
```

```
     column category    # of rows
    ----------------------------------
1.     USER                2
2.     NONUSER             2
```

```
Press RETURN to continue>>>
```

```
     row name
    ----------------------------------
1. MALE
2. FEMALE
```

```
Do you want to
1. Go on to the next step
2. Examine your data specifications
   again
```

(continued)

Table 17.7-2.   (Continued)

```
3.  Change your column category name(s)
4.  Change your # of rows?
5.  Change the # of column categories
6.  Change row names?
Select one: 1

Now you will be prompted for
the individual frequencies

USER     1: 32
USER     2: 96
NONUSER 1: 8
NONUSER 2: 64


Press RETURN to continue>>>
```

Would you like to check the data for
errors? (Y/N) N

If you have a second disk drive, you can
store your data on a separate diskette.
Would you like to store the data? (Y/N) N

Please wait a few moments.

Do you want the results sent to the

  1. Screen

  2. Printer?

Select a number: 1

NOTE: Use CTRL and S simultaneously to
stop and start the scrolling of
the screen.

```
Press RETURN to continue>>>

Test for independence

Chi-square Value        df
***************************
  5.556                 1

***********
  p < 0.018
***********

Press RETURN to continue>>>

Category Key
```

(continued)

Table 17.7-2.   (Continued)

Columns

1C = USER
2C = NONUSER

Rows

1R = MALE
2R = FEMALE

O: Observed frequency

E: Expected frequency

CONTINGENCY TABLE

		1C	2C	ROW FREQ
O	1R	32.00	8.00	40.
E		25.60	14.40	
O	2R	96.00	64.00	160.
E		102.40	57.60	

COLUMN
FREQ 128.   72.

Total frequency = 200.

Press RETURN to continue>>>

Would you like to have Cramer's
V computed (Y/N)? Y

Cramer's V = 0.167

Press RETURN to continue>>>

Would you like to

1. Perform statistical analyses

2. See a demonstration of the
   Central Limit Theorem

3. End the session?

Select a number: 3

Just turn off the computer, remove your
diskettes, and have a nice day!!

18

Statistical Inference for Rank Data

Section 18.1 Introduction to Assumption-Freer Tests[1]

So far we have presented procedures for testing hypotheses about a variety of population parameters. In each case it was necessary to assume that the sampled population had a probability distribution of a particular form—usually normal or binomial. For many variables, such an assumption seems warranted. For others, however, we may not know the form of the underlying distribution and may be unwilling to make an assumption about it. The procedures we have presented also required other assumptions. For example, the independent-samples t statistic for testing the hypothesis that $\mu_1 - \mu_2 = 0$ assumes that the population variances are equal. To avoid having to make such assumptions, nonparametric and distribution-free tests have been developed that are assumption-freer—that is, they require less stringent assumptions. Before describing them, let us review the three kinds of distributions about which assumptions are made: (1) the sampled population (for example, the population of the observation statistic X), (2) the sampling distribution of the descriptive statistic used in the test (for example, $\bar{X}$), and (3) the sampling distribution of the test statistic (for example, t). Nonparametric and distribution-free tests are assumption-freer with respect to the distribution of the sampled population. *A statistical test is* **nonparametric** *if it does not test a hypothesis concerning one of the parameters of the sampled population. It is* **distribution-free** *if it makes no assumptions about the precise form of the sampled population.* Most distribution-free tests, however, do assume that the sampled population is continuous. Although the distinction between nonparametric and distribution-free tests seems clear enough, in practice the distinction is frequently blurred. Consequently, many statisticians use the terms interchangeably. We will follow Ury's (1967) lead and denote both kinds of tests by the more descriptive label *assumption-freer tests.*

Several assumption-freer tests have already been described. The chi-square tests in Chapter 17, for example, are appropriately called assumption-freer tests for frequency data. The two tests to be described in this chapter are assumption-freer tests for rank data.

Assumption-freer tests differ from parametric tests in a number of important respects. We will mention one difference now and defer a complete discussion to

---

[1] For a more detailed introduction see Bradley (in Kirk, 1972a). Sections of this chapter were influenced by Bradley's article.

Section 18.4. Parametric tests utilize the magnitude information contained in observations (scores), but assumption-freer tests ignore this information. Instead, they use either the frequency with which observations occur, as in the case of chi-square, or their rank (ordinal position). One consequence of focusing on the categorical or ordinal information contained in observations has already been mentioned: we can avoid having to make assumptions regarding the shape of the sampled population. But as we will see, this freedom is bought at a price. Assumption-freer tests tend to be less efficient than parametric tests when the assumptions of the parametric tests are fulfilled. A happier consequence is that assumption-freer tests require less sophisticated measurement procedures. One of the simplest measurement procedures is ranking people or objects with respect to a characteristic. We observe, for example, that John is a better quarterback than Fred, who is better than Elmer; this piece of pie looks better than that one; or Jane is more resourceful than Dennis. It is convenient to use numbers to denote rank order. For example, John is number one; Fred is two; and Elmer is three. However, the numbers don't reflect the magnitude of the difference in quarterbacking skill between John and Fred or whether the difference is the same as that between Fred and Elmer. Presumably, measuring instruments could be devised that would assign numbers that reflect the magnitude of differences in quarterbacking skill, pie attractiveness, and resourcefulness. This has been done in the area of intellectual assessment; the measuring instrument is an intelligence test. But such instruments are not easy to develop, which is why experimenters frequently resort to counting or ranking.

Two test statistics that utilize only the ordinal information contained in observations are described in Sections 18.2 and 18.3. They are regarded as assumption-freer alternatives to the independent and dependent two-sample t tests for means. There are many other assumption-freer tests. The interested reader should refer to Bradley (1968) and Marascuilo and McSweeney (1977).

Section 18.2   Mann–Whitney U Test for Two Independent Samples

The Mann–Whitney U test[2] is used to test the hypothesis that two population distributions are identical. It assumes that the populations are continuous and that random samples have been drawn from each. The test statistic is based on the ranks of observations rather than on their numerical values, and hence it is appropriate for most data in the behavioral sciences and education. Because of the U test's modest assumptions, it is used widely as an assumption-freer alternative to the two-sample t test for independent samples.

Computational Procedure for Small Samples

The computational procedure described here for the U test can be used when each sample size is 20 or less; for larger samples, a normal approximation procedure illustrated later can be used. Suppose an experiment was performed to deter-

---

[2] The test was originally developed by Frank Wilcoxon in 1945 and was called the Wilcoxon rank-sum test. Since then, various forms of the test have appeared—a form by Festinger in 1946, the Mann–Whitney form in 1947, and a form by White in 1952.

mine whether observing aggression affects amount of aggressive behavior. A sample of $n = 21$ six-year-old girls was randomly assigned to one of two conditions: viewing a simulated television program containing numerous aggressive acts, the experimental condition, and viewing a program without aggression, the control condition. Following the television viewing, each girl was observed at play, and her aggressive acts were counted.

Table 18.2-1.   Computational procedure for Mann–Whitney U test

(i) *Data:* Samples are treated as one combined sample, and the scores are ranked from 1 to n, with the smallest receiving a rank of 1. Two or more scores with the same value (tied scores) are assigned the mean of the ranks they would have received. For example, the two scores of 13 would have received ranks of 12 and 13; instead, they both receive the mean rank of 12.5.

Number of aggressive acts for experimental group	R_1	Number of aggressive acts for control group	R_2
2	3	8	8
19	17	1	2
13	12.5	0	1
9	9	10	10
17	15	20	18
18	16	5	6
24	21	11	11
15	14	7	7
22	20	3	4
21	19	13	12.5
		4	5
	$\sum R_1 = 146.5$		$\sum R_2 = 84.5$

Computational check:

$$\sum R_1 + \sum R_2 = \frac{n(n + 1)}{2}$$

$$146.5 + 84.5 = \frac{21(21 + 1)}{2} = 231$$

(ii) *Computation of U:*

$$U = \text{Smaller of} \begin{bmatrix} n_1 n_2 + \dfrac{n_1(n_1 + 1)}{2} - \sum R_1 \\[2mm] n_1 n_2 + \dfrac{n_2(n_2 + 1)}{2} - \sum R_2 \end{bmatrix}$$

$$U = \text{Smaller of} \begin{bmatrix} (10)(11) + \dfrac{10(10 + 1)}{2} - 146.5 = 18.5 \\[2mm] (10)(11) + \dfrac{11(11 + 1)}{2} - 84.5 = 91.5 \end{bmatrix} = 18.5$$

To be significant at α level of significance, U must be less than the critical value, $U_{\alpha/2;n_1,n_2}$, in Table D.10. This value is $U_{.05/2;10,11} = 26$. Since $U < U_{\alpha/2;n_1,n_2}$, the null hypothesis is rejected.

The experimental data are given in Table 18.2-1, along with the computational procedure for the U test. For purposes of the analysis, the data for the experimental and control groups are treated as one sample. The scores, frequency of aggressive acts for each girl, are ordered from the smallest to the largest. The first n integers are then substituted for the scores, with the smallest score receiving a rank of 1 and the largest, a rank of n. If two or more observations have the same value (tied scores), they are assigned the mean of the ranks they would have occupied. The ranks associated with the experimental and control groups are then added separately. The Mann–Whitney test statistic, U, is based on the smaller of the two sums of ranks as indicated in part (ii) of the table.

No assumptions regarding the shape of the populations are required, because the test statistic is based not on scores, but instead on their ranks. The null hypothesis tested is that the distribution of aggressive acts for girls in the experimental population is identical to that for girls in the control population. The statistical hypotheses for the experiment can be stated as follows.[3]

H_0: · Population distributions for the experimental and control groups are identical.

H_1: Population distributions are not identical.

The .05 level of significance is adopted. To be significant at the .05 level, the value of U must be less than the critical value, $U_{\alpha/2;n_1,n_2}$, in Appendix Table D.10. For the data in Table 18.2-1, $U = 18.5$ and is less than $U_{.05/2;10,11} = 26$. Hence, we can conclude that the two populations are not identical. Inspection of the data indicates that the girls who watched a television program containing aggression engaged in more aggressive acts than those who didn't.

The U test also can be used to test directional hypotheses if the population distributions are symmetrical. The test statistic U is computed as before. However, U must be less than the one-tailed critical value from Table D.10. In addition, the relative position of the sample distributions must be consistent with the alternative hypothesis. For example, if the alternative hypothesis states that the population distribution for the experimental group is displaced (shifted) above that for the control group, the sample distributions must exhibit a similar displacement.

Computational Procedures when One or Both n's Exceed 20

When one of the samples contains more than 20 scores, a normal approximation procedure can be used. The test statistic is

$$z = \frac{(U + c) - E(U)}{\sigma_U} = \frac{(U + c) - n_1 n_2/2}{\sqrt{(n_1 n_2/12)(n_1 + n_2 + 1)}},$$

where U is defined in Table 18.2-1 and n_1 and n_2 are the two sample sizes. The term c in the formula is a correction for continuity and is equal to 0.5. The decision rule for the z test is as follows: Reject H_0 if z falls in the critical region of the

---

[3] If it can be assumed that the two population distributions are symmetrical, the Mann–Whitney U test is a test of the hypothesis that the population medians are equal.

normal distribution; otherwise, don't reject H_0. Because of the way U is defined, this region is always in the lower tail of the normal distribution regardless of whether the test is one-tailed or two-tailed. To be significant, the z test statistic must be equal to or less than $-z_{\alpha/2}$ for a two-tailed test or $-z_\alpha$ for a one-tailed test.

As noted earlier, if two or more observations have the same value (tied scores), they are assigned the mean of the ranks they would have occupied. The denominator, σ_U, of the z statistic can be corrected for ties; the corrected formula is

$$\sigma_U = \sqrt{\frac{n_1 n_2}{12}(n_1 + n_2 + 1)\left[1 - \frac{\sum(t_i^3 - t_i)}{(n_1 + n_2)^3 - (n_1 + n_2)}\right]},$$

where t_i is the number of tied observations in a particular set. The term $(t_i^3 - t_i)$ is computed for each set and then is summed for the sets. If $n_1 + n_2$ is large and the number of ties is small, the correction can be ignored.

The normal distribution approximation will be illustrated using the data in Table 18.2-1. This is done for comparison purposes; the approximate test isn't recommended unless n_1 or n_2 exceeds 20. The test statistic is

$$z = \frac{(U + c) - n_1 n_2/2}{\sqrt{(n_1 n_2/12)(n_1 + n_2 + 1)\{1 - (\sum(t_i^3 - t_i)/[(n_1 + n_2)^3 - (n_1 + n_2)])\}}}$$

$$= \frac{(18.5 + 0.5) - (10)(11)/2}{\sqrt{[(10)(11)/12](10 + 11 + 1)\left\{1 - \dfrac{[(2)^3 - 2]}{[(10 + 11)^3 - (10 + 11)]}\right\}}}$$

$$= \frac{-36.00}{\sqrt{201.667(0.999)}} = -2.54.$$

The critical value of z for a two-tailed test at $\alpha = .05$ is -1.96. Since $z < -z_{.05/2}$, the null hypothesis is rejected. The absolute value of the z test statistic is large enough to be significant at the .02 level. A similar conclusion would be reached using the small-sample exact test procedure. As expected, the correction for ties (0.999) had virtually no effect on the test.

We mentioned earlier that assumption-freer tests tend to be less efficient than parametric tests when the assumptions of the parametric tests are fulfilled. Two statistics for comparing the relative efficiency of different tests will be described now.

Two Indexes of Efficiency

We saw in Section 11.4 that power is determined by four factors: (1) level of significance, (2) sample size, (3) population dispersion, and (4) magnitude of the difference between the true and hypothesized parameters. Any desired power can be achieved for a given significance level and true alternative hypothesis by obtaining a sufficiently large sample. If one test statistic requires a smaller sample size to achieve a desired power than another statistic, it is said to be more efficient. *One index for comparing the efficiency of two test statistics when both are* used to test the same null hypothesis at α significance level against the same alternative hypothesis is called **power efficiency.** It is given by

$$Power\ efficiency = \frac{100(n_S)}{n_L},$$

where n_L is the sample size required by test L to equal the power of the more efficient test S, based on n_S observations. Suppose, for example, that test S requires 40 subjects to reject H_0 in favor of H_1 at α significance level with power equal to .90, and that test L requires 80 subjects. The power efficiency, PE, of test L relative to S is

$$PE = \frac{100(40)}{80} = 50\%.$$

The PE index has one serious disadvantage. Its value is determined by the particular values of α, power, H_0, H_1, and the specified sample size of the more efficient comparison test statistic. Mathematicians prefer an index called *asymptotic relative efficiency*, ARE, which doesn't depend on qualifying conditions that vary from one situation to the next. This index assumes that the sampling distributions of the test statistics being compared are normal when their n's are infinitely large and that the power of each test approaches 1 as the sample n's approach infinity. The index is independent of α but is computed for a hypothetical situation, where the qualifying conditions specify that the sample size is infinite, H_1 is essentially identical to H_0, and H_1 is one-tailed. The rationale underlying ARE is beyond the scope of this book.[4] ARE has the advantage of comparing the efficiencies of two test statistics under standard conditions. However, in interpreting a particular value of the index, one should keep in mind its unrealistic qualifying conditions—infinitely large samples and a negligible difference between H_0 and H_1. It turns out that when the two indexes PE and ARE are used to rank order various test statistics, the results are almost identical. The ARE index is preferred for comparing the efficiencies of more than two test statistics, because all comparisons are made under a standard set of conditions. Comparisons among power efficiencies are meaningful only if the qualifying conditions are held constant.

Asymptotic Relative Efficiency of the Mann–Whitney U Test

In general, when the assumptions of parametric tests are met, they are more efficient than assumption-freer tests. This is true for the two-sample t test for independent samples when compared with the Mann–Whitney U test. The two provide tests of the same hypothesis if observations are randomly sampled from a normal population, since in that case $\mu = Mdn_{Pop}$. The asymptotic relative efficiency of the U test in comparison with the t test is 95.5%. Thus, the U test is an excellent alternative to the t test and has two compensating advantages: it is applicable to rank data and it assumes only that random samples have been obtained from continuous populations.

---

[4] Bradley (1968, pp. 56–62) provides a good introduction.

Review Exercises for Sections 18.1 and 18.2

*1. Recognizing that in practice the distinction between nonparametric and distribution-free tests is often blurred, indicate the principal differences between them. How do these tests differ from parametric tests?

2. Compare the assumptions of the Mann–Whitney U test with those for the two-sample t test for independent samples. What are the relative merits of the tests?

*3. The effect on exploratory behavior of administering at random intervals a noxious stimulus, electric shock, to gerbils during infancy was investigated. Animals were randomly assigned to the experimental condition, shock, or the control condition, nonshock. The dependent variable, duration in minutes of exploratory behavior during one day, was measured when the animals were 6 months old. For the data in the table, test the hypothesis that the experimental and control populations are identical. Let $\alpha = .05$.

Control group				Experimental group		
40	42	25	30	32	26	30
33	31	34		16	24	20

4. A paired-associates learning task is one in which subjects are presented with stimulus–response paired items and must learn to give the second item in each pair when the first is presented. The effect on learning of having as the stimulus item a dirty word versus a neutral word was investigated. Subjects were randomly assigned to the conditions. For the recall scores listed in the table, test the hypothesis that the populations are identical. Let $\alpha = .05$.

Dirty stimulus item				Neutral stimulus item			
12	14	16	13	11	10	7	5
10	18	19		8	13	9	

*5. The effect on motor skill development of playing with educational toys for 6 months was investigated. The subjects were 4- and 5-year-olds. Half of the subjects were randomly assigned to the play group; the remaining half didn't play with the toys. For the data in the table, test the hypothesis that the two populations are identical. Let $\alpha = .01$. Perform the test with and without a correction for ties.

Motor skill scores for play group				Motor skill scores for control group			
28	22	19	15	27	18	17	13
26	21	18	14	25	18	17	10
25	20	18	12	23	18	16	8
24	19	17	11	21	17	16	7
23	19	16	9	20	17	16	6

6. The experiment described in Exercise 5 was repeated with 8- and 9-year-old children. For the data in the table, test the hypothesis that the two populations are identical. Let $\alpha = .05$. Perform the test with and without a correction for ties.

Motor skill scores for play group				Motor skill scores for control group			
47	36	43	32	36	23	26	16
46	35	39	31	34	21	25	15
44	34	37	30	32	20	25	14
45	33	37	29	29	19	24	11
44	32	36	27	27	18	24	12
48							

7. In order to measure the effect of time of day on test performance, a multiple-choice general knowledge test was administered to a sample of 24 college students. Half of the students were tested at 7:30 A.M. on Saturday; the other half were tested at 4:00 P.M. on the same day. The two testing times were randomly assigned to the subjects, with the restriction that the subsample n's were equal. The following data representing number of correct answers on the test were obtained. (Suggested by Hughey, Arron Wilson. Effects of scheduling test administration on the academic performance of college students. *Psychological Reports*, 1982, *50*, 1346.)

7:30 A.M. group				4:00 P.M. group			
90	80	89	94	88	91	85	80
90	97	92	95	96	88	86	77
93	82	87	83	84	79	81	78

  a. Compute the medians for the two samples.
  b. Use the Mann–Whitney U statistic to test the hypothesis that the two population medians are equal. Let $\alpha = .05$.
  c. What assumption must be tenable in order to use the Mann–Whitney U statistic to test the hypothesis of equal medians?
*8. What are the qualifying conditions associated with PE and ARE?
 9. Terms to remember:
  a. Nonparametric tests
  b. Distribution-free tests
  c. Assumption-freer tests
  d. Power efficiency
  e. Asymptotic relative efficiency

Section 18.3 Wilcoxon T Test for Dependent Samples

The Wilcoxon matched-pairs signed-ranks test is used to test the hypothesis that two population distributions are identical. It is appropriate for dependent samples. Such samples can result from (1) obtaining repeated measures on the same subjects, (2) using subjects matched on a variable that is known to be correlated with the dependent variable, (3) using identical twins or litter mates, or (4) obtaining pairs of subjects who are matched by mutual selection.[5] The Wilcoxon test assumes that the populations are continuous and that a random sample of paired elements has been obtained. Its ARE, in comparison with the two-sample t test for dependent

---

[5] Procedures for obtaining dependent samples are discussed in Section 13.5.

samples, is 95.5%, when the latter test's assumptions are met. Thus, Wilcoxon's test is an excellent alternative to the t test.

The Wilcoxon test statistic, denoted by T, is based on the rank of the absolute difference between paired observations rather than on the numerical value of the difference. Consequently, the test is appropriate for observations that reflect at least ordinal information.

Computational Procedure for Small Samples

The computational procedure for the T test described here can be used for samples containing 50 or fewer pairs of scores; for larger samples, a normal approximation procedure illustrated later can be used.

Suppose that a test of assertiveness was administered to female college students and the scores were used to form 16 pairs of women matched on assertiveness. One woman in each pair was randomly assigned to participate in an assertiveness training group; the other member of the pair participated in a psychology seminar, the control condition. It was hypothesized that women in the assertiveness training group would become more assertive relative to those in the control group. The statistical hypotheses are the following:

H_0: The population distributions of assertiveness scores are identical for the two groups.

H_1: The population distribution of scores for women in the training group is displaced (shifted) above that for the control group.

The alternative hypothesis calls for a one-tailed test in which the sample distribution for the training group is displaced above that for the control group. The .05 level of significance is adopted. The experimental data are given in Table 18.3-1, along with the computational procedure.

The computations in Table 18.3-1 required for Wilcoxon's T test are easy to perform. The magnitude of the difference between each pair of observations is determined (see column 4). These differences are rank ordered in column 5 in terms of their absolute size; that is, their signs are ignored. The smallest difference receives a rank of 1; the largest is ranked n. Finally, two columns, 6 and 7, containing the ranks, respectively, for the positive and negative differences are formed. The test statistic T is the smaller of the sum of the positive ranks and the absolute value of the negative ranks.

The presence of zero differences requires a computational adjustment. If the number of zero differences is even, each zero difference is assigned the average rank for the set, and then half are arbitrarily given a positive sign and half, a negative sign. If an odd number of zero differences occurs, one randomly selected difference is discarded, and the procedure for an even number of zero differences is followed. When a score is discarded, the sample size, n, is reduced by 1.

To be significant, the T for the data in Table 18.3-1 must be less than the one-tailed critical value $T_{\alpha,n}$ in Appendix Table D.11, and the training group must be displaced above the control group. Inspection of the data indicates that the

Table 18.3-1.   Computational procedure for Wilcoxon T test

(i) *Data:*

(1) Pair	(2) Training group	(3) Control group	(4) Difference	(5) Rank of difference, ignoring sign	(6) Rank associated with positive difference, R_+	(7) Rank associated with negative difference, R_-
1	34	32	2	6.5	6.5	
2	36	26	10	16.0	16.0	
3	31	28	3	8.5	8.5	
4	42	41	1	4.0	4.0	
5	47	47	0	1.5		-1.5^a
6	32	33	-1	4.0		-4.0
7	33	29	4	10.0	10.0	
8	39	41	-2	6.5		-6.5
9	31	26	5	11.0	11.0	
10	34	28	6	12.0	12.0	
11	32	29	3	8.5	8.5	
12	41	41	0	1.5	1.5	
13	35	28	7	13.0	13.0	
14	45	44	1	4.0	4.0	
15	31	23	8	14.0	14.0	
16	33	24	9	15.0	15.0	
	$\sum X_T = 576$	$\sum X_C = 520$			$\sum R_+ = 124.0$	$|\sum R_-| = 12.0$

Computational check:

$$\sum R_+ + |\sum R_-| = \frac{n(n+1)}{2}$$

$$124 + 12 = \frac{16(16+1)}{2} = 136$$

(ii) *Test statistic:*

$$T = \text{Smaller of } \sum R_+ \text{ and } |\sum R_-| = 12.0$$

To be significant at α level of significance, T must be less than the one-tailed critical value, $T_{\alpha,n}$, in Appendix Table D.11 and the training group must be displaced above the control group. From Table D.11, $T_{.05,16} = 35$. Since $T < T_{\alpha,n}$ and the training group is displaced above the control group, the null hypothesis is rejected.

^{a} See text for explanation of why this zero difference is assigned a negative sign.

latter condition is satisfied. The value of the test statistic, T, in Table 18.3-1 is 12 and is less than $T_{.05,16} = 35$. Hence, we can conclude that the two populations are not identical; the training group is displaced above the control group.

Computational Procedures when n Is Greater than 50

A normal approximation procedure can be used when a sample contains more than 50 pairs of scores. The test statistic is

$$z = \frac{(T+c) - E(T)}{\sigma_T} = \frac{(T+c) - n(n+1)/4}{\sqrt{n(n+1)(2n+1)/24}},$$

where T is defined in Table 18.3-1 and n is the number of pairs of scores. The term c in the formula is a correction for continuity and is equal to 0.5. The decision rule for the z test is as follows: Reject H_0 if z is less than $-z_\alpha$; otherwise, don't reject H_0. Because of the way T is defined, the critical region always will be in the lower tail of the normal distribution, whether the test is one-tailed or two-tailed.

If two or more ranks have the same value, they are assigned the mean of the ranks they would have occupied. The denominator, σ_T, of the z statistic can be corrected for ties; the corrected formula is

$$\sigma_T = \sqrt{\frac{n(n + 1)(2n + 1)}{24} - \frac{1}{48} \sum (t_i^3 - t_i)},$$

where t_i is the number of tied observations in a particular set. The term $(t_i^3 - t_i)$ is computed for each set and then is summed for the sets. If n is large and the number of ties is small, the correction can be ignored.

The normal distribution approximation will be illustrated using the data in Table 18.3-1. The procedure provides a satisfactory approximation for $n \geq 8$. The test statistic is

$$z = \frac{(T + c) - n(n + 1)/4}{\sqrt{n(n + 1)(2n + 1)/24 - (1/48) \sum (t_i^3 - t_i)}}$$

$$= \frac{(12 + 0.5) - (16)(16 + 1)/4}{\sqrt{16(16 + 1)[(2)(16) + 1]/24 - (1/48)[(2^3 - 2) + (2^3 - 2) + (2^3 - 2)]}}$$

$$= \frac{-55.5}{\sqrt{374 - 0.375}} = \frac{-55.5}{19.33} = -2.87.$$

The critical value of z for a one-tailed test at $\alpha = .05$ is -1.645. Since $z < -z_{.05}$ and the training group is displaced above the control group, the null hypothesis is rejected. The absolute value of the z test statistic is large enough to be significant at the .003 level. The table of critical values for T doesn't have significance levels beyond .005, but the observed T is small enough to have been significant at this level.

Review Exercises for Section 18.3

10. Compare the assumptions of the Wilcoxon T test with those for the two-sample t test for dependent samples. What are the relative merits of the tests?

*11. The effect on one's sense of well-being of participating in a transactional analysis group was investigated. Subjects completed a questionnaire before and after participating in the group. For the data in the table, test the hypothesis that the two populations are identical. Let $\alpha = .05$. The higher the score, the higher the individual's sense of well-being.

Subject	Score before participation	Score after participation	Subject	Score before participation	Score after participation
1	50	56	8	36	40
2	46	50	9	35	34
3	45	50	10	35	34
4	43	48	11	34	34
5	40	44	12	34	32
6	37	40	13	33	33
7	36	38	14	31	31

12. An experiment was performed to determine the effects of sustained physical activity on hand steadiness. For the data in the table, test the hypothesis that the two populations are identical. Let $\alpha = .01$.

Subject	Steadiness before activity	Steadiness after activity	Subject	Steadiness before activity	Steadiness after activity
1	14	12	9	13	9
2	12	11	10	11	10
3	16	13	11	14	12
4	6	6	12	13	11
5	13	14	13	9	6
6	15	10	14	11	9
7	14	10	15	13	12
8	12	12			

*13. The absolute threshold for a 1000-Hertz tone was investigated under the effects of an hallucinogen, hashish, and of a placebo. The order of administration of the conditions was randomized independently for each subject. For the data in the table, test the hypothesis that the two populations are identical versus the alternative that the placebo population is displaced above the hallucinogen population. Let $\alpha = .05$. Scores are dB re. 0.0002 dyne/cm^2.

Subject	Hashish	Placebo	Subject	Hashish	Placebo
1	4	6	9	0	0
2	0	0	10	5	8
3	1	1	11	-1	-2
4	0	-1	12	6	7
5	0	1	13	2	2
6	1	2	14	1	1
7	3	3	15	4	5
8	-1	0	16	2	4

*14. It has been claimed that college students' grades improve following marriage. To test the hypothesis, the variables of college aptitude, sex, and size of high school attended were used to form matched pairs of students; one had been married for at least two semesters and the other was unmarried. For the data in the table, test the hypothesis that the populations are identical versus the alternative that the married population is

Pair	Married	Unmarried	Pair	Married	Unmarried
1	3.7	3.8	19	3.1	3.1
2	3.4	3.2	20	3.8	3.6
3	2.9	3.1	21	3.6	3.3
4	2.9	2.7	22	3.6	3.8
5	3.0	2.8	23	3.5	3.1
6	2.2	2.3	24	3.1	3.0
7	3.1	2.7	25	2.8	2.9
8	3.2	2.7	26	2.6	2.8
9	3.4	3.4	27	2.5	2.3
10	3.5	3.4	28	3.6	3.4
11	3.3	2.6	29	3.5	3.0
12	2.7	3.1	30	3.4	3.5
13	3.2	3.1	31	3.3	3.6
14	1.8	2.3	32	3.2	3.0
15	3.4	3.0	33	3.5	2.2
16	3.9	3.4	34	3.4	2.6
17	3.3	3.2	35	2.0	2.4
18	3.2	2.6			

displaced above the unmarried population. Use the approximate z test to analyze the data. Let $\alpha = .05$. Don't use the correction for ties.

15. Use the approximate z test to analyze the data in Exercise 12. Use the correction for ties.

*16. Use the approximate z test to analyze the data in Exercise 13. Use the correction for ties.

17. Suppose the two-sample t test required 57 subjects to reject the nondirectional null hypothesis at the .05 level of significance with power equal to .80, and the Wilcoxon T test required 60 subjects. (a) What is the PE of the Wilcoxon test? (b) What are the qualifying conditions associated with your estimate?

Section 18.4 *Comparison of Parametric Tests and Assumption-Freer Tests for Rank Data*

When assumption-freer tests first appeared, they were regarded as no more than quick and dirty substitutes for parametric tests, since their power efficiency was thought to be inferior. Now we have a clearer understanding of the differences between the two kinds of tests, which mainly involve (1) their assumptions, (2) the level of mathematics necessary to understand their rationale, (3) their computational simplicity, and (4) the nature of the hypotheses they test. These differences are examined below.

Most parametric test statistics assume that (1) population elements are randomly sampled and/or experimental conditions are randomly assigned to the elements, (2) the population is normally distributed, and (3) the null hypothesis is true. Of course, the null hypothesis is advanced provisionally in the hope that it can be rejected. A fourth assumption is required by some test statistics if the null hypothesis concerns two or more populations—that the population variances are equal.

Assumption-freer test statistics are much less demanding and hence can be used in situations for which parametric methods are not appropriate. Most assumption-freer procedures assume that (1) the population elements are randomly sampled and/or experimental conditions are randomly assigned to the elements, (2) the sampled population is continuous, which implies that no two population elements have the same value (that is, no tied values)[6], and (3) the null hypothesis is true. It is relatively easy to determine whether the assumptions of assumption-freer tests are satisfied. For example, a sampling procedure is under an experimenter's control; the experimenter knows whether it is random. And one can decide on logical grounds whether or not the population is continuous.

On the other hand, the parametric assumptions of normality and equal variances are more difficult to check, since in any practical situation the population is not available for examination. Statistical tests can be applied to sample data to test these assumptions. However, for the small samples typically used in the behavioral sciences and education, the tests may fail to detect departures from normality and equal variances.

If all the assumptions for parametric tests are met, these tests are more efficient than, or equally efficient as, their assumption-freer counterparts. If, however, the assumptions aren't met, parametric tests are only approximate. This means the probability of making a type I error is not α as specified by the experimenter. Fortunately, parametric tests are relatively insensitive to violation of the normality and equal variances assumptions.[7] Nevertheless, one may prefer to use an assumption-freer test for which the probability of making a type I error is known rather than to rely on an inexact parametric test.

The second major way in which parametric and assumption-freer tests differ is in the level of mathematics necessary to understand their rational. The derivation of parametric tests involves mathematics beyond the training of most researchers in the behavioral sciences and education. Many assumption-freer tests, however, can be derived using high school algebra and elementary probability and counting rules. This is a real plus, since most researchers want to understand the rationale for the procedures they use rather than having to accept their validity and appropriateness on faith.

Third, assumption-freer tests differ from parametric tests in being easier to apply. For example, the chi-square tests discussed in Chapter 17 use the simplest kind of measurement—counting the number of observations in categories—and a test statistic that is easy to compute. The tests in the present chapter also use a simple measuring procedure—ranking—and statistics that are easy to compute.

The fourth difference is in the nature of the hypothesis tested. Two-sample parametric methods, for example, test hypotheses about particular population parameters; most assumption-freer methods test hypotheses about equality of population distributions. As we have seen, populations can differ in a number of ways, such as central tendency, dispersion, skewness, and kurtosis. To test the hypothesis $\mu_1 - \mu_2 = 0$ using a t statistic, one must assume that the populations have equal variances and are symmetrical and mesokurtic. Thus, to test a hypothesis about one population parameter, we must be willing to make assumptions

---

[6] This follows, since the probability of randomly drawing the same value twice in a finite sample from a continuous population is zero. However, even if the population is continuous, the same value may occur twice in one's sample, because the measuring instrument is calibrated in discrete units.

[7] This point is developed in Sections 10.3 and 12.2.

about other parameters. Assumption-freer tests don't require such assumptions and, accordingly, are less specific in what they tell us.

Section 18.5   Summary

Test statistics are often classified according to whether they are parametric, non-parametric, or distribution-free. The classification scheme is not entirely satis-factory because some tests fall into more than one category. A test is parametric if it tests a hypothesis concerning one of the parameters of the sampled popula-tion and if it requires stringent assumptions regarding the precise form of the sampled population; if not, it is nonparametric. A test is distribution-free if it makes no assumptions about the precise form of the sampled population. The Mann–Whitney U test and the Wilcoxon T test fit into both categories, since they don't require assumptions about the precise form of the sampled population nor do they test hypotheses about parameters of the sampled population. Certainly, a classification scheme is less useful if its categories are not mutually exclusive. Even the parametric–distribution-free distinction is not always clear. For example, many parametric tests that assume the sampled population is normally distributed are approximately distribution-free for very large samples. Little is gained by trying to distinguish between nonparametric and distribution-free tests; it is more accurate to label the two categories collectively as assumption-freer.

The Mann–Whitney U test is used to test the hypothesis that two population distributions are identical. It is an excellent alternative to the two-sample t test for independent samples, since its ARE is 95.5%. The Wilcoxon T test often is used in place of the two-sample t test for dependent samples when the assump-tions of the latter aren't tenable. Although it involves less stringent assumptions, Wilcoxon's T is nearly as efficient as the two-sample t—its ARE is 95.5%. Like the U test, it tests the hypothesis that two population distributions are identical.

The major differences between assumption-freer and parametric tests can be summarized as follows: The assumption-freer methods based on ranks (1) require less stringent assumptions, (2) involve assumptions that are easier to verify, (3) are usually less powerful when the assumptions of corresponding parametric tests are satisfied, (4) are easier to compute, (5) require simpler mathematical proce-dures for their derivation and understanding, (6) usually test hypotheses about population distributions instead of parameters, and (7) utilize information regard-ing rank order instead of the numerical value of individual observations.

Review of
Basic Mathematics

This appendix provides a brief review of selected arithmetic and algebraic concepts.[1] You have, no doubt, been exposed to this material in the past, but chances are you have forgotten some of it. If so, this review should help refresh your memory.

The following test is designed to appraise your knowledge of basic mathematics and help you pinpoint concepts that you should review. Answers are given at the end of the test, along with references to relevant review sections.

Section A.1   Test of Mathematical Skills

Round the following numbers to three digits.

1. 2.576 _____   2. 100.4 _____
3. 1.645 _____   4. 2.328 _____
5. 15.35 _____   6. 16.25 _____

Perform the following basic operations.

7. $|-3| + |3|$ = _____
8. $-5 + 2$ = _____
9. $3 - 2 + 4 - 8$ = _____
10. $-6 - 3$ = _____
11. $5 - (-1)$ = _____
12. $-9 - (-4)$ = _____
13. $(-2)(-6)$ = _____
14. $10/(-2)$ = _____
15. $0/6$ = _____
16. $9/0$ = _____
17. $(a/b)(n/n)$ = _____
18. $(a/b)^2$ = _____
19. $(2/5)(3/6)$ = _____
20. $(\frac{3}{4})/2$ = _____

[1] For a more detailed review see Baggaley (1969), Bradshaw (1969), Clark and Tarter (1968), Kearney (1970), and Walker (1951)

21. $3/(\frac{4}{2})$ = _____

22. 2^0 = _____

23. $(X)(X^2)$ = _____

24. $(X^2)^3$ = _____

25. 2^{-1} = ( )/( ) _____

26. 3^{-2} = ( )/( )( ) _____

27. $3^2/3^4$ = ( )( ) _____

28. $3\sqrt{15}$ = $\sqrt{(\ )}15$ _____

29. Factor $X^2 - 2XY + Y^2$ = _____

30. Factor $pn - p$ = _____

31. $3!$ = _____

32. $0!$ = _____

Remove the parentheses.

33. $X + (Y - Z)$ _____   34. $X - (Y + C)$ _____

35. $nS(1 - R)^2$ _____   36. $(X - Y)S + M$ _____

Solve the equations and inequalities.

37. $3X - 6 = 12$ $X =$ _____   38. $2a/3 = 6$ $a =$ _____

39. $X = \dfrac{a + b}{n}$ $a =$ _____   40. $z = (X - Y)/S$ $X =$ _____

41. $Y = a + bX$ $X =$ _____   42. $S = R/\sqrt{2n}$ $R =$ _____

43. $S = a\sqrt{1 - b^2}$ $b =$ _____   44. $X = [(n - 1)S]/b$ $S =$ _____

45. $2X - 1 < 3$ $X <$ _____   46. $-3 \le \dfrac{16 - X}{7} \le 5$ $\le X \le$ _____

47. $-z \le \dfrac{X - M}{S} \le z$ $\le M \le$ _____   48. $[(n - 1)S]/b \le X$ $b \ge$ _____

The answers to the skills test follow. The numbers in parentheses refer to the review sections that discuss the principle involved.

Principles discussed in Section A.2 (Rounding Numbers)

1. 2.58   (1)   2. 100   (2)   3. 1.64   (3)
4. 2.33   (1)   5. 15.4   (3)   6. 16.2   (3)

Principles discussed in Section A.3 (Basic Operations)

7. 6   (5b, 6a)   8. -3 (6b)   9. -3 (6c)
10. -9 (7)   11. 6   (7)   12. -5 (7)
13. 12   (8)   14. -5 (8)   15. 0   (9c)
16. Undefined   (9d)   17. a/b (10c-iii)   18. a^2/b^2 (10c-iv)
19. $\frac{6}{30}$ (10c-i)   20. $\frac{3}{8}$ (10d-i)   21. $\frac{6}{4}$ (10d-ii)

22. 1   (11d)   23. X^3 (11b-i)   24. X^6 (11b-ii)
25. $\frac{1}{2}$ or 0.5   (11c-i)   26. $1/3^2$ (11c-i)   27. 3^{-2} (11c-iv)
28. $\sqrt{9(15)}$ (12c)   29. $(X - Y)^2$ (13)   30. $p(n - 1)$ (13)
31. 6   (14a)   32. 1   (14b)

Principles discussed in Section A.4 (Order of Performing Operations)
33. $X + Y - Z$ (17a)   34. $X - Y - C$ (17b)
35. $nS - nSR^2$ (17c)   36. $XS - YS + M$ (17c)

Principles discussed in Section A.5 (Equations)
37. 6   (19c, 20, 21, 22)   38. 9   (19c, 20, 22)
39. $nX - b$ (19c, 20, 21, 22)   40. $zS + Y$ (19c, 20, 21, 22)
41. $(Y - a)/b$ (19c, 20, 21, 22)   42. $S\sqrt{2n}$ (19c, 20, 22)
43. $\sqrt{1 - S^2/a^2}$ (19c, 20, 21, 22)   44. $Xb/(n - 1)$ (19c, 20, 22)

Principles discussed in Section A.6 (Inequalities)
45. 2   (25)   46. $-19 \leq X \leq 37$ (25, 26)
47. $X - zS \leq M \leq X + zS$ (25, 26)   48. $[(n - 1)S]/X$ (25, 27)

Section A.2   Rounding Numbers

The number of significant digits in a number (all digits except 0 when it is used only to position the decimal point) should reflect the precision of a measurement. Therefore, numbers should be rounded to give a correct impression of the measurement precision actually achieved. Rounding involves dropping digits if they are to the right of the decimal or replacing them by 0 if they are to the left of the decimal.

1. When the digit to be dropped is greater than 5 or is a 5 with nonzero digits to the right, the digit to the left of it is increased by 1.
   Examples 246.36 rounded to four significant digits becomes 246.4
   386 rounded to two significant digits becomes 390
   0.0068 rounded to one significant digit becomes 0.007
   6.51 rounded to one significant digit becomes 7
2. When the digit to be dropped is less than 5, no change is made in the digit to the left of it.
   Examples 246.31 rounded to four significant digits becomes 246.3
   384 rounded to two significant digits becomes 380
   0.0063 rounded to one significant digit becomes 0.006
3. When the digit to be dropped is 5 or is 5 with only zeros to the right, the digit to the left of 5 is increased by 1 if it is odd and is not changed if it is even. (Not all hand-held calculators follow this convention; some always increase the digit to the left of 5.)
   Examples 75 rounded to one significant digit becomes 80
   935.35 rounded to four significant digits becomes 935.4
   674.5 rounded to three significant digits becomes 674
   912.5 rounded to three significant digits becomes 912

4. If the final result of a computation is to be rounded to s digits to the right of the decimal, at least $s + 1$ digits to the right should be retained in intermediate computational steps.

   Example $(115 - 100)/(14/\sqrt{26}) = 5/(14/5.099) = 5/2.746 = 1.82$

Section A.3   Basic Operations

5. Numbers
   a. A signed number, for example, 4, -2, 9, -11, indicates (1) direction and (2) size. The sign, $+$ or $-$, indicates the direction of movement away from a starting point, 0. Zero has no direction. The number part of the signed number indicates the extent or size of movement away from 0.

   Example The size and direction of movement for -2 and 4 are shown here.

   b. The absolute value of a number, denoted by $|\ \ |$, indicates size but not direction. The absolute value is always positive for nonzero numbers. The general rule is

$$|a| = \begin{cases} a & \text{if} \quad a \geq 0 \\ -a & \text{if} \quad a < 0. \end{cases}$$

   The second part of the rule appears contradictory but isn't. If a is less than 0, it is a negative number, and a minus sign in front of a negative number makes it positive.

   Examples $|3| = 3$; $|-3| = 3$; $|0| = 0$; $|-1.96| = 1.96$;
   $|X| = X$ if X is a positive number, and $|X| = -X$ if X is a negative number.

6. Addition
   a. Two numbers of like sign: add the absolute values of the numbers and attach the common sign to the sum.

   Examples $3 + 2 = 5$; $-3 + (-2) = -5$

   b. Two numbers of unlike sign: determine the difference between their absolute values and attach the sign of the larger number.

   Examples $3 + (-2) = 1$; $-3 + 2 = -1$

   c. More than two numbers with unlike signs: add the absolute values of the positive numbers, as in Rule 6a, and do the same for the negative numbers; then determine the difference between their absolute values and attach the sign of the larger of the two, as in Rule 6b.

   Example $3 + 2 + (-5) + (-3) + (-1) = 5 + (-9) = -4$

7. Subtraction

   To subtract one number from another, change the sign of the number to be subtracted and proceed as in addition.

   $$\begin{aligned} Examples \quad 3 - (2) &= \quad 3 + (-2) = \quad 1 \\ 3 - (-2) &= \quad 3 + 2 \quad = \quad 5 \\ -3 - (2) &= -3 + (-2) = -5 \\ -3 - (-2) &= -3 + 2 \quad = -1 \end{aligned}$$

8. Multiplication and division

   Multiplying two numbers together results in another number called a *product*; dividing one number by another results in another number called a *quotient*. When two numbers have like signs, their product and quotient are positive; when they have unlike signs, their product and quotient are negative.

   $$\begin{aligned} Examples \quad (6)(3) &= \quad 18 & 6/3 &= \quad 2 \\ (6)(-3) &= -18 & 6/-3 &= -2 \\ (-6)(3) &= -18 & -6/3 &= -2 \\ (-6)(-3) &= \quad 18 & -6/-3 &= \quad 2 \end{aligned}$$

9. Operations with zero

   a. If 0 is added to or subtracted from any number, the result is the number itself.

      Examples $3 + 0 = 3$; $9 - 0 = 9$

   b. The product of 0 and any other number is equal to 0.

      Examples $(3)(0) = 0$; $(2)(3)(0) = 0$

   c. $0/a = 0$ for all nonzero values of a.

      Examples $\frac{0}{3} = 0$; $\frac{0}{7} = 0$

   d. The use of 0 as a divisor results in a fraction that cannot be evaluated.

      Example $\frac{5}{0}$ is undefined

10. Fractions

    a. A fraction, for example, 6/2 or a/b, is the result of dividing one number or expression by another. The upper part of the fraction is called the *numerator*; the lower part (the divisor) is called the *denominator*.

    b. Addition and subtraction

       i. To add or subtract fractions with the same denominator, perform the indicated operation on the numerator and leave the denominator unchanged.

          $$Examples \quad \frac{a}{c} + \frac{b}{c} = \frac{a+b}{c}; \quad \frac{a}{c} - \frac{b}{c} = \frac{a-b}{c}; \quad \frac{2}{3} + \frac{4}{3} = \frac{6}{3}; \quad \frac{2}{3} - \frac{4}{3} = \frac{-2}{3}$$

       ii. To add or subtract fractions with different denominators, find a common denominator, change all fractions accordingly, and proceed as above. Some multiple of all the original denominators is selected as the common denominator; each new numerator is formed by multiplying the original numerator by the number of times the original denominator divides into the common denominator.

          $$Examples \quad \frac{a}{b} + \frac{c}{b} = \frac{ad}{bd} + \frac{bc}{bd} = \frac{ad+bc}{bd}; \quad \frac{1}{2} + \frac{2}{3} = \frac{3}{6} + \frac{4}{6} = \frac{7}{6}$$

iii. In general, if the same quantity is added to or subtracted from both the numerator and the denominator, the value of the fraction is changed.

Examples $\dfrac{a}{b} \neq \dfrac{a+n}{b+n}$ unless $a = b$ or $n = 0$;

$$\frac{3}{4} \neq \frac{3+2}{4+2} = \frac{5}{6}$$

c. Multiplication
  i. To multiply two or more fractions, multiply their numerators together and their denominators together to obtain, respectively, the numerator and the denominator of the product.

Examples $\left(\dfrac{a}{b}\right)\left(\dfrac{c}{d}\right) = \dfrac{ac}{bd}$; $\left(\dfrac{2}{3}\right)\left(\dfrac{3}{4}\right) = \dfrac{(2)(3)}{(3)(4)} = \dfrac{6}{12}$

  ii. Multiplying anything by 1 leaves its value unchanged.

Examples $\dfrac{a}{b}(1) = \dfrac{a}{b}$; $\dfrac{3}{2}(1) = \dfrac{3}{2}$

  iii. Multiplying both the numerator and the denominator of a fraction by the same quantity other than 0 does not change its value.

Examples $\left(\dfrac{a}{b}\right)\left(\dfrac{n}{n}\right) = \dfrac{an}{bn} = \dfrac{a}{b}$; $\left(\dfrac{3}{4}\right)\left(\dfrac{2}{2}\right) = \dfrac{(3)(2)}{(4)(2)} = \dfrac{3}{4}$

  iv. In general, squaring a fraction or taking its square root changes its value.

Examples $\dfrac{a}{b} \neq \left(\dfrac{a}{b}\right)^2$ and $\dfrac{a}{b} \neq \dfrac{\sqrt{a}}{\sqrt{b}}$ unless $a = b$;

$$\frac{4}{9} \neq \left(\frac{4}{9}\right)^2 = \frac{16}{81} \quad \text{and} \quad \frac{4}{9} \neq \frac{\sqrt{4}}{\sqrt{9}} = \frac{2}{3}$$

d. Division
  i. To divide a fraction by a quantity, multiply the denominator of the fraction by that quantity.

Examples $\dfrac{a/b}{c} = \dfrac{a}{bc}$; $\dfrac{2/3}{4} = \dfrac{2}{(3)(4)} = \dfrac{2}{12}$

  ii. To divide a quantity by a fraction, invert the fraction and multiply.

Examples $\dfrac{a}{b/c} = a\,\dfrac{c}{b} = \dfrac{ac}{b}$; $\dfrac{2}{3/4} = 2\left(\dfrac{4}{3}\right) = \dfrac{8}{3}$

iii. To divide a fraction by another fraction, invert the second fraction and multiply.

$$\text{Examples} \quad \frac{a/b}{c/d} = \left(\frac{a}{b}\right)\left(\frac{d}{c}\right) = \frac{ad}{bc}; \quad \frac{2/3}{4/5} = \left(\frac{2}{3}\right)\left(\frac{5}{4}\right) = \frac{10}{12}$$

iv. Any quantity (other than 0) divided by itself equals 1.

   Examples $a/a = 1$ if $a \neq 0$; $3/3 = 1$

v. Dividing any quantity by 1 leaves its value unchanged.

$$\text{Examples} \quad \frac{a/b}{1} = \frac{a}{b}; \quad \frac{3/2}{1} = \frac{3}{2}$$

11. **Exponents**

 a. The number of times a number, the *base*, is multiplied by itself is denoted by a superscript, the *exponent*.

   Examples $a^1 = a$; $a^2 = (a)(a)$; $a^3 = (a)(a)(a)$

 b. Laws of positive exponents

  i. $a^n a^m = a^{n+m}$ Example $(2)^2(2)^3 = 2^{2+3} = 2^5$

  ii. $(a^n)^m = a^{nm}$ Example $(2^2)^3 = 2^{(2)(3)} = 2^6$

  iii. $a^n b^n = (ab)^n$ Example $(2)^2(4)^2 = [(2)(4)]^2 = 8^2$

 c. Laws of negative exponents and mixed exponents

  i. $a^{-1} = \dfrac{1}{a}$ Example $2^{-1} = \dfrac{1}{2}$

   $a^{-2} = \dfrac{1}{a^2}$ Example $2^{-2} = \dfrac{1}{2^2}$

   $a^{-n} = \dfrac{1}{a^n}$

  ii. $(a^n)^{-1} = (a^{-1})^n = a^{(-1)(n)} = a^{-n}$

   Example $(2^3)^{-1} = 2^{(3)(-1)} = 2^{-3}$

  iii. $\left(\dfrac{a}{b}\right)^n = a^n\left(\dfrac{1}{b}\right)^n = a^n(b^{-1})^n = a^n b^{-n} = \dfrac{a^n}{b^n}$

   Example $\left(\dfrac{2}{4}\right)^2 = \dfrac{(2)^2}{(4)^2}$

  iv. $\dfrac{a^n}{a^m} = a^n a^{-m} = a^{n-m}$

   Example $\dfrac{(2)^2}{(2)^3} = 2^{2-3} = 2^{-1}$

 d. A number not equal to 0 with a zero exponent, a^0, is equal to 1. This follows, since according to 11c-iv, $a^n/a^n = a^{n-n} = a^0$, and according to 10d-iv, $a^n/a^n = 1$; therefore $a^0 = 1$.

   Examples $a^0 = 1$ for $a \neq 0$; $2^0 = 1$, $5^0 = 1$

e. Do not confuse exponents and coefficients. When multiplying terms with coefficients and exponents, add exponents and multiply coefficients.
Examples $(3X^2)(2X^3) = 6X^5$; $(3X)(5X^2) = 15X^3$

12. Radicals

a. A radical is used to indicate a specific root of a quantity, as, for example, in the expression $b = \sqrt[n]{a}$; b is said to be the nth root of a, $\sqrt{}$ is a radical sign, and a is called the radicand. If n is not specified, it is understood to equal 2, in which case $b = \sqrt{a}$ is called the square root of a.

b. To multiply two radicals if both radicands are positive, multiply their radicands under one radical. Similarly, to divide one radical by another, divide (under one radical) the radicand of the first by the radicand of the second.

Examples $\sqrt{a}\sqrt{b} = \sqrt{ab}$; $\sqrt{2}\sqrt{3} = \sqrt{6}$
$\sqrt{a}/\sqrt{b} = \sqrt{a/b}$; $\sqrt{2}/\sqrt{3} = \sqrt{2/3}$

c. To multiply or divide a radical of the nth order by a number a not equal to 0, place the number raised to the nth power under the radical and multiply or divide the radicand by it.

Examples $a\sqrt{b} = \sqrt{a^2b}$; $2\sqrt{3} = \sqrt{4(3)}$; $2\sqrt[3]{3} = \sqrt[3]{(2)^3(3)}$
$\sqrt{a}/c = \sqrt{a/c^2}$; $\sqrt{3}/2 = \sqrt{3/4}$

d. Note that the nth root of a sum is not equal to the sum of the respective nth roots.

Examples $\sqrt{a^2 + b^2} \neq \sqrt{a^2} + \sqrt{b^2}$; $\sqrt{9 + 16} = 5 \neq \sqrt{9} + \sqrt{16} = 7$

13. Factoring

Factoring an expression consists of dividing it into smaller terms or expressions that, when multiplied, will yield the original expression.
Examples $ab - a = a(b - 1)$
$a^2 - b^2 = (a + b)(a - b)$
$a^2 - 5a - 6 = (a + 1)(a - 6)$
$a^2 - 5a + 6 = (a - 2)(a - 3)$

14. Factorials

a. The product of the first n natural numbers (positive integers) is called n *factorial* and is denoted by $n!$, which equals $n(n - 1)(n - 2) \cdots (3)(2)(1)$.
Example $4! = 4(4 - 1)(4 - 2)(4 - 3) = (4)(3)(2)(1) = 24$

b. For $n = 0$, $0!$ is defined as 1.

Section A.4   Order of Performing Operations

15. The order in which numbers are added does not affect the result.
Examples $a + b = b + a$; $2 + 3 = 3 + 2 = 5$;
$(a + b) + c = a + (b + c)$; $(2 + 3) + 5 = 2 + (3 + 5) = 10$

$a + b = b + a$ illustrates the *commutative law of addition.*

$(a + b) + c = a + (b + c)$ illustrates the *associative law of addition.*

16. The order in which numbers are multiplied does not affect the result.

    Examples $ab = ba$; $(2)(3) = (3)(2) = 6$; $(ab)c = a(bc)$;

    $\quad\quad\quad [(2)(3)]5 = 2[(3)(5)] = 30$

    $ab = ba$ illustrates the *commutative law of multiplication.*

    $(ab)c = a(bc)$ illustrates the *associative law of multiplication.*

17. Parentheses ( ), braces { }, brackets [ ], and the radical $\sqrt{\quad}$ indicate that the enclosed expression is to be treated as a single number. The bar of a fraction has a similar effect; the numerator and the denominator are treated as single numbers.

    Examples $10(16 - 14) = 10(2) = 20$; $(3 - 1)4.21 = (2)4.21 = 8.42$;

    $$\frac{2 + 4}{3 - 1} = \frac{6}{2} = 3$$

    a. When a plus sign precedes parentheses, the parentheses may be removed without changing the signs of terms within the parentheses.

       Examples $a + (b + c) = a + b + c$; $2 + (3 + 5) = 2 + 3 + 5 = 10$

    b. If a minus sign precedes parentheses and the parentheses are removed, the sign of every term within the parentheses must be changed.

       Examples $(a + b) - (c + d) = a + b - c - d$; $2 - (3 + 5) = 2 - 3 - 5 = -6$

    c. When a quantity within parentheses is to be multiplied by a number, each term within the parentheses must be so multiplied.

       Examples $a(b + c) = ab + ac$;

       $\quad 10(16 - 14) = 10(16) - 10(14) = 160 - 140 = 20$;

       $$10\left(\frac{1}{20} + \frac{6}{24}\right) = \frac{10}{20} + \frac{60}{24} = \frac{1}{2} + \frac{5}{2} = 3$$

    $a(b + c) = ab + ac$ illustrates the *distributive law.*

18. Unless specifically altered, for example by parentheses, the order for performing operations is as follows: first, exponentiation (raising a number to a power); next, multiplication and division; and last, addition and subtraction.

    Examples $Mdn = 44.5 + 10\left(\dfrac{30/2 - 14}{3}\right) = 44.5 + 10\left(\dfrac{15 - 14}{3}\right)$

    $$= 44.5 + 10\left(\frac{1}{3}\right) = 44.5 + 3.33 = 47.8;$$

    $$S = \sqrt{\frac{10(32) - (16)^2}{(10)^2}} = \sqrt{\frac{320 - 256}{100}} = \sqrt{\frac{64}{100}} = 0.8;$$

    $$F = \frac{320/(3 - 1)}{1350/[3(10) - 3]} = \frac{320/2}{1350/27} = \frac{160}{50} = 3.2$$

Section A.5   Equations

19. An equation is a statement asserting that what is on the left side of the equals sign is equal to what is on the right.

    Examples $2 + 4 = 6$ is an example of an *arithmetic equation*, since it contains only numbers.

    $2X - 5 = 7$ is an example of an *algebraic equation*, since it contains a symbol.

    a. An equation in which both sides have the same numerical value or one that is true for all values of the variables employed is called an *identity*.

    Examples $2 + 4 = 6$; $3a + 4a = 7a$

    One that is true only when certain values are substituted for variables is called a *conditional equation*.

    Examples $2X = 6$; $3X - 4 = X + 6$

    b. To solve for an unknown in an algebraic equation, find the set of values (called *roots*) that, when substituted for the unknown, makes the two sides of the equation numerically equal. For $2X - 5 = 7$, the root is 6, since $2(6) - 5 = 7$—that is, $7 = 7$.

    c. To find the roots of an equation, perform a series of manipulations that place the unknown alone on the left side (see Rules 20–23).

    Example $2X - 5 = 7$

    $2X - 5 + 5 = 7 + 5$ Add 5 to both sides (see Rule 21)

    $$\frac{2X}{2} = \frac{7 + 5}{2}$$ Divide both sides by 2 (see Rule 22)

    $$X = \frac{12}{2}$$

    $$X = 6$$

20. An operation performed on one side of an equation must also be performed on the other. The condition of equality is not affected by
    a. adding the same quantity to both sides,
    b. subtracting the same quantity from both sides,
    c. multiplying both sides by the same quantity,
    d. dividing both sides by the same nonzero quantity,
    e. raising both sides to the same power if both sides have the same sign,
    f. taking the same root of both sides if both sides have the same sign.

21. Any term on one side of an equation may be transposed to the other side by changing its sign. In essence, the term to be transposed is either added to or subtracted from both sides of the equation.

    Example Solving for *a*:

    $$a + b = c$$
    $$a + b - b = c - b$$
    $$a = c - b$$

22. A quantity that multiplies one side of an equation may be transposed to divide the other side or vice versa. In essence, both sides of the equation are subjected to the same operation—either multiplication or division.

Examples Solving for *a*:   Solving for *X*:

$$ab = c$$

$$\left(\frac{X - a}{S}\right)b + c = z$$

$$\frac{ab}{b} = \frac{c}{b}$$

$$\left(\frac{X - a}{S}\right)b = z - c$$

$$a = \frac{b}{c}$$

$$\frac{X - a}{S} = \frac{z - c}{b}$$

$$X - a = \left(\frac{z - c}{b}\right)S$$

$$X = \left(\frac{z - c}{b}\right)S + a$$

23. When each side of an equation consists of a fraction, the fractions can be removed by cross-multiplying as follows.

Example $\dfrac{a}{b} = \dfrac{c}{d}$

$$ad = bc$$

Section A.6   Inequalities

24. Two or more expressions connected by one of the ordering symbols $<$, $>$, $\leq$, or $\geq$ is an *inequality*.
Examples $a < b$; $-1.96 \leq t \leq 1.96$; $z > 2.576$

25. The solutions to an inequality are not affected by the following operations:
    a. Adding the same quantity to both sides.
    b. Subtracting the same quantity from both sides.
    c. Multiplying both sides by a positive quantity.
    d. Dividing both sides by a positive quantity.
       Example $4X + 2 \geq 10$
       $$4X \geq 10 - 2$$
       $$X \geq 2$$

26. If both sides of an inequality are multiplied or divided by the same negative number, a new inequality is formed, with direction opposite to that of the original.
    Examples $-a < b$ $-3 < 2$
                $-1(-a < b)$ $-1(-3 < 2)$
                   $a > -b$ $3 > -2$

Find μ if $-3 \le \dfrac{2 - \mu}{5} \le 8$.

$$-3(5) \le 2 - \mu \le 8(5)$$ Multiply each member by 5

$$-15 - 2 \le \quad -\mu \le 40 - 2$$ Subtract 2 from each member

$$-1(-17 \le \quad -\mu \le 38)$$ Multiply by -1

$$17 \ge \quad \mu \ge -38$$

$$-38 \le \quad \mu \le 17$$ Rearrange terms

27. Taking the reciprocal of, or inverting, all expressions in an inequality results in a new inequality with direction opposite to that of the original.

Examples $a > b$ implies that $\dfrac{1}{a} < \dfrac{1}{b}$ if $ab > 0$.

$$a \le \frac{(n-1)b}{c} \le d$$

$$\frac{1}{a} \ge \frac{c}{(n-1)b} \ge \frac{1}{d}$$

Glossary of Symbols

Mathematical Symbols

Symbol	Example	Meaning[1]
$+$	$X + Y$	X and Y are added (A.3)
$-$	$X - Y$	Y is subtracted from X (A.3)
( )( ), $\times$	$(X)(Y)$, $X \times Y$, or XY	X and Y are multiplied (A.3)
$/$, $\div$	X/Y or $X \div Y$	X is divided by Y (A.3)
$=$	$X = Y$	X is equal to Y (A.5)
$\neq$	$X \neq Y$	X is not equal to Y
$\simeq$, $\cong$	$X \simeq Y$, $X \cong Y$	X is approximately equal to Y
$>$	$X > Y$	X is greater than Y (A.6)
$\geq$	$X \geq Y$	X is greater than or equal to Y (A.6)
$<$	$X < Y$	X is less than Y (A.6)
$\leq$	$X \leq Y$	X is less than or equal to Y (A.6)
$\leq \leq$	$W \leq X \leq Y$	X is greater than or equal to W and less than or equal to Y (A.6)
$\sum\limits_{i=1}^{n}$	$\sum\limits_{i=1}^{n} X_i$	Sum of X_i, letting i equal $1-n$ (3.3, 3.9)
$\ldots$	$1, 2, 3, \ldots, 6$	Continue the pattern—that is, 1, 2, 3, 4, 5, 6 in this case
$\vert\ \vert$	$\vert X \vert$	Absolute value of X; for $X = 0$, $\vert X \vert = 0$, and for $X \neq 0$, $\vert X \vert$ is equal to the positive member of the couple X, $-X$ (A.3)
$\sqrt{\ }$	$\sqrt{X}$	Square root of X (A.3)
$!$	$n!$	n factorial, $n(n-1)(n-2)\cdots(3)(2)(1)$ (A.3)

Greek Letters

Symbol	Meaning
α (alpha)	Probability of a type I error (11.4); significance level (11.2)
β (beta)	Probability of a type II error (11.4)

[1] The letter/number in parentheses refers to the section in which the symbol is discussed.

Symbol	*Meaning*
α_j	Treatment effect in the jth population (16.2)
δ_0 (delta)	Value of the difference between two population parameters specified by the null hypothesis (13.2)
ε_{ij} (epsilon)	Error effect for subject i in treatment level j (16.2)
η^2 (eta)	Correlation ratio (5.6)
θ (theta)	Population parameter (10.3)
μ (mu)	Population mean (3.3)
$\mu_j, \mu_{j'}$	Means for populations j and j', $j \neq j'$ (16.1)
μ_0	Value of the population mean specified by the null hypothesis (11.2)
$\mu_{\bar{X}}$	Mean of means (10.3)
v (nu)	Degrees of freedom (12.2)
π (pi)	Ratio of the circumference of a circle to the diameter; approximately 3.1416 (10.1)
ρ (rho)	Pearson product-moment population correlation parameter (5.2)
$\sum$ (sigma)	Summation (3.3)
$\sum R_1$ and $\sum R_2$	Sum of ranks, respectively, for variables 1 and 2 (18.2)
$\sum R_+$ and $\sum R_-$	Sum of ranks, respectively, for positive and negative ranks (18.3)
σ (*sigma*)	Population standard deviation (4.2); a caret over the symbol, $\hat{\sigma}$, denotes an estimate of σ (4.2).
σ_{Mdn}	Standard error of a median (10.3)
σ_S	Standard error of a standard deviation (10.3)
$\sigma_{\bar{X}}$	Standard error of a mean (10.3); a caret over the symbol, $\hat{\sigma}_{\bar{X}}$, denotes an estimate of $\sigma_{\bar{X}}$ (12.2)
$\hat{\sigma}_{\bar{X}_D}$	Estimate of the standard error of the mean of difference scores (13.5)
$\sigma_{\bar{X}_1 - \bar{X}_2}$	Standard error of the difference between means (13.2)
$\hat{\sigma}_{Y \cdot X}$	Estimate of the population standard error of estimate (6.3)
σ^2	Population variance (4.2); a caret over the symbol, $\hat{\sigma}^2$, denotes an estimate of σ^2 (4.2)
σ_0^2	Value of the population variance specified by the null hypothesis (12.3)
σ_ε^2	Population error variance (16.2)
$\hat{\sigma}_{est}^2$	Estimate of the population variance (10.5)
σ_U	Standard error of U (18.2)
σ_T	Standard error of T (18.3)
$\hat{\sigma}_{Pooled}^2, \hat{\sigma}_{Pd}^2$	Weighted estimate of the population variance (13.4)
ϕ' (phi)	Cramer's measure of association; also denoted by V (17.3); a caret over the symbol, $\hat{\phi}'$, denotes an estimate of ϕ' (17.3)
χ^2 (chi)	Chi-square random variable (12.3); Pearson's chi-square random variable (17.2)

Symbol	*Meaning*
$\chi^2_{\alpha,v}$ and $\chi^2_{1-\alpha,v}$	Values that cut off, respectively, the upper and lower α regions of the sampling distribution of χ^2 for v degrees of freedom (12.3)
ψ (psi)	Contrast among population means (16.5); a caret over the symbol, $\hat{\psi}$, denotes an estimate of ψ (16.5)

English Letters

Symbol	*Meaning*
ARE	Asymptotic relative efficiency (18.2)
$a_{Y \cdot X}$	Y intercept of a line (6.2)
$b_{Y \cdot X}$	Sample coefficient of linear regression of Y on X (6.2)
c_j	Coefficient of a linear contrast (16.5)
$_nC_r$	Combination of n objects taken r at a time (8.1)
c	A constant (3.9); number of qualitative categories (4.2, 17.4); correction for continuity (18.2)
Cum f	Cumulative frequency (2.2)
Cum $\% f$	Cumulative percentage frequency (2.2)
Cum prop. f	Cumulative proportionate frequency (2.2)
D	Index of dispersion (4.2)
D_i	Difference between scores for the ith pair of elements (13.5)
DP	Number of distinguishable pairs (4.2)
DP_{max}	Number of distinguishable pairs when observations are equally divided among categories (4.2)
d	Effect size (11.4)
df	Degrees of freedom; also denoted by v (12.2)
E	Index of forecasting efficiency (6.4)
E_i	ith event (7.2); sample point for the ith event (7.2)
E_j	Expected frequency in the jth category (17.1)
$E(\chi_v^2)$	Expected value of a chi-square random variable (12.3)
$E(\theta)$	Expected value of an estimator (10.3)
$E(\hat{\sigma}^2)$	Expected value of $\hat{\sigma}^2$ (10.5)
$E(S^2)$	Expected value of S^2 (10.5)
$E(T)$	Expected value of T in the Wilcoxon T statistic (18.3)
$E(MS_{BG})$ and $E(MS_{WG})$	Expected value of MS_{BG} and MS_{WG} (16.2)
$E(U)$	Expected value of U in the Mann–Whitney U statistic (18.2)
$E(X)$	Expected value of X (9.2)
e	Base of the system of natural logarithms; approximately 2.7183 (10.1); number of distribution parameters estimated (16.2)

Symbol	*Meaning*
e_i	Prediction error for the ith element; difference between the observed and predicted scores (6.2)
F	F random variable (14.1); value of Scheffé's test statistic (16.5)
$F_{\alpha; v_1, v_2}$ and $F_{1-\alpha; v_1, v_2}$	Values that cut off, respectively, the upper and lower α regions of the sampling distribution of F (14.1)
F'	Critical value of Scheffé's test statistic (16.5)
f_a and f_b	Number of scores, respectively, above the real upper limit of a class interval and below the real lower limit of a class interval (3.4)
f, f_i, f_j	Frequency of a measurement or event class (2.2)
f_i	Number of scores in the class interval containing a particular statistic (3.4)
$\%f$	Percentage frequency (2.2)
H_0	Null hypothesis (11.1)
H_1	Alternative hypothesis (11.1)
i	Class interval size (2.2); index of summation (3.3)
k	Number of class intervals in a frequency distribution (3.3); number of standard deviation units in Tchebycheff's theorem (4.5); coefficient of alienation (6.4)
k^2	Coefficient of nondetermination (5.4)
Kur	Kurtosis index (4.6)
Mdn	Sample median (3.4)
Mdn_{Pop}	Population median (18.2)
Mo	Sample mode (3.2)
MS_{BG}	Between-groups mean square (16.2)
MS_{WG}	Within-groups mean square (16.2)
n	Number of observations in a sample (2.2); number of trials in a binomial experiment (9.3)
$n!$	n factorial (8.1)
n_A	Number of equally likely events favoring A (7.1)
n_L and n_S	Sample size of tests L and S (18.2)
n_S	Total number of equally likely events (7.1)
O_j	Number of observations in the jth category (17.1)
$_nP_n$	Permutation of n objects taken n at a time (8.1)
$_nP_r$	Permutation of n objects taken r at a time (8.1)
$_nP_{r_1, r_2, \ldots, r_k}$	Permutation of n objects in which $r_1, r_2, \ldots, r_k$ are alike (8.1)
PE	Power efficiency (18.2)
p	Probability of a success (9.3); population proportion (12.4); a caret over the symbol, $\hat{p}$, denotes an estimate of p (12.4); number of levels of treatment A (16.4)

Symbol	Meaning
p_j	Population proportion of observations in the jth category (17.1); a caret over the symbol, $\hat{p}_j$, denotes an estimate of p_j (17.2)
p'_j	Value of the population proportion specified by the null hypothesis (17.2)
p_0	Value of the population proportion specified by the null hypothesis (12.4)
$\hat{p}_{Pooled}, \hat{p}_{Pd}$	Weighted mean of two population proportion estimates (14.2)
$p(A)$	Probability of event A (7.1)
$p(A \mid B)$	Conditional probability of A given B (7.3)
$p(A \text{ and } B)$	Probability of the intersection of events A and B (7.3)
$p(A \text{ or } B)$	Probability of the union of events A and B (7.3)
$p(E_i)$	Probability of event E_i (7.2)
$p(X = r)$	Probability that X is equal to r (9.2)
Prop f	Proportionate frequency (2.2)
Q	Semi-interquartile range (4.2)
Q_1, Q_2, Q_3	First, second, and third quartile points, respectively (4.2)
q	Probability of a failure (9.3); number of levels of treatment B (16.6)
q_0	Value of the population proportion specified by the null hypothesis (12.4)
R	Sample range (4.2)
R_X and R_Y	Ranks, respectively, on variables X and Y (5.7)
r	Sample product-moment correlation coefficient (5.2); number of successes in a binomial experiment (9.3)
r_s	Spearman rank correlation coefficient for a sample (5.7)
r^2	Sample coefficient of determination (5.4)
S	Sample standard deviation (4.2)
S^2	Sample variance (4.2)
S'	Desired sample standard deviation (10.2)
S_{cX}	Standard deviation that has been altered by multiplying each score by a constant (4.8)
Sk	Skewness index (4.6)
S_X and S_Y	Sample standard deviations, respectively, of X and Y (6.2)
$S_X{}^2$ and $S_Y{}^2$	Sample variances, respectively, of X and Y (5.4)
S_{X+c}	Standard deviation that has been altered by adding a constant to each score (4.8)
S_{XY}	Sample covariance (5.3)
$S_{Y \cdot X}$	Standard error of estimate for predicting Y from X (6.3)

Symbol	Meaning
SS_{BG}	Between-groups sum of squares (16.2)
SS_{Total}	Total sum of squares (16.2)
SS_{WG}	Within-groups sum of squares (16.2)
T	Wilcoxon's T random variable (18.3)
$T_{\alpha,n}$ and $T_{\alpha/2,n}$	Values that cut off, respectively, the α and $\alpha/2$ regions of the sampling distributions of Wilcoxon's T for n pairs of observations (18.3)
t	Student's t random variable (12.1)
t_i	Number of tied observations in a set (18.2)
$t_{\alpha,v}$, and $t_{\alpha/2,v}$	Values that cut off, respectively, the upper α and $\alpha/2$ regions of the sampling distribution of t for v degrees of freedom (12.2)
U	Mann–Whitney U random variable (18.2)
$U_{\alpha;n_1,n_2}$ and $U_{\alpha/2;n_1,n_2}$	Values that cut off, respectively, the α and $\alpha/2$ regions of the sampling distribution of U for n_1 and n_2 observations in samples 1 and 2 (18.2)
V	A variable (3.9); Cramer's measure of association; also denoted by ϕ' (17.3); a caret over the symbol, $\hat{V}$, denotes an estimate of V (17.3)
$\text{Var}(t)$	Variance of Student's t random variable (12.2)
$\text{Var}(\bar{X})$	Variance of sample means (10.3)
$\text{Var}(\chi_v^2)$	Variance of a chi-square random variable (12.3)
$\text{Var}(Mdn)$	Variance of sample medians (10.3)
$\text{Var}(\hat{\sigma}^2)$	Variance of $\hat{\sigma}^2$ (10.5)
$\text{Var}(\hat{\sigma}_{est}^2)$	Variance of $\hat{\sigma}_{est}^2$ (10.5)
W	A variable (3.9)
X	A score (1.4); the independent variable in an experiment (5.1)
X_i	A score for the ith measurement or event class (3.3)
X_j	Midpoint of the jth class interval (3.3)
X_{ij}	A score for the ith subject in the jth treatment condition (16.2)
X_{ll}	Real lower limit of a score (4.2) or class interval (3.4)
X_{ul}	Real upper limit of a score (4.2) or class interval (3.4)
$\bar{X}$	Sample arithmetic mean (3.3)
$\bar{X}_D$	Mean of difference scores (13.5)
$\bar{X}'$	Desired sample mean (10.2)
$\bar{X}_{..}$	Arithmetic mean of all scores (16.2)
$\bar{X}_{\cdot j}$	Arithmetic mean of scores in the jth treatment condition (16.2)
$\bar{X}_{..k}$	Arithmetic mean of the kth level of treatment B (16.6)

Symbol	*Meaning*
$\bar{X}_{\cdot jk}$	Arithmetic mean of the jth and kth treatment combination (16.6)
$\bar{X}_W$	Weighted mean of two or more sample means (3.7)
Y	A score (1.4); the dependent variable in an experiment (5.1)
Y_i'	A predicted Y score (6.2)
$\bar{Y}$	Sample arithmetic mean (3.3)
Z'	Fisher's transformation of r (12.5)
Z_0'	Value of population Z' specified by the null hypothesis (12.5)
Z_{Pop}'	Fisher's transformation of ρ (14.6)
z	Standard score (10.1); z random variable (11.2)
z_α and $z_{\alpha/2}$	Values that cut off, respectively, the upper α and $\alpha/2$ regions of the sampling distribution of z (11.4)

Answers to Starred Exercises

Chapter 1

1. (a) Those who must be able to understand statistical presentations of findings in their fields; those who select, apply, and interpret statistical procedures in their work; professional (applied) statisticians; and mathematical statisticians.
4. (a) white female students in this university, a female student, a measure of career ambivalence
   (b) all drivers in Tequila Tech students' age group, a driver in that age group, the number of automobile accidents for that driver
   (c) homes in Chickasha, Oklahoma; a home in Chickasha, Oklahoma; presence or absence of a color TV in the home
   (d) students at Ginebra University, a student at Ginebra University, grade point average of that student
6. one and all but one of the population elements
9. (a) R (b) NR (c) R (d) NR
13. (a) D (b) U (c) C (d) O
15. (a) D (b) U (c) D (d) O
17. (a) ratio (b) nominal (c) between ordinal and interval (d) ordinal
19. nominal: one-to-one substitution, ordinal: monotonic, interval: positive linear, ratio: multiplication by a positive constant
20. (a) interval (b) between ordinal and interval
23. A change of three points on the measurement scale represents the same empirical change from 62 to 65 as from 68 to 71 and from 71 to 74.
25. (a) national statistics, probability theory, and experimental statistics
27. The modern era uses exact inductive procedures appropriate for both large and small samples; the previous period relied on large-sample procedures.

Chapter 2

1.

X	f	X	f
13	1	6	2
12	0	5	4
11	0	4	6
10	1	3	3
9	1	2	2
8	0	1	1
7	1	0	1

$n = 23$

3.

X	f	X	f
25	1	13	1
24	0	12	2
23	0	11	4
22	1	10	4
21	1	9	5
20	0	8	8
19	0	7	10
18	0	6	9
17	1	5	7
16	0	4	3
15	2	3	1
14	0		

$$n = 60$$

4.   see list in Section 2.2
5.   (a)   49.5–54.5, 5   (b)   73.5–74.5, 1   (d)   17.95–19.95, 2
6.   (a)   16, 3, 21–23   (b)   11, 15, 105–119   (c)   17, 10, 90–99

7.

X	f	X	f
95–99	1	60–64	2
90–94	1	55–59	4
85–89	4	50–54	5
80–84	5	45–49	3
75–79	4	40–44	2
70–74	2	35–39	1
65–69	1		

$$n = 35$$

9.   (a)

X	f	X	f
210–212	1	192–194	3
207–209	0	189–191	5
204–206	1	186–188	6
201–203	2	183–185	4
198–200	2	180–182	1
195–197	2		

$$n = 27$$

X	f	X	f
210–211	1	194–195	1
208–209	0	192–193	3
206–207	0	190–191	3
204–205	1	188–189	4
202–203	2	186–187	4
200–201	0	184–185	4
198–199	2	182–183	0
196–197	1	180–181	1

$$n = 27$$

(b)   The distribution with $i = 3$ is better than the one with $i = 2$ because the number of scores is relatively small and because it provides a clearer picture of the distribution of scores.

11.

X	Prop. f	X	Prop. f
95–99	0.03	60–64	0.06
90–94	0.03	55–59	0.11
85–89	0.11	50–54	0.14
80–84	0.14	45–49	0.09
75–79	0.11	40–44	0.06
70–74	0.06	35–39	0.03
65–69	0.03		

15.

X	f	Cum f	X	f	Cum f
16	1	32	10	7	17
15	0	31	9	4	10
14	1	31	8	3	6
13	2	30	7	2	3
12	5	28	6	1	1
11	6	23			

$$n = 32$$

16.

X	f	Prop. f	Cum prop. f	X	f	Prop. f	Cum prop. f
13	1	0.04	0.98	6	2	0.09	0.82
12	0	0.00	0.94	5	4	0.17	0.73
11	0	0.00	0.94	4	6	0.26	0.56
10	1	0.04	0.94	3	3	0.13	0.30
9	1	0.04	0.90	2	2	0.09	0.17
8	0	0.00	0.86	1	1	0.04	0.08
7	1	0.04	0.86	0	1	0.04	0.04

$$n = 23$$

19.

X	f
Graduate	1
Senior	8
Junior	10
Sophomore	6
Freshman	4

$$n = 29$$

26.

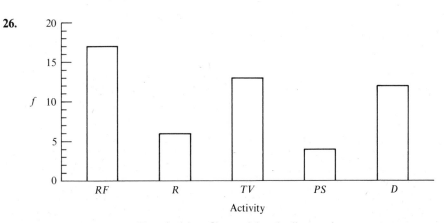

Favorite leisure time activity of college students

28.

Attitudes toward research

33.

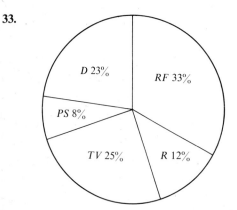

Favorite leisure time activity of college students

39.

Histogram for number of cigarettes smoked per day by mothers
whose first babies were stillborn

44. (a) 22   (c) 132.5   (e) 22

45.

Frequency polygon for number of cigarettes smoked by mothers
whose first babies were stillborn

50. (a)

Cumulative percentage frequency polygon for number of cigarettes
smoked per day by mothers whose first babies were stillborn

(b)   22.5

53. A cumulative polygon will have an "S" shape if there are more scores in the middle of the
corresponding frequency distribution than at the extremes.

54.

Stems (class intervals)	Leaves (scores)
30–32	0 0 0 1 1 2
27–29	7 7 7 8 8 9
24–26	5 5 6
21–23	1 1 2 3
18–20	8 9 9 0
15–17	5 6 6
12–14	2 3 4
9–11	9 0
6–8	
3–5	3
0–2	0

57. (a) true   (b) true

58. (a)

(b)

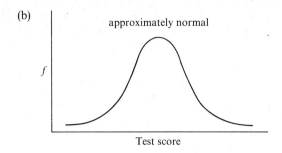

approximately normal

f

Test score

60.

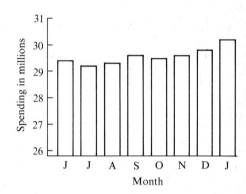

Chapter 3

1. (a) $Mo = S$ (b) ordered qualitative variable
4. (a) The distribution is bimodal; the two maximum values occur at 1 and 3.
   (b) unordered qualitative variable
6. (a) X score for element one   (c) mean of population one   (h) X score for element i
   (o) X score for element j
7. (a) $X_1 + X_2 + \cdots + X_n$ (c) $(Z_1 + Z_2 + Z_4)/3$ (d) $(f_1 X_1 + f_2 X_2 + \cdots + f_k X_k)/n$
8. $\bar{X} = 4.59$
10. $\bar{X} = 9.44$

12.

X_j	f_j	$f_j X_j$
9	1	9
8	0	0
7	2	14
6	2	12
5	4	20
4	3	12
3	2	6
2	2	4
1	1	1

$$n = 17 \qquad \sum_{j=1}^{k} f_j X_j = 78$$

$\bar{X} = 78/17 = 4.59$

14.

X_j	f_j	f_jX_j
22	1	22
21	0	0
20	0	0
19	0	0
18	0	0
17	0	0
16	1	16
15	0	0
14	2	28
13	1	13
12	3	36
11	4	44
10	6	60
9	5	45
8	5	40
7	3	21
6	0	0
5	2	10
4	0	0
3	1	3
2	1	2
1	0	0
0	1	0

$$n = 36 \qquad \sum_{j=1}^{k} f_jX_j = 340$$

$\bar{X} = 340/36 = 9.44$

18. (a) 9   (c) 18   (e) 3.25   (f) 3.75

19.

X_j	f_j	Cum f
22	1	36
21	0	35
20	0	35
19	0	35
18	0	35
17	0	35
16	1	35
15	0	34
14	2	34
13	1	32
12	3	31
11	4	28
10	6	24
9	5	18
8	5	13
7	3	8
6	0	5
5	2	5
4	0	3
3	1	3
2	1	2
1	0	1
0	1	1

$$n = 36$$

$Mdn = 8.5 + 1[(18 - 13)/5] = 9.5$

21. X_{ul} = real upper limit of class interval containing the median, i = class interval size, n = number of scores, $\sum f_a$ = number of scores above X_{ul}, f_i = number of scores in the class interval containing the median

24. (a) mean, quantitative data and the distribution is relatively symmetrical
    (c) mean, quantitative data and the distribution is relatively symmetrical

25. (a) $\bar{X} = 1$, $Mdn = 2$, $Mo = 3$ (e) only Mo is appropriate

28. (a) positively skewed   (e) positively skewed

29. (a) 43.33

30. (a) $X_1 + X_2 + X_3$ (b) $Y_1 + Y_2 + Y_3 + Y_4$ (d) $f_1 X_1 + f_2 X_2 + f_3 X_3$
    (e) $f_1 X_1 + f_2 X_2 + \cdots + f_k X_k$ (g) $aX_1 + aX_2 + aX_3 = a(X_1 + X_2 + X_3)$
    (i) $(X_1 + a) + (X_2 + a) + \cdots + (X_n + a) = (X_1 + X_2 + \cdots + X_n) + na$

31. (a) $(3)(2) = 6$ (b) $(4)(3) = 12$ (e) 9   (f) 16   (g) 5
    (i) $(2)(2 + 3 + 4) = 18$ (k) $(2 + 3 + 4) + (3)(2) = 15$
    (m) $(1 + 2 + 4 + 9) + (4)(2) - (4)(3) = 12$ (n) $(2 + 3) + (2)(2) = 9$

32. (a) $\displaystyle\sum_{i=1}^{n} (X_i + c) = \sum_{i=1}^{n} X_i + \sum_{i=1}^{n} c$ Rule 3.9-1-4

$$\sum_{i=1}^{n} c = nc \quad \text{Rule 3.9-1-1}$$

Chapter 4

1. (a) 16   (b) 16
4. (a) $Mdn = 27.33$, $Q = 1.33$ (b) $P_{10} = 24.70$, $P_{90} = 30.30$
6. 87.50
9. $\bar{X} = 2.467$, $S = 1.258$
10. $\sqrt{[(115) - (37)^2/(15)]/15} = 1.258$
14. (a) Mo = moderately desire career, $D = 0.97$

17. (a) $\displaystyle\sum_{i=1}^{n} (X_i + c)/n = \sum_{i=1}^{n} X_i/n + \sum_{i=1}^{n} c/n$ Rule 3.9-1-4

$$\sum_{i=1}^{n} c/n = nc/n \quad \text{Rule 3.9-1-1}$$

18. (a) Approximately 68% of the scores fall between 85 and 115.
20. (a) Compute the mean and the standard deviation because the variable is quantitative and the distribution is relatively symmetrical.
    (c) Compute the median and the semi-interquartile range because the distribution is skewed.
    (d) Compute the mode and the index of dispersion because the variable is qualitative.
21. (a) 84.14%
22. (a) 3.37   (b) 20
24. $S_{min} = 3$, $S_{max} = 15$
26. (a) correct
28. (a) 67.35%
29. (a) Minimum is 0%; probable percent is 68.27%.
    (b) Minimum is 75%; probable percent is 95.45%.
30. (a) $Sk = -1.02$; distribution is negatively skewed.

(b)

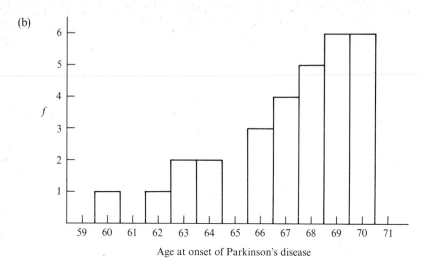

Age at onset of Parkinson's disease

(c) yes
32. Since $Kur = -0.04$, the data do not support the prediction.
33. $Kur = 0.23$; distribution is leptokurtic.

Chapter 5

1. (a)

Test A	Test B									
	30	31	32	33	34	35	36	37	38	39
33										1
32										1
31								1		
30									1	
29							1	1		
28						1	1			
27					1	1	1			
26			1		2	1	1			
25				1	2		1			
24			1		1					
23			1							
22				1						
21			1							
20	1									

(b) linear
3. The term means that extreme scores for one variable, that is, scores that differ considerably from their mean, are likely to be paired with less extreme scores for the other variable.
6. (a) 1   (e) -0.9
7. (a) positive   (b) positive
9. 0.86
12. (a) Data appear to be linearly related.   (b) $r = -0.53$
    (c) The faster the music tempo, the slower the rate of sipping.
13. 0.90
15. (a) 18, quadrants 1 and 3, variables are positively related
    (b) 0, quadrants 1, 2, 3 and 4, variables are not related

18. 1, −1, 0

21. (a)   The proportion of variance in English grades explained by variation in grades in bowling is 0.048; the proportion that is not explained is 0.952.

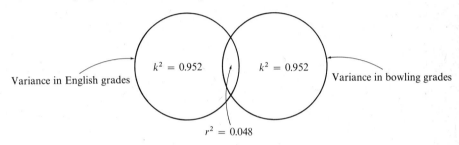

23. (a)   Correct interpretation
(c)   This interpretation is incorrect, since it uses an arbitrary descriptive label, medium, to denote r's between 0.30 and 0.69.
25. The mean IQ should increase because of regression toward the mean.
28. (a) r underestimates the magnitude of the relationship.

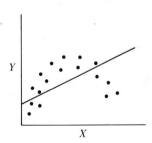

(d) r overestimates the magnitude of the relationship.

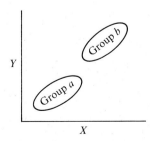

(e)   *r* underestimates the magnitude of the relationship.

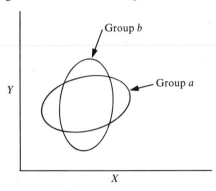

(g)   *r* underestimates the magnitude of the relationship for small values of *X* and large values of *Y*; the converse is true for large values of *X* and small values of *Y*.

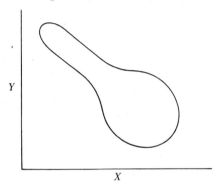

35.   0.76
39.   An estimate of r_s using the Pearson product-moment correlation formula is -0.30.
40.   (a)   strictly monotonic   (b)   nonmonotonic

Chapter 6

1.   0.5
3.   (a)

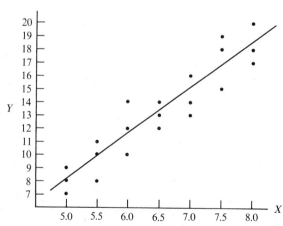

Data appear to be linearly related.

(b) $a_{Y \cdot X} = -9.356$
$b_{Y \cdot X} = 3.476$
$r = 0.93$

(c) Estimate based on the regression equation is 11.5; estimate based on the line of best fit is 11.5.

5. They are best fitting lines in the sense that they minimize the sum of the squared prediction errors.

8. $b_{X \cdot Y} = \dfrac{\sum(X - \bar{X})(Y - \bar{Y})}{\sum(X - \bar{X})^2} = \dfrac{\sum(XY - X\bar{Y} - Y\bar{X} + \bar{X}\bar{Y})}{\sum(X^2 - 2X\bar{X} + \bar{X}^2)}$

$= \dfrac{\sum XY - \bar{Y}\sum X - \bar{X}\sum Y + n\bar{X}\bar{Y}}{\sum X^2 - 2\bar{X}\sum X + n\bar{X}^2} = \dfrac{\sum XY - \dfrac{\sum Y}{n}\sum X - \dfrac{\sum X}{n}\sum Y + n\dfrac{\sum X}{n}\dfrac{\sum Y}{n}}{\sum X^2 - 2\dfrac{\sum X}{n}\sum X + n\dfrac{(\sum X)^2}{n^2}}$

$= \dfrac{\sum XY - 2\dfrac{\sum X \sum Y}{n} + \dfrac{\sum X \sum Y}{n}}{\sum X^2 - 2\dfrac{(\sum X)^2}{n} + \dfrac{(\sum X)^2}{n}}$

$= \dfrac{\sum XY - \dfrac{\sum X \sum Y}{n}}{\sum X^2 - \dfrac{(\sum X)^2}{n}}$

10. (a) $Y_i' = \bar{Y} - b\bar{X} + bX_i$

$\sum Y_i' = \sum(\bar{Y} - b\bar{X} + bX_i)$

$= n\bar{Y} - nb\bar{X} + b\sum X_i$

$= n\dfrac{\sum Y_i}{n} - nb\dfrac{\sum X_i}{n} + b\sum X_i$

$= \sum Y_i$

(b) r is equal to either -1 or 1.

12. (a) 1.42
(b) $13.2 \pm 1.42 = 11.78$ and 14.62
(c) For $r = 0$, $S_{Y \cdot X} = 2.97$; for $r = 1$, $S_{Y \cdot X} = 0$
(d) $k = 0.48$ and $E = 0.52$. The standard error of estimate is 48% of the size it would be if the correlation were zero. By taking the value of X_i into account in predicting Y_i, we have reduced the magnitude of the prediction error by 52%.

Chapter 7

3. (a) $\frac{1}{2}$
(b) It is assumed that there are six possible outcomes, that they are equally likely, and that the number of outcomes favoring an odd number is three.

5. (a) 0.52   (b) 0.5

6. (a)

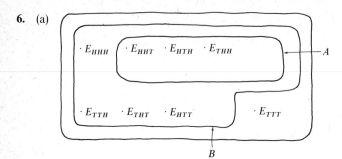

 (b) $p(A) = \frac{3}{8}$ (c) $p(B) = \frac{7}{8}$

9. (a) 3/10   (b) 6/10,000

13. (a) $\frac{4}{52} = \frac{1}{13}$ (b) $\frac{13}{52}$ (c) $\frac{16}{52} = \frac{4}{13}$ (d) $\frac{26}{52} = \frac{1}{2}$ (e) $\frac{12}{52} = \frac{3}{13}$ (f) $\frac{16}{52} = \frac{4}{13}$

14. (c) $\frac{2}{3}$

15. (a) 0.48   (b) $(0.4)(0.2) = 0.08$ (c) $0.80 + 0.60 - 0.48 = 0.92$

16. (a)

	F	*Not F*	
D	$p(D \text{ and } F)$ $= 0.002$	$p(D \text{ and } Not F)$ $= 0.098$	$p(D) = 0.10$
O	$p(O \text{ and } F)$ $= 0.002$	$p(O \text{ and } Not F)$ $= 0.898$	$p(O) = 0.90$

       $p(F) = 0.004$ $p(Not F) = 0.996$

 (b) 0.002

17. $p(Not D \text{ and } M) = 0.147$

21. (a) 0.4

Chapter 8

1. (a) 8   (b) 1296   (c) 12

4. 24

6. 5040

10. (a) $(9)(8) = 72$ (b) $9[8!/(8-2)!] = 504$ (c) $9\{8!/[2!(8-2)!]\} = 252$

Chapter 9

1. Identify the population, decide whether to sample with or without replacement, and select elements using a random sampling procedure.

3. (a) 2118760

4. (a) $8!/[4!(8-4)!] = 70$ (b) 0.0143   (c) $(8)(8)(8)(8) = 4096$

9. (a)

r	$p(X = r)$
0	0.0625
1	0.2500
2	0.3750
3	0.2500
4	0.0625

10. (a) 0.93   (b) 0.07   (c) 0.77   (d) 1.24   (e) 0.88
13. $22
16. (a) 1/1000   (b) $E(X) = (1/1000)[750 + (-1.00)] + (999/1000)(-1.00) = -.25$
(c) no   (d) $(1/1000)[750 + (-T)] + (999/1000)(-T) = 0$

$$0.75 - \frac{T}{1000} - \frac{999T}{1000} = 0$$

$$T = 0.75$$

20. (a) $p(X = 0) = 0.16$, $p(X = 1) = 0.48$, $p(X = 2) = 0.36$

(b)

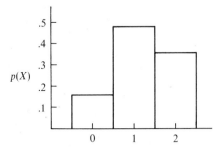

Number of males

(c) $E(X) = 1.2$, $\sigma = 0.693$
22. (a) 0.149   (b) $E(X) = 3$, $\sigma = 1.449$
24. (a) 250   (b) 25   (c) 500
26. 0.071
27. 0.091

Chapter 10

2. (a) normal   (b) normal   (c)   (d) normal

Income

(e)   (f)

Age   Arrival time

3. (a) 2   (b) -1.6   (c) -1
4. (a) 0.0668   (b) 0.0228   (c) 0.4987   (e) 0.1359
5. (a) 0.6826   (b) 0.8990   (c) 0.9500
6. (a) 190   (b) 120   (c) 212
7. (a) 0   (b) 1.645   (c) 0.25

9. The number eligible will decrease by 9.94%.
12. (a) 0.25 (b) 0.13
13. (a) 0.06 (b) 0.92 (c) 0.04
17. Performance on test 3 was best, and performance on test 2 was worst.
19. Yes.
20. (a) 88 (b) 42 (c) 8.4

22. (a)

Sample no.	Sample values	$\bar{X}_j$	Sample no.	Sample values	$\bar{X}_j$
1	0, 0	0.0	9	2, 0	1.0
2	0, 1	0.5	10	2, 1	1.5
3	0, 2	1.0	11	2, 2	2.0
4	0, 3	1.5	12	2, 3	2.5
5	1, 0	0.5	13	3, 0	1.5
6	1, 1	1.0	14	3, 1	2.0
7	1, 2	1.5	15	3, 2	2.5
8	1, 3	2.0	16	3, 3	3.0

   (b) $\mu = 6/4 = 1.5$, $\sigma_{\bar{x}} = 1.11803/\sqrt{2} = 0.7906$ (c) $\mu_{\bar{x}} = 24/16 = 1.5$, $\sigma_{\bar{x}} = 0.7906$
26. (a) 7.07 (b) 5.00
27. $(115 - 120)/(10/\sqrt{25}) = -2.5$, $p = 0.0062$
28. $(182 - 165)/(15/\sqrt{9}) = 3.4$, $p = 0.0003$

Chapter 11

1. (a) scientific hypothesis (b) scientific hypothesis (c) not a scientific hypothesis
   (d) scientific hypothesis
3. (a) yes (b) no (c) yes (d) no (e) yes (f) no (g) yes (h) no
   (i) yes (j) yes
7. $H_0: \mu \le 8$, $H_1: \mu > 8$
11. $H_0: \mu \le 15$
15. When σ^2 is known, the population distribution of X is normal or n is very large, and H_0 is true.
17. (a) $z = (55.1667 - 50)/(10/\sqrt{30}) = 2.83$
    (b) The H_0 is rejected, and this supports the truth of the scientific hypothesis.

20. (a)

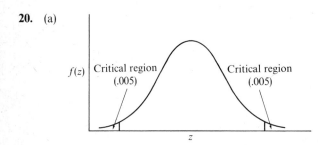

22. (a) H_0 is exact; H_1 is inexact.
    (b) H_0 is inexact; H_1 is inexact.
    (c) H_0 is inexact; H_1 is inexact.
24. (a) 0.67 (b) $[1.645 - (-1.28)]^2/[(118 - 115)^2/(15)^2] = 214$

Chapter 12

3. (a) $t = (88.75 - 100)/(10.1620/\sqrt{16}) = -4.428$. Reject H_0 since $t < -t_{.05,15} = -1.753$; this year's classes are below average in intelligence.

   (b) $n = 12$

8. $\chi^2 = (16 - 1)(10.1620)^2/(15)^2 = 6.884$. Reject H_0 since $\chi^2 < \chi^2_{1-10/2,\,15}$.

11. n should be at least 26 <u>since in that case</u> both $n(0.2)$ and $n(1 - 0.2)$ are greater than 5.

12. $z = [(12 - \frac{1}{2}) - 10]/\sqrt{20(0.5)(0.5)} = 0.67$. The hypothesis that $p = 0.50$ cannot be rejected.

17. $z = (0.24\sqrt{26-2})/\sqrt{1-(0.24)^2} = 1.21$. Do not reject the null hypothesis.

18. (a) 0.497   (b) -0.234

19. (a) 0.50   (b) -0.20

21. (a) No

    (b) The value of r is less than 0.50, which means that it is in the wrong tail of the sampling distribution.

Chapter 13

4. $z = (112 - 116)/\sqrt{(15)^2/50 + (15)^2/50} = -1.333$; $z > -z_{.05} = -1.645$. Do not reject the null hypothesis.

8. The significant result indicates only that chance is an unlikely explanation for the observed difference.

11. (a) random assignment

14. Pooling is appropriate when the population variances are equal.

15. $t = (24.731 - 18.000)/0.712 = 9.45$; $t > t_{.05,47} \simeq 1.68$. The data support the experimenter's scientific hypothesis.

18. (a) $t = (0.072800 - 0.148200)/0.013299 = -5.67$; $|t| > t_{.001/2,\,26} = 3.707$. Reject the null hypothesis.

    (b) Early language experience in Spanish-speaking homes resulted in better discrimination of the Spanish contrasts.

20. The order of presentation of the conditions should be randomized independently for each subject.

22. (a) The larger the positive correlation between samples, the smaller is the standard error of the difference between means.

    (b) The larger the positive correlation, the higher is the probability of rejecting a false null hypothesis.

23. (a) $t = -2.1875/0.2772 = -7.891$; $t < -t_{.05,15} = -1.753$. Reject the null hypothesis; viewing the film does result in more favorable attitudes toward legalization of marijuana.

    (b) 11

25. (a) $t = -9.3182/1.2916 = -7.214$; $|t| > t_{.01/2,21} = 2.831$. Reject the null hypothesis.

    (b) $n = 20$

    (c) $t = (25.6818 - 35.0000)/1.9209 = -4.851$; $|t| > t_{.01/2,42} \simeq 2.7$.
    The absolute value of the dependent samples t statistic is approximately 2.5 times larger than the critical value; the independent samples t statistic is only 1.8 times larger than the critical value. The use of repeated measures was an effective strategy.

    (d) In the present experiment the difference $\bar{X}_1 - \bar{X}_2$ reflects the effects of the ten-day workshop as well as other effects, such as (1) improved test-taking skills due to taking the pretest, (2) increased sensitivity during the workshop to the kinds of material on the test due to taking the pretest, and (3) acquisition of cancer knowledge from sources other than the workshop such as professional journals and colleagues. The following design, with subjects randomly divided into experimental and control groups, would enable the experimenter to measure the combined magnitude of the effects of 1 and 3.

	Pretest	*Workshop*	*Posttest*
Experimental group	X	X	X
Control group	X		X

The use of two experimental and two control groups as follows would enable the experimenter to measure the effects of 1, 2, and 3 above.

	Pretest	Workshop	Posttest
Experimental group 1	X	X	X
Experimental group 2		X	X
Control group 1	X		X
Control group 2			X

27. (a) $n_1 + n_2 = 100$ (b) $n_1 + n_2 = 1492$ (d) $n = 21$ (e) $n = 225$

Chapter 14

1. No
4. $F = 1.24$ or 0.81, depending on which sample variance is made the numerator; the corresponding critical values are $F_{.20/2;22,25} \simeq 1.7$ and $F_{1-.20/2;25,22} \simeq 0.59$. Do not reject the null hypothesis.
7. $z = (0.04 - 0.11)/0.01563 = -4.48$; reject the null hypothesis.
10. $z = (0.78 - 0.85)/0.01468 = -4.77$; reject the null hypothesis.
11. $z = (16 - 27)/\sqrt{16 + 27} = -1.68$. Do not reject the null hypothesis; the data do not provide evidence that the population proportions differ.

Chapter 15

2. $z = (\bar{X} - \mu)/(\sigma/\sqrt{n})$; hence $-z_{\alpha/2} \le (\bar{X} - \mu)/(\sigma/\sqrt{n}) \le z_{\alpha/2} = -\bar{X} - z_{\alpha/2}(\sigma/\sqrt{n}) \le -\mu \le -\bar{X} + z_{\alpha/2}(\sigma/\sqrt{n}) = \bar{X} - z_{\alpha/2}(\sigma/\sqrt{n}) \le \mu \le \bar{X} + z_{\alpha/2}(\sigma/\sqrt{n})$.
4. L_1 and L_2 would be equal if all of the scores were equal or if the sample exhausted all of the elements in the population, in which case $\bar{X} = \mu$.
5. $7.05 \le \mu \le 7.35$
6. (a) I   (b) C   (c) I   (d) C   (e) C   (f) I
7. $392.77 \le \mu \le 409.33$
9. $0.94 \le \mu_1 - \mu_2 \le 4.06$
11. $2.15 \le \mu_D \le 4.97$
12. (a) The larger σ, the larger the interval.   (b) The larger n, the smaller the interval.
    (c) The larger $1 - \alpha$, the larger the interval.
14. $164.38 \le \sigma^2 \le 779.37$
16. The population variances could be equal because the confidence interval includes 1.
17. (a) $(5.76/6.76)(1/2.25) \le \sigma_1{}^2/\sigma_2{}^2 \le (5.76/6.76)(2.25) = 0.38 \le \sigma_1{}^2/\sigma_2{}^2 \le 1.92$
    (b) yes   (c) 0.25
19. $0.391 \le p \le 0.529$
21. $0.004 \le p \le 0.136$
24. $0.095 \le p_1 - p_2 \le 0.293$
28. $0.37 \le \rho \le 0.68$
31. $-0.24 \le \rho_1 - \rho_2 \le 0.43$

Chapter 16

1. (a) $H_0: \mu_1 = \mu_2 = \cdots = \mu_5$ (b) 10
2. It means that at least two of the population means are not equal.
6. (b) The number of levels of the independent variables is equal to 2.
7. (a) level 2 of treatment A
    (b) score for subject 2 in treatment level 4
    (c) score for subject 16 in treatment level 1

(d) mean of treatment level 4

(e) grand mean

(f) linear model equation for subject 7 in treatment level 3

(g) treatment effect of population 2

(h) error effect for subject 1 in treatment level 3

9. (a) correct   (b) incorrect   (c) incorrect   (d) incorrect   (e) correct

10. (a) $X_{83} = \bar{X}.. + (\bar{X}._3 - \bar{X}..) + (X_{83} - \bar{X}._3)$

12. (a) Total = 83, $BG = 3$, $WG = 80$ (b) Total = 54, $BG = 4$, $WG = 50$

(d) Total = 16, $BG = 2$, $WG = 14$

13. Both MS_{BG} and MS_{WG} estimate $\sigma_\varepsilon{}^2$ when random samples are drawn from normally distributed populations having equal means and equal variances.

14. MS_{BG}/MS_{WG} is distributed as the F distribution when p random samples of size n are drawn from normally distributed populations having equal means and equal variances.

15. MS_{BG}/MS_{WG} tends to be larger than 1 when random samples are drawn from normally distributed populations having two or more unequal means but equal variances.

18.
$$n \sum_{j=1}^{p} (\bar{X}._j - \bar{X}..)^2 = n \sum_{j=1}^{p} (\bar{X}_{.j}^2 - 2\bar{X}..\bar{X}._j + \bar{X}..^2)$$

$$= n \sum_{j=1}^{p} \frac{\left(\sum_{i=1}^{n} X_{ij}\right)^2}{n^2} - 2n \frac{\left(\sum_{j=1}^{p}\sum_{i=1}^{n} X_{ij}\right) \sum_{j=1}^{p}\left(\sum_{i=1}^{n} X_{ij}\right)}{pn} + np \frac{\left(\sum_{j=1}^{p}\sum_{i=1}^{n} X_{ij}\right)^2}{p^2 n^2}$$

$$= \sum_{j=1}^{p} \frac{\left(\sum_{i=1}^{n} X_{ij}\right)^2}{n} - 2 \frac{\left(\sum_{j=1}^{p}\sum_{i=1}^{n} X_{ij}\right)^2}{pn} + \frac{\left(\sum_{j=1}^{p}\sum_{i=1}^{n} X_{ij}\right)^2}{pn}$$

$$= \sum_{j=1}^{p} \frac{\left(\sum_{i=1}^{n} X_{ij}\right)^2}{n} - \frac{\left(\sum_{j=1}^{p}\sum_{i=1}^{n} X_{ij}\right)^2}{pn}$$

$$\sum_{j=1}^{p}\sum_{i=1}^{n} (X_{ij} - \bar{X}._j)^2 = \sum_{j=1}^{p}\sum_{i=1}^{n} (X_{ij}^2 - 2\bar{X}._j X_{ij} + \bar{X}._j^2)$$

$$= \sum_{j=1}^{p}\sum_{i=1}^{n} X_{ij}^2 - 2 \sum_{j=1}^{p} \frac{\left(\sum_{i=1}^{n} X_{ij}\right)}{n} \sum_{i=1}^{n} (X_{ij}) + n \sum_{j=1}^{p} \frac{\left(\sum_{i=1}^{n} X_{ij}\right)^2}{n^2}$$

$$= \sum_{j=1}^{p}\sum_{i=1}^{n} X_{ij}^2 - \sum_{j=1}^{p} \frac{\left(\sum_{i=1}^{n} X_{ij}\right)^2}{n}$$

20. (a) Descriptive statistics for time required to learn nonsense syllables

	Association value of nonsense syllables			
	25%	50%	75%	100%
	a_1	a_2	a_3	a_4
$\bar{X}._j$	21.875	20.000	17.875	17.000
$\hat{\sigma}_j$	1.246	1.309	1.246	1.309

(b) Yes

(c)

Source	SS	df	MS	F
1 Between groups (association value)	115.1250	3	38.3750	$[\frac{1}{2}]$ 23.49*
2 Within groups	45.7500	28	1.6339	
3 Total	160.8750	31		

* $p < .01$
Reject the null hypothesis.

20. (d) According to the ANOVA, the learning-time population means for the four levels of association value are not all equal, $F(3, 28) = 23.49$, $p < .01$.

23. (a) Descriptive statistics for
reaction time in seconds to
three colors of warning lights

	Color of warning light		
	Yellow a_1	Red a_2	Green a_3
$\bar{X}_{.j}$	20.300	21.700	21.000
$\hat{\sigma}_j$	0.949	0.949	1.183

(b) Yes

(c)

Source	SS	df	MS	F
Between groups (color of light)	9.8000	2	4.9000	$[\frac{1}{2}]$ 4.54*
Within groups	30.2000	28	1.0786	
Total	40.0000	30		

* $p < .05$
Reject the null hypothesis.

(d) According to the ANOVA, the reaction-time population means for the three colors of instrument panel warning lights are not equal, $F(2, 28) = 4.54$, $p < .05$.

26. The test is robust if the treatment populations have the same shape.
27. (a) Assumption is tenable.
33. (a) $1 - 1$ (b) $1 - \frac{1}{2} - \frac{1}{2}$ (f) $1 - \frac{2}{3} - \frac{1}{3}$
34. (a) meets requirements   (b) meets requirements
35. (a) satisfies   (b) does not satisfy
36. (a) 12.93   (c) 7.10
37. Do not reject $\mu_1 - \mu_2 = 0$, $F = 0.612$. Reject $\mu_1 - \mu_3 = 0$, $F = 16.200$. Reject $\mu_2 - \mu_3 = 0$, $F = 10.512$. Reject $(\mu_1 + \mu_2) - \mu_3 = 0$, $F = 17.604$. $(3 - 1)F_{.01;2,120} = (3 - 1)4.79 = 9.58$.
38. Do not reject $\mu_1 - \mu_2 = 0$, $F = 0.707$. Reject $\mu_1 - \mu_3 = 0$, $F = 12.500$. Reject $\mu_2 - \mu_3 = 0$, $F = 8.000$. $(3 - 1)F_{.05;2,12} = (3 - 1)3.89 = 7.78$.

Chapter 17

2. (a) State the statistical hypotheses—H_0: $p_1 = 0.50$, H_1: $p_1 \neq 0.50$. Specify the test statistic—$\chi^2 = \sum (O - E)^2/E$. Specify the sample size—$n = 50$, and the sampling distribution—chi-square distribution. Specify the level of significance—$\alpha = .05$. Obtain a random sample of size 50, compute χ^2 and make a decision.

    (b) $\chi^2 = 5.12$. Reject H_0; the data suggest that opinion is not equally divided on the issue.

4. $\chi^2 = 0.120$; the data do not suggest that there has been a change in party affiliation.

6. $\chi^2 = 9.280$; the data do not suggest that the die is not fair.

11. Both r and $\hat{V}$ are measures of association, but, unlike r, $\hat{V}$ is appropriate for unordered qualitative variables and ranges over values of 0 to 1. Further, $\hat{V}$ does not have a simple interpretation in terms of proportion of explained variance.

13. (a) State the statistical hypotheses—H_0: $p(A \text{ and } B) = p(A)p(B)$, H_1: H_0 is false. Specify the test statistic—$\chi^2 = \sum (O - E)^2/E$. Specify the sample size—$n = 200$, and the sampling distribution—chi-square distribution. Specify the level of significance—$\alpha = .001$. Obtain a random sample of size 200, compute χ^2 and make a decision.

    (b) $\chi^2 = 32.28$. Reject the null hypothesis; the variables are not independent.

    (c) $\hat{V} = 0.28$

18. H_0: $p_1 = p_2$; $\chi^2 = 8.864$. Reject the null hypothesis.

20. The population proportions in the three categories of variable A, 0.30, 0.60, 0.10, are equal across the three categories of variable B.

21. (a) H_0:

$$\begin{bmatrix} p_{A_1|B_1} \\ p_{A_2|B_1} \\ p_{A_3|B_1} \\ p_{A_4|B_1} \end{bmatrix} = \begin{bmatrix} p_{A_1|B_2} \\ p_{A_2|B_2} \\ p_{A_3|B_2} \\ p_{A_4|B_2} \end{bmatrix} = \begin{bmatrix} p_{A_1|B_3} \\ p_{A_2|B_3} \\ p_{A_3|B_3} \\ p_{A_4|B_3} \end{bmatrix} = \begin{bmatrix} p_{A_1|B_4} \\ p_{A_2|B_4} \\ p_{A_3|B_4} \\ p_{A_4|B_4} \end{bmatrix}$$

$$H_1: \begin{bmatrix} p_{A_1|B_j} \\ p_{A_2|B_j} \\ p_{A_3|B_j} \\ p_{A_4|B_j} \end{bmatrix} \neq \begin{bmatrix} p_{A_1|B_{j'}} \\ p_{A_2|B_{j'}} \\ p_{A_3|B_{j'}} \\ p_{A_4|B_{j'}} \end{bmatrix} \quad \text{for some } j \text{ and } j'$$

    (b) $\chi^2 = 17.560$; reject the null hypothesis.

Chapter 18

1. A test is nonparametric if it does not test a hypothesis concerning one of the parameters of the sampled population. It is distribution free if it makes no assumptions about the precise form of the sampled population. Parametric tests test hypotheses about one of the parameters of the sampled population, and they make assumptions about the precise form of the sampled population.

3. $U = 5.5$; reject the null hypothesis that the populations are identical.

5. Without the correction, $z = -1.38$; with the correction, -1.38. Do not reject the null hypothesis.

8. PE is dependent on α, $1 - \beta$, H_0, and H_1, and the sample size of the more efficient comparison test statistic; ARE is a theoretical value that applies when the sample size is infinite, H_1 is essentially identical to H_0, and H_1 is directional.

11. $T = 16.5$; reject the null hypothesis.

13. $T = 30.5$; reject the null hypothesis.

14. $z = -2.28$; reject the null hypothesis.

16. $z = -1.94$; reject the null hypothesis.

Tables

Table D.1. Random Numbers[a]

	1 2 3 4 5	6 7 8 9 10	11 12 13 14 15	16 17 18 19 20	21 22 23 24 25
1	10 27 53 96 23	71 50 54 36 23	54 31 04 82 98	04 14 12 15 09	26 78 25 47 47
2	28 41 50 61 88	64 85 27 20 18	83 36 36 05 56	39 71 65 09 62	94 76 62 11 89
3	34 21 42 57 02	59 19 18 97 48	80 30 03 30 98	05 24 67 70 07	84 97 50 87 46
4	61 81 77 23 23	82 82 11 54 08	53 28 70 58 96	44 07 39 55 43	42 34 43 39 28
5	61 15 18 13 54	16 86 20 26 88	90 74 80 55 09	14 53 90 51 17	52 01 63 01 59
6	91 76 21 64 64	44 91 13 32 97	75 31 62 66 54	84 80 32 75 77	56 08 25 70 29
7	00 97 79 08 06	37 30 28 59 85	53 56 68 53 40	01 74 39 59 73	30 19 99 85 48
8	36 46 18 34 94	75 20 80 27 77	78 91 69 16 00	08 43 18 73 68	67 69 61 34 25
9	88 98 99 60 50	65 95 79 42 94	93 62 40 89 96	43 56 47 71 66	46 76 29 67 02
10	04 37 59 87 21	05 02 03 24 17	47 97 81 56 51	92 34 86 01 82	55 51 33 12 91
11	63 62 06 34 41	94 21 78 55 09	72 76 45 16 94	29 95 81 83 83	79 88 01 97 30
12	78 47 23 53 90	34 41 92 45 71	09 23 70 70 07	12 38 92 79 43	14 85 11 47 23
13	87 68 62 15 43	53 14 36 59 25	54 47 33 70 15	59 24 48 40 35	50 03 42 99 36
14	47 60 92 10 77	88 59 53 11 52	66 25 69 07 04	48 68 64 71 06	61 65 70 22 12
15	56 88 87 59 41	65 28 04 67 53	95 79 88 37 31	50 41 06 94 76	81 83 17 16 33
16	02 57 45 86 67	73 43 07 34 48	44 26 87 93 29	77 09 61 67 84	06 69 44 77 75
17	31 54 14 13 17	48 62 11 90 60	68 12 93 64 28	46 24 79 16 76	14 60 25 51 01
18	28 50 16 43 36	28 97 85 58 99	67 22 52 76 23	24 70 36 54 54	59 28 61 71 96
19	63 29 62 66 50	02 63 45 52 38	67 63 47 54 75	83 24 78 43 20	92 63 13 47 48
20	45 65 58 26 51	76 96 59 38 72	86 57 45 71 46	44 67 76 14 55	44 88 01 62 12
21	39 65 36 63 70	77 45 85 50 51	74 13 39 35 22	30 53 36 02 95	49 34 88 73 61
22	73 71 98 16 04	29 18 94 51 23	76 51 94 84 86	79 93 96 38 63	08 58 25 58 94
23	72 20 56 20 11	72 65 71 08 86	79 57 95 13 91	97 48 72 66 48	09 71 17 24 89
24	75 17 26 99 76	89 37 20 70 01	77 31 61 95 46	26 97 05 73 51	53 33 18 72 87
25	37 48 60 82 29	81 30 15 39 14	48 38 75 93 29	06 87 37 78 48	45 56 00 84 47
26	68 08 02 80 72	83 71 46 30 49	89 17 95 88 29	02 39 56 03 46	97 74 06 56 17
27	14 23 98 61 67	70 52 85 01 50	01 84 02 78 43	10 62 98 19 41	18 83 99 47 99
28	49 08 96 21 44	25 27 99 41 28	07 41 08 34 66	19 42 74 39 91	41 96 53 78 72
29	78 37 06 08 43	63 61 62 42 29	39 68 95 10 96	09 24 23 00 62	56 12 80 73 16
30	37 21 34 17 68	68 96 83 23 56	32 84 60 15 31	44 73 67 34 77	91 15 79 74 58
31	14 29 09 34 04	87 83 07 55 07	76 58 30 83 64	87 29 25 58 84	86 50 60 00 25
32	58 43 28 06 36	49 52 83 51 14	47 56 91 29 34	05 87 31 06 95	12 45 57 09 09
33	10 43 67 29 70	80 62 80 03 42	10 80 21 38 84	90 56 35 03 09	43 12 74 49 14
34	44 38 88 39 54	86 97 37 44 22	00 95 01 31 76	17 16 29 56 63	38 78 94 49 81
35	90 69 59 19 51	85 39 52 85 13	07 28 37 07 61	11 16 36 27 03	78 86 72 04 95
36	41 47 10 25 62	97 05 31 03 61	20 26 36 31 62	68 69 86 95 44	84 95 48 46 45
37	91 94 14 63 19	75 89 11 47 11	31 56 34 19 09	79 57 92 36 59	14 93 87 81 40
38	80 06 54 18 66	09 18 94 06 19	98 40 07 17 81	22 45 44 84 11	24 62 20 42 31
39	67 72 77 63 48	84 08 31 55 58	24 33 45 77 58	80 45 67 93 82	75 70 16 08 24
40	59 40 24 13 27	79 26 88 86 30	01 31 60 10 39	53 58 47 70 93	85 81 56 39 38
41	05 90 35 89 95	01 61 16 96 94	50 78 13 69 36	37 68 53 37 31	71 26 35 03 71
42	44 43 80 69 98	46 68 05 14 82	90 78 50 05 62	77 79 13 57 44	59 60 10 39 66
43	61 81 31 96 82	00 57 25 60 59	46 72 60 18 77	55 66 12 62 11	08 99 55 64 57
44	42 88 07 10 05	24 98 65 63 21	47 21 61 88 32	27 80 30 21 60	10 92 35 36 12
45	77 94 30 05 39	28 10 99 00 27	12 73 73 99 12	49 99 57 94 82	96 88 57 17 91
46	78 83 19 76 16	94 11 68 84 26	23 54 20 86 85	23 86 66 99 07	36 37 34 92 09
47	87 76 59 61 81	43 63 64 61 61	65 76 36 95 90	18 48 27 45 68	27 23 65 30 72
48	91 43 05 96 47	55 78 99 95 24	37 55 85 78 78	01 48 41 19 10	35 19 54 07 73
49	84 97 77 72 73	09 62 06 65 72	87 12 49 03 60	41 15 20 76 27	50 47 02 29 16
50	87 41 60 76 83	44 88 96 07 80	83 05 83 38 96	73 70 66 81 90	30 56 10 48 59

Table D.1 is taken from Table XXXIII of Fisher and Yates: *Statistical Tables for Biological, Agricultural and Medical Research*, published by Longman Group Ltd., London (previously published by Oliver & Boyd, Edinburgh), and reprinted by permission of the authors and publishers.

[a] Discussed in Section 9.1.

Table D.1.   (Continued)

	1 2 3 4 5	6 7 8 9 10	11 12 13 14 15	16 17 18 19 20	21 22 23 24 25
1	22 17 68 65 84	68 95 23 92 35	87 02 22 57 51	61 09 43 95 06	58 24 82 03 47
2	19 36 27 59 46	13 79 93 37 55	39 77 32 77 09	85 52 05 30 62	47 83 51 62 74
3	16 77 23 02 77	09 61 87 25 21	28 06 24 25 93	16 71 13 59 78	23 05 47 47 25
4	78 43 76 71 61	20 44 90 32 64	97 67 63 99 61	46 38 03 93 22	69 81 21 99 21
5	03 28 28 26 08	73 37 32 04 05	69 30 16 09 05	88 69 58 28 99	35 07 44 75 47
6	93 22 53 64 39	07 10 63 76 35	87 03 04 79 88	08 13 13 85 51	55 34 57 72 69
7	78 76 58 54 74	92 38 70 96 92	52 06 79 79 45	82 63 18 27 44	69 66 92 19 09
8	23 68 35 26 00	99 53 93 61 28	52 70 05 48 34	56 65 05 61 86	90 92 10 70 80
9	15 39 25 70 99	93 86 52 77 65	15 33 59 05 28	22 87 26 07 47	86 96 98 29 06
10	58 71 96 30 24	18 46 23 34 27	85 13 99 24 44	49 18 09 79 49	74 16 32 23 02
11	57 35 27 33 72	24 53 63 94 09	41 10 76 47 91	44 04 95 49 66	39 60 04 59 81
12	48 50 86 54 48	22 06 34 72 52	82 21 15 65 20	33 29 94 71 11	15 91 29 12 03
13	61 96 48 95 03	07 16 39 33 66	98 56 10 56 79	77 21 30 27 12	90 49 22 23 62
14	36 93 89 41 26	29 70 83 63 51	99 74 20 52 36	87 09 41 15 09	98 60 16 03 03
15	18 87 00 42 31	57 90 12 02 07	23 47 37 17 31	54 08 01 88 63	39 41 88 92 10
16	88 56 53 27 59	33 35 72 67 47	77 34 55 45 70	08 18 27 38 90	16 95 86 70 75
17	09 72 95 84 29	49 41 31 06 70	42 38 06 45 18	64 84 73 31 65	52 53 37 97 15
18	12 96 88 17 31	65 19 69 02 83	60 75 86 90 68	24 64 19 35 51	56 61 87 39 12
19	85 94 57 24 16	92 09 84 38 76	22 00 27 69 85	29 81 94 78 70	21 94 47 90 12
20	38 64 43 59 98	98 77 87 68 07	91 51 67 62 44	40 98 05 93 78	23 32 65 41 18
21	53 44 09 42 72	00 41 86 79 79	68 47 22 00 20	35 55 31 51 51	00 83 63 22 55
22	40 76 66 26 84	57 99 99 90 37	36 63 32 08 58	37 40 13 68 97	87 64 81 07 83
23	02 17 79 18 05	12 59 52 57 02	22 07 90 47 03	28 14 11 30 79	20 69 22 40 98
24	95 17 82 06 53	31 51 10 96 46	92 06 88 07 77	56 11 50 81 69	40 23 72 51 39
25	35 76 22 42 92	96 11 83 44 80	34 68 35 48 77	33 42 40 90 60	73 96 53 97 86
26	26 29 13 56 41	85 47 04 66 08	34 72 57 59 13	82 43 80 46 15	38 26 61 70 04
27	77 80 20 75 82	72 82 32 99 90	63 95 73 76 63	89 73 44 99 05	48 67 26 43 18
28	46 40 66 44 52	91 36 74 43 53	30 82 13 54 00	78 45 63 98 35	55 03 36 67 68
29	37 56 08 18 09	77 53 84 46 47	31 91 18 95 58	24 16 74 11 53	44 10 13 85 57
30	61 65 61 68 66	37 27 47 39 19	84 83 70 07 48	53 21 40 06 71	95 06 79 88 54
31	93 43 69 64 07	34 18 04 52 35	56 27 09 24 86	61 85 53 83 45	19 90 70 99 00
32	21 96 60 12 99	11 20 99 45 18	48 13 93 55 34	18 37 79 49 90	65 97 38 20 46
33	95 20 47 97 97	27 37 83 28 71	00 06 41 41 74	45 89 09 39 84	51 67 11 52 49
34	97 86 21 78 73	10 65 81 92 59	58 76 17 14 97	04 76 62 16 17	17 95 70 45 80
35	69 92 06 34 13	59 71 74 17 32	27 55 10 24 19	23 71 82 13 74	63 52 52 01 41
36	04 31 17 21 56	33 73 99 19 87	26 72 39 27 67	53 77 57 68 93	60 61 97 22 61
37	61 06 98 03 91	87 14 77 43 96	43 00 65 98 50	45 60 33 01 07	98 99 46 50 47
38	85 93 85 86 88	72 87 08 62 40	16 06 10 89 20	23 21 34 74 97	76 38 03 29 63
39	21 74 32 47 45	73 96 07 94 52	09 65 90 77 47	25 76 16 19 33	53 05 70 53 30
40	15 69 53 82 80	79 96 23 53 10	65 39 07 16 29	45 33 02 43 70	02 87 40 41 45
41	02 89 08 04 49	20 21 14 68 86	87 63 93 95 17	11 29 01 95 80	35 14 97 35 33
42	87 18 15 89 79	85 43 01 72 73	08 61 74 51 69	89 74 39 82 15	94 51 33 41 67
43	98 83 71 94 22	59 97 50 99 52	08 52 85 08 40	87 80 61 65 31	91 51 80 32 44
44	10 08 58 21 66	72 68 49 29 31	89 85 84 46 06	59 73 19 85 23	65 09 29 75 63
45	47 90 56 10 08	88 02 84 27 83	42 29 72 23 19	66 56 45 65 79	20 71 53 20 25
46	22 85 61 68 90	49 64 92 85 44	16 40 12 89 88	50 14 49 81 06	01 82 77 45 12
47	67 80 43 79 33	12 83 11 41 16	25 58 19 68 70	77 02 54 00 52	53 43 37 15 26
48	27 62 50 96 72	79 44 61 40 15	14 53 40 65 39	27 31 58 50 28	11 39 03 34 25
49	33 78 80 87 15	38 30 06 38 21	14 47 47 07 26	54 96 87 53 32	40 36 40 96 76
50	13 13 92 66 99	47 24 49 57 74	32 25 43 62 17	10 97 11 69 84	99 63 22 32 98

Table D.2. **Areas under the Standard Normal Distribution**[a]

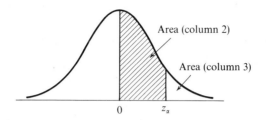

Area (column 2)

Area (column 3)

0   z_α

(1)	(2) Area between Mean and	(3) Area beyond	(1)	(2) Area between Mean and	(3) Area beyond	(1)	(2) Area between Mean and	(3) Area beyond
z_α	z_α	z_α	z_α	z_α	z_α	z_α	z_α	z_α
0.00	0.0000	0.5000	0.30	0.1179	0.3821	0.60	0.2257	0.2743
0.01	0.0040	0.4960	0.31	0.1217	0.3783	0.61	0.2291	0.2709
0.02	0.0080	0.4920	0.32	0.1255	0.3745	0.62	0.2324	0.2676
0.03	0.0120	0.4880	0.33	0.1293	0.3707	0.63	0.2357	0.2643
0.04	0.0160	0.4840	0.34	0.1331	0.3669	0.64	0.2389	0.2611
0.05	0.0199	0.4801	0.35	0.1368	0.3632	0.65	0.2422	0.2578
0.06	0.0239	0.4761	0.36	0.1406	0.3594	0.66	0.2454	0.2546
0.07	0.0279	0.4721	0.37	0.1443	0.3557	0.67	0.2486	0.2514
0.08	0.0319	0.4681	0.38	0.1480	0.3520	0.68	0.2517	0.2483
0.09	0.0359	0.4641	0.39	0.1517	0.3483	0.69	0.2549	0.2451
0.10	0.0398	0.4602	0.40	0.1554	0.3446	0.70	0.2580	0.2420
0.11	0.0438	0.4562	0.41	0.1591	0.3409	0.71	0.2611	0.2389
0.12	0.0478	0.4522	0.42	0.1628	0.3372	0.72	0.2642	0.2358
0.13	0.0517	0.4483	0.43	0.1664	0.3336	0.73	0.2673	0.2327
0.14	0.0557	0.4443	0.44	0.1700	0.3300	0.74	0.2704	0.2296
0.15	0.0596	0.4404	0.45	0.1736	0.3264	0.75	0.2734	0.2266
0.16	0.0636	0.4364	0.46	0.1772	0.3228	0.76	0.2764	0.2236
0.17	0.0675	0.4325	0.47	0.1808	0.3192	0.77	0.2794	0.2206
0.18	0.0714	0.4286	0.48	0.1844	0.3156	0.78	0.2823	0.2177
0.19	0.0753	0.4247	0.49	0.1879	0.3121	0.79	0.2852	0.2148
0.20	0.0793	0.4207	0.50	0.1915	0.3085	0.80	0.2881	0.2119
0.21	0.0832	0.4168	0.51	0.1950	0.3050	0.81	0.2910	0.2090
0.22	0.0871	0.4129	0.52	0.1985	0.3015	0.82	0.2939	0.2061
0.23	0.0910	0.4090	0.53	0.2019	0.2981	0.83	0.2967	0.2033
0.24	0.0948	0.4052	0.54	0.2054	0.2946	0.84	0.2995	0.2005
0.25	0.0987	0.4013	0.55	0.2088	0.2912	0.85	0.3023	0.1977
0.26	0.1026	0.3974	0.56	0.2123	0.2877	0.86	0.3051	0.1949
0.27	0.1064	0.3936	0.57	0.2157	0.2843	0.87	0.3078	0.1922
0.28	0.1103	0.3897	0.58	0.2190	0.2810	0.88	0.3106	0.1894
0.29	0.1141	0.3859	0.59	0.2224	0.2776	0.89	0.3133	0.1867

Table D.2 is abridged from Table IIi of Fisher and Yates: *Statistical Tables for Biological, Agricultural and Medical Research*, published by Longman Group Ltd., London (previously published by Oliver & Boyd, Edinburgh), and reprinted by permission of the authors and publishers.

[a] Discussed in Section 10.1.

Table D.2.   (Continued)

(1) z_α	(2) Area between Mean and z_α	(3) Area beyond z_α	(1) z_α	(2) Area between Mean and z_α	(3) Area beyond z_α	(1) z_α	(2) Area between Mean and z_α	(3) Area beyond z_α
0.90	0.3159	0.1841	1.35	0.4115	0.0885	1.80	0.4641	0.0359
0.91	0.3186	0.1814	1.36	0.4131	0.0869	1.81	0.4649	0.0351
0.92	0.3212	0.1788	1.37	0.4147	0.0853	1.82	0.4656	0.0344
0.93	0.3238	0.1762	1.38	0.4162	0.0838	1.83	0.4664	0.0336
0.94	0.3264	0.1736	1.39	0.4177	0.0823	1.84	0.4671	0.0329
0.95	0.3289	0.1711	1.40	0.4192	0.0808	1.85	0.4678	0.0322
0.96	0.3315	0.1685	1.41	0.4207	0.0793	1.86	0.4686	0.0314
0.97	0.3340	0.1660	1.42	0.4222	0.0778	1.87	0.4693	0.0307
0.98	0.3365	0.1635	1.43	0.4236	0.0764	1.88	0.4699	0.0301
0.99	0.3389	0.1611	1.44	0.4251	0.0749	1.89	0.4706	0.0294
1.00	0.3413	0.1587	1.45	0.4265	0.0735	1.90	0.4713	0.0287
1.01	0.3438	0.1562	1.46	0.4279	0.0721	1.91	0.4719	0.0281
1.02	0.3461	0.1539	1.47	0.4292	0.0708	1.92	0.4726	0.0274
1.03	0.3485	0.1515	1.48	0.4306	0.0694	1.93	0.4732	0.0268
1.04	0.3508	0.1492	1.49	0.4319	0.0681	1.94	0.4738	0.0262
1.05	0.3531	0.1469	1.50	0.4332	0.0668	1.95	0.4744	0.0256
1.06	0.3554	0.1446	1.51	0.4345	0.0655	1.96	0.4750	0.0250
1.07	0.3577	0.1423	1.52	0.4357	0.0643	1.97	0.4756	0.0244
1.08	0.3599	0.1401	1.53	0.4370	0.0630	1.98	0.4761	0.0239
1.09	0.3621	0.1379	1.54	0.4382	0.0618	1.99	0.4767	0.0233
1.10	0.3643	0.1357	1.55	0.4394	0.0606	2.00	0.4772	0.0228
1.11	0.3665	0.1335	1.56	0.4406	0.0594	2.01	0.4778	0.0222
1.12	0.3686	0.1314	1.57	0.4418	0.0582	2.02	0.4783	0.0217
1.13	0.3708	0.1292	1.58	0.4429	0.0571	2.03	0.4788	0.0212
1.14	0.3729	0.1271	1.59	0.4441	0.0559	2.04	0.4793	0.0207
1.15	0.3749	0.1251	1.60	0.4452	0.0548	2.05	0.4798	0.0202
1.16	0.3770	0.1230	1.61	0.4463	0.0537	2.06	0.4803	0.0197
1.17	0.3790	0.1210	1.62	0.4474	0.0526	2.07	0.4808	0.0192
1.18	0.3810	0.1190	1.63	0.4484	0.0516	2.08	0.4812	0.0188
1.19	0.3830	0.1170	1.64	0.4495	0.0505	2.09	0.4817	0.0183
			1.645	0.4500	0.0500			
1.20	0.3849	0.1151	1.65	0.4505	0.0495	2.10	0.4821	0.0179
1.21	0.3869	0.1131	1.66	0.4515	0.0485	2.11	0.4826	0.0174
1.22	0.3888	0.1112	1.67	0.4525	0.0475	2.12	0.4830	0.0170
1.23	0.3907	0.1093	1.68	0.4535	0.0465	2.13	0.4834	0.0166
1.24	0.3925	0.1075	1.69	0.4545	0.0455	2.14	0.4838	0.0162
1.25	0.3944	0.1056	1.70	0.4554	0.0446	2.15	0.4842	0.0158
1.26	0.3962	0.1038	1.71	0.4564	0.0436	2.16	0.4846	0.0154
1.27	0.3980	0.1020	1.72	0.4573	0.0427	2.17	0.4850	0.0150
1.28	0.3997	0.1003	1.73	0.4582	0.0418	2.18	0.4854	0.0146
1.29	0.4015	0.0985	1.74	0.4591	0.0409	2.19	0.4857	0.0143
1.30	0.4032	0.0968	1.75	0.4599	0.0401	2.20	0.4861	0.0139
1.31	0.4049	0.0951	1.76	0.4608	0.0392	2.21	0.4864	0.0136
1.32	0.4066	0.0934	1.77	0.4616	0.0384	2.22	0.4868	0.0132
1.33	0.4082	0.0918	1.78	0.4625	0.0375	2.23	0.4871	0.0129
1.34	0.4099	0.0901	1.79	0.4633	0.0367	2.24	0.4875	0.0125

Table D.2.   (Continued)

(1) z_α	(2) Area between Mean and z_α	(3) Area beyond z_α	(1) z_α	(2) Area between Mean and z_α	(3) Area beyond z_α	(1) z_α	(2) Area between Mean and z_α	(3) Area beyond z_α
2.25	0.4878	0.0122	2.64	0.4959	0.0041	3.00	0.4987	0.0013
2.26	0.4881	0.0119	2.65	0.4960	0.0040	3.01	0.4987	0.0013
2.27	0.4884	0.0116	2.66	0.4961	0.0039	3.02	0.4987	0.0013
2.28	0.4887	0.0113	2.67	0.4962	0.0038	3.03	0.4988	0.0012
2.29	0.4890	0.0110	2.68	0.4963	0.0037	3.04	0.4988	0.0012
2.30	0.4893	0.0107	2.69	0.4964	0.0036	3.05	0.4989	0.0011
2.31	0.4896	0.0104	2.70	0.4965	0.0035	3.06	0.4989	0.0011
2.32	0.4898	0.0102	2.71	0.4966	0.0034	3.07	0.4989	0.0011
2.33	0.4901	0.0099	2.72	0.4967	0.0033	3.08	0.4990	0.0010
2.34	0.4904	0.0096	2.73	0.4968	0.0032	3.09	0.4990	0.0010
2.35	0.4906	0.0094	2.74	0.4969	0.0031	3.10	0.4990	0.0010
2.36	0.4909	0.0091	2.75	0.4970	0.0030	3.11	0.4991	0.0009
2.37	0.4911	0.0089	2.76	0.4971	0.0029	3.12	0.4991	0.0009
2.38	0.4913	0.0087	2.77	0.4972	0.0028	3.13	0.4991	0.0009
2.39	0.4916	0.0084	2.78	0.4973	0.0027	3.14	0.4992	0.0008
2.40	0.4918	0.0082	2.79	0.4974	0.0026	3.15	0.4992	0.0008
2.41	0.4920	0.0080	2.80	0.4974	0.0026	3.16	0.4992	0.0008
2.42	0.4922	0.0078	2.81	0.4975	0.0025	3.17	0.4992	0.0008
2.43	0.4925	0.0075	2.82	0.4976	0.0024	3.18	0.4993	0.0007
2.44	0.4927	0.0073	2.83	0.4977	0.0023	3.19	0.4993	0.0007
2.45	0.4929	0.0071	2.84	0.4977	0.0023	3.20	0.4993	0.0007
2.46	0.4931	0.0069	2.85	0.4978	0.0022	3.21	0.4993	0.0007
2.47	0.4932	0.0068	2.86	0.4979	0.0021	3.22	0.4994	0.0006
2.48	0.4934	0.0066	2.87	0.4979	0.0021	3.23	0.4994	0.0006
2.49	0.4936	0.0064	2.88	0.4980	0.0020	3.24	0.4994	0.0006
2.50	0.4938	0.0062	2.89	0.4981	0.0019	3.25	0.4994	0.0006
2.51	0.4940	0.0060	2.90	0.4981	0.0019	3.30	0.4995	0.0005
2.52	0.4941	0.0059	2.91	0.4982	0.0018	3.35	0.4996	0.0004
2.53	0.4943	0.0057	2.92	0.4982	0.0018	3.40	0.4997	0.0003
2.54	0.4945	0.0055	2.93	0.4983	0.0017	3.45	0.4997	0.0003
2.55	0.4946	0.0054	2.94	0.4984	0.0016	3.50	0.4998	0.0002
2.56	0.4948	0.0052	2.95	0.4984	0.0016	3.60	0.4998	0.0002
2.57	0.4949	0.0051	2.96	0.4985	0.0015	3.70	0.4999	0.0001
2.576	0.4950	0.0050	2.97	0.4985	0.0015	3.80	0.4999	0.0001
2.58	0.4951	0.0049	2.98	0.4986	0.0014	3.90	0.49995	0.00005
2.59	0.4952	0.0048	2.99	0.4986	0.0014	4.00	0.49997	0.00003
2.60	0.4953	0.0047						
2.61	0.4955	0.0045						
2.62	0.4956	0.0044						
2.63	0.4957	0.0043						

Table D.3. Percentage Points of Student's t Distribution[a]

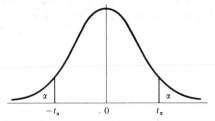

Degrees of Freedom, v	Level of Significance for a One-Tailed Test								
	.25	.20	.15	.10	.05	.025	.01	.005	.0005
	Level of Significance for a Two-Tailed Test								
	.50	.40	.30	.20	.10	.05	.02	.01	.001
1	1.000	1.376	1.963	3.078	6.314	12.706	31.821	63.657	636.619
2	.816	1.061	1.386	1.886	2.920	4.303	6.965	9.925	31.598
3	.765	.978	1.250	1.638	2.353	3.182	4.541	5.841	12.924
4	.741	.941	1.190	1.533	2.132	2.776	3.747	4.604	8.610
5	.727	.920	1.156	1.476	2.015	2.571	3.365	4.032	6.869
6	.718	.906	1.134	1.440	1.943	2.447	3.143	3.707	5.959
7	.711	.896	1.119	1.415	1.895	2.365	2.998	3.499	5.408
8	.706	.889	1.108	1.397	1.860	2.306	2.896	3.355	5.041
9	.703	.883	1.100	1.383	1.833	2.262	2.821	3.250	4.781
10	.700	.879	1.093	1.372	1.812	2.228	2.764	3.169	4.587
11	.697	.876	1.088	1.363	1.796	2.201	2.718	3.106	4.437
12	.695	.873	1.083	1.356	1.782	2.179	2.681	3.055	4.318
13	.694	.870	1.079	1.350	1.771	2.160	2.650	3.012	4.221
14	.692	.868	1.076	1.345	1.761	2.145	2.624	2.977	4.140
15	.691	.866	1.074	1.341	1.753	2.131	2.602	2.947	4.073
16	.690	.865	1.071	1.337	1.746	2.120	2.583	2.921	4.015
17	.689	.863	1.069	1.333	1.740	2.110	2.567	2.898	3.965
18	.688	.862	1.067	1.330	1.734	2.101	2.552	2.878	3.922
19	.688	.861	1.066	1.328	1.729	2.093	2.539	2.861	3.883
20	.687	.860	1.064	1.325	1.725	2.086	2.528	2.845	3.850
21	.686	.859	1.063	1.323	1.721	2.080	2.518	2.831	3.819
22	.686	.858	1.061	1.321	1.717	2.074	2.508	2.819	3.792
23	.685	.858	1.060	1.319	1.714	2.069	2.500	2.807	3.767
24	.685	.857	1.059	1.318	1.711	2.064	2.492	2.797	3.745
25	.684	.856	1.058	1.316	1.708	2.060	2.485	2.787	3.725
26	.684	.856	1.058	1.315	1.706	2.056	2.479	2.779	3.707
27	.684	.855	1.057	1.314	1.703	2.052	2.473	2.771	3.690
28	.683	.855	1.056	1.313	1.701	2.048	2.467	2.763	3.674
29	.683	.854	1.055	1.311	1.699	2.045	2.462	2.756	3.659
30	.683	.854	1.055	1.310	1.697	2.042	2.457	2.750	3.646
40	.681	.851	1.050	1.303	1.684	2.021	2.423	2.704	3.551
60	.679	.848	1.046	1.296	1.671	2.000	2.390	2.660	3.460
120	.677	.845	1.041	1.289	1.658	1.980	2.358	2.617	3.373
∞	.674	.842	1.036	1.282	1.645	1.960	2.326	2.576	3.291

Table D.3 is taken from Table III of Fisher and Yates: *Statistical Tables for Biological, Agricultural and Medical Research*, published by Longman Group Ltd., London (previously published by Oliver & Boyd, Edinburgh), and reprinted by permission of the authors and publishers.

[a] Discussed in Section 12.2.

Table D.4.   **Upper Percentage Points of the Chi-Square Distribution**[a]

ν^b	.99	.98	.95	.90	.80	.70	.50	.30	.20	.10	.05	.02	.01	.001
1	.0³157	.0³628	.00393	.0158	.0642	.148	.455	1.074	1.642	2.706	3.841	5.412	6.635	10.827
2	.0201	.0404	.103	.211	.446	.713	1.386	2.408	3.219	4.605	5.991	7.824	9.210	13.815
3	.115	.185	.352	.584	1.005	1.424	2.366	3.665	4.642	6.251	7.815	9.837	11.345	16.266
4	.297	.429	.711	1.064	1.649	2.195	3.357	4.878	5.989	7.779	9.488	11.668	13.277	18.467
5	.554	.752	1.145	1.610	2.343	3.000	4.351	6.064	7.289	9.236	11.070	13.388	15.086	20.515
6	.872	1.134	1.635	2.204	3.070	3.828	5.348	7.231	8.558	10.645	12.592	15.033	16.812	22.457
7	1.239	1.564	2.167	2.833	3.822	4.671	6.346	8.383	9.803	12.017	14.067	16.622	18.475	24.322
8	1.646	2.032	2.733	3.490	4.594	5.527	7.344	9.524	11.030	13.362	15.507	18.168	20.090	26.125
9	2.088	2.532	3.325	4.168	5.380	6.393	8.343	10.656	12.242	14.684	16.919	19.679	21.666	27.877
10	2.558	3.059	3.940	4.865	6.179	7.267	9.342	11.781	13.442	15.987	18.307	21.161	23.209	29.588
11	3.053	3.609	4.575	5.578	6.989	8.148	10.341	12.899	14.631	17.275	19.675	22.618	24.725	31.264
12	3.571	4.178	5.226	6.304	7.807	9.034	11.340	14.011	15.812	18.549	21.026	24.054	26.217	32.909
13	4.107	4.765	5.892	7.042	8.634	9.926	12.340	15.119	16.985	19.812	22.362	25.472	27.688	34.528
14	4.660	5.368	6.571	7.790	9.467	10.821	13.339	16.222	18.151	21.064	23.685	26.873	29.141	36.123
15	5.229	5.985	7.261	8.547	10.307	11.721	14.339	17.322	19.311	22.307	24.996	28.259	30.578	37.697
16	5.812	6.614	7.962	9.312	11.152	12.624	15.338	18.418	20.465	23.542	26.296	29.633	32.000	39.252
17	6.408	7.255	8.672	10.085	12.002	13.531	16.338	19.511	21.615	24.769	27.587	30.995	33.409	40.790
18	7.015	7.906	9.390	10.865	12.857	14.440	17.338	20.601	22.760	25.989	28.869	32.346	34.805	42.312
19	7.633	8.567	10.117	11.651	13.716	15.352	18.338	21.689	23.900	27.204	30.144	33.687	36.191	43.820
20	8.260	9.237	10.851	12.443	14.578	16.266	19.337	22.775	25.038	28.412	31.410	35.020	37.566	45.315
21	8.897	9.915	11.591	13.240	15.445	17.182	20.337	23.858	26.171	29.615	32.671	36.343	38.932	46.797
22	9.542	10.600	12.338	14.041	16.314	18.101	21.337	24.939	27.301	30.813	33.924	37.659	40.289	48.268
23	10.196	11.293	13.091	14.848	17.187	19.021	22.337	26.018	28.429	32.007	35.172	38.968	41.638	49.728
24	10.856	11.992	13.848	15.659	18.062	19.943	23.337	27.096	29.553	33.196	36.415	40.270	42.980	51.179
25	11.524	12.697	14.611	16.473	18.940	20.867	24.337	28.172	30.675	34.382	37.652	41.566	44.314	52.620
26	12.198	13.409	15.379	17.292	19.820	21.792	25.336	29.246	31.795	35.563	38.885	42.856	45.642	54.052
27	12.879	14.125	16.151	18.114	20.703	22.719	26.336	30.319	32.912	36.741	40.113	44.140	46.963	55.476
28	13.565	14.847	16.928	18.939	21.588	23.647	27.336	31.391	34.027	37.916	41.337	45.419	43.278	56.893
29	14.256	15.574	17.708	19.768	22.475	24.577	28.336	32.461	35.139	39.087	42.557	46.693	49.588	58.302
30	14.953	16.306	18.493	20.599	23.364	25.508	29.336	33.530	36.250	40.256	43.773	47.962	50.892	59.703

Table D.4 is taken from Table IV of Fisher and Yates: *Statistical Tables for Biological, Agricultural and Medical Research*, published by Longman Group Ltd., London (previously published by Oliver & Boyd, Edinburgh), and reprinted by permission of the authors and publishers.

[a] Discussed in Section 12.3.

[b] For $v > 30$, the expression $\sqrt{2\chi^2} - \sqrt{2v - 1}$ may be referred to the standard normal distribution, Table D.2.

Table D.5.   **Upper Percentage Points of the *F* Distribution**[a]

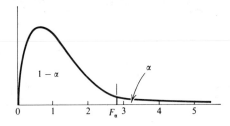

Degrees of Freedom for Denominator, v_2	α	Degrees of Freedom for Numerator, v_1											
		1	*2*	*3*	*4*	*5*	*6*	*7*	*8*	*9*	*10*	*11*	*12*
1	.25	5.83	7.50	8.20	8.58	8.82	8.98	9.10	9.19	9.26	9.32	9.36	9.41
	.10	39.9	49.5	53.6	55.8	57.2	58.2	58.9	59.4	59.9	60.2	60.5	60.7
	.05	161	200	216	225	230	234	237	239	241	242	243	244
2	.25	2.57	3.00	3.15	3.23	3.28	3.31	3.34	3.35	3.37	3.38	3.39	3.39
	.10	8.53	9.00	9.16	9.24	9.29	9.33	9.35	9.37	9.38	9.39	9.40	9.41
	.05	18.5	19.0	19.2	19.2	19.3	19.3	19.4	19.4	19.4	19.4	19.4	19.4
	.01	98.5	99.0	99.2	99.2	99.3	99.3	99.4	99.4	99.4	99.4	99.4	99.4
3	.25	2.02	2.28	2.36	2.39	2.41	2.42	2.43	2.44	2.44	2.44	2.45	2.45
	.10	5.54	5.46	5.39	5.34	5.31	5.28	5.27	5.25	5.24	5.23	5.22	5.22
	.05	10.1	9.55	9.28	9.12	9.01	8.94	8.89	8.85	8.81	8.79	8.76	8.74
	.01	34.1	30.8	29.5	28.7	28.2	27.9	27.7	27.5	27.3	27.2	27.1	27.1
4	.25	1.81	2.00	2.05	2.06	2.07	2.08	2.08	2.08	2.08	2.08	2.08	2.08
	.10	4.54	4.32	4.19	4.11	4.05	4.01	3.98	3.95	3.94	3.92	3.91	3.90
	.05	7.71	6.94	6.59	6.39	6.26	6.16	6.09	6.04	6.00	5.96	5.94	5.91
	.01	21.2	18.0	16.7	16.0	15.5	15.2	15.0	14.8	14.7	14.5	14.4	14.4
5	.25	1.69	1.85	1.88	1.89	1.89	1.89	1.89	1.89	1.89	1.89	1.89	1.89
	.10	4.06	3.78	3.62	3.52	3.45	3.40	3.37	3.34	3.32	3.30	3.28	3.27
	.05	6.61	5.79	5.41	5.19	5.05	4.95	4.88	4.82	4.77	4.74	4.71	4.68
	.01	16.3	13.3	12.1	11.4	11.0	10.7	10.5	10.3	10.2	10.1	9.96	9.89
6	.25	1.62	1.76	1.78	1.79	1.79	1.78	1.78	1.78	1.77	1.77	1.77	1.77
	.10	3.78	3.46	3.29	3.18	3.11	3.05	3.01	2.98	2.96	2.94	2.92	2.90
	.05	5.99	5.14	4.76	4.53	4.39	4.28	4.21	4.15	4.10	4.06	4.03	4.00
	.01	13.7	10.9	9.78	9.15	8.75	8.47	8.26	8.10	7.98	7.87	7.79	7.72
7	.25	1.57	1.70	1.72	1.72	1.71	1.71	1.70	1.70	1.69	1.69	1.69	1.68
	.10	3.59	3.26	3.07	2.96	2.88	2.83	2.78	2.75	2.72	2.70	2.68	2.67
	.05	5.59	4.74	4.35	4.12	3.97	3.87	3.79	3.73	3.68	3.64	3.60	3.57
	.01	12.2	9.55	8.45	7.85	7.46	7.19	6.99	6.84	6.72	6.62	6.54	6.47
8	.25	1.54	1.66	1.67	1.66	1.66	1.65	1.64	1.64	1.63	1.63	1.63	1.62
	.10	3.46	3.11	2.92	2.81	2.73	2.67	2.62	2.59	2.56	2.54	2.52	2.50
	.05	5.32	4.46	4.07	3.84	3.69	3.58	3.50	3.44	3.39	3.35	3.31	3.28
	.01	11.3	8.65	7.59	7.01	6.63	6.37	6.18	6.03	5.91	5.81	5.73	5.67
9	.25	1.51	1.62	1.63	1.63	1.62	1.61	1.60	1.60	1.59	1.59	1.58	1.58
	.10	3.36	3.01	2.81	2.69	2.61	2.55	2.51	2.47	2.44	2.42	2.40	2.38
	.05	5.12	4.26	3.86	3.63	3.48	3.37	3.29	3.23	3.18	3.14	3.10	3.07
	.01	10.6	8.02	6.99	6.42	6.06	5.80	5.61	5.47	5.35	5.26	5.18	5.11

[a] Discussed in Section 14.1.

Table D.5.   (Continued)

				Degrees of Freedom for Numerator, v_1									Degrees of Freedom for Denominator,
15	20	24	30	40	50	60	100	120	200	500	∞	α	v_2
9.49	9.58	9.63	9.67	9.71	9.74	9.76	9.78	9.80	9.82	9.84	9.85	.25	
61.2	61.7	62.0	62.3	62.5	62.7	62.8	63.0	63.1	63.2	63.3	63.3	.10	1
246	248	249	250	251	252	252	253	253	254	254	254	.05	
3.41	3.43	3.43	3.44	3.45	3.45	3.46	3.47	3.47	3.48	3.48	3.48	.25	
9.42	9.44	9.45	9.46	9.47	9.47	9.47	9.48	9.48	9.49	9.49	9.49	.10	2
19.4	19.4	19.5	19.5	19.5	19.5	19.5	19.5	19.5	19.5	19.5	19.5	.05	
99.4	99.4	99.5	99.5	99.5	99.5	99.5	99.5	99.5	99.5	99.5	99.5	.01	
2.46	2.46	2.46	2.47	2.47	2.47	2.47	2.47	2.47	2.47	2.47	2.47	.25	
5.20	5.18	5.18	5.17	5.16	5.15	5.15	5.14	5.14	5.14	5.14	5.13	.10	3
8.70	8.66	8.64	8.62	8.59	8.58	8.57	8.55	8.55	8.54	8.53	8.53	.05	
26.9	26.7	26.6	26.5	26.4	26.4	26.3	26.2	26.2	26.2	26.1	26.1	.01	
2.08	2.08	2.08	2.08	2.08	2.08	2.08	2.08	2.08	2.08	2.08	2.08	.25	
3.87	3.84	3.83	3.82	3.80	3.80	3.79	3.78	3.78	3.77	3.76	3.76	.10	4
5.86	5.80	5.77	5.75	5.72	5.70	5.69	5.66	5.66	5.65	5.64	5.63	.05	
14.2	14.0	13.9	13.8	13.7	13.7	13.7	13.6	13.6	13.5	13.5	13.5	.01	
1.89	1.88	1.88	1.88	1.88	1.88	1.87	1.87	1.87	1.87	1.87	1.87	.25	
3.24	3.21	3.19	3.17	3.16	3.15	3.14	3.13	3.12	3.12	3.11	3.10	.10	5
4.62	4.56	4.53	4.50	4.46	4.44	4.43	4.41	4.40	4.39	4.37	4.36	.05	
9.72	9.55	9.47	9.38	9.29	9.24	9.20	9.13	9.11	9.08	9.04	9.02	.01	
1.76	1.76	1.75	1.75	1.75	1.75	1.74	1.74	1.74	1.74	1.74	1.74	.25	
2.87	2.84	2.82	2.80	2.78	2.77	2.76	2.75	2.74	2.73	2.73	2.72	.10	6
3.94	3.87	3.84	3.81	3.77	3.75	3.74	3.71	3.70	3.69	3.68	3.67	.05	
7.56	7.40	7.31	7.23	7.14	7.09	7.06	6.99	6.97	6.93	6.90	6.88	.01	
1.68	1.67	1.67	1.66	1.66	1.66	1.65	1.65	1.65	1.65	1.65	1.65	.25	
2.63	2.59	2.58	2.56	2.54	2.52	2.51	2.50	2.49	2.48	2.48	2.47	.10	7
3.51	3.44	3.41	3.38	3.34	3.32	3.30	3.27	3.27	3.25	3.24	3.23	.05	
6.31	6.16	6.07	5.99	5.91	5.86	5.82	5.75	5.74	5.70	5.67	5.65	.10	
1.62	1.61	1.60	1.60	1.59	1.59	1.59	1.58	1.58	1.58	1.58	1.58	.25	
2.46	2.42	2.40	2.38	2.36	2.35	2.34	2.32	2.32	2.31	2.30	2.29	.10	8
3.22	3.15	3.12	3.08	3.04	3.02	3.01	2.97	2.97	2.95	2.94	2.93	.05	
5.52	5.36	5.28	5.20	5.12	5.07	5.03	4.96	4.95	4.91	4.88	4.86	.01	
1.57	1.56	1.56	1.55	1.55	1.54	1.54	1.53	1.53	1.53	1.53	1.53	.25	
2.34	2.30	2.28	2.25	2.23	2.22	2.21	2.19	2.18	2.17	2.17	2.16	.10	9
3.01	2.94	2.90	2.86	2.83	2.80	2.79	2.76	2.75	2.73	2.72	2.71	.05	
4.96	4.81	4.73	4.65	4.57	4.52	4.48	4.42	4.40	4.36	4.33	4.31	.01	

Table D.5. (Continued)

Degrees of Freedom for Denominator, v_2	α	\multicolumn{12}{c}{Degrees of Freedom for Numerator, v_1}
		1	2	3	4	5	6	7	8	9	10	11	12
10	.25	1.49	1.60	1.60	1.59	1.59	1.58	1.57	1.56	1.56	1.55	1.55	1.54
	.10	3.29	2.92	2.73	2.61	2.52	2.46	2.41	2.38	2.35	2.32	2.30	2.28
	.05	4.96	4.10	3.71	3.48	3.33	3.22	3.14	3.07	3.02	2.98	2.94	2.91
	.01	10.0	7.56	6.55	5.99	5.64	5.39	5.20	5.06	4.94	4.85	4.77	4.71
11	.25	1.47	1.58	1.58	1.57	1.56	1.55	1.54	1.53	1.53	1.52	1.52	1.51
	.10	3.23	2.86	2.66	2.54	2.45	2.39	2.34	2.30	2.27	2.25	2.23	2.21
	.05	4.84	3.98	3.59	3.36	3.20	3.09	3.01	2.95	2.90	2.85	2.82	2.79
	.01	9.65	7.21	6.22	5.67	5.32	5.07	4.89	4.74	4.63	4.54	4.46	4.40
12	.25	1.46	1.56	1.56	1.55	1.54	1.53	1.52	1.51	1.51	1.50	1.50	1.49
	.10	3.18	2.81	2.61	2.48	2.39	2.33	2.28	2.24	2.21	2.19	2.17	2.15
	.05	4.75	3.89	3.49	3.26	3.11	3.00	2.91	2.85	2.80	2.75	2.72	2.69
	.01	9.33	6.93	5.95	5.41	5.06	4.82	4.64	4.50	4.39	4.30	4.22	4.16
13	.25	1.45	1.55	1.55	1.53	1.52	1.51	1.50	1.49	1.49	1.48	1.47	1.47
	.10	3.14	2.76	2.56	2.43	2.35	2.28	2.23	2.20	2.16	2.14	2.12	2.10
	.05	4.67	3.81	3.41	3.18	3.03	2.92	2.83	2.77	2.71	2.67	2.63	2.60
	.01	9.07	6.70	5.74	5.21	4.86	4.62	4.44	4.30	4.19	4.10	4.02	3.96
14	.25	1.44	1.53	1.53	1.52	1.51	1.50	1.49	1.48	1.47	1.46	1.46	1.45
	.10	3.10	2.73	2.52	2.39	2.31	2.24	2.19	2.15	2.12	2.10	2.08	2.05
	.05	4.60	3.74	3.34	3.11	2.96	2.85	2.76	2.70	2.65	2.60	2.57	2.53
	.01	8.86	6.51	5.56	5.04	4.69	4.46	4.28	4.14	4.03	3.94	3.86	3.80
15	.25	1.43	1.52	1.52	1.51	1.49	1.48	1.47	1.46	1.46	1.45	1.44	1.44
	.10	3.07	2.70	2.49	2.36	2.27	2.21	2.16	2.12	2.09	2.06	2.04	2.02
	.05	4.54	3.68	3.29	3.06	2.90	2.79	2.71	2.64	2.59	2.54	2.51	2.48
	.01	8.68	6.36	5.42	4.89	4.56	4.32	4.14	4.00	3.89	3.80	3.73	3.67
16	.25	1.42	1.51	1.51	1.50	1.48	1.47	1.46	1.45	1.44	1.44	1.44	1.43
	.10	3.05	2.67	2.46	2.33	2.24	2.18	2.13	2.09	2.06	2.03	2.01	1.99
	.05	4.49	3.63	3.24	3.01	2.85	2.74	2.66	2.59	2.54	2.49	2.46	2.42
	.01	8.53	6.23	5.29	4.77	4.44	4.20	4.03	3.89	3.78	3.69	3.62	3.55
17	.25	1.42	1.51	1.50	1.49	1.47	1.46	1.45	1.44	1.43	1.43	1.42	1.41
	.10	3.03	2.64	2.44	2.31	2.22	2.15	2.10	2.06	2.03	2.00	1.98	1.96
	.05	4.45	3.59	3.20	2.96	2.81	2.70	2.61	2.55	2.49	2.45	2.41	2.38
	.01	8.40	6.11	5.18	4.67	4.34	4.10	3.93	3.79	3.68	3.59	3.52	3.46
18	.25	1.41	1.50	1.49	1.48	1.46	1.45	1.44	1.43	1.42	1.42	1.41	1.40
	.10	3.01	2.62	2.42	2.29	2.20	2.13	2.08	2.04	2.00	1.98	1.96	1.93
	.05	4.41	3.55	3.16	2.93	2.77	2.66	2.58	2.51	2.46	2.41	2.37	2.34
	.01	8.29	6.01	5.09	4.58	4.25	4.01	3.84	3.71	3.60	3.51	3.43	3.37
19	.25	1.41	1.49	1.49	1.47	1.46	1.44	1.43	1.42	1.41	1.41	1.40	1.40
	.10	2.99	2.61	2.40	2.27	2.18	2.11	2.06	2.02	1.98	1.96	1.94	1.91
	.05	4.38	3.52	3.13	2.90	2.74	2.63	2.54	2.48	2.42	2.38	2.34	2.31
	.01	8.18	5.93	5.01	4.50	4.17	3.94	3.77	3.63	3.52	3.43	3.36	3.30
20	.25	1.40	1.49	1.48	1.46	1.45	1.44	1.43	1.42	1.41	1.40	1.39	1.39
	.10	2.97	2.59	2.38	2.25	2.16	2.09	2.04	2.00	1.96	1.94	1.92	1.89
	.05	4.35	3.49	3.10	2.87	2.71	2.60	2.51	2.45	2.39	2.35	2.31	2.28
	.01	8.10	5.85	4.94	4.43	4.10	3.87	3.70	3.56	3.46	3.37	3.29	3.23

Table D.5. (Continued)

15	20	24	30	40	50	60	100	120	200	500	∞	α	Degrees of Freedom for Denominator, v_2
1.53	1.52	1.52	1.51	1.51	1.50	1.50	1.49	1.49	1.49	1.48	1.48	.25	
2.24	2.20	2.18	2.16	2.13	2.12	2.11	2.09	2.08	2.07	2.06	2.06	.10	10
2.85	2.77	2.74	2.70	2.66	2.64	2.62	2.59	2.58	2.56	2.55	2.54	.05	
4.56	4.41	4.33	4.25	4.17	4.12	4.08	4.01	4.00	3.96	3.93	3.91	.01	
1.50	1.49	1.49	1.48	1.47	1.47	1.47	1.46	1.46	1.46	1.45	1.45	.25	
2.17	2.12	2.10	2.08	2.05	2.04	2.03	2.00	2.00	1.99	1.98	1.97	.10	11
2.72	2.65	2.61	2.57	2.53	2.51	2.49	2.46	2.45	2.43	2.42	2.40	.05	
4.25	4.10	4.02	3.94	3.86	3.81	3.78	3.71	3.69	3.66	3.62	3.60	.01	
1.48	1.47	1.46	1.45	1.45	1.44	1.44	1.43	1.43	1.43	1.42	1.42	.25	
2.10	2.06	2.04	2.01	1.99	1.97	1.96	1.94	1.93	1.92	1.91	1.90	.10	12
2.62	2.54	2.51	2.47	2.43	2.40	2.38	2.35	2.34	2.32	2.31	2.30	.05	
4.01	3.86	3.78	3.70	3.62	3.57	3.54	3.47	3.45	3.41	3.38	3.36	.01	
1.46	1.45	1.44	1.43	1.42	1.42	1.42	1.41	1.41	1.40	1.40	1.40	.25	
2.05	2.01	1.98	1.96	1.93	1.92	1.90	1.88	1.88	1.86	1.85	1.85	.10	13
2.53	2.46	2.42	2.38	2.34	2.31	2.30	2.26	2.25	2.23	2.22	2.21	.05	
3.82	3.66	3.59	3.51	3.43	3.38	3.34	3.27	3.25	3.22	3.19	3.17	.01	
1.44	1.43	1.42	1.41	1.41	1.40	1.40	1.39	1.39	1.39	1.38	1.38	.25	
2.01	1.96	1.94	1.91	1.89	1.87	1.86	1.83	1.83	1.82	1.80	1.80	.10	14
2.46	2.39	2.35	2.31	2.27	2.24	2.22	2.19	2.18	2.16	2.14	2.13	.05	
3.66	3.51	3.43	3.35	3.27	3.22	3.18	3.11	3.09	3.06	3.03	3.00	.01	
1.43	1.41	1.41	1.40	1.39	1.39	1.38	1.38	1.37	1.37	1.36	1.36	.25	
1.97	1.92	1.90	1.87	1.85	1.83	1.82	1.79	1.79	1.77	1.76	1.76	.10	15
2.40	2.33	2.29	2.25	2.20	2.18	2.16	2.12	2.11	2.10	2.08	2.07	.05	
3.52	3.37	3.29	3.21	3.13	3.08	3.05	2.98	2.96	2.92	2.89	2.87	.01	
1.41	1.40	1.39	1.38	1.37	1.37	1.36	1.36	1.35	1.35	1.34	1.34	.25	
1.94	1.89	1.87	1.84	1.81	1.79	1.78	1.76	1.75	1.74	1.73	1.72	.10	16
2.35	2.28	2.24	2.19	2.15	2.12	2.11	2.07	2.06	2.04	2.02	2.01	.05	
3.41	3.26	3.18	3.10	3.02	2.97	2.93	2.86	2.84	2.81	2.78	2.75	.01	
1.40	1.39	1.38	1.37	1.36	1.35	1.35	1.34	1.34	1.34	1.33	1.33	.25	
1.91	1.86	1.84	1.81	1.78	1.76	1.75	1.73	1.72	1.71	1.69	1.69	.10	17
2.31	2.23	2.19	2.15	2.10	2.08	2.06	2.02	2.01	1.99	1.97	1.96	.05	
3.31	3.16	3.08	3.00	2.92	2.87	2.83	2.76	2.75	2.71	2.68	2.65	.01	
1.39	1.38	1.37	1.36	1.35	1.34	1.34	1.33	1.33	1.32	1.32	1.32	.25	
1.89	1.84	1.81	1.78	1.75	1.74	1.72	1.70	1.69	1.68	1.67	1.66	.10	18
2.27	2.19	2.15	2.11	2.06	2.04	2.02	1.98	1.97	1.95	1.93	1.92	.05	
3.23	3.08	3.00	2.92	2.84	2.78	2.75	2.68	2.66	2.62	2.59	2.57	.01	
1.38	1.37	1.36	1.35	1.34	1.33	1.33	1.32	1.32	1.31	1.31	1.30	.25	
1.86	1.81	1.79	1.76	1.73	1.71	1.70	1.67	1.67	1.65	1.64	1.63	.10	19
2.23	2.16	2.11	2.07	2.03	2.00	1.98	1.94	1.93	1.91	1.89	1.88	.05	
3.15	3.00	2.92	2.84	2.76	2.71	2.67	2.60	2.58	2.55	2.51	2.49	.01	
1.37	1.36	1.35	1.34	1.33	1.33	1.32	1.31	1.31	1.30	1.30	1.29	.25	
1.84	1.79	1.77	1.74	1.71	1.69	1.68	1.65	1.64	1.63	1.62	1.61	.10	20
2.20	2.12	2.08	2.04	1.99	1.97	1.95	1.91	1.90	1.88	1.86	1.84	.05	
3.09	2.94	2.86	2.78	2.69	2.64	2.61	2.54	2.52	2.48	2.44	2.42	.01	

Degrees of Freedom for Numerator, v_1

Table D.5. (*Continued*)

Degrees of Freedom for Denominator, v_2	α	Degrees of Freedom for Numerator, v_1											
		1	2	3	4	5	6	7	8	9	10	11	12
22	.25	1.40	1.48	1.47	1.45	1.44	1.42	1.41	1.40	1.39	1.39	1.38	1.37
	.10	2.95	2.56	2.35	2.22	2.13	2.06	2.01	1.97	1.93	1.90	1.88	1.86
	.05	4.30	3.44	3.05	2.82	2.66	2.55	2.46	2.40	2.34	2.30	2.26	2.23
	.01	7.95	5.72	4.82	4.31	3.99	3.76	3.59	3.45	3.35	3.26	3.18	3.12
24	.25	1.39	1.47	1 46	1.44	1.43	1.41	1.40	1.39	1.38	1.38	1.37	1.36
	.10	2.93	2.54	2.33	2.19	2.10	2.04	1.98	1.94	1.91	1.88	1.85	1.83
	.05	4.26	3.40	3.01	2.78	2.62	2.51	2.42	2.36	2.30	2.25	2.21	2.18
	.01	7.82	5.61	4.72	4.22	3.90	3.67	3.50	3.36	3.26	3.17	3.09	3.03
26	.25	1.38	1.46	1.45	1.44	1.42	1.41	1.39	1.38	1.37	1.37	1.36	1.35
	.10	2.91	2.52	2.31	2.17	2.08	2.01	1.96	1.92	1.88	1.86	1.84	1.81
	.05	4.23	3.37	2.98	2.74	2.59	2.47	2.39	2.32	2.27	2.22	2.18	2.15
	.01	7.72	5.53	4.64	4.14	3.82	3.59	3.42	3.29	3.18	3.09	3.02	2.96
28	.25	1.38	1.46	1.45	1.43	1.41	1.40	1.39	1.38	1.37	1.36	1.35	1.34
	.10	2.89	2.50	2.29	2.16	2.06	2.00	1.94	1.90	1.87	1.84	1.81	1.79
	.05	4.20	3.34	2.95	2.71	2.56	2.45	2.36	2.29	2.24	2.19	2.15	2.12
	.01	7.64	5.45	4.57	4.07	3.75	3.53	3.36	3.23	3.12	3.03	2.96	2.90
30	.25	1.38	1.45	1.44	1.42	1.41	1.39	1.38	1.37	1.36	1.35	1.35	1.34
	.10	2.88	2.49	2.28	2.14	2.05	1.98	1.93	1.88	1.85	1.82	1.79	1.77
	.05	4.17	3.32	2.92	2.69	2.53	2.42	2.33	2.27	2.21	2.16	2.13	2.09
	.01	7.56	5.39	4.51	4.02	3.70	3.47	3.30	3.17	3.07	2.98	2.91	2.84
40	.25	1.36	1.44	1.42	1.40	1.39	1.37	1.36	1.35	1.34	1.33	1.32	1.31
	.10	2.84	2.44	2.23	2.09	2.00	1.93	1.87	1.83	1.79	1.76	1.73	1.71
	.05	4.08	3.23	2.84	2.61	2.45	2.34	2.25	2.18	2.12	2.08	2.04	2.00
	.01	7.31	5.18	4.31	3.83	3.51	3.29	3.12	2.99	2.89	2.80	2.73	2.66
60	.25	1.35	1.42	1.41	1.38	1.37	1.35	1.33	1.32	1.31	1.30	1.29	1.29
	.10	2.79	2.39	2.18	2.04	1.95	1.87	1.82	1.77	1.74	1.71	1.68	1.66
	.05	4.00	3.15	2.76	2.53	2.37	2.25	2.17	2.10	2.04	1.99	1.95	1.92
	.01	7.08	4.98	4.13	3.65	3.34	3.12	2.95	2.82	2.72	2.63	2.56	2.50
120	.25	1.34	1.40	1.39	1.37	1.35	1.33	1.31	1.30	1.29	1.28	1.27	1.26
	.10	2.75	2.35	2.13	1.99	1.90	1.82	1.77	1.72	1.68	1.65	1.62	1.60
	.05	3.92	3.07	2.68	2.45	2.29	2.17	2.09	2.02	1.96	1.91	1.87	1.83
	.01	6.85	4.79	3.95	3.48	3.17	2.96	2.79	2.66	2.56	2.47	2.40	2.34
200	.25	1.33	1.39	1.38	1.36	1.34	1.32	1.31	1.29	1.28	1.27	1.26	1.25
	.10	2.73	2.33	2.11	1.97	1.88	1.80	1.75	1.70	1.66	1.63	1.60	1.57
	.05	3.89	3.04	2.65	2.42	2.26	2.14	2.06	1.98	1.93	1.88	1.84	1.80
	.01	6.76	4.71	3.88	3.41	3.11	2.89	2.73	2.60	2.50	2.41	2.34	2.27
∞	.25	1.32	1.39	1.37	1.35	1.33	1.31	1.29	1.28	1.27	1.25	1.24	1.24
	.10	2.71	2.30	2.08	1.94	1.85	1.77	1.72	1.67	1.63	1.60	1.57	1.55
	.05	3.84	3.00	2.60	2.37	2.21	2.10	2.01	1.94	1.88	1.83	1.79	1.75
	.01	6.63	4.61	3.78	3.32	3.02	2.80	2.64	2.51	2.41	2.32	2.25	2.18

Table D.5. (Continued)

15	20	24	30	40	50	60	100	120	200	500	∞	α	v_2
				Degrees of Freedom for Numerator, v_1									Degrees of Freedom for Denominator,
1.36	1.34	1.33	1.32	1.31	1.31	1.30	1.30	1.30	1.29	1.29	1.28	.25	
1.81	1.76	1.73	1.70	1.67	1.65	1.64	1.61	1.60	1.59	1.58	1.57	.10	22
2.15	2.07	2.03	1.98	1.94	1.91	1.89	1.85	1.84	1.82	1.80	1.78	.05	
2.98	2.83	2.75	2.67	2.58	2.53	2.50	2.42	2.40	2.36	2.33	2.31	.01	
1.35	1.33	1.32	1.31	1.30	1.29	1.29	1.28	1.28	1.27	1.27	1.26	.25	
1.78	1.73	1.70	1.67	1.64	1.62	1.61	1.58	1.57	1.56	1.54	1.53	.10	24
2.11	2.03	1.98	1.94	1.89	1.86	1.84	1.80	1.79	1.77	1.75	1.73	.05	
2.89	2.74	2.66	2.58	2.49	2.44	2.40	2.33	2.31	2.27	2.24	2.21	.01	
1.34	1.32	1.31	1.30	1.29	1.28	1.28	1.26	1.26	1.26	1.25	1.25	.25	
1.76	1.71	1.68	1.65	1.61	1.59	1.58	1.55	1.54	1.53	1.51	1.50	.10	26
2.07	1.99	1.95	1.90	1.85	1.82	1.80	1.76	1.75	1.73	1.71	1.69	.05	
2.81	2.66	2.58	2.50	2.42	2.36	2.33	2.25	2.23	2.19	2.16	2.13	.01	
1.33	1.31	1.30	1.29	1.28	1.27	1.27	1.26	1.25	1.25	1.24	1.24	.25	
1.74	1.69	1.66	1.63	1.59	1.57	1.56	1.53	1.52	1.50	1.49	1.48	.10	28
2.04	1.96	1.91	1.87	1.82	1.79	1.77	1.73	1.71	1.69	1.67	1.65	.05	
2.75	2.60	2.52	2.44	2.35	2.30	2.26	2.19	2.17	2.13	2.09	2.06	.01	
1.32	1.30	1.29	1.28	1.27	1.26	1.26	1.25	1.24	1.24	1.23	1.23	.25	
1.72	1.67	1.64	1.61	1.57	1.55	1.54	1.51	1.50	1.48	1.47	1.46	.10	30
2.01	1.93	1.89	1.84	1.79	1.76	1.74	1.70	1.68	1.66	1.64	1.62	.05	
2.70	2.55	2.47	2.39	2.30	2.25	2.21	2.13	2.11	2.07	2.03	2.01	.01	
1.30	1.28	1.26	1.25	1.24	1.23	1.22	1.21	1.21	1.20	1.19	1.19	.25	
1.66	1.61	1.57	1.54	1.51	1.48	1.47	1.43	1.42	1.41	1.39	1.38	.10	40
1.92	1.84	1.79	1.74	1.69	1.66	1.64	1.59	1.58	1.55	1.53	1.51	.05	
2.52	2.37	2.29	2.20	2.11	2.06	2.02	1.94	1.92	1.87	1.83	1.80	.01	
1.27	1.25	1.24	1.22	1 21	1 20	1.19	1.17	1.17	1.16	1.15	1.15	.25	
1.60	1.54	1.51	1.48	1.44	1.41	1.40	1.36	1.35	1.33	1.31	1.29	.10	60
1.84	1.75	1.70	1.65	1.59	1.56	1.53	1.48	1.47	1.44	1.41	1.39	.05	
2.35	2.20	2.12	2.03	1.94	1.88	1.84	1.75	1.73	1.68	1.63	1.60	.01	
1.24	1.22	1.21	1.19	1.18	1.17	1.16	1.14	1.13	1.12	1.11	1.10	.25	
1.55	1.48	1.45	1.41	1.37	1.34	1.32	1.27	1.26	1.24	1.21	1.19	.10	120
1.75	1.66	1.61	1.55	1.50	1.46	1.43	1.37	1.35	1.32	1.28	1.25	.05	
2.19	2.03	1.95	1.86	1.76	1.70	1.66	1.56	1.53	1.48	1.42	1.38	.01	
1.23	1.21	1.20	1.18	1.16	1.14	1.12	1.11	1.10	1.09	1.08	1.06	.25	
1.52	1.46	1.42	1.38	1.34	1.31	1.28	1.24	1.22	1.20	1.17	1.14	.10	200
1.72	1.62	1.57	1.52	1.46	1.41	1.39	1.32	1.29	1.26	1.22	1.19	.05	
2.13	1.97	1.89	1.79	1.69	1.63	1.58	1.48	1.44	1.39	1.33	1.28	.01	
1.22	1.19	1.18	1.16	1.14	1.13	1.12	1.09	1.08	1.07	1.04	1.00	.25	
1.49	1.42	1.38	1.34	1.30	1.26	1.24	1.18	1.17	1.13	1.08	1.00	.10	∞
1.67	1.57	1.52	1.46	1.39	1.35	1.32	1.24	1.22	1.17	1.11	1.00	.05	
2.04	1.88	1.79	1.70	1.59	1.52	1.47	1.36	1.32	1.25	1.15	1.00	.01	

Table D.6.   Critical Values of the Pearson r^a

Degrees of Freedom, $v = n - 2$ [b]	Level of Significance for a One-Tailed Test			
	.05	.025	.01	.005
	Level of Significance for a Two-Tailed Test			
	.10	.05	.02	.01
1	0.988	0.997	0.9995	0.9999
2	0.900	0.950	0.980	0.990
3	0.805	0.378	0.934	0.959
4	0.729	0.811	0.882	0.917
5	0.669	0.754	0.833	0.874
6	0.622	0.707	0.789	0.834
7	0.582	0.666	0.750	0.798
8	0.549	0.632	0.716	0.765
9	0.521	0.602	0.685	0.735
10	0.497	0.576	0.658	0.708
11	0.476	0.553	0.634	0.684
12	0.458	0.532	0.612	0.661
13	0.441	0.514	0.592	0.641
14	0.426	0.497	0.574	0.623
15	0.412	0.482	0.558	0.606
16	0.400	0.468	0.542	0.590
17	0.389	0.456	0.528	0.575
18	0.378	0.444	0.516	0.561
19	0.369	0.433	0.503	0.549
20	0.360	0.423	0.492	0.537
21	0.352	0.413	0.482	0.526
22	0.344	0.404	0.472	0.515
23	0.337	0.396	0.462	0.505
24	0.330	0.388	0.453	0.496
25	0.323	0.381	0.445	0.487
26	0.317	0.374	0.437	0.479
27	0.311	0.367	0.430	0.471
28	0.306	0.361	0.423	0.463
29	0.301	0.355	0.416	0.456
30	0.296	0.349	0.409	0.449
35	0.275	0.325	0.381	0.418
40	0.257	0.304	0.358	0.393
45	0.243	0.288	0.338	0.372
50	0.231	0.273	0.322	0.354
60	0.211	0.250	0.295	0.325
70	0.195	0.232	0.274	0.302
80	0.183	0.217	0.256	0.283
90	0.173	0.205	0.242	0.267
100	0.164	0.195	0.230	0.254
120	0.150	0.178	0.210	0.232
150	0.134	0.159	0.189	0.208
200	0.116	0.138	0.164	0.181
300	0.095	0.113	0.134	0.148
400	0.082	0.098	0.116	0.128
500	0.073	0.088	0.104	0.115

Table D.6 is taken from Table VII of Fisher and Yates: *Statistical Tables for Biological, Agricultural and Medical Research*, published by Longman Group Ltd., London (previously published by Oliver & Boyd, Edinburgh), and reprinted by permission of the authors and publishers.

[a] Discussed in Section 12.5.
[b] n is the number of pairs.

Table D.7. **Critical Values of r_s (Spearman Rank Correlation Coefficient)^a**

Number of Pairs, n	Level of Significance for a One-Tailed Test			
	.05	.025	.01	.005
	Level of Significance for a Two-Tailed Test			
	.10	.05	.02	.01
5	0.900	1.000	1.000	—
6	0.829	0.886	0.943	1.000
7	0.714	0.786	0.893	0.929
8	0.643	0.738	0.833	0.881
9	0.600	0.683	0.783	0.833
10	0.564	0.648	0.746	0.794
12	0.506	0.591	0.712	0.777
14	0.456	0.544	0.645	0.715
16	0.425	0.506	0.601	0.665
18	0.399	0.475	0.564	0.625
20	0.377	0.450	0.534	0.591
22	0.359	0.428	0.508	0.562
24	0.343	0.409	0.485	0.537
26	0.329	0.392	0.465	0.515
28	0.317	0.377	0.448	0.496
30	0.306	0.364	0.432	0.478

From "Distributions of Sums of Squares of Rank Differences for Small Numbers of Individuals," by E. G. Olds, *Annals of Mathematical Statistics*, 1938, *9*, 133–148, and from "The 5 Per Cent Significance Levels for Sums of Squares of Rank Differences and a Correction," by E. G. Olds, *Annals of Mathematical Statistics*, 1949, *20*, 117–118. Reprinted by permission.

^{a} Discussed in Section 12.5.

Table D.8.   Transformation of r to Z'[a]

r	Z'	r	Z'	r	Z'	r	Z'	r	Z'
0.000	0.000	0.200	0.203	0.400	0.424	0.600	0.693	0.800	1.099
0.005	0.005	0.205	0.208	0.405	0.430	0.605	0.701	0.805	1.113
0.010	0.010	0.210	0.213	0.410	0.436	0.610	0.709	0.810	1.127
0.015	0.015	0.215	0.218	0.415	0.442	0.615	0.717	0.815	1.142
0.020	0.020	0.220	0.224	0.420	0.448	0.620	0.725	0.820	1.157
0.025	0.025	0.225	0.229	0.425	0.454	0.625	0.733	0.825	1.172
0.030	0.030	0.230	0.234	0.430	0.460	0.630	0.741	0.830	1.188
0.035	0.035	0.235	0.239	0.435	0.466	0.635	0.750	0.835	1.204
0.040	0.040	0.240	0.245	0.440	0.472	0.640	0.758	0.840	1.221
0.045	0.045	0.245	0.250	0.445	0.478	0.645	0.767	0.845	1.238
0.050	0.050	0.250	0.255	0.450	0.485	0.650	0.775	0.850	1.256
0.055	0.055	0.255	0.261	0.455	0.491	0.655	0.784	0.855	1.274
0.060	0.060	0.260	0.266	0.460	0.497	0.660	0.793	0.860	1.293
0.065	0.065	0.265	0.271	0.465	0.504	0.665	0.802	0.865	1.313
0.070	0.070	0.270	0.277	0.470	0.510	0.670	0.811	0.870	1.333
0.075	0.075	0.275	0.282	0.475	0.517	0.675	0.820	0.875	1.354
0.080	0.080	0.280	0.288	0.480	0.523	0.680	0.829	0.880	1.376
0.085	0.085	0.285	0.293	0.485	0.530	0.685	0.838	0.885	1.398
0.090	0.090	0.290	0.299	0.490	0.536	0.690	0.848	0.890	1.422
0.095	0.095	0.295	0.304	0.495	0.543	0.695	0.858	0.895	1.447
0.100	0.100	0.300	0.310	0.500	0.549	0.700	0.867	0.900	1.472
0.105	0.105	0.305	0.315	0.505	0.556	0.705	0.877	0.905	1.499
0.110	0.110	0.310	0.321	0.510	0.563	0.710	0.887	0.910	1.528
0.115	0.116	0.315	0.326	0.515	0.570	0.715	0.897	0.915	1.557
0.120	0.121	0.320	0.332	0.520	0.576	0.720	0.908	0.920	1.589
0.125	0.126	0.325	0.337	0.525	0.583	0.725	0.918	0.925	1.623
0.130	0.131	0.330	0.343	0.530	0.590	0.730	0.929	0.930	1.658
0.135	0.136	0.335	0.348	0.535	0.597	0.735	0.940	0.935	1.697
0.140	0.141	0.340	0.354	0.540	0.604	0.740	0.950	0.940	1.738
0.145	0.146	0.345	0.360	0.545	0.611	0.745	0.962	0.945	1.783
0.150	0.151	0.350	0.365	0.550	0.618	0.750	0.973	0.950	1.832
0.155	0.156	0.355	0.371	0.555	0.626	0.755	0.984	0.955	1.886
0.160	0.161	0.360	0.377	0.560	0.633	0.760	0.996	0.960	1.946
0.165	0.167	0.365	0.383	0.565	0.640	0.765	1.008	0.965	2.014
0.170	0.172	0.370	0.388	0.570	0.648	0.770	1.020	0.970	2.092
0.175	0.177	0.375	0.394	0.575	0.655	0.775	1.033	0.975	2.185
0.180	0.182	0.380	0.400	0.580	0.662	0.780	1.045	0.980	2.298
0.185	0.187	0.385	0.406	0.585	0.670	0.785	1.058	0.985	2.443
0.190	0.192	0.390	0.412	0.590	0.678	0.790	1.071	0.990	2.647
0.195	0.198	0.395	0.418	0.595	0.685	0.795	1.085	0.995	2.994

Table D.8 is taken from Table VIIi of Fisher and Yates: *Statistical Tables for Biological, Agricultural and Medical Research*, published by Longman Group Ltd., London (previously published by Oliver & Boyd, Edinburgh), and reprinted by permission of the authors and publishers.

[a] Discussed in Section 12.5.

Table D.9.   **Approximate *n* Required for Testing Hypotheses about Means**[a]

One-Sample Test

Effect Size, d	α	One-Tailed Hypothesis, $1-\beta$ .80	.90	.95	Two-Tailed Hypothesis, $1-\beta$ .80	.90	.95
0.2	.05	156	215	272	198	264	326
	.01	253	328	396	294	374	447
0.5	.05	27	36	45	34	44	54
	.01	43	55	66	51	63	75
0.8	.05	12	15	19	15	19	22
	.01	19	24	28	22	27	32

Two-Sample Test (Independent Samples)

Effect Size, d	α	.80	.90	.95	.80	.90	.95
0.2	.05	310	429	542	393	526	651
	.01	503	652	790	586	746	892
0.5	.05	50	69	87	64	85	105
	.01	82	105	128	95	120	144
0.8	.05	21	27	35	26	34	42
	.01	33	42	51	38	48	57

Two-Sample Test (Dependent Samples)

d	α	ρ	.80	.90	.95	.80	.90	.95
0.2	.05	0.4	187	258	326	237	317	391
		0.5	156	215	272	198	264	326
		0.6	125	172	218	159	212	261
		0.7	94	130	164	119	159	197
		0.8	63	87	109	80	107	131
		0.9	32	44	55	41	54	66
	.01	0.4	303	393	475	353	449	537
		0.5	253	328	396	294	374	447
		0.6	203	262	317	236	300	358
		0.7	153	197	239	177	225	269
		0.8	102	132	160	119	151	180
		0.9	52	67	81	60	76	91
0.5	.05	0.4	31	42	53	39	52	63
		0.5	26	35	44	33	44	53
		0.6	21	29	36	27	35	43
		0.7	16	22	27	20	27	33
		0.8	11	15	19	14	18	22
		0.9	6	8	10	8	10	12

[a] Discussed in Sections 12.2 and 13.5. For the two-sample test (independent samples), it is assumed that $\sigma_1^2 = \sigma_2^2$ and $n_1 = n_2$; the values in the table are for each of the samples. If dependent samples are used, the values in the table are for the number of pairs of dependent elements.

Table D.9.   (Continued)

			Two-Sample Test (Dependent Samples)					
Effect Size, *d*	α		*One-Tailed Hypothesis, $1 - \beta$*			*Two-Tailed Hypothesis, $1 - \beta$*		
			.80	.90	.95	.80	.90	.95
		ρ						
	.01	0.4	50	65	78	58	73	88
		0.5	42	54	65	49	62	73
		0.6	34	44	52	39	50	59
		0.7	26	33	40	30	38	45
		0.8	18	23	27	21	26	31
		0.9	10	12	15	11	14	16
0.8	.05	0.4	13	17	21	16	21	26
		0.5	11	15	18	13	18	22
		0.6	9	12	15	11	15	18
		0.7	7	9	11	9	11	14
		0.8	5	7	8	6	8	10
		0.9	3	4	5	4	5	6
	.01	0.4	21	26	32	24	30	35
		0.5	18	22	27	20	25	30
		0.6	15	18	22	17	21	24
		0.7	11	14	17	13	16	19
		0.8	8	10	12	9	11	13
		0.9	5	6	7	6	7	8

Table D.10. **Critical Values of the Mann–Whitney U.**[a] **For a one-tailed test at $\alpha = .01$ (roman type) and $\alpha = .005$ (boldface type) and for a two-tailed test at $\alpha = .02$ (roman type) and $\alpha = .01$ (boldface type).**

n_2 \ n_1	1	2	3	4	5	6	7	8	9	10	11	12	13	14	15	16	17	18	19	20
1	—[b]	—	—	—	—	—	—	—	—	—	—	—	—	—	—	—	—	—	—	—
2	—	—	—	—	—	—	—	—	—	—	—	—	0	0	0	0	0	0	1	1
													—	—	—	—	—	—	**0**	**0**
3	—	—	—	—	—	—	—	0	0	1	1	1	2	2	2	3	3	4	4	5
									0	**0**	**0**	**1**	**1**	**1**	**2**	**2**	**2**	**2**	**3**	**3**
4	—	—	—	—	0	1	1	2	3	3	4	5	5	6	7	7	8	9	9	10
						0	**0**	**1**	**1**	**2**	**2**	**3**	**3**	**4**	**5**	**5**	**6**	**6**	**7**	**8**
5	—	—	—	0	1	2	3	4	5	6	7	8	9	10	11	12	13	14	15	16
					0	**1**	**1**	**2**	**3**	**4**	**5**	**6**	**7**	**7**	**8**	**9**	**10**	**11**	**12**	**13**
6	—	—	—	1	2	3	4	6	7	8	9	11	12	13	15	16	18	19	20	22
				0	**1**	**2**	**3**	**4**	**5**	**6**	**7**	**9**	**10**	**11**	**12**	**13**	**15**	**16**	**17**	**18**
7	—	—	0	1	3	4	6	7	9	11	12	14	16	17	19	21	23	24	26	28
				0	**1**	**3**	**4**	**6**	**7**	**9**	**10**	**12**	**13**	**15**	**16**	**18**	**19**	**21**	**22**	**24**
8	—	—	0	2	4	6	7	9	11	13	15	17	20	22	24	26	28	30	32	34
				1	**2**	**4**	**6**	**7**	**9**	**11**	**13**	**15**	**17**	**18**	**20**	**22**	**24**	**26**	**28**	**30**
9	—	—	1	3	5	7	9	11	14	16	18	21	23	26	28	31	33	36	38	40
			0	**1**	**3**	**5**	**7**	**9**	**11**	**13**	**16**	**18**	**20**	**22**	**24**	**27**	**29**	**31**	**33**	**36**
10	—	—	1	3	6	8	11	13	16	19	22	24	27	30	33	36	38	41	44	47
			0	**2**	**4**	**6**	**9**	**11**	**13**	**16**	**18**	**21**	**24**	**26**	**29**	**31**	**34**	**37**	**39**	**42**
11	—	—	1	4	7	9	12	15	18	22	25	28	31	34	37	41	44	47	50	53
			0	**2**	**5**	**7**	**10**	**13**	**16**	**18**	**21**	**24**	**27**	**30**	**33**	**36**	**39**	**42**	**45**	**48**
12	—	—	2	5	8	11	14	17	21	24	28	31	35	38	42	46	49	53	56	60
			1	**3**	**6**	**9**	**12**	**15**	**18**	**21**	**24**	**27**	**31**	**34**	**37**	**41**	**44**	**47**	**51**	**54**
13	—	0	2	5	9	12	16	20	23	27	31	35	39	43	47	51	55	59	63	67
		—	**1**	**3**	**7**	**10**	**13**	**17**	**20**	**24**	**27**	**31**	**34**	**38**	**42**	**45**	**49**	**53**	**56**	**60**
14	—	0	2	6	10	13	17	22	26	30	34	38	43	47	51	56	60	65	69	73
		—	**1**	**4**	**7**	**11**	**15**	**18**	**22**	**26**	**30**	**34**	**38**	**42**	**46**	**50**	**54**	**58**	**63**	**67**
15	—	0	3	7	11	15	19	24	28	33	37	42	47	51	56	61	66	70	75	80
		—	**2**	**5**	**8**	**12**	**16**	**20**	**24**	**29**	**33**	**37**	**42**	**46**	**51**	**55**	**60**	**64**	**69**	**73**
16	—	0	3	7	12	16	21	26	31	36	41	46	51	56	61	66	71	76	82	87
		—	**2**	**5**	**9**	**13**	**18**	**22**	**27**	**31**	**36**	**41**	**45**	**50**	**55**	**60**	**65**	**70**	**74**	**79**
17	—	0	4	8	13	18	23	28	33	38	44	49	55	60	66	71	77	82	88	93
		—	**2**	**6**	**10**	**15**	**19**	**24**	**29**	**34**	**39**	**44**	**49**	**54**	**60**	**65**	**70**	**75**	**81**	**86**
18	—	0	4	9	14	19	24	30	36	41	47	53	59	65	70	76	82	88	94	100
		—	**2**	**6**	**11**	**16**	**21**	**26**	**31**	**37**	**42**	**47**	**53**	**58**	**64**	**70**	**75**	**81**	**87**	**92**
19	—	1	4	9	15	20	26	32	38	44	50	56	63	69	75	82	88	94	101	107
		0	**3**	**7**	**12**	**17**	**22**	**28**	**33**	**39**	**45**	**51**	**56**	**63**	**69**	**74**	**81**	**87**	**93**	**99**
20	—	1	5	10	16	22	28	34	40	47	53	60	67	73	80	87	93	100	107	114
		0	**3**	**8**	**13**	**18**	**24**	**30**	**36**	**42**	**48**	**54**	**60**	**67**	**73**	**79**	**86**	**92**	**99**	**105**

[a] Discussed in Section 18.2. To be significant for any given n_1 and n_2, obtained U must be *equal to* or *less than* the value shown in the table.

[b] Dashes in the body of the table indicate that no decision is possible at the stated level of significance.

Table D.10 (Continued). Critical values for a one-tailed test at α = .05 (roman type) and α = .025 (boldface type) and for a two-tailed test at α = .10 (roman type) and α = .05 (boldface type).

n_2 \ n_1	1	2	3	4	5	6	7	8	9	10	11	12	13	14	15	16	17	18	19	20
1	—	—	—	—	—	—	—	—	—	—	—	—	—	—	—	—	—	—	0	0
																		—	—	—
2	—	—	—	—	0	0	0	1	1	1	1	2	2	2	3	3	3	4	4	4
	—	—	—	—	**0**	**0**	**0**	**0**	**1**	**1**	**1**	**1**	**1**	**1**	**1**	**2**	**2**	**2**	**2**	**2**
3	—	—	0	0	1	2	2	3	3	4	5	5	6	7	7	8	9	9	10	11
	—	—	—	—	**0**	**1**	**1**	**2**	**2**	**3**	**3**	**4**	**4**	**5**	**5**	**6**	**6**	**7**	**7**	**8**
4	—	—	0	1	2	3	4	5	6	7	8	9	10	11	12	14	15	16	17	18
	—	—	—	**0**	**1**	**2**	**3**	**4**	**4**	**5**	**6**	**7**	**8**	**9**	**10**	**11**	**11**	**12**	**13**	**13**
5	—	0	1	2	4	5	6	8	9	11	12	13	15	16	18	19	20	22	23	25
	—	—	**0**	**1**	**2**	**3**	**5**	**6**	**7**	**8**	**9**	**11**	**12**	**13**	**14**	**15**	**17**	**18**	**19**	**20**
6	—	0	2	3	5	7	8	10	12	14	16	17	19	21	23	25	26	28	30	32
	—	—	**1**	**2**	**3**	**5**	**6**	**8**	**10**	**11**	**13**	**14**	**16**	**17**	**19**	**21**	**22**	**24**	**25**	**27**
7	—	0	2	4	6	8	11	13	15	17	19	21	24	26	28	30	33	35	37	39
	—	—	**1**	**3**	**5**	**6**	**8**	**10**	**12**	**14**	**16**	**18**	**20**	**22**	**24**	**26**	**28**	**30**	**32**	**34**
8	—	1	3	5	8	10	13	15	18	20	23	26	28	31	33	36	39	41	44	47
	—	**0**	**2**	**4**	**6**	**8**	**10**	**13**	**15**	**17**	**19**	**22**	**24**	**26**	**29**	**31**	**34**	**36**	**38**	**41**
9	—	1	3	6	9	12	15	18	21	24	27	30	33	36	39	42	45	48	51	54
	—	**0**	**2**	**4**	**7**	**10**	**12**	**15**	**17**	**20**	**23**	**26**	**28**	**31**	**34**	**37**	**39**	**42**	**45**	**48**
10	—	1	4	7	11	14	17	20	24	27	31	34	37	41	44	48	51	55	58	62
	—	**0**	**3**	**5**	**8**	**11**	**14**	**17**	**20**	**23**	**26**	**29**	**33**	**36**	**39**	**42**	**45**	**48**	**52**	**55**
11	—	1	5	8	12	16	19	23	27	31	34	38	42	46	50	54	57	61	65	69
	—	**0**	**3**	**6**	**9**	**13**	**16**	**19**	**23**	**26**	**30**	**33**	**37**	**40**	**44**	**47**	**51**	**55**	**58**	**62**
12	—	2	5	9	13	17	21	26	30	34	38	42	47	51	55	60	64	68	72	77
	—	**1**	**4**	**7**	**11**	**14**	**18**	**22**	**26**	**29**	**33**	**37**	**41**	**45**	**49**	**53**	**57**	**61**	**65**	**69**
13	—	2	6	10	15	19	24	28	33	37	42	47	51	56	61	65	70	75	80	84
	—	**1**	**4**	**8**	**12**	**16**	**20**	**24**	**28**	**33**	**37**	**41**	**45**	**50**	**54**	**59**	**63**	**67**	**72**	**76**
14	—	2	7	11	16	21	26	31	36	41	46	51	56	61	66	71	77	82	87	92
	—	**1**	**5**	**9**	**13**	**17**	**22**	**26**	**31**	**36**	**40**	**45**	**50**	**55**	**59**	**64**	**67**	**74**	**78**	**83**
15	—	3	7	12	18	23	28	33	39	44	50	55	61	66	72	77	83	88	94	100
	—	**1**	**5**	**10**	**14**	**19**	**24**	**29**	**34**	**39**	**44**	**49**	**54**	**59**	**64**	**70**	**75**	**80**	**85**	**90**
16	—	3	8	14	19	25	30	36	42	48	54	60	65	71	77	83	89	95	101	107
	—	**1**	**6**	**11**	**15**	**21**	**26**	**31**	**37**	**42**	**47**	**53**	**59**	**64**	**70**	**75**	**81**	**86**	**92**	**98**
17	—	3	9	15	20	26	33	39	45	51	57	64	70	77	83	89	96	102	109	115
	—	**2**	**6**	**11**	**17**	**22**	**28**	**34**	**39**	**45**	**51**	**57**	**63**	**67**	**75**	**81**	**87**	**93**	**99**	**105**
18	—	4	9	16	22	28	35	41	48	55	61	68	75	82	88	95	102	109	116	123
	—	**2**	**7**	**12**	**18**	**24**	**30**	**36**	**42**	**48**	**55**	**61**	**67**	**74**	**80**	**86**	**93**	**99**	**106**	**112**
19	0	4	10	17	23	30	37	44	51	58	65	72	80	87	94	101	109	116	123	130
	—	**2**	**7**	**13**	**19**	**25**	**32**	**38**	**45**	**52**	**58**	**65**	**72**	**78**	**85**	**92**	**99**	**106**	**113**	**119**
20	0	4	11	18	25	32	39	47	54	62	69	77	84	92	100	107	115	123	130	138
	—	**2**	**8**	**13**	**20**	**27**	**34**	**41**	**48**	**55**	**62**	**69**	**76**	**83**	**90**	**98**	**105**	**112**	**119**	**127**

***Table D.11.*   Critical Values of the Wilcoxon T^a**

	Level of Significance for a One-Tailed Test					Level of Significance for a One-Tailed Test			
	.05	.025	.01	.005		.05	.025	.01	.005
	Level of Significance for a Two-Tailed Test					Level of Significance for a Two-Tailed Test			
n	.10	.05	.02	.01	n	.10	.05	.02	.01
5	0	—	—	—	28	130	116	101	91
6	2	0	—	—	29	140	126	110	100
7	3	2	0	—	30	151	137	120	109
8	5	3	1	0	31	163	147	130	118
9	8	5	3	1	32	175	159	140	128
10	10	8	5	3	33	187	170	151	138
11	13	10	7	5	34	200	182	162	148
12	17	13	9	7	35	213	195	173	159
13	21	17	12	9	36	227	208	185	171
14	25	21	15	12	37	241	221	198	182
15	30	25	19	15	38	256	235	211	194
16	35	29	23	19	39	271	249	224	207
17	41	34	27	23	40	286	264	238	220
18	47	40	32	27	41	302	279	252	233
19	53	46	37	32	42	319	294	266	247
20	60	52	43	37	43	336	310	281	261
21	67	58	49	42	44	353	327	296	276
22	75	65	55	48	45	371	343	312	291
23	83	73	62	54	46	389	361	328	307
24	91	81	69	61	47	407	378	345	322
25	100	89	76	68	48	426	396	362	339
26	110	98	84	75	49	446	415	379	355
27	119	107	92	83	50	466	434	397	373

^{a} Discussed in Section 18.3. The symbol T denotes the smaller sum of ranks associated with differences that are all of the same sign. For any given n (number of ranked differences), the obtained T is significant at a given level if it is *equal to* or *less than* the value shown in the table.

References

Anderson, N. H. Scales and statistics: Parametric and nonparametric. *Psychological Bulletin*, 1961, *58*, 305–316. [16]*

Arken, A., & Colton, R. *Graphs: How to make and use them* (2nd ed.). New York: Harper, 1938. [34]

Baggaley, A. R. *Mathematics for introductory statistics*. New York: Wiley, 1969. [416]

Bakan, D. The test of significance in psychological research. *Psychological Bulletin*, 1966, *66*, 423–437. [255]

Binder, A. Further considerations on testing the null hypothesis and the strategy and tactics of investigating theoretical models. *Psychological Review*, 1963, *70*, 107–115. [237]

Boneau, C. A. The effects of violations of assumptions underlying the *t* test. *Psychological Bulletin*, 1960, *57*, 49–64. [266]

Boneau, C. A. A note on measurement scales and statistical tests. *American Psychologist*, 1961, *16*, 160–161. [16]

Boneau, C. A., & Pennypacker, H. S. Group matching as research strategy: How not to get significant results. *Psychological Reports*, 1961, *8*, 143–147. [298]

Box, G. E. P., & Anderson, S. L. Permutation theory in the derivation of robust criteria and the study of departures from assumptions. *Journal of the Royal Statistical Society, Series B*, 1955, *17*, 1–34. [358]

Bradley, J. V. *Distribution-free statistical tests*. Englewood Cliffs, N.J.: Prentice-Hall, 1968. [402, 406]

Bradley, J. V. Non-parametric statistics. In R. E. Kirk (Ed.), *Statistical issues*. Monterey, Calif.: Brooks/Cole, 1972. [401]

Bradshaw, W. L. *Mathematics for statistics*. New York: Wiley, 1969. [416]

Bresnahan, J. L., & Shapiro, M. M. A general equation and technique for the exact partitioning of chi-square contingency tables. *Psychological Bulletin*, 1966, *66*, 252–262. [382]

Campbell, D. T. Factors relevant to the validity of experiments in social settings. *Psychological Bulletin*, 1957, *54*, 297–312. [265, 358]

Campbell, D. T., & Stanley, J. C. *Experimental and quasi-experimental designs for research*. Chicago: Rand McNally, 1963. [358]

Careers in statistics. Washington, D.C.: American Statistical Association. [3]

Castellan, N. J., Jr. On the partitioning of contingency tables. *Psychological Bulletin*, 1965, *64*, 330–338. [382]

Chissom, B. S. Interpretation of the kurtosis statistic. *American Statistician*, 1970. *24*(4), 19–22. [108]

Clark, V. A., & Tarter, M. E. *Preparation for basic statistics*. New York: McGraw-Hill, 1968. [416]

Cochran, W. G. Some consequences when the assumptions for the analysis of variance are not satisfied. *Biometrics*, 1947, *3*, 22–38. [359]

* *Page on which reference is cited.*

Cochran, W. G. Some methods for strengthening the common chi-square tests. *Biometrics*, 1954, *10*, 417–451. [377]

Cohen, Jacob. *Statistical power analysis for the behavioral sciences*. New York: Academic Press, 1969. [254, 265, 271, 274, 351]

Conover, W. J. Rejoinder. *Journal of the American Statistical Association*, 1974, *69*, 382. (a) [380]

Conover, W. J. Some reasons for not using the Yates continuity correction on 2 × 2 contingency tables. *Journal of the American Statistical Association*, 1974, *69*, 374–376. (b) [380]

Darlington, R. B. Is kurtosis really "peakedness?" *American Statistician*, 1970, *24*(2), 19–22. [108]

Dixon, W. J., & Massey, F. J. *Introduction to statistical analysis*. New York: McGraw-Hill, 1957. [265]

Dixon, W. J., Brown, M. B., Engelman, L., Frane, J. W., Hill, M. A., Jennrich, R. I., & Toporek, J. D. *BMDP statistical software 1981*. Berkeley: University of California Press, 1981. [57, 111]

Dudycha, A. L., & Dudycha, L. W. Behavioral statistics: An historical perspective. In R. E. Kirk (Ed.), *Statistical issues*. Monterey, Calif.: Brooks/Cole, 1972. [18, 235]

Edgington, E. S. Statistical inference and nonrandom samples. *Psychological Bulletin*, 1966, *66*, 485–487. [286]

Edgington, E. S. A new tabulation of statistical procedures used in APA journals. *American Psychologist*, 1974, *29*, 25–26. [338]

Edwards, A. L. On "the use and misuse of the chi-square test"—the case of the 2 × 2 contingency table. *Psychological Bulletin*, 1950, *47*, 347–355. [373]

Edwards, A. L. *Experimental design in psychological research* (4th ed.). New York: Holt, Rinehart and Winston, 1972. [339, 343, 357, 364]

Edwards, W. Tactical note on the relation between scientific and statistical hypotheses. *Psychological Bulletin*, 1965, *63*, 400–402. [237]

Fisher, R. A., & Yates, F. *Statistical tables for biological, agricultural and medical research*. London: Longman, 1974. Previously, Edinburgh: Oliver & Boyd, 1963. [458, 461, 464, 465, 472, 474]

Gaito, J. Scale classification and statistics. *Psychological Review*, 1960, *67*, 277–278. [16]

Galton, F. *Natural inheritance*. London and New York: Macmillan, 1889. [117]

Games, P. A. Multiple comparisons of means. *American Educational Research Journal*, 1971, *8*, 531–565. [360, 362]

Gardner, P. L. Scales and statistics. *Review of Educational Research*, 1975, *45*, 43–57. [16]

Glass, G. V., & Stanley, J. C. *Statistical methods in education and psychology*. Englewood Cliffs, N.J.: Prentice-Hall, 1970. [118, 133, 144]

Godard, R. H., & Lindquist, E. F. An empirical study of the effect of heterogeneous within-groups variance upon certain *F*-tests of significance in analysis of variance. *Psychometrika*, 1940, *5*, 263–274. [359]

Grant, D. A. Testing the null hypothesis and the strategy and tactics of investigating theoretical models. *Psychological Review*, 1962, *69*, 54–61. [237]

Grizzle, J. E. Continuity correction in the χ^2-test for 2 × 2 tables. *American Statistician*, 1967, *21*(4), 28–32. [380]

Halperin, M., Hartley, H. O., & Hoel, P. G. Recommended standards for statistical symbols and notation. *American Statistician*, 1965, *19*(3), 12–14. [61]

Hammond, K. R., Householder, J. E., & Castellan, N. J., Jr. *Introduction to the statistical method* (2nd ed.). New York: Knopf, 1970. [94]

Hays, W. L. *Statistics for the social sciences* (3rd ed.). New York: Holt, Rinehart and Winston, 1981. [133, 144, 172, 201, 205, 206, 359, 383, 384]

Helwig, J. T., and Council, K. A. *SAS user's guide 1979*. Cary, N. C.: SAS Institute, Inc., 1979. [394]

Horsnell, G. The effect of unequal group variances on the *F*-test for the homogeneity of group means. *Biometrika*, 1953, *40*, 128–136. [359]

Horst, P. A proof that the point from which the absolute deviations is a minimum is the median. *Journal of Educational Psychology*, 1931, *22*, 463–464. [70]

Howell, D. C. *Statistical methods for psychology*. Boston: Duxbury Press, 1982. [144]

Huff, D. *How to lie with statistics*. New York: Norton, 1954. [45]

Kearney, P. A. *Programmed review of fundamental mathematics for elementary statistics*. Englewood Cliffs, N.J.: Prentice-Hall, 1970. [416]

Keppel, G. *Design and analysis: A researcher's handbook*. Englewood Cliffs, N.J.: Prentice-Hall, 1973. [343, 364]

Kirk, R. E. (Ed.). *Statistical issues*. Monterey, Calif.: Brooks/Cole, 1972. (a) [16, 235, 237, 250, 265, 286, 298, 318, 358, 401]

Kirk, R. E. Classification of ANOVA designs. In R. E. Kirk (Ed.), *Statistical issues*. Monterey, Calif.: Brooks/Cole, 1972. (b) [343]

Kirk, R. E. *Introductory statistics*. Monterey, Calif.: Brooks/Cole, 1978. [95, 105, 128, 141, 160, 161, 253]

Kirk, R. E. *Experimental design: Procedures for the behavioral sciences* (2nd ed.). Monterey, Calif.: Brooks/Cole, 1982. [118, 133, 339, 343, 348, 350, 351, 357, 359, 360, 362, 364, 366]

Korin, B. P. *Statistical concepts for the social sciences*. Cambridge, Mass.: Winthrop, 1975. [377]

Lee, W. *Experimental design and analysis*. San Francisco: W. H. Freeman, 1975. [343, 364]

Lewis, D., & Burke, C. J. The use and misuse of the chi-square test. *Psychological Bulletin*, 1949, *46*, 433–489. [373]

Lewis, D., & Burke, C. J. Further discussion of the use and misuse of the chi-square test. *Psychological Bulletin*, 1950, *47*, 347–355. [373]

Lindman, H. R. *Analysis of variance in complex experimental designs*. San Francisco: W. H. Freeman, 1974. [343, 364]

Lindquist, E. F. *Design and analysis of experiments in psychology and education*. Boston: Houghton Mifflin, 1953. [358, 359]

Loether, H. J., & McTavish, D. G. *Descriptive statistics for sociologists*. Boston: Allyn & Bacon, 1974. [383]

Mantel, N. Comment and suggestion. *Journal of the American Statistical Association*, 1974, *69*, 378–380. [380]

Marascuilo, L. A. Large sample multiple comparisons. *Psychological Bulletin*, 1966, *65*, 280–290. [382]

Marascuilo, L. A. *Statistical methods for behavioral science research*. New York: McGraw-Hill, 1971. [376]

Marascuilo, L. A., & McSweeney, M. *Nonparametric and distribution-free methods for the social sciences*. Monterey, Calif.: Brooks/Cole, 1977. [376, 382, 385, 402]

McGee, V. E. *Principles of statistics*. New York: Appleton-Century-Crofts, 1971. [172]

McNemar, Q. Note on the sampling error of the difference between correlated proportions or percentages. *Psychometrika*, 1947, *12*, 153–157. [312]

McNemar, Q. *Psychological statistics* (4th ed.). New York: Wiley, 1969. [153]

Miettinen, O. S. Comment. *Journal of the American Statistical Association*, 1974, *69*, 380–382. [380]

Minium, E. W. *Statistical reasoning in psychology and education*. New York: Wiley, 1970. [265]

Mueller, J. H., & Schuessler, K. F. *Statistical reasoning in sociology*. Boston: Houghton Mifflin, 1961. [94]

Myers, J. L. *Fundamentals of experimental design* (2nd ed.). Boston: Allyn & Bacon, 1979. [339, 343, 357, 364]

Natrella, M. G. The relation between confidence intervals and tests of significance. *American Statistician*, 1960, *14*(1), 20–22, 33. [318]

Nie, N. H., Hull, C. H., Franklin, M. N., Jenkins, J. G., Sours, K. J., Norušis, M. J., & Beadle, V. *A user's guide to the SCSS conversational system*. New York: McGraw-Hill, 1980. [56]

Nie, N. H., Hull, C. H., Jenkins, J. G., Steinbrenner, K., & Bent, D. H. *Statistical package for the social sciences* (2nd ed.). New York: McGraw-Hill, 1975. [80]

Novick, M. R., & Jackson, P. H. *Statistical methods for educational and psychological research*. New York: McGraw-Hill, 1974. [172]

Nunnally, J. The place of statistics in psychology. *Educational and Psychological Measurement*, 1960, *20*, 641–650. [255]

Olds, E. G. Distributions of sums of squares of rank differences for small numbers of individuals. *Annals of Mathematical Statistics*, 1938, *9*, 133–148. [473]

Olds, E. G. The 5 per cent significance levels for sums of squares of rank differences and a correction. *Annals of Mathematical Statistics*, 1949, *20*, 117–118. [473]

Pastore, N. Some comments on "The use and misuse of the chi-square test." *Psychological Bulletin*, 1950, *47*, 338–340. [373]

Pearson, E. S. The analysis of variance in cases of non-normal variation. *Biometrika*, 1931, *23*, 114–133. [358]

Pearson, E. S., & Hartley, H. O. *Biometrika tables for statisticians* (Vol. 1, 3rd ed.). New York: Cambridge, 1966. [466]

Peters, C. C. The misuse of chi-square—a reply to Lewis and Burke. *Psychological Bulletin*, 1950, *47*, 331–337. [373]

Plackett, R. L. The continuity correction in 2 × 2 tables. *Biometrika*, 1964, *51*, 327–337. [380]

Rozeboom, W. W. The fallacy of the null-hypothesis significance test. *Psychological Bulletin*, 1960, *57*, 416–428. [255]

Ryan, T. A., Joiner, B. L., & Ryan, B. J. *Minitab student handbook*. North Scituate, Mass: Duxbury Press, 1976. [335]

Senders, V. L. *Measurement and statistics*. New York: Oxford, 1958. [16]

Siegel, S. *Nonparametric statistics for the behavioral sciences*. New York: McGraw-Hill, 1956. [16]

Sockloff, A. Behavior of the product-moment correlation coefficient when two heterogeneous subgroups are pooled. *Educational and Psychological Measurement*, 1975, *35*, 267–276. [134]

Starmer, C. J., Grizzle, J. E., & Sen, P. K. Comment. *Journal of the American Statistical Association*, 1974, *69*, 376–378. [380]

Steinmetz, J. E., Romano, A. G., & Patterson, M. M. Statistical programs for the Apple II computer. *Behavioral Research Methods and Instrumentation*, 1981, *13*(5), 702. [147, 369]

Stevens, S. S. On the theory of scales of measurement. *Science*, 1946, *103*, 667–680. [12, 16]

Stevens, S. S. Mathematics, measurement, and psychophysics. In S. S. Stevens (Ed.), *Handbook of experimental psychology*. New York: Wiley, 1951. [16]

Stevens, S. S. Measurement, statistics, and the schemapiric view. *Science*, 1968, *161*, 849–856. [16]

Stilson, D. W. *Probability and statistics in psychological research*. San Francisco: Holden-Day, 1966. [205]

Taylor, P. A. *An introduction to statistical methods*. Itasca, Ill.: J. E. Peacock, 1972. [377, 384]

Tippett, L. H. C. On the extreme individuals and the range of samples from a population. *Biometrika*, 1925, *17*, 386. [105]

Tukey, J. W. *Exploratory data analysis*. Reading, Mass.: Addison-Wesley, 1977. [40]

Ury, H. In response to Noether's letter, "Needed—a new name." *American Statistician*, 1967, *21*(4), 53. [401]

Wainer, H. Estimating coefficients in linear models: It don't make no nevermind. *Psychological Bulletin*, 1976, *83*, 213–217. [16]

Walker, H. *Mathematics essential for elementary statistics*. New York: Holt, Rinehart and Winston, 1951. [416]

Walpole, R. E. *Introduction to statistics*. New York: Macmillan, 1968. [205]

Wilson, B. *A user-oriented computer program in introductory statistics*. Unpublished master's thesis, Baylor University, Waco, Tx., 1984. [81, 166, 277, 394]

Wilson, W. R., & Miller, H. A note on the inconclusiveness of accepting the null hypothesis. *Psychological Review*, 1964, *71*, 238–242. [237]

Wilson, W., Miller, H. L., & Lower, J. S. Much ado about the null hypothesis. *Psychological Bulletin*, 1967, *67*, 188–196. [237]

Winer, B. J. *Statistical principles in experimental design* (2nd ed.). New York: McGraw-Hill, 1971. [339, 343, 357, 364]

Index